Harold Brown

Physical Characteristics of Early Films as Aids to Identification

NEW, EXPANDED EDITION

Edited by Camille Blot-Wellens

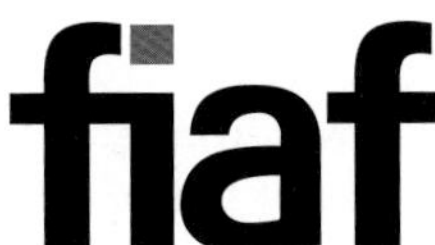

Second edition, published in 2020 by FIAF.
ISBN: 9782960029697

FIAF / Fédération internationale des archives du film
Rue Blanche 42
1060 Brussels
Belgium
T +32 2 538 3065
F +32 2 534 4774
info@fiafnet.org

Author of the original edition: Harold Brown
Editor: Camille Blot-Wellens
Graphic Design: Lara Denil
Copy-editor: Catherine A. Surowiec
Translations: Itzíar Gómez Carrasco, Aymeric Leroy
Executive Publisher: Christophe Dupin
Printed and bound in Ghent (Belgium) by Graphius

Cover image: Harold Brown examining a film.
Back cover image: Éclair film on 1910 film stock.
Courtesy of the BFI National Archive.

Table of contents

PART II

Foreword

Martin Koerber

In 1992, thanks to a kind letter of recommendation by Eva Orbanz, then curator of the film archive at the Deutsche Kinemathek, I was accepted as one of perhaps two dozen students of the FIAF Summer School. This school, which had been created in the early 1970s in (then East-) Berlin, was about to be revived at the British Film Institute's J. Paul Getty Jr. Conservation Centre in Berkhamsted, near London. There I joined a very diverse and interesting group of people from film archives the world over, and was very excited to spend several weeks at an institution which most of us saw as "the mother of all film archives", at least as far as technical expertise was concerned.

As we were shown around the facilities in Berkhamsted, everything seemed to be fantastically well organized, and intimidatingly superior to the places most of us came from. Having vaults filled with hundreds of thousands of carefully numbered cans is one thing, having them properly climatized another, but to have one's own laboratory for the duplication of film is still unheard of today for most film archives, and remains a utopian proposition.

Apart from sitting in with every department of the laboratory in order to understand how everything worked hand-in-hand to form a "factory of preservation", so to speak, we were exposed to a rigorous scheme of lectures informing us about all aspects of film preservation, delivered by presenters who were either the leaders of the various departments in the BFI itself, or outside experts from film laboratories, other archives in the UK, or well-known authors of reference books some of us had at least heard about.

Again and again in the lectures, a certain "Mr. Brown" was mentioned, as someone who had said, or found, or installed this-or-that procedure decades ago. The senior staff also referred to him repeatedly in conversation, thus creating an almost mythological, omnipresent figure, who knew it all before anybody had even thought of proper film preservation. After a few days, it dawned on me that "Mr. Brown" was none other than the elderly gentleman who was sitting in on most of the lectures, always in the back of the room, without ever saying a word, but taking notes, and seeming as much interested as he seemed pleased with what he saw being presented.

The name "Harold Brown" was not unknown to me, since the publication of *Physical Characteristics of Early Film as Aids to Identification* by FIAF in 1990. Once it had reached the library of the Kinemathek I had immediately photocopied every page to take home and study. This slender publication (it couldn't really be called a book, *per se*) was a tremendous eye-opener for someone like me, who had worked in film but so far had considered the perforation area only for writing up lists of numbers to give to the negative cutters. Learning about the wealth of useful information that was hiding outside the image area next to those tedious numbers opened a field of investigation for years to come. One might say this was my own private "material turn". No longer was the projected image the only thing to think about. In my mind any film element became a three-dimensional object, which would reveal its inseparable components of recorded picture, duplication status, emulsion type, and photographic quality, as a result of all these factors – components which only when understood as a whole make you fully understand the film element you have in front of you. And then there is a fourth dimension: Time. Think of all the events of handling, projecting, or even re-editing an element might have seen during its active life, and which have inscribed themselves not only into the surface of the film, but which might perhaps have been re-photographed several times into other elements of the same film, every time leaving traces which make the riddle more complex...

Seen through a potentially ever-stronger magnifying glass, for me every film thus became a galaxy to dive into, filled with molecular silver clouds, and numbers and letters on the margins as indications, which were now finally explained in Mr. Brown's "star map", which would guide one through all those nebulae.

So here I was, in the same room with Mr. Brown, and it took not much thinking to figure out that it might be a good idea to talk to him about his "method", and perhaps get a glimpse of the impressive collection of snippets from the most exotic pieces of film he might have kept...

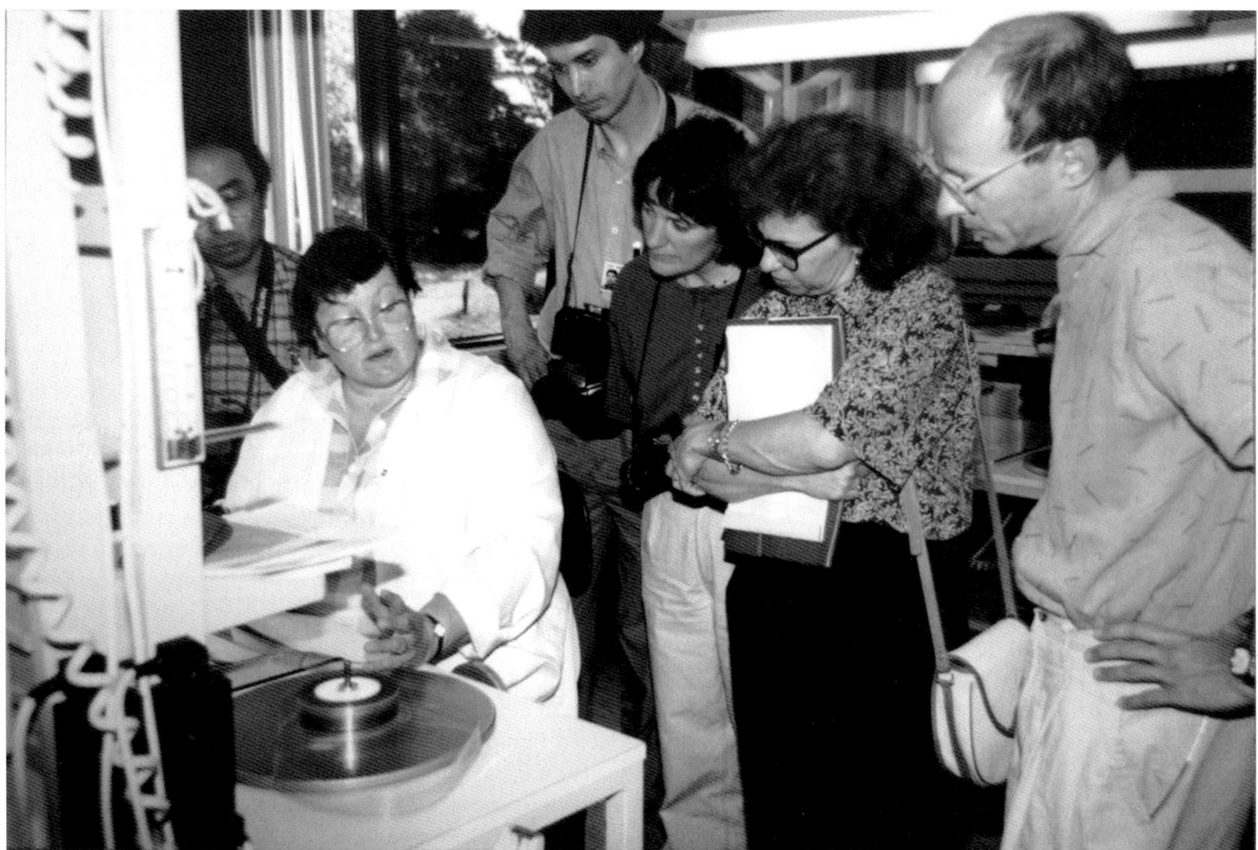

Jan Wilkinson, film repairer at the National Film Archive's J. Paul Getty Jr. Conservation Centre in Berkhamsted, teaches students of the FIAF Summer School 1992 (among them Eric Loné, Marilyn Koolik, and Martin Koerber). Harold Brown Special Collection, BFI National Archive.

When yet another excursion to yet another film vault was scheduled for us, I dared to ask our minders whether it would be OK if instead I could stay in Berkhamsted and ask Mr. Brown some questions about his reference collection. My wish was granted. And so when the others were put on the bus one morning I was driven to Berkhamsted, to spend my day with Mr. Brown. When I arrived he was already there, wearing his lab smock as he had done for decades while working, and I was given a likeness of it to wear before we sat down for business.

A number of folders had been brought out, and off we went, going through "specimens" page by page. Mr. Brown was sometimes explaining, and at other times silently watching me turning pages, and then he would sometimes test me with questions to be sure I understood what I was looking at.

When opening a film can, and before even looking at an image, it has since become my practice to slowly wind through the reel on the bench, equipped with a magnifying glass, deciphering print-throughs, reading stock marks, and establishing the age and generation of the element I am about to look at in projection.

Harold Brown's work was essential in preparing the "material turn" that film historians would experience in the 1990s. However, it still seems strange to me that no one else came forward in the same way with tools and methods that linked film research and very basic facts about film production and film duplication. They were, after all, more or less readily available to many people who were working in film-stock production and film laboratories. The gap between the places where films were made and the film archives was huge, and I still wonder why. No art historian would be taken seriously without proven knowledge about pigments, image carriers, and painting techniques, and an intimate acquaintance with originals. No serious museum would hire restorers who never looked at the back of a painting for traces of provenance. But for decades it seems to have been possible to forge a career as a film historian without ever touching a piece of film.

When I learned that Harold Brown's principal research was initially published internally at the FIAF Congress in (East-) Berlin in the form of a lecture and a mimeographed handout as early as 1967, I was shocked. Why did it take FIAF almost a quarter of a century to make all this worthwhile information available? And why did nobody ever answer Brown's request to come forward and pool information of the same nature that others may have found? That there *were* others is certain, but where was their research recorded or kept? Gerhard Lamprecht, the founder of the Deutsche Kinemathek in 1963 and an ardent film collector from before World War I onwards, is cited by Brown with dating information about Pathé's film stock. Lamprecht deduced this

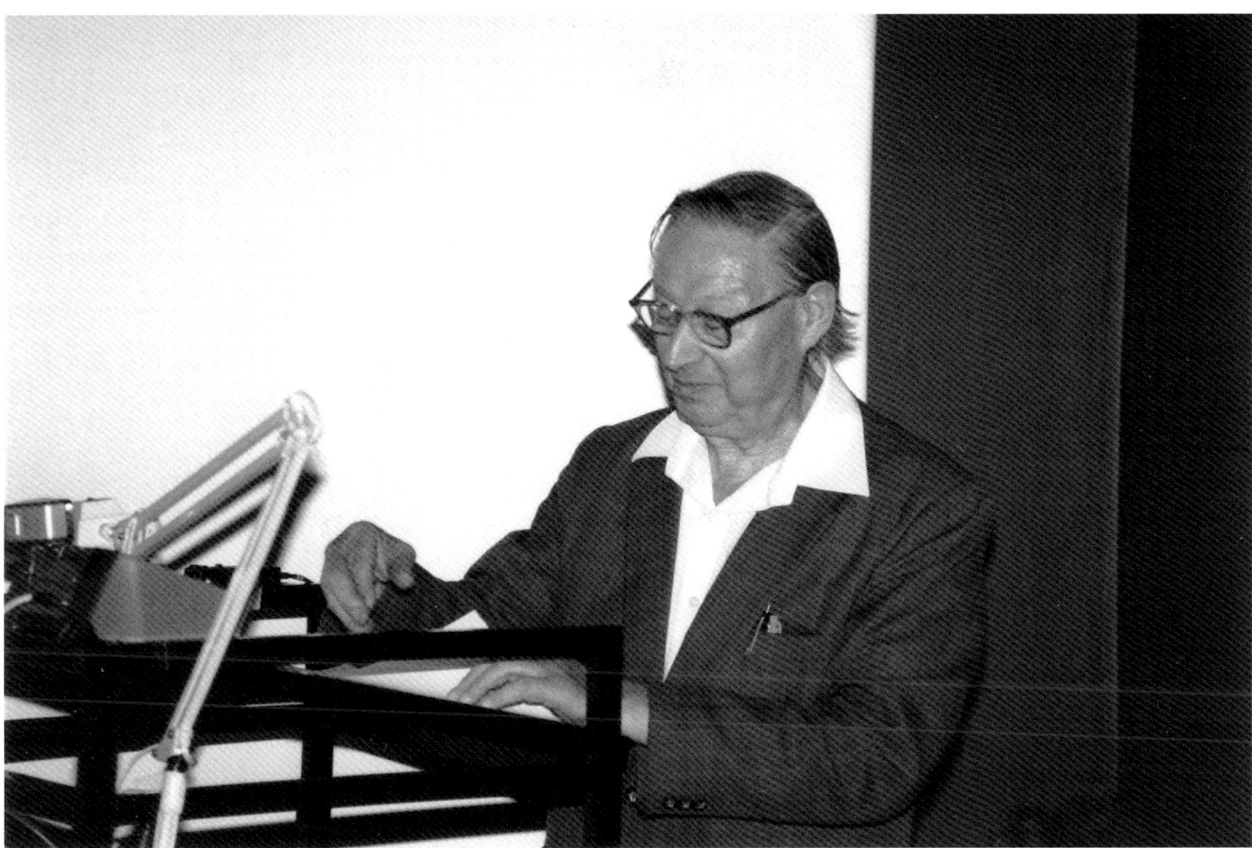

Harold Brown teaching at the 1992 FIAF Summer School, hosted by the National Film Archive in Berkhamsted. Courtesy of Heather Davies.

from the nitrate prints in his collection, which are lost, alas, since a Russian grenade went into his film vault in April 1945. We still have, however, the wonderful index cards which he compiled describing the prints in his collection, which recorded information about what was written in the perforation area, as well as remarks about title styles, tinting and toning, and how the prints were acquired, and when and where they were duplicated in the 1930s. Ironically, most of this information never made it into the catalogue of the Deutsche Kinemathek. This situation was finally corrected when Eva Orbanz retired in 2007, and thankfully used a large part of her free time to transcribe the index cards at last.

We have to thank the endurance of Camille Blot-Wellens, and the enthusiasm of the other authors for the new and very enlarged edition of Harold Brown's *Physical Characteristics...* Instead of a few tables and some sentences explaining a film stock, we now find extensive essays on the most important stocks, explaining company history as well as the technical development of manufacturing film, information difficult to find elsewhere. Also, the explanations are no longer limited to "Early Film" – one might rather wish to change the title of this book to "Physical Characteristics as Aids to Identification of ANY Film"! In many essays we also learn about the end of production of this precious material, alas. So it is all the more important that this information has been gathered now, while archives have not been dispersed or discarded, and while people who remember how film was made are still around and can pass on their knowledge.

Thus, what is published here is not only deduced from collected prints – apart from studying "specimens", most authors did extensive research in the paper archives of film manufacturers, as far as they survive. This represents an important change in methodology. At the time of Brown's initial internal publication in the 1960s, only Kodak had revealed, somewhat reluctantly, their dating symbols to the British Film Institute. And this was due, most probably, to a single person's understanding that leaking this information was crucial for film archives, and in the end would not endanger Kodak's interests. For many years these Kodak symbols seemed to be the only ones we would ever know. More serious research seemed to be too far-fetched even to undertake, and would also have asked too much from a single person who had many other duties to fulfil. A more scholarly approach has since taken hold, and more people are "on the job" now – which is very good news indeed.

Despite the wealth of information that can be found in this new and expanded edition of *Physical Characteristics...*, I would like to see it also as an invitation to dig up even more information, and to – as Harold Brown put it – more "pooling of information" in the future.

Harold Brown,
a True Pioneer of Film Preservation

Christophe Dupin

Harold is the authentic pioneer of the accepted techniques of film preservation and restoration. Every film archive in the world acknowledges him as such, bows to his expertise in these fields, and sends its own technicians to Aston Clinton [then the National Film Archive's conservation centre] to learn direct from the master. He is a longstanding member of FIAF's Preservation Commission and is invariably the one turned to for final arbitration on and interpretation of its findings and recommendations. His contribution to our present-day knowledge of cinematographic film, its properties, its identification, and above all its conservation, is unique and profound; and yet, thanks to an innate gift for clarity of expression, it is also entirely accessible, as his various writings on the subject amply demonstrate.

Clyde Jeavons, 25 March 1982[1]

Few personalities in the film archiving field (or perhaps any human activity) have had their entire career (life, even) so intrinsically intertwined with the technological development of that field as Harold Brown. Hired by the British Film Institute aged 15 in March 1935 – only a few months before the official inauguration of the National Film Library (later National Film Archive, today the BFI National Archive), and at a time when the concept of film preservation hardly existed at all – he retired from the same archive as Chief Preservation Officer in August 1984, a few months short of his golden jubilee on the BFI payroll, and his active contribution to the technical developments of film archiving went on for many years after that.

Harold Godart Brown, of French Huguenot ancestry, was born on 15 August 1919 in Walthamstow, a district in northeast London, "just within the sound of Bow Bells when the wind blows in the right direction",[2] which technically made him a Cockney, as he liked stating. He was raised in the London East End borough of Redbridge and left school at the age of fourteen after failing to obtain a scholarship to go to the County Secondary School. He worked for seven months as an office boy for an export merchant in the City of London, but lack of advancement opportunities made him look for another job. In March 1935, he came across an advertisement for an office boy/learner typist position at the recently formed British Film Institute in *The Daily Telegraph*. He applied for the job, was interviewed twice – first by the BFI's then book librarian, a young man called Ernest Lindgren, and later by the General Manager J.W. Brown. He was offered the job and started work in the BFI's general office at 4 Great Russell Street in London's West End on 18 March 1935. Much of his work at the start consisted of "addressing envelopes, transcribing film reviews from the trade papers, and running errands".[3] Among the other eight staff members were Rose Taylor, the General Manager's Secretary, and Joan Gardener, Ernest Lindgren's Secretary. The former would become Mrs. Lindgren and the latter Mrs. Brown. As soon as he started, the young Harold was faced with a difficulty over his name. As he shared his family name with the General Manager, he was instructed to use his middle name when working in the office. For that reason, he was known as "Godart" (even in official documents) for many years – he would still sign internal memos with that name decades later.

1 Clyde Jeavons, "Harold Brown", BFI internal memorandum to David Francis and Scott Meek, 25 March 1982, p.1. Harold Brown Collection, BFI Special Collections. In this 3-page memorandum, Jeavons strongly argued in favour of campaigning for the presentation of the BAFTA Outstanding British Contribution to Cinema Award to Harold Brown. BAFTA, however, decided otherwise.

2 Harold Brown interviewed by the author, 21 March 2005, at his home in Aylesbury, Buckinghamshire. Historically, "Cockneys" are people born within earshot of Bow Bells, the bells of St. Mary-le-Bow in the East End of London.

3 Harold Brown interviewed by the author, 21 March 2005.

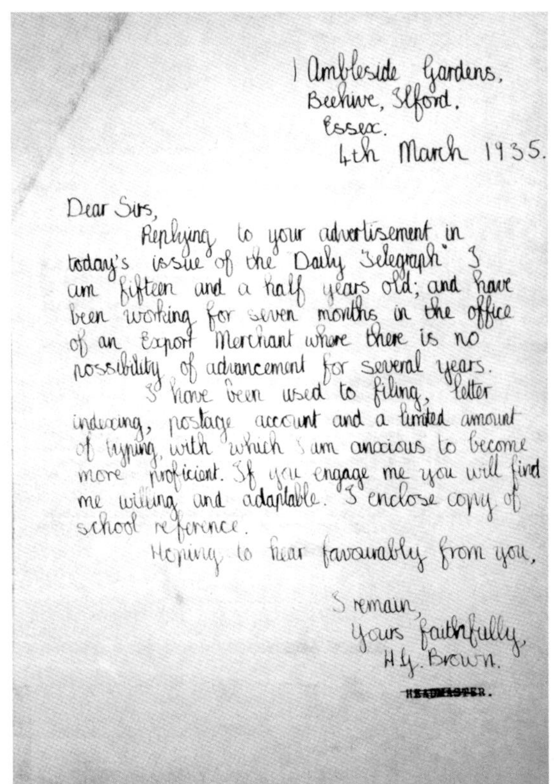

1 Ambleside Gardens,
Beehive, Ilford.
Essex.
4th March 1935.

Dear Sirs,

Replying to your advertisement in today's issue of the "Daily Telegraph" I am fifteen and a half years old; and have been working for seven months in the office of an Export Merchant where there is no possibility of advancement for several years.

I have been used to filing, letter indexing, postage account and a limited amount of typing, with which I am anxious to become more proficient. If you engage me you will find me willing and adaptable. I enclose copy of school reference.

Hoping to hear favourably from you,

I remain,
Yours faithfully,
H.G. Brown.

~~HEADMASTER.~~

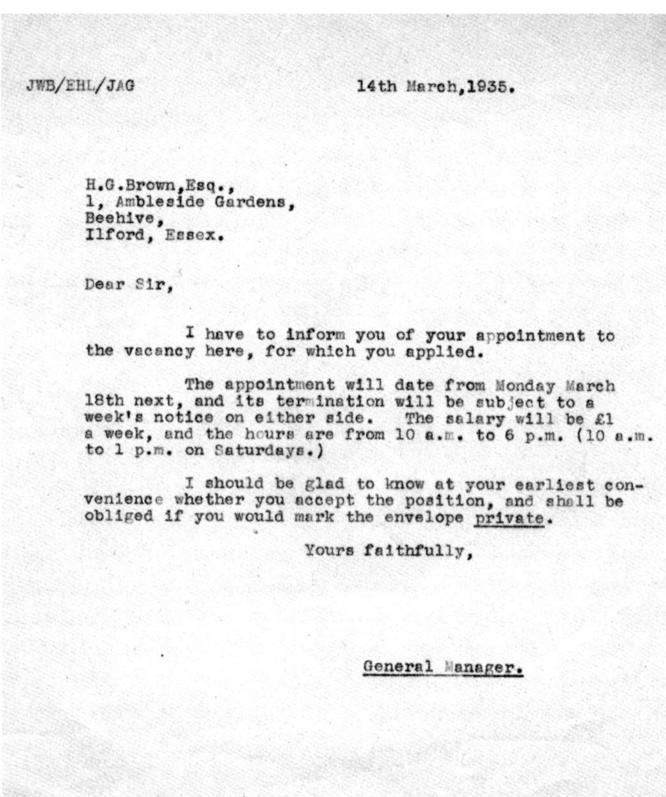

JWB/EHL/JAG 14th March,1935.

H.G.Brown,Esq.,
1, Ambleside Gardens,
Beehive,
Ilford, Essex.

Dear Sir,

I have to inform you of your appointment to the vacancy here, for which you applied.

The appointment will date from Monday March 18th next, and its termination will be subject to a week's notice on either side. The salary will be £1 a week, and the hours are from 10 a.m. to 6 p.m. (10 a.m. to 1 p.m. on Saturdays.)

I should be glad to know at your earliest convenience whether you accept the position, and shall be obliged if you would mark the envelope <u>private</u>.

Yours faithfully,

<u>General Manager.</u>

Correspondence between Harold Brown and the British Film Institute regarding his appointment in March 1935.
Harold Brown Special Collection, BFI National Archive.

Official Remit No. 6 of the young Film Institute (founded in September 1933) was to "establish a national repository of films of permanent value".[4] In May 1935, just two months after Harold Brown was hired, the National Film Library was born and Ernest Lindgren was appointed as its Curator. He soon took the young Harold with him to collect the first film acquisitions, a few early trick films donated by the famous British psychic researcher Harry Price, who had been appointed Chair of the National Film Library. "Soon I was typing fewer envelopes and film reviews in the office, and instead learning how to handle these reels," Brown later remembered.[5] He did not receive any formal training, since there was none available in these early days of film preservation, but he quickly picked up basic technical skills from H.D. Waley, the BFI's Technical Director, from the weekly lectures he attended at the British Kinematograph Society, from the leaflet on film preservation that the BKS produced in 1934 for the BFI[6], and from the projection booth of the Forum Cinema located under the arches below Charing Cross station, where he was sent to learn to project film for the BFI's small preview theatre, but also to handle (and splice) film. In the late 1930s, under the leadership of Ernest Lindgren, who too was learning everything about the new activity of film preservation, he worked more and more on film projecting[7] and servicing the two separate sections of the NFL, Preservation and Lending.

In the early years of the NFL, the fast-growing collections[8] (including many nitrate films) were stored in the basement vaults located at 5 Denmark Street, a few blocks away from the BFI headquarters in central London. When war broke in September 1939, the Government immediately issued a new regulation stating that all nitrate films should be moved out of central London. Temporary storage was found in Rudgwick, Sussex, and Brown helped move the collection there. Meanwhile, Lindgren went looking for a permanent home for the collection, "out of London, away from the continent, outside the target zone, but near enough for communication".[9] He found a stable in the village of Aston Clinton, Buckinghamshire, obtained permission to purchase it, and by the end of 1939 the first six film vaults had been built. Harold Brown lodged with the Lindgrens in Aston Clinton, where they retreated for the duration of the war, and cycled home at weekends. Between 1939

4 British Film Institute: *First Annual Report, Year Ended 30 September 1934*, London: BFI, p.9.

5 Harold Brown, "Trying to Save Frames", in Roger Smither and Catherine A. Surowiec (eds.), *This Film Is Dangerous: A Celebration of Nitrate Film*, FIAF, 2002, p.98.

6 *Report of Special Committee Set Up by the British Kinematograph Society to Consider Means that Should Be Adopted to Preserve Cinematograph Films for an Indefinite Period*, London: British Film Institute, Leaflet No. 4, August 1934, 8 pp. This is the seminal text on which Ernest Lindgren and Harold Brown based their entire outlook on film preservation. See the article devoted to this text in the *Journal of Film Preservation*, No. 101, October 2019, pp.86-90: Kieron Webb, "The British Kinematograph Society's 1934 Report on Methods of Film Preservation".

7 A BFI staff list of May 1939 gives "projectionist" as his official job title at that time.

8 In May 1939, the NFL announced to the press that its collections had reached 1,000 films and two million feet of film.

9 Patience Coster, "Harold Brown's Half Century at the National Film Archive", *Three Sixty: British Film Institute News*, August 1984, p.13.

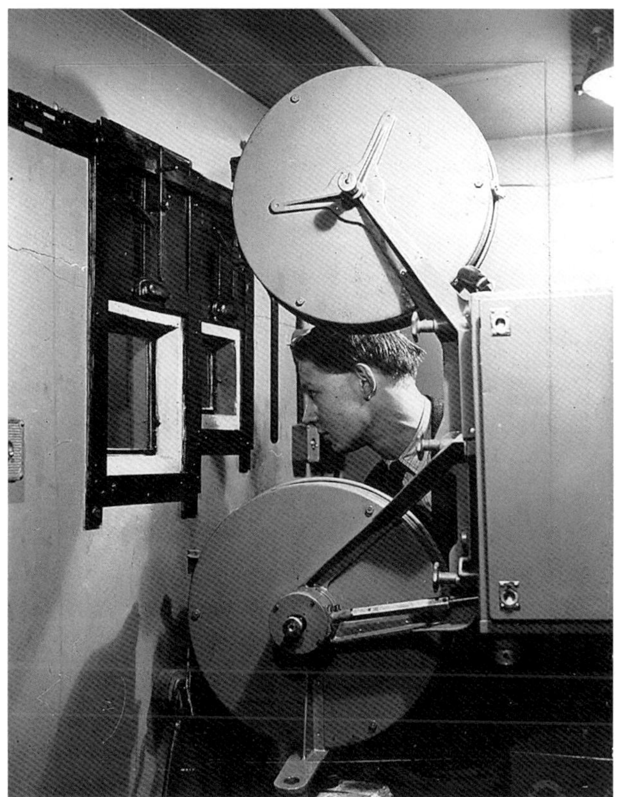

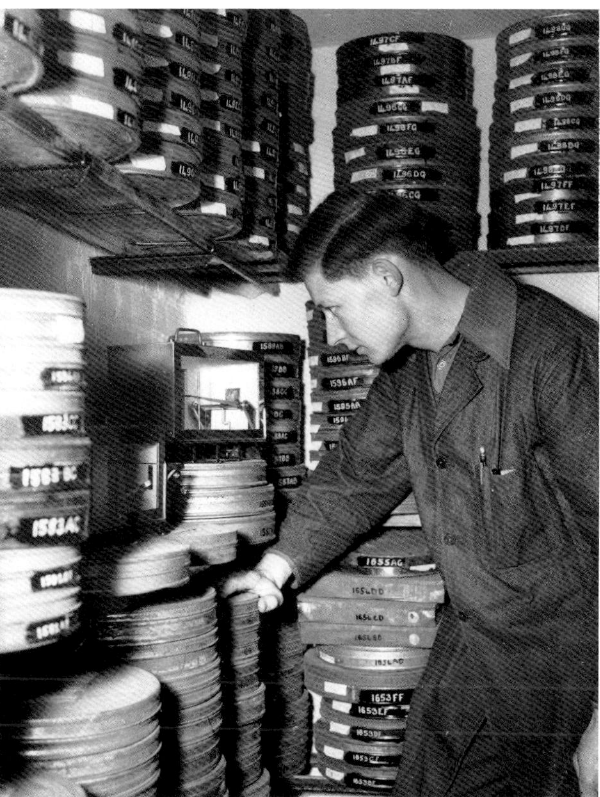

A young Harold Brown in the projection booth, and in a vault at Aston Clinton. Courtesy of Harold Brown's Estate.

and 1942, one of his key tasks was to do the legwork on the production of the compilation film *Film and Reality*, a montage of 58 extracts illustrating the history of documentary film and realism in cinema, produced by the NFL and assembled by the documentary filmmaker Alberto Cavalcanti and Ernest Lindgren.[10]

At the start of the war the young Brown had declared himself a conscientious objector, just as his father had been in the First World War. He was given exemption from military service on condition that he continue to work for the BFI, which he did until 1942. However, he eventually felt that it was not right to be almost unaffected by the war, so he joined the Friends Ambulance Unit as a medical orderly, working in hospitals and field surgical units. Two weeks after D-Day, he crossed to France where he served with the Unit until the end of the war. He married Joan Gardener on 22 April 1945 during a quick leave. By the end of July 1945 Brown was back in Aston Clinton. The Lindgrens moved back to London and the Browns moved into the archive's house.

In July 1942, Lindgren and Brown had realized that some of the early films in their collection were showing signs of stickiness. The following year, with the help of the Kodak Research Department, a system of artificial ageing tests was devised to predict the imminent disintegration of nitrate film (which, until its decomposition starts, shows no sign of the imminence of that condition).[11] The test was improved by two Government chemists at the end of the war. In the following years, carrying out this test, organizing the copying of soon unstable films, as well as inspecting new acquisitions and recording their condition, were Harold Brown's main tasks.

In 1948, the BFI decided to hire a new Technical Officer, C.R. Gibbs. However, within two years he was gone, and Harold Brown was then promoted to the post of Film Preservation Officer in recognition of the technical experience which he had accumulated in his first fifteen years at the archive. He did not carry out his film preservation work in total isolation, as he enjoyed the regular advice of the NFL Technical Committee.[12]

10 The 58 extracts were selected not only from the NFL's own collection but also from those of the other three founding members of FIAF, which made *Film and Reality* the first truly collaborative project among FIAF members. The film was met with great critical acclaim and proved popular in the British non-commercial sector during 1942, and it was acquired by a number of FIAF archives after the war.

11 A small disc of film was punched out of the tested film reel and dropped into a small test tube along with an indicator paper of Alizarin Red, then heated at 134 C. The time at which the indicator paper changed colour indicated a certain minimum remaining life. This result determined whether a film should be submitted for immediate copying or not.

12 Appointed by the National Film Library Committee (the NFL's governing body), the Technical Committee included scientists from Kodak, Ilford, Brent Laboratories, the British Museum, the Science Museum, and the British Government.

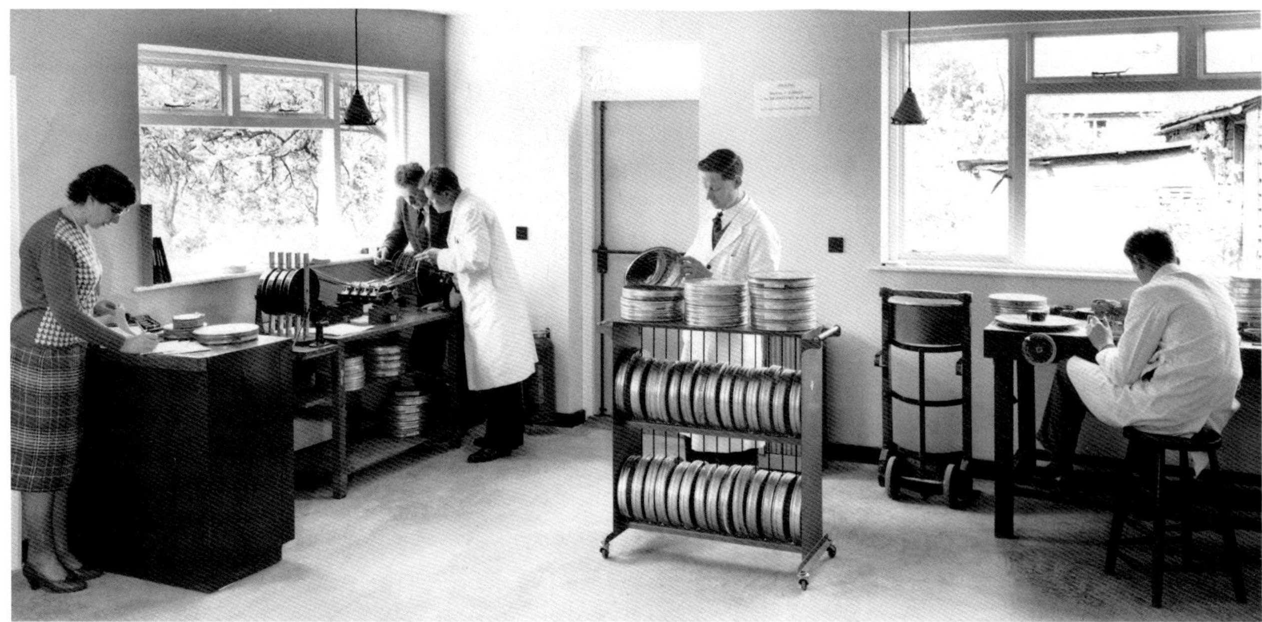

Harold Brown and his team at Aston Clinton in the 1950s. Courtesy of the BFI National Archive.

In 1951, at the invitation of Malcolm Hoare, the Chief Technician at Kay Laboratories (one of the two labs in which the NFL had its films duplicated), Brown was able to spend a month there, observing and learning various aspect of film processing – a key experience for him. In early 1952, the NFL was invited by the British Kinematograph Society to deliver one of their renowned Wednesday evening lectures. Brown later remembered: "The natural thing would have been for the curator to do this lecture. But he didn't. He put me onto it. This is an indication of the sort of person he was."[13] They wrote the lecture together and Brown delivered it at the BKS on 5 March. The paper, "Problems of Storing Film for Archive Purposes", was published in *British Kinematography* a couple of months later.[14] This 14-page text went beyond the mere question of storage. It examined the fragile and hazardous nature of nitrate film, which then constituted the vast majority of the archive's collections, and provided a detailed overview of the NFL's film preservation work, including its artificial ageing tests, film repair and duplication, and the specific problems of tinted and toned films, and colour films.

In his article, Brown highlighted the particular difficulty of duping early films, which constituted a significant proportion of the NFL's collections. Many of these did not have subsequent standard perforations and had suffered significant damage and defect over time (shrinkage especially), which meant that commercial printing laboratories could not duplicate them using their usual printers. Brown mentioned early attempts by the BFI's H.D. Waley to construct home-made optical printers using parts of early cinema projectors, but, as he noted, "there is still a field for further trial and experiment". It was Henri Langlois who gave him the opportunity to engage in this new endeavour during one of his visits to the BFI in 1954. Langlois asked Lindgren whether the NFL was able to dupe single-perforation Lumière films. Lindgren turned to Brown, who spent the next few years building a unique step-printer using parts of his childhood Meccano set, plywood, glass, pieces of tin can, "knicker elastic" (as he liked putting it), a car headlamp bulb, and an old projector sprocket, all put together using basic carpentry tools and a soldering iron. The printer, eventually known as "Mark IV" (the reason being that there had been three failed versions), never copied a Lumière film for Langlois, as by the time it operated successfully (the late 1950s) the relationship between Langlois and Lindgren had become very strained. The Mark IV printer, however, did copy the National Film Archive's own Lumière films (the first one was *A Heavy Load*), as well as Lumière films for other FIAF archives. It could not only handle Lumière (as well as other early films with non-standard perforations), but also badly damaged and heavily shrunken film which could not be dealt with in commercial labs. In 1984, as he was about to retire, he said with his usual modesty: "its only virtue is particularly shrunk, non-standard, fragile film, and the footage it has done is very small, but nothing is too shrunk or damaged to go through it."[15]

13 Harold Brown interviewed by the author, 21 March 2005.
14 Harold Brown, "Problems of Storing Film for Archive Purposes", *British Kinematography*, Vol. 20, No. 5, May 1952. It was also reprinted as a 14-page stand-alone booklet.
15 Patience Coster, "Harold Brown's Half Century at the National Film Archive", *Three Sixty: British Film Institute News*, August 1984, p.13.

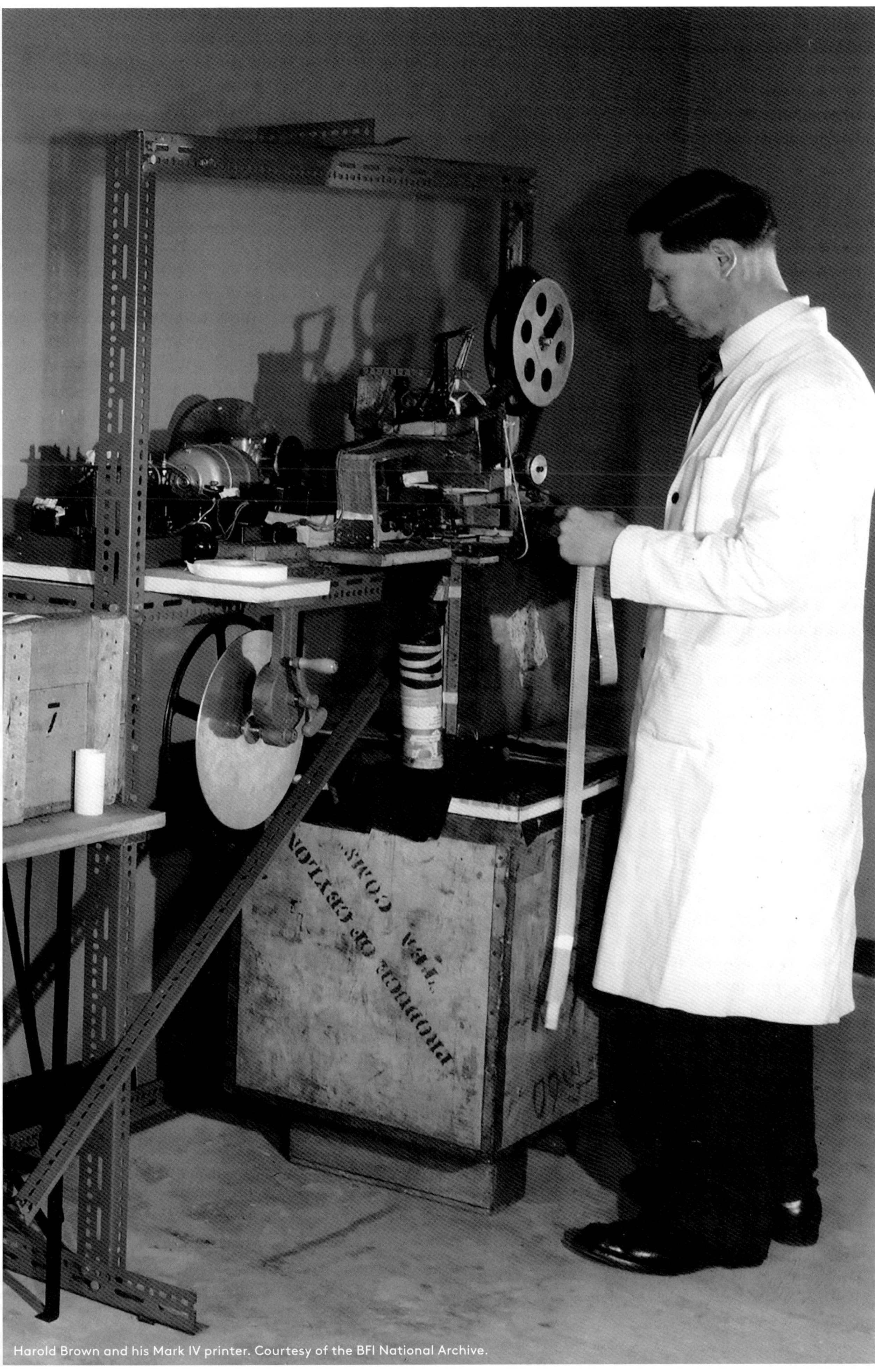

Harold Brown and his Mark IV printer. Courtesy of the BFI National Archive.

The Mark IV printer remained operational for nearly three decades. Over the years Brown improved it and adapted it to the archive's needs. Initially the printer took five seconds per frame, but this was later reduced to one second per frame. It was originally operated by hand, pulling each individual frame through, until Brown could afford to add a motor. The printer was also used to copy 28mm film onto 35mm, and to copy 9.5mm onto 16mm, or even to copy some two-colour Technicolor film. Although he later conceived other, perhaps more sophisticated, printing machines to deal with a variety of film formats,[16] the Mark IV was still used for years for the most difficult 35mm films. It even progressively acquired an almost mythical status. At the 1967 FIAF Congress in East Berlin, Brown, who was attending his first FIAF Congress, introduced the Mark IV, and screened a film showing the printer at work and several examples of early films it copied. The demonstration was a huge success, all the delegates present warmly congratulating him for his "ingenuity and hard work".[17] 23 years later, in November 1990, the Mark IV printer, which had then gone into retirement, was shipped to Bologna, where it was used by the Cineteca di Bologna for training purposes, and was then exhibited at L'Immagine Ritrovata lab for several years. In 2010, it was the centrepiece of the display celebrating 75 years of the BFI National Archive at BFI Southbank.

Harold Brown's work on the Mark IV in the 1950s showed his keen interest and fast-growing expertise in early films. In 1955, he wrote in the *British Film Institute News Sheet*: "I am sometimes able to assist the cataloguing department in cases where technical information can help to identify films; thus, for instance, if we have a film of the one-reel period which has lost all its titles; it is probable that I can determine the producing company which made it and, within a year or two, its date."[18] Brown was undoubtedly the first person to approach the key issue of film identification scientifically, by carefully examining both the image and the film material. He himself acknowledged that film identification "can sometimes require a skill and experience in research comparable to that applied to ancient manuscripts or prehistoric monuments".[19] In the 1950s, as his expertise in that field grew, he learned to recognize a number of characteristics particular to most early film producers, and his advice to the cataloguing department of the archive saved his colleagues a lot of research. Soon Ernest Lindgren started pressing him to write down his findings for the benefit of all in the film archiving field. In 1967, with his boss's assistance (and under his close supervision), he finally wrote a 30-page paper with two pages of photographed film frames, "Film Identification by Examination of Film Copies" (the title was suggested by Lindgren), which he delivered to the delegates of the 23rd FIAF Congress in East Berlin. His presentation was highly praised by his peers. As it turned out, other film archives were also working on the issue of film identification, notably the archive of the Czechoslovak Film Institute, who offered to host a symposium on film identification the following year. The week-long symposium took place in Gottwaldov (today Zlín) in south-eastern Moravia, in March 1968 (in the midst of the Prague Spring, as Brown later remembered), and Brown delivered one of the keynote lectures. Those were followed by screenings of unidentified films, which gave the experts present a chance to exchange views on the various methods of film identification. Although Brown and the other experts promised one another to keep the project alive by continuing their exchanges and appealing to other film archivists to record their discoveries and pool them for the benefit of the whole FIAF community, things didn't really go according to plan. By the time Brown retired in 1984, he confessed that he had never found enough time since 1968 to take the matter further. He said, "maybe now that I'm not here [at the BFI] officially, I might come back and spend more time on it. I hope so, because all it needs in a sense is the films and a bench to wind them on."[20] Taking his research on film identification a step further and writing it up was indeed one of the many things he could finally work on in his early retirement. In 1990, FIAF published a much expanded version of his 1967 paper under the title *Physical Characteristics of Early Films as Aids to Identification*. FIAF's famous "red book" has been one of FIAF's best-sellers ever since, just as this new, expanded edition, 30 years later, gives it a new lease of life.

Harold Brown's contribution to the field of film preservation was of course not limited to film identification, nor to the mere premises of the National Film Archive. Although his first public appearance on the international stage was at the 1967 FIAF Congress, he had been involved in FIAF's early technical discussions earlier in that decade. In 1965, he was involved in the translation into English of FIAF's very first technical manual, *Film Preservation*, written by the East German Herbert Volkmann, the first Head of the FIAF Preservation Commission. Following the success of his presentations at the East Berlin Congress, his became a name to reckon with in FIAF's technical circles. With Lindgren's blessing, he joined the Preservation Commission in 1970, and attended his first Commission meeting in Stockholm in October that year – the first of many on the theme of "colour

16 In particular optical printers that copied obsolete gauges such as 17.5mm or 60mm films. See Harold Brown, "Copying Small Amounts of Non-Standard Film Gauges", in Eva Orbanz (ed.), *Archiving the Audio-Visual Heritage: A Joint Technical Symposium*, Berlin: Stiftung Deutsche Kinemathek, 1988, pp.99-101.

17 *Minutes of the 23rd FIAF Congress in Berlin* (GDR), 8-13 June 1967, p.20. FIAF Historical Archive.

18 Harold Brown, "My Job", *British Film Institute News Sheet*, No. 5, 29 March 1956, p.2.

19 Harold Brown, "Film Identification by Examination of Film Copies", typewritten paper, 1967 FIAF Congress East Berlin, p.6.

20 Harold Brown, quoted in Elizabeth Sussex, "Preserving: Harold Brown, Joining Boy of the Archive", *Sight and Sound*, Autumn 1984, p.238.

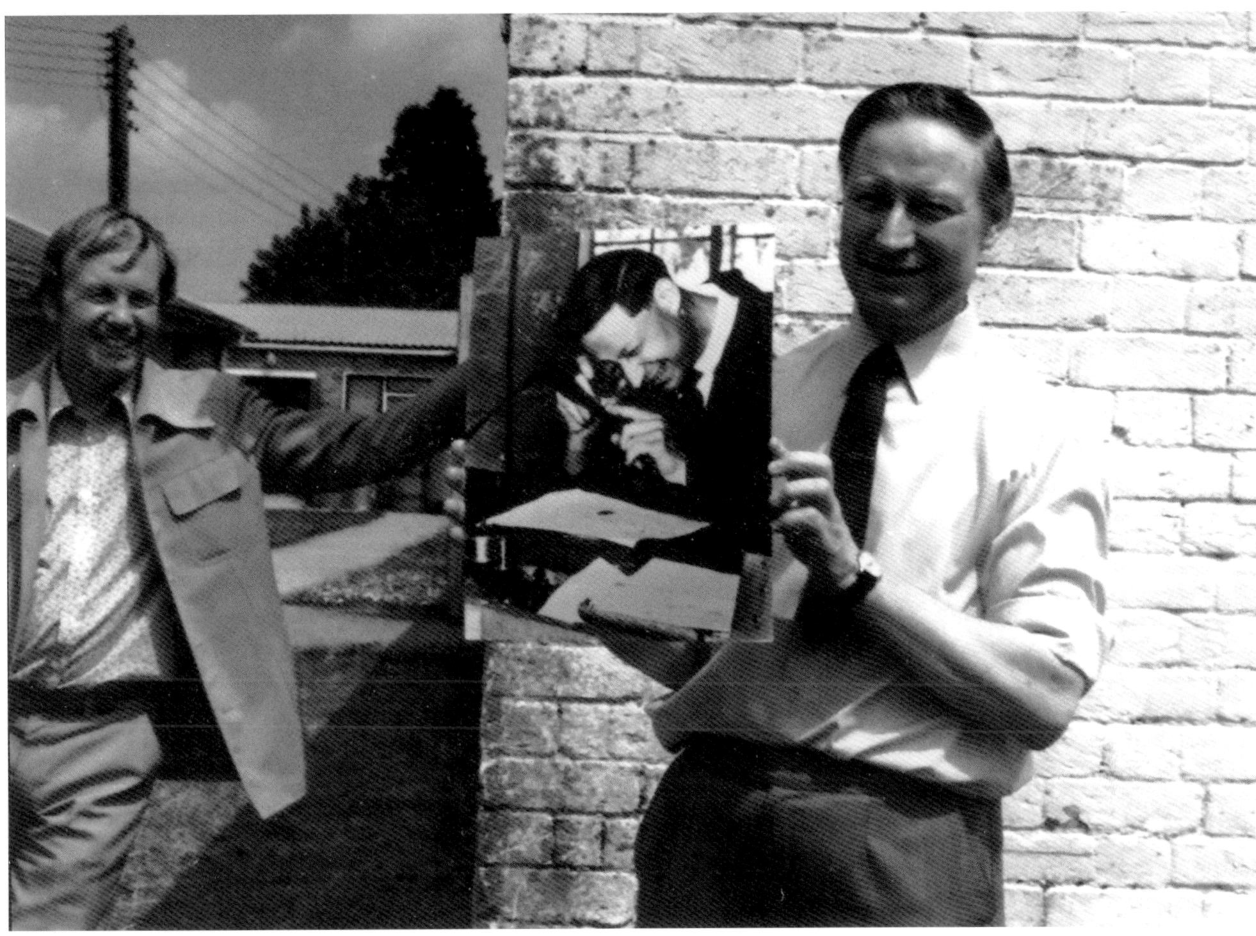

Harold Brown celebrated by the FIAF community in Aston Clinton during the 1978 Brighton Congress (left: David Francis, then Curator of the NFA). Courtesy of Heather Davies.

film preservation", as throughout the 1970s the Preservation Commission focused its efforts almost entirely on that theme, and Brown's expertise was once again very useful. He contributed to the discussions that led to the publication in 1977 of *The Preservation and Restoration of Colour and Sound in Films*, the Preservation Commission's second manual, which he also helped translate from German to English.[21] Clyde Jeavons later wrote: "Harold Brown was a prime mover in establishing the accepted procedures for colour preservation [...]. Even more significant, perhaps, have been his achievements with the accurate copying of obsolete colour systems, including early hand- and stencil-colouring, tinting and toning (by recreating the original dyes), Dufaycolor, Gasparcolor, and two-colour Technicolor."[22]

Brown also started playing an active role in a number of FIAF Congresses from the late 1970s. The first one was the Brighton Congress (1978), and its legendary symposium "Cinema 1900-1906", which owed a great debt to Brown in many ways. First, it has been rightly argued by various historians that the event – the marathon screening of 548 early films over five days, from 22 to 27 May 1978, to a small audience of film historians and archivists – was partly the result of his long-time interest in the rescue and restoration of the cinema's earliest films. As Sabine Lenk wrote, "Thanks to Brown's preliminary work, a certain number of archives started investing preservation time and money into films from a period they had often neglected."[23] Secondly, Harold Brown led the very complex technical preparations for the Symposium, handling the nearly 600 films received by the NFA from film archives around the world, repairing them, having them copied, and mounting them on reels in a strict chronological order. Finally, it was he who had the very physical task of operating the projector during these five demanding days of film screenings at the Brighton Film Theatre. His efforts were rewarded by the presentation by his boss David Francis, on behalf of the National Film Archive and FIAF, with a silver plaque "in deep appreciation and admiration of his pioneering work and many years of service in the cause of film preservation" during a special ceremony in Aston Clinton attended by all the Congress delegates.[24]

---

21 In the 1970s, the main language of the Preservation Commission was German, because of the influence of the then-Head Herbert Volkmann, and the many East Europeans of the Commission.
22 Clyde Jeavons, "Harold Brown", BFI internal memorandum, 25 March 1982, p.1.
23 Sabine Lenk, "Harold G. Brown and the Identification of Early Films", *The Moving Image*, Vol. 16, No. 1, Spring 2016, p.47. Lenk argues that without Brown's 1967 film identification paper, the Brighton Symposium would have taken place much later.
24 The plaque, along with other trophies he received throughout his career, was recently donated to the BFI National Archive by his daughter Gillian.

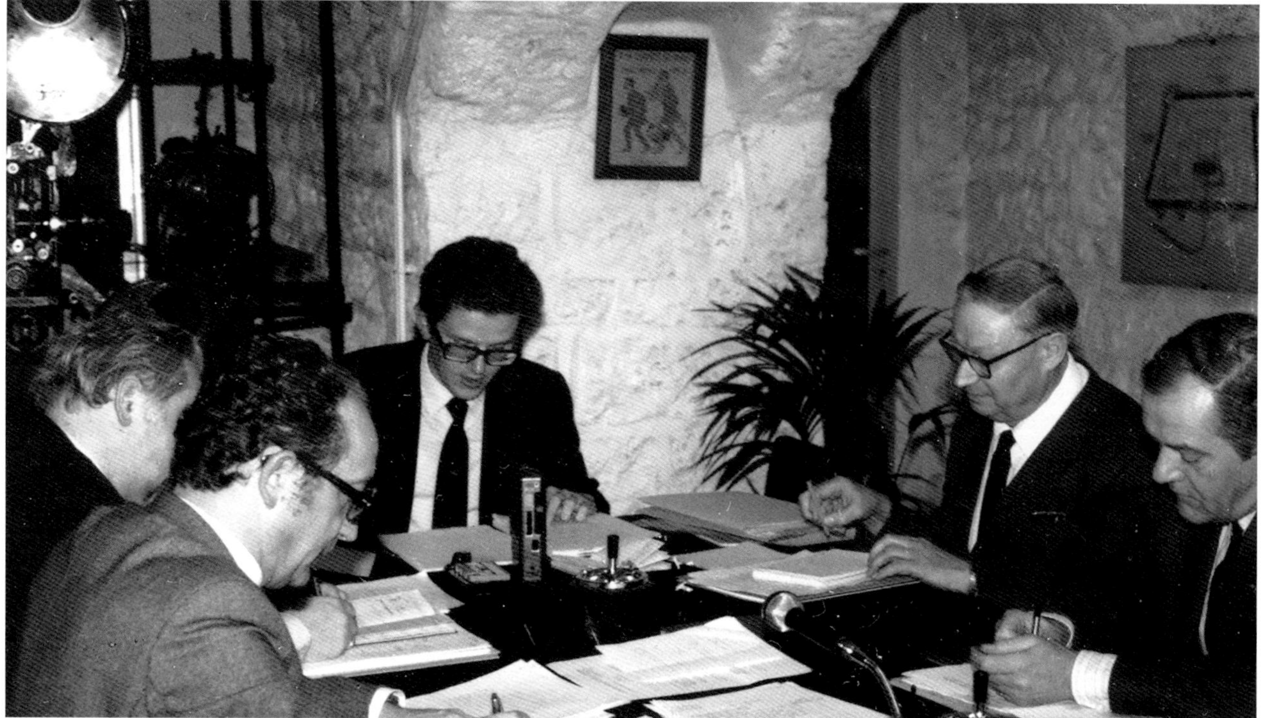

The FIAF Preservation Commission at work in the 1980s (left to right: Peter Konlechner, Hans Karnstadt, Henning Schou, Harold Brown, and Frantz Schmitt). Courtesy of Harold Brown's Estate.

Three years later, he also played a key role in the Rapallo Symposium on "The Problem of Colour Fading: A Statement of the Problem and of the Nature of Some Approaches to Solutions". Brown provided an analysis of the results of a survey on "colour films held in archives" he had organized prior to the Symposium, and delivered a paper entitled "The Problem of Colour Fading".

In 1983, he took part in the first Joint Technical Symposium held in Stockholm as part of the Congress, and gave a lecture on "Basic Film Handling", a largely expanded version of which was published as a Commission paper in April 1985, and soon became a much-used handbook by film archives around the world.

Brown retired from the BFI in August 1984. In his honour, the National Film Theatre organized a fitting tribute screening of the Douglas Fairbanks Technicolor epic *The Black Pirate* (1926), which he had famously restored and reprinted almost frame by frame. His professional retirement was not, however, the end of his contribution to the field of film preservation. Widely praised for his unique generosity, he devoted the next decade of his life to disseminating the expertise and knowledge he had accumulated throughout his unique career. He remained a very active member of the Preservation Commission, for which he wrote up a number of papers included in the Commission's *Technical Manual*, and continued to give presentations at FIAF Congresses (Canberra, 1986; West Berlin, 1987; Lisbon, 1989). He also became a tireless teacher of archive technical practices. He trained students at FIAF Summer Schools (1984 in Berlin; 1992 and 1996 in London) and other courses (the University of East Anglia's MA in Film Studies and Film Archiving; the Cineteca di Bologna's training programme), and travelled to many new or established film archives around the world, providing advice and training to fellow archivists (in the first two years of retirement alone he spent weeks in Manila, Canberra, Wellington, and Bangkok). It was during a visit to the vaults of the film archive in the Philippines that he famously coined the term "vinegar syndrome" to describe the chemical degradation of acetate film and its characteristic smell.[25]

He remained on the Preservation Commission until the mid-1990s. By that time, it was deemed time to rename it the "Technical Commission", redefine its remit to fit the reality of the 1990s film archiving world, and call upon a new generation of experts to serve on it. After 60 years of dedicated service to the cause of film preservation, he finally deserved a little rest. However, he kept himself informed of the latest developments in the field, and met his former international colleagues again at the 2000 Congress in London. When I interviewed him at his home in 2005, as part of the BFI history project I was involved in, he happily reminisced about a lifetime of passionate commitment to the cause of film preservation, looking back on it with the same generosity and modesty as ever.

25 Harold Brown, "Trying to Save Frames", in *This Film Is Dangerous: A Celebration of Nitrate Film* (FIAF, 2002), p.102.

His unique contribution to our field was not left unrewarded. He was awarded an MBE (Member of the Order of the British Empire) in 1967, was made an Honorary Fellow of the British Kinematograph, Sound and Television Society in 1984, received the Jean Mitry Award at the Giornate del Cinema Muto in Pordenone in 1987, and was elected a FIAF Honorary Member in 1992. He died on 14 November 2008, aged 89. A celebration of his life, attended by many of his friends and former colleagues, took place at the Alfred Rose Community Centre in Aylesbury on 8 December 2008.

Further Reading

Articles about Harold Brown

Patience Coster, "Harold Brown's Half Century at the National Film Archive", *Three Sixty: British Film Institute News*, August 1984, pp.12-14

Clyde Jeavons, "In Memoriam: Harold Brown (1919-2008)", *Journal of Film Preservation*, No. 79-80, May 2009, pp.106-109

Sabine Lenk, "Harold G. Brown and the Identification of Early Films", *The Moving Image*, Vol. 16, No. 1, Spring 2016, pp.35-56

Elizabeth Sussex, "Preserving: Harold Brown, Joining Boy of the Archive", *Sight and Sound*, Fall 1984, pp.237-238

Texts published by Harold Brown

Harold Brown, "Problems of Storing Film for Archive Purposes", *British Kinematography*, Vol. 20, No. 5, May 1952, pp.150-162

Harold Brown, "Film Identification by Examination of Film Copies", typewritten paper, 1967 FIAF Congress East Berlin

Harold Brown, "Technical Problems of Preservation", *Journal of the Society of Film and Television Arts*, No. 39, Spring 1970, pp.8-15

Harold Brown, "The Problem of Colour Fading: A Statement of the Problem and of the Nature of Some Approaches to Solutions", in *The Preservation of Colour Films: A Simple Examination of the Problem and the Solutions Currently Available*, FIAF Symposium, Rapallo, 1981, Appendix 2

Harold Brown, "Basic Film Handling", April 1985

Harold Brown, "Copying Small Amounts of Non-Standard Film Gauges", in Eva Orbanz (ed.), *Archiving the Audio-Visual Heritage: A Joint Technical Symposium*, Berlin: Stiftung Deutsche Kinemathek, 1988, pp.99-101

Harold Brown, "A Discussion Paper on Methods of Copying Tinted, Toned and Stencil-Coloured Films for Preservation and Presentation", FIAF, January 1993

Harold Brown, "Film Joins (Splices): Comments on Cement and Tape Splicers", May 1993

Harold Brown, "Survey of Printing Machines", FIAF, 1993

Harold Brown, "Trying to Save Frames", in Roger Smither and Catherine A. Surowiec (eds.), *This Film Is Dangerous: A Celebration of Nitrate Film*, Brussels: FIAF, 2002, pp.98-102

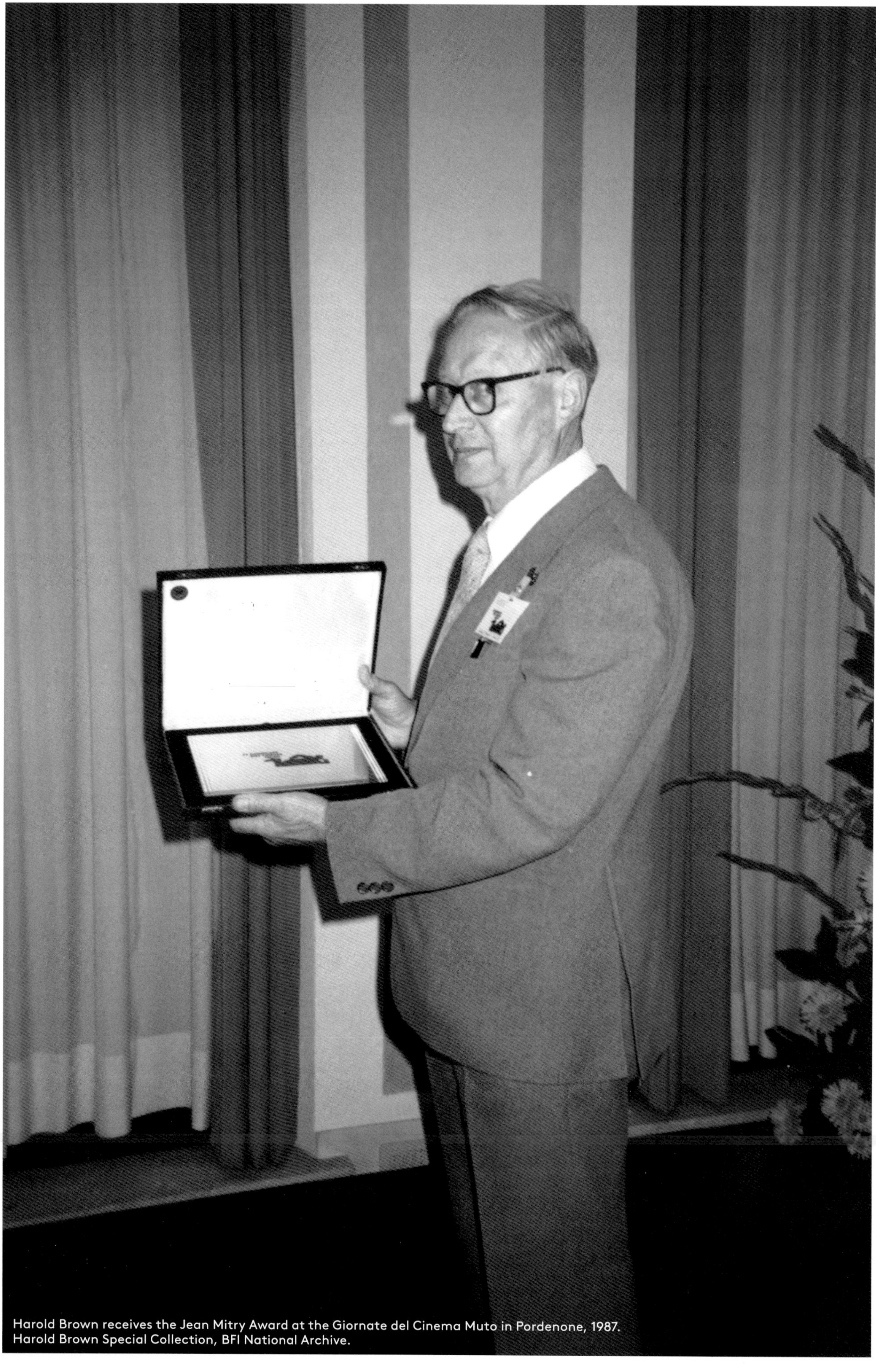

Harold Brown receives the Jean Mitry Award at the Giornate del Cinema Muto in Pordenone, 1987.
Harold Brown Special Collection, BFI National Archive.

Editor's Notes on the New Edition

Camille Blot-Wellens

In the process of working on this new, expanded edition of *Physical Characteristics of Early Films as Aids to Identification*, I tried to understand its genesis, its chronology, and its methodology — as well as the reactions of the FIAF community to this project. This study was made possible thanks to three collections: the personal archives of Harold Brown at the British Film Institute (referred to in the text and footnotes as HB-BFI), the archives of Einar Lauritzen, former Treasurer of FIAF and later Honorary Member, at the Swedish Film Institute (EL-SFI), and the FIAF Historical Archive, partly digitized by the Secretariat (FIAF).[1]

Genesis

Harold Brown started working at the National Film Library on 18 March 1935, as an office boy. He was then not even 16 years old. He soon became increasingly involved with film projection and loans. In 1939 the National Film Library moved to Aston Clinton, and after the war Brown worked with the NFL's Assistant Film Examiner Cedric Pheasant on the physical diagnostics and inspection of the films which entered the Library's collections. Among them were many films from the silent period.

In 1952, Brown, then Preservation Officer, published a study on film storage in *British Kinematography*, the journal of the British Kinematograph Society. This text is also very interesting to understand his relation to the materials: "for many years at the beginning of the history of the industry there was almost no standard. Perforation shapes and sizes varied from maker to maker, and even among the products of a manufacturer".[2] So, in the early 1950s Brown had already noticed the importance of the physical characteristics of films, not only for duplication, but also for identification and dating.

In that same period, the archive's curator Ernest Lindgren contacted Kodak regarding edge codes observed by Brown on the film materials. Kodak answered Lindgren on 27 July 1953, providing the key to understanding them. But Kodak asked that the information remain confidential: "the list of symbols is marked confidential, and I understand you will treat it as such in the sense that the list will be only known to whose working within your organisation."[3] It would not be revealed for many years, since Lindgren would never agree to violate Kodak's request, as a memo sent by Lindgren to Brown in January 1967 reveals: "Since these symbols are regarded as confidential it is impossible for us to disclose the systems of the various manufacturers (...). FIAF members must approach manufacturers or their representatives in their own countries and try to obtain this information individually."[4]

Nevertheless, "probably around the end of the 1950s, (...) Lindgren (...) encouraged the author [Brown] to try to describe the tangible features by which he recognised the product of the different makers, and to write it down and illustrate it for the benefit of other film archivists".[5]

---

1 The author warmly thanks Carolyne Bevan and Nathalie Morris (British Film Institute Special Collections) for giving her access to the Harold Brown Collection in Berkhamsted in early June 2016, and Ola Törjas (Swedish Film Institute) for all his help with the Einar Lauritzen Archives at the Film House in Stockholm in February and March 2019.

2 *British Kinematography*, Vol. 20, No. 5 (May 1952), p.158.

3 Letter from C. J. Craig (Technical Service – Motion Picture Film Division, Kodak Limited) to Ernest Lindgren (27 July 1953) – HB-BFI.

4 Memorandum (17 January 1967) – HB-BFI.

5 Harold Brown, *Physical Characteristics of Early Films as Aids to Identification*, Brussels: FIAF (1990), p.2. Quoted phrasing as in the 2020 edition: see p.39 for clarification on the use of "tangible".

Ernest Lindgren would be decisive in this work. In the Minutes of the FIAF Executive Committee meeting held in Paris in early November 1965, the project was then referred to as a "Method of ascertaining to which 'generation' a copy belongs", and a full report was promised for the following meeting, at the end of February 1966. However, Lindgren would have no news to announce at the next meeting, and had no other option than to suggest a "plan or a proposal" for the FIAF Congress in Sofia in May 1966. The situation bothered him, as we see in this exasperated memo: "At the FIAF Committee meeting last week-end, I was asked if there was any hope of seeing your projected paper on methods of 'reading' the history of a film copy by examining it. Is there? Even a short, preliminary report, listing, for example, the various forms of evidence – sprocket holes, edge-marks, etc. – would be most welcome, and better than nothing at all. If you see no hope of doing it, I will ask FIAF to delete it from their Agenda. This would be better than continuing to hold unfulfilled hope."[6]

Brown began to devote himself to writing the document, and was able to send a first draft to Lindgren that spring, consisting of a text of almost 45 pages (double-spaced), which Lindgren extensively edited, deciding its title and its structure. Lindgren also wrote the introduction.[7]

The text was ready in May 1967, and the following month was presented and distributed to all FIAF members attending the Congress in East Berlin, as a 30-page paper, whose title, "Film Identification by Examination of Film Copies", was suggested (or decided) by Lindgren. Brown not only presented the information he had gathered and the methodology he had used, but took the opportunity to make an appeal to the FIAF community: "Much more work could be usefully done to work out these, and other ideas, in greater detail, and I very much hope that other FIAF members will be encouraged by this paper to record their own discoveries and their own experience, so that they can be pooled for the benefit of us all."

What was the reaction of the FIAF community? According to the Minutes of the East Berlin Congress, Jacques Ledoux (curator of the Royal Film Archive of Belgium) suggested that the efforts of Harold Brown and Myrtil Frida (historian and film archivist in Prague) be combined into a joint publication. Stanislav Zvoniček, director of the Czechoslovak Film Institute, and its curator Bohumil Brejcha offered to invite Brown to a symposium on film identification, to be held soon. Eventually, a few months later (March 1968), the first symposium on the "Identification of old films" was organized by the Czechoslovak Archive in Gottwaldov.

Brown's impressions after the event were not entirely positive[8]:

- "Many films in bad condition were projected and some damage thereby inflicted."
- "To be useful to archivists around the world, this is not enough. It is surely necessary that the details, on which the comments, convictions are founded, be written down, as the data in the Preservation Commission report are written down. Unfortunately in this sphere there is a vital difference between the natures of the matter of the two Commissions[9]; for temperatures and vault sizes and such like are susceptible to measurement, and precise statement and they can be written in a form easily consulted; but styles of set decoration, recognition of actors and such clues to identity do not lend themselves simply to precise written communication."
- "There was plenty of opportunity at the lecture sessions for any participants to put questions to the lecturers about their methods of identification, and to state any means of identification which they themselves used, but there was almost no such questioning or contribution."

The Symposium also raised an interesting thought in Brown's mind: "What are we to regard as constituting 'Identification'? In the case of some films there is on record almost all that there is to be known. In very many cases we can, perhaps, never hope to know more than a little." It seemed to him that it was not appropriate to think of and refer to a film as "Identified" or "Unidentified", but rather to think in terms of a statement of known data.

6 Inter-Office Memo from Ernest Lindgren to Harold Brown (4 March 1966) – HB-BFI.
7 Letter from Ernest Lindgren to Harold Brown (28 April 1967) – HB-BFI.
8 *Notes on the Symposium on Identification of Early Films* (no date) – HB-BFI.
9 Here Brown seems to refer to the Preservation Commission and to the attendees of the Symposium as another "Commission".

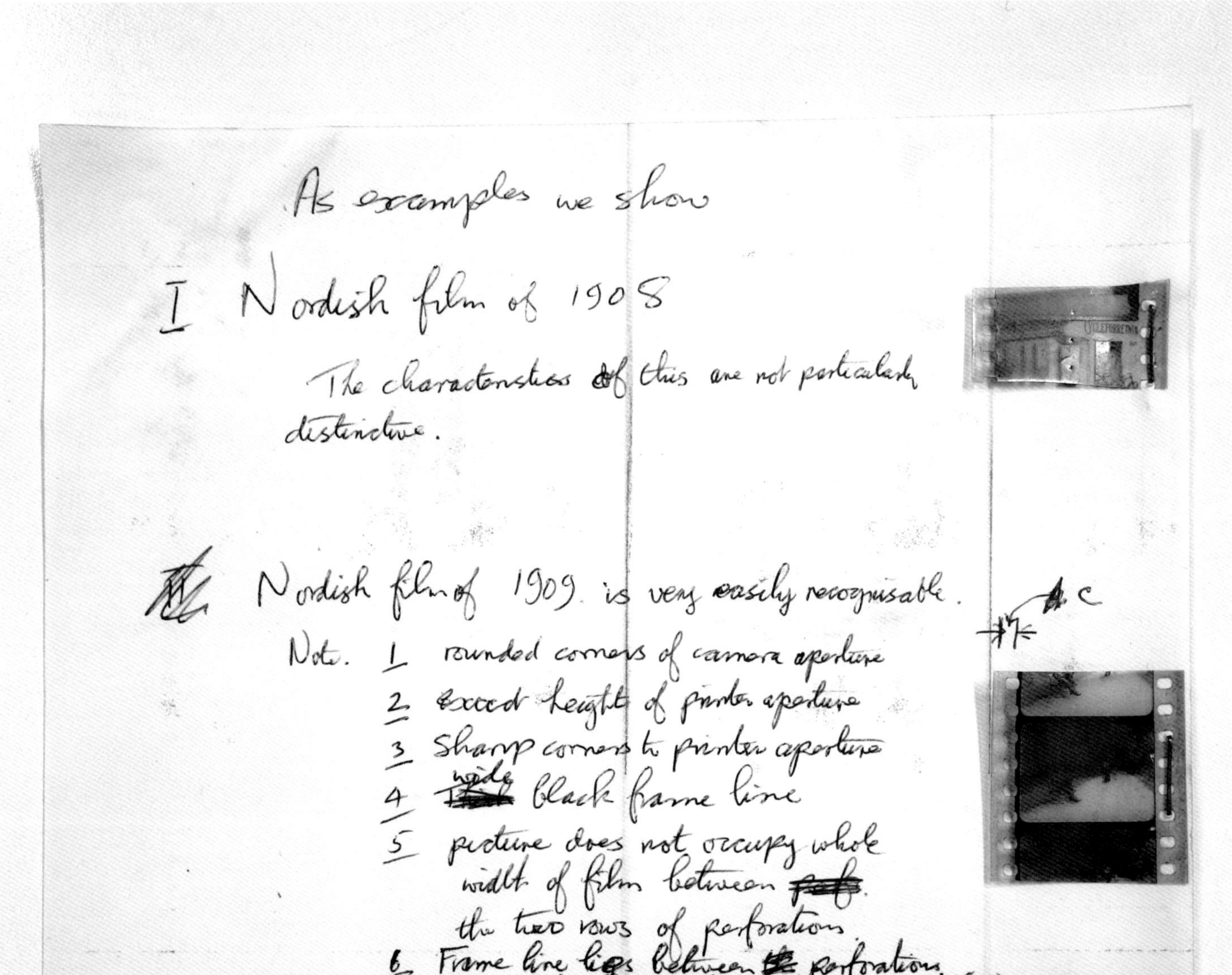

As examples we show

I Nordish film of 1908

The characteristics of this are not particularly distinctive.

II Nordish film of 1909 is very easily recognisable.

Note.
1 rounded corners of camera aperture
2 exceed height of printer aperture
3 Sharp corners to printer aperture
4 ~~black~~ wide black frame line
5 picture does not occupy whole width of film between ~~perf~~ the two rows of perforation
6 Frame line lies between ~~the~~ perforations
7 Note also the ~~sharp~~ perforation; the dimension 'C' is less than in the other samples

Nordish film of 1911–14

note.
1 Wide frame line
2 picture encroaches slightly onto the two rows of perfs.
3 Frame line lies between perforations

Handwritten draft of Harold Brown's 1967 paper "Film Identification by Examination of Film Copies", distributed to the delegates of the 1967 FIAF Congress in East Berlin. Harold Brown Special Collection, BFI National Archive.

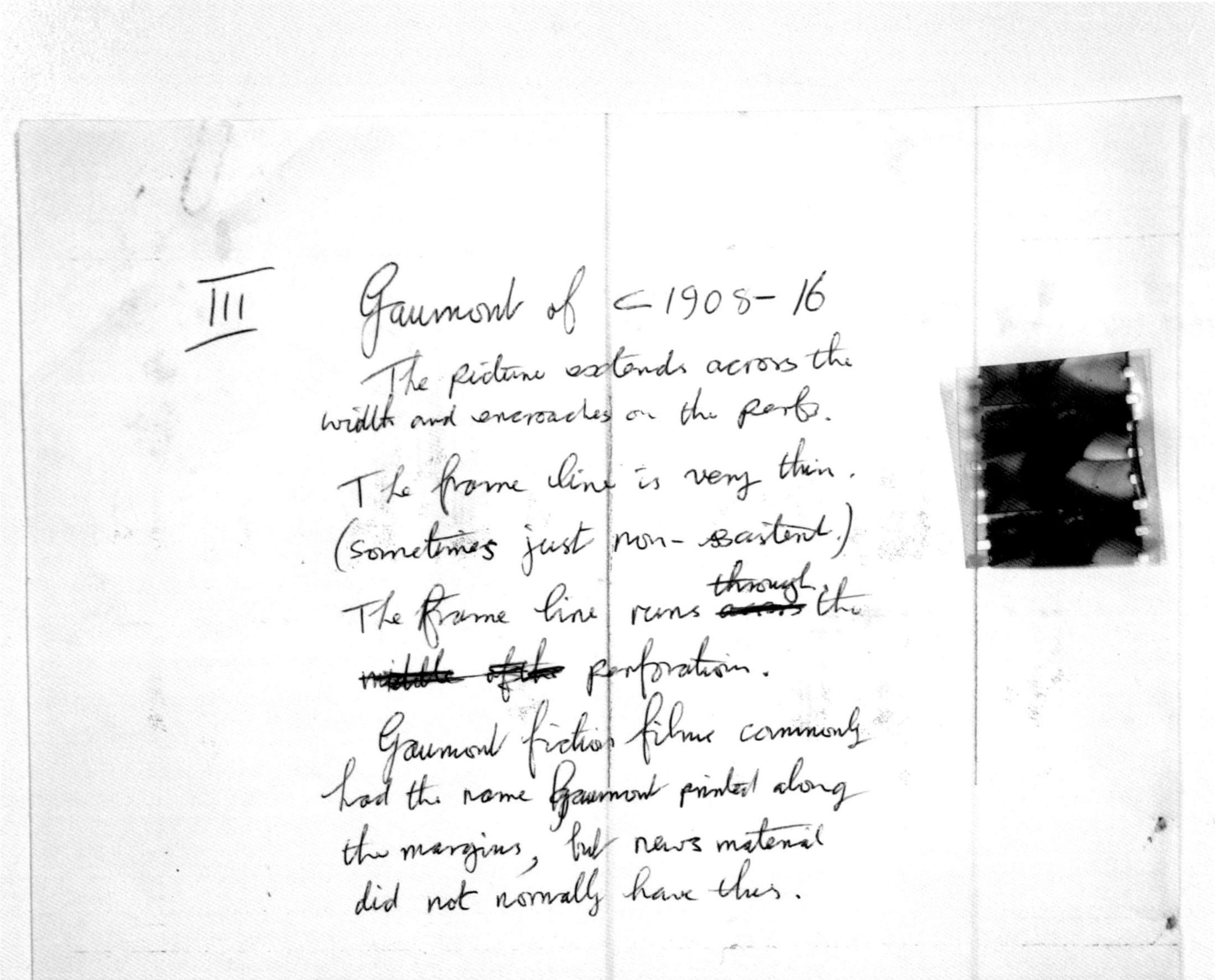

III Gaumont of c1908–16

The picture extends across the width and encroaches on the perfs.

The frame line is very thin. (sometimes just non-existent.)

The frame line runs ~~across~~ through the ~~middle of the~~ perforation.

Gaumont fiction films commonly had the name Gaumont printed along the margins, but news material did not normally have this.

IV Pathe of c1908–16

Note the thickness of the frame line.

slightly rounded corners to picture.

~~the centre of the frame line is darker~~

There is a darker line within the frame line.

Frame line lies between ~~the~~ perforations

Normally the picture does not encroach on the perfs. ~~Pa~~

Pathe fiction material ~~had~~ normally had edge marks as described in section I.

This was not normally printed on news ~~material~~ films

Handwritten draft of Harold Brown's 1967 paper "Film Identification by Examination of Film Copies", distributed to the delegates of the 1967 FIAF Congress in East Berlin. Harold Brown Special Collection, BFI National Archive.

Nevertheless, during the following FIAF Congress, which took place in London a few weeks later (May 1968), the participants of the symposium suggested the creation of a Permanent Commission on the Identification of Old Films.[10] Brejcha even offered to host the Commission's meetings, but such a commission would never be created.

Soon after, in 1970, Harold Brown joined the Preservation Commission. From this moment on, for 25 years, he became a very active member of the Commission.

But during the FIAF Congress held in Lyon in May 1970, Herbert Volkmann (ex-director of the East German State Film Archive, and Head of the Commission) announced that in November 1969 the Commission had decided to focus exclusively on the problems of the preservation of colour film.[11]

And things "got worse" for Film Identification when, at the Executive Committee meeting in Kleinmachnow in November 1970, Lindgren communicated that Brown did not wish to participate in the Symposium on Film Identification planned for spring 1972 in Prague. Even Myrtil Frida expressed his "disappointment about the lack of interest that the big archives showed in that subject".[12]

From this time onwards, the enthusiasm expressed after Brown's 1967 presentation in East Berlin started to fade away, and rather soon afterwards the subject of Film Identification would totally disappear from the ongoing projects listed by the Executive Committee.

Nevertheless, Brown would go on working on this topic, studying materials in the National Film Archive. His work on the films of the early years would be crucial. According to Clyde Jeavons, "a culmination of Harold's work on the cinema's first moving images was his overseeing of the preparation and projection of the 548 films made between 1900 and 1906 shown and studied at the seminal FIAF Congress symposium held in Brighton in 1978".[13]

Around the same time, Brown went on gathering information on film stock, notably Kodak, for which he would update the information in June 1981 and August 1983. Later, towards 1985, Brown, then retired from the NFA but still an active member of the Preservation Commission, helped Frantz Schmitt (Head of the archives of the CNC) on a project on film stocks, establishing, among other things, an English "interpretation" of the document (originally in French).[14]

On 2 October 1985, Brown delivered the annual Ernest Lindgren Memorial Lecture at the British Film Institute. The subject was, naturally, "Physical Characteristics of Early Films". During his talk Brown sadly noted that "It was pointed out in the [1967] booklet that much more information could be obtained by further study of the physical features of these early films. Since then a little more information has been extracted."[15]

Almost 20 years after the 1967 East Berlin Congress, which was followed by unsatisfying initiatives, and 7 years after the 1978 Brighton Symposium, which brought to light the early years of cinema, knowledge on the identification of early films was still limited, and to a large extent limited to Brown's knowledge, besides some isolated works by others. This was a sad and disturbing situation for Brown, who repeated his call to the FIAF community, now almost a warning: "Much more could usefully be done, given the opportunity, before the nitrate originals are finally lost to us."

Did the 1985 Ernest Lindgren Lecture revive interest in the identification of early films? It seems so. In November 1986, the Preservation Commission (then headed by Henning Schou, curator at the National Film & Sound Archive in Australia) offered to "publish an update of Brown's 1967 paper on identification of early film stock illustrated with quality photographs".[16]

---

10 Bohumil Brejcha, *Proposal made by the participants of the Symposium on Identification of old films at Gottwaldov, Czechoslovakia, March 1968* (25 May 1968) – EL-SFI.
11 Report of the Preservation Commission, *Minutes of the FIAF Congress* (May 1970) – EL-SFI.
12 *Minutes of the Executive Committee Meeting* (Kleinmachnow, November 1970), p.21 – EL-SFI.
13 Clyde Jeavons, "Harold Brown (1919-2008)", *Journal of Film Preservation,* No. 79-80 (April 2009), p.108.
14 English translation of *Étude récapitulative sommaire des principales émulsions cinématographiques disponibles sur le marché mondial* (Catalogue of All Major Film Stocks) – HB-BFI.
15 *Physical Characteristics of Early Films* (Unpublished typescript of the 1985 Lindgren Memorial Lecture) – HB-BFI. "Little" was underlined by Brown.
16 *Minutes of the Executive Committee Meeting* (Glasgow, November 1986), p.23 – FIAF.

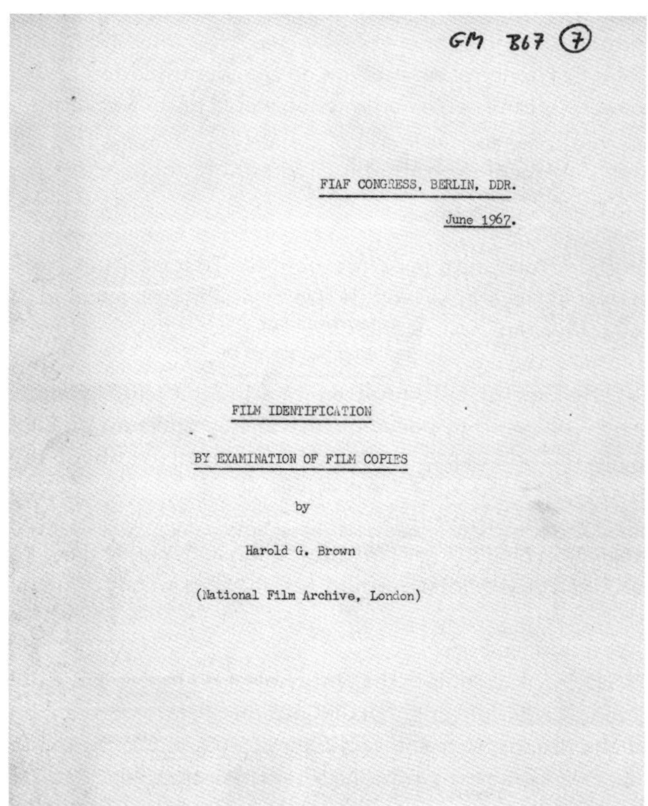

FIAF CONGRESS, BERLIN, DDR.

June 1967.

FILM IDENTIFICATION

BY EXAMINATION OF FILM COPIES

by

Harold G. Brown

(National Film Archive, London)

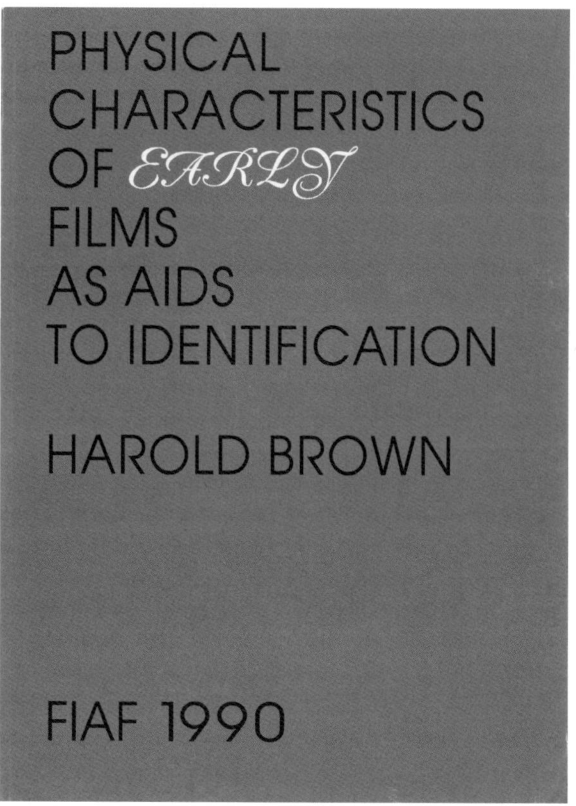

Covers of Harold Brown's 1967 paper "Film Identification by Examination of Film Copies", and 1990 book *Physical Characteristics of Early Films as Aids to Identification*. FIAF Historical Archive.

Therefore, in 1987 Brown started working on this project, studying more materials in London, but also in Canberra and Prague. In the same period, during the Berlin Congress of May 1987, Jiři Levy from Prague offered to organize a new symposium on film identification, "like the one they held some 20 years ago. He thought it was an important part of the work of the archives".[17]

Brown's new publication, *Physical Characteristics of Early Films as Aids to Identification*, was ready at the end of 1989, and was sent to Seoul to be printed by the Korean Film Archive. The 30-page paper presented in 1967 had become a book with 81 pages of text and almost 250 illustrations, a work which has since been used on a daily basis by most archives worldwide.

Methodology

In the 1967 paper, Brown described his methodology "to provide a complete and accurate identification and description of a film for cataloguing". The paper focused on the films produced during the silent era, more specifically before World War I.

The features he enumerated and described were of very different kinds, and required very diverse approaches and knowledge. The most obvious ones were the marks introduced on the edges, on the titles, or as internal evidence, like logos/trademarks appearing in the sets onscreen, introduced by the producers in order to be seen by the customers and the audience when a film was purchased or screened. But there was also technical data related to the manufacture of film stocks or the making of the films: frame-lines and aperture shapes, exposure of the edges, edge printings, perforations, and other identifiers introduced by the film stock manufacturers, as well as, for instance, specific practices of producers that could be observed in the editing or the use of tinting. Other information that can help in dating a film can be, for instance, changes in the address of the board of censors (as in Britain).

17 *Minutes of the FIAF 43rd General Assembly* (Berlin, May 1987), pp.41-42 – EL-SFI.

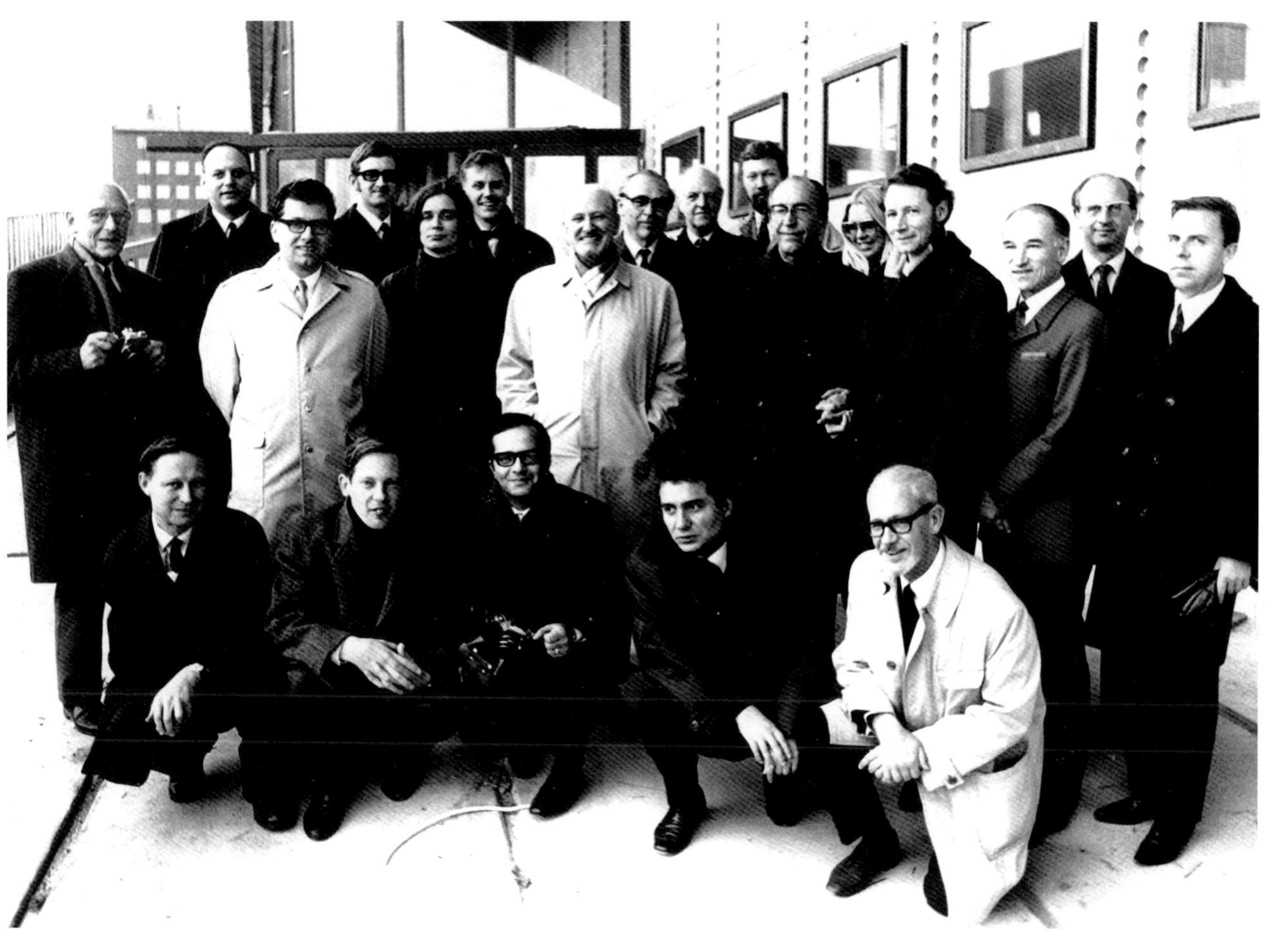

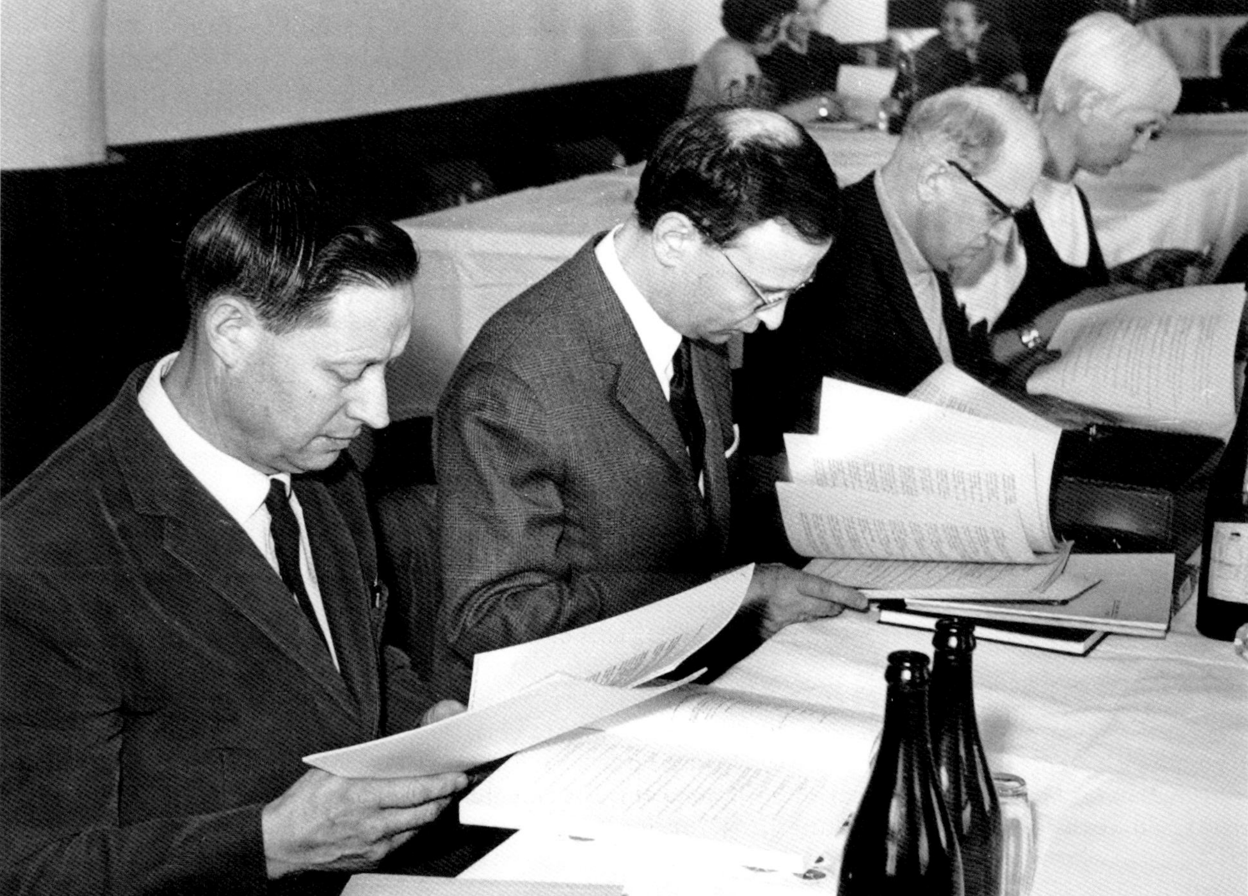

Top photo: Group photo of the first FIAF Preservation Commission meeting attended by Harold Brown (bottom left) at the Swedish Film Institute in Stockholm in October 1970. FIAF Historical Archive. Bottom photo: Harold Brown at the symposium on "Identification of old films" held in Gottwaldov (Czechoslovakia) in March 1968. Harold Brown Special Collection, BFI National Archive.

The 1990 edition of *Physical Characteristics of Early Films as Aids to Identification* is not only a unique tool to identify early films and several film stocks, it also – and this aspect may be in some way the most relevant and important – suggests a methodology based on the study of the materials and their physical features, what they mean and how they can help us identify, date, and understand films. Brown gathered under the umbrella "physical characteristics" all the concrete attributes introduced by manufacturers or production companies (that is, all the visual features observable on film materials), whether they pertain to the manufacture of film stock, or the making, editing, printing, or processing of the film. This approach puts the film back into its technological context, inseparable from its cultural context, and thus guides the identification of the film as a cultural artefact, an interaction highlighted by Brown: "You cannot fully identify a mysterious film by use of physical characteristics alone. Reference to written sources of information is vital for this purpose. What the physical characteristics can do is to materially narrow down the area within which it is necessary to search the literature".[18]

Brown's writings are an invitation to archivists to study the film element as deeply as possible, and to try to determine the period and company of its production by combining all the information available. To study, describe, write down, and compare in a consistent and relevant way the materials in collections may sometimes be the only way to improve not only our knowledge of the films, but also of the history and the techniques of the early years of film.

Since 1967, Harold Brown had insisted that "[it] is important for readers to realise that this paper is by no means exhaustive. Study of as many as possible of other producers' films would yield more information of the same kind. It is to be hoped that other workers who have access to relevant films will contribute to FIAF any information which they can find."[19] To gather information that has been found since then is precisely the idea behind this new expanded edition. Beware, not all the information, but the information that film archivists and researchers have been able to find, and have agreed to share "for the benefit of us all".

Notes on the Present Edition

Chronology

The origins of this new edition reside in the fact that for years film archivists and researchers have studied the materials and their characteristics without always making the information available to a broader audience. It was therefore necessary to share the information as extensively as possible, but also very important to conduct further research while persons who had the knowledge were still around (especially among film manufacturers). In autumn 2014, as a member of the Technical Commission, I submitted the project to the Executive Committee, and first presented it to the FIAF community during the Congress held in Canberra in April 2015.

The first step was to contact archivists and researchers who, as far as I knew, had some new information. The first reactions were very enthusiastic, and I was able to gather information provided by several archivists and researchers (Luciano Berriatúa, Martin Koerber, Egbert Koppe, Brian Pritchard, Akira Tochigi, and Nikolaus Wostry).

The second step was to decide which materials to focus on, since it was impossible to consider studying all the film manufacturers or production companies within the scope of this project. It seemed important from the beginning to research and study production companies or film stocks that might be represented in many archives. It is equally important to underline that better knowledge comes from the study of the actual materials whenever possible. Therefore, the new contributions of this edition are linked to the collections studied by the authors.

The next step was to contact film manufacturers or corporate archives to try to find more information. Unfortunately, in some cases the paper archives of the firms had disappeared, and it was impossible to get more information (as in the case of Ilford). In other instances the research proved to be very productive: a project developed at the Swedish Film Institute (2014-2015) and the notebooks of the Pathé engineers from CECIL

18 Brown, *Physical Characteristics of Early Films as Aids to Identification*, p.41 (of the current edition).
19 Brown: *Physical Characteristics of Early Films as Aids to identification*, p.40 (of the current edition).

(Cercle des Conservateurs de l'Image Latente) made available by the Fondation Jérôme Seydoux-Pathé were very useful in gaining better knowledge of Pathé materials (autumn 2015). Frank Böhme from FilmoTec (May 2015 and June 2019), and Uwe Holtz and Manfred Gill from the Industrie- und Filmmuseum Wolfen for Agfa and Orwo, were very generous during my visit (May 2016). Marc Sutherland from Agfa was very helpful regarding the edge printing system adopted by Gevaert and Agfa-Gevaert (January 2016). Between 2016 and 2018, I was invited by Shivendra Singh Dungarpur to be part of the Film Preservation and Restoration Workshops organized every year by the Film Heritage Foundation in India, which allowed me to study more Orwo and Indu film stock.

Also, in order to better understand the genesis of Harold Brown's methodology, I spent a few days at the British Film Institute's J. Paul Getty Jr. Conservation Centre in Berkhamsted in June 2016, studying his personal archives.

Among the persons contacted, Luciano Berriatúa was the first to accept to contribute to this new edition, sharing his study on Agfa film stocks of the 1920s and 1930s. In June 2016 James Layton was the next to approach me about contributing to the project, by studying Kodak materials from MoMA, and in October 2016 Peter Bagrov accepted to help with Soviet film stock.

During the spring of 2018 I worked on the Madrid Project – developed in the early 2000s by Alfonso del Amo, then Head of the Technical Commission – in order to make it available to the FIAF community on the FIAF website. These two projects may be different, but they are both motivated by the need for a better knowledge of film stocks. The Madrid Project gathers more than 1200 documents, between 1912 and 2005, mainly technical sheets from film manufacturers, but also manuals and articles provided by archives, laboratories, and researchers.

Entering into the final phase (2018), I confirmed the contributions of several more archivists and researchers: Christophe Dupin offered to write a biographical essay on Harold Brown; in June, Jacques Malthête agreed to write a text on Méliès; in October, Eric Loné shared several essays on French film companies prepared by the archivists of the CNC; in November, Hidenori Okada accepted to write a text on Fuji, and Brian Pritchard authorized the insertion of the information he had sent on various film stock manufacturers. During the Berlinale in February 2019, Martin Koerber agreed to write the Foreword.

This new edition may be incomplete, but it is the result of years of research and analysis of film stocks and production companies conducted by archivists in different countries. And it is also the first attempt (to my knowledge) to gather all the information into one publication. I want to believe that, in a way, the wish expressed so many times by Harold Brown, of seeing film archivists and researchers working on collections in order to better know and understand, and share their knowledge and understanding with the FIAF community, has been fulfilled.

Structure

After working for 30 years with the 1990 edition of *Physical Characteristics of Early Films as Aids to Identification*, readers will notice some changes in the layout and organization of the sections in this new edition. There are multiple reasons for this. We hope that our explanations of these changes make sense to readers and will help them to find their way.

One reason for this new edition was to provide better illustrations in a more comprehensive way, integrated with the text. Today Harold Brown's 1990 system to refer to the illustrations in a separate section is no longer relevant, as they are now reproduced in black & white directly within the text.

A special Colour Section has now been added to illustrate the title styles and tints. For practical reasons, it was decided to group the reproductions of all the colour illustrations together.

All the separate essays about companies (Gaumont, Cines, Vitagraph, and Éclair) in the 1990 first edition are now incorporated in the Section "Title Styles". By doing this, we do not feel we are doing anything wrong, since Harold Brown indicated references to these essays in his text. As for the tables of film listings by company, it was considered relevant to reproduce them at the end of Brown's text.

Brown's essays about Pasquali and Selig illustrate the use of company production numbers perfectly, so for this reason they are now incorporated in the Section "Production Serial Numbers".

Brown's texts on Thanhouser and Hepworth have been left as independent essays, as their contents are too miscellaneous to be incorporated under specific topic headings.

The numbering of the paragraphs has been revised in an attempt to harmonize the system used by Brown and respect the hierarchy of the text.

When new information has been added to Brown's text (such as that provided by Eric Loné on Gaumont), it is introduced in footnotes, and in the cases that they are in the main text (such as in the Section "Frame Characteristics"), they are indicated with an asterisk (*).

When relevant, a film's original title has been added, as well as the English-language title used for British or American distribution. The sources for these are as follows:

Books and theses:

Richard Abel: *The Ciné Goes to Town: French Cinema, 1896-1914*, Berkeley: University of California Press (1994)

Aldo Bernardini & Vittorio Martinelli: *Il Cinema muto italiano*, Roma: Centro Sperimentale di Cinematografia / Nuova Eri (1991-1996)

Aldo Bernardini: *Cinema muto italiano. I film "dal vero" 1895-1914*, Gemona: La Cineteca del Friuli (2002)

Frédéric Delmeulle: *Contribution à l'histoire du cinéma documentaire en France. Le cas de l'Encyclopédie Gaumont (1909-1929)*, Villeneuve d'Ascq: Le Septentrion (1999)

Philippe d'Hugues & Dominique Muller (eds.): *Gaumont. 90 ans de cinéma*, Paris: Ramsay – La Cinémathèque française (1986)

Francis Lacassin: *Louis Feuillade. Maître des lions et des vampires*, Paris: Pierre Bordas et Fils (1995)

Laurent Le Forestier: *Les Films comiques produits par Gaumont entre 1907 et 1914. Mémoire de maîtrise sous la direction de M. Michel Marie*, Université Paris 3 (1992-1993)

Bénédicte Salomon: *Filmographie de la Société française des films et cinématographes Éclair (1907-1918). Mémoire de maîtrise sous la direction de M. Jean Tulard*, Université Paris IV (1987)

Corporate journals of the period:

Ciné-Journal; The Cinema News and Property Gazette; The Film Index; Motography; Moving Picture News; Moving Picture World; The New York Clipper; The Nickelodeon; Reel Life; Views and Film Index.

Websites:

British Film Institute catalogue online; Internet Movie Database (IMDb); the Pathé catalogues reconstituted by Henri Bousquet and made available online by the Fondation Jérôme Seydoux-Pathé; and the Media History Digital Library, an invaluable resource where many original publications can be consulted.

This edition's new contributions are presented in Part II, after Brown's text. These new chapters deal with early film companies as well as film stock manufacturers, since these two aspects were also addressed by Harold Brown. Some more recent manufacturers of film stocks have now also been included.

The texts delivered in French were translated into English by Aymeric Leroy, and the text in Spanish was translated by Itzíar Gómez Carrasco.

Illustrations

Insofar as possible, I have tried to use the same illustrations as Harold Brown. The negatives used for the reproductions in 1990 are kept at the FIAF offices in Brussels, and I thank Christophe Dupin for giving me access to them. A great part of these documents were scanned in 2010 by Sabine Lenk, who kindly allowed me to use them. Some illustrations were not reproduced in the 1990 edition; these are credited "HB" in the current edition.

Finally, in order to reproduce some of the images in colour (in the "Title Styles" section), I spent several days at the BFI National Archive's collections in Berkhamsted in April 2019, to take new photographs of the materials that could be located. This would not have been possible without the invaluable help of Bryony Dixon, Jane

Fernandes, and Kieron Webb. I also was able to take photographs of Cines films from the Cineteca di Bologna at the laboratory L'Immagine Ritrovata in June 2019. Many thanks there to Davide Pozzi, Andrea Meneghelli, Marianna De Sanctis, and Maura Pischedda.

In some instances, it was necessary to replace the original illustrations with new ones. In these cases, the name of the original collection is mentioned (see below).

All additional illustrations not in the original 1990 edition are indicated with an asterisk (*).

I wish to take this opportunity to warmly thank not only all the archives who gave me access to their collections (for almost 20 years), but also the archivists who helped by sharing images or by scanning frames especially for this new edition. The illustrations in the new essays have been provided by their respective authors.

When the archive source is known, it is indicated by its acronym or initials, as follows:

AFA: Academy Film Archive (Los Angeles, USA). Photographs by Joe Lindner

BA: Bundesarchiv (Berlin, Germany). Photographs by Florian Wrobel

BFI: British Film Institute (London, UK). Photographs by Camille Blot-Wellens and Jane Fernandes

CdB: Cineteca di Bologna (Bologna, Italy). Photographs by Camille Blot-Wellens

CF: Cinémathèque française (Paris, France). Photographs by Camille Blot-Wellens

CNC: Direction du Patrimoine du Centre national de la Cinématographie (Bois d'Arcy, France). Photographs by Eric Loné

CRB: Cinémathèque royale de Belgique / Koninklijk Belgisch Filmarchief (Brussels, Belgium). Photographs by David Gruwez

CS: Cinémathèque suisse (Lausanne, Switzerland). Photographs by Noémie Jean and Luciano Berriatúa

FE: Filmoteca Española (Madrid, Spain). Photographs by Camille Blot-Wellens and Encarni Rus

FINA: Filmoteka narodowa – Instytut audiowizualny (Warsaw, Poland). Photographs by Camille Blot-Wellens

GEM: George Eastman Museum (Rochester, USA). Photographs from the Turconi and Joye collections provided by Nancy Kauffman

GF: Gosfilmofond (Moscow, Russia). Photographs by Peter Bagrov, Olga Dereviankina, Alisa Nasrtdinova, and Artiom Sopin

GTsMK: State Central Film Museum (Moscow, Russia). Photographs by Peter Bagrov and Marianna Kushnerova

IFM-W: Industrie- und Filmmuseum Wolfen (Wolfen, Germany). Scans by Uwe Holz

MoMA: Museum of Modern Art (New York, USA). Photographs by James Layton

NFA: Národni filmový archiv (Prague, Czech Republic). Photographs by Jeanne Pommeau

NFAI: National Film Archive of India (Pune, India). Photographs by Camille Blot-Wellens

NFAJ: National Film Archive of Japan (Tokyo, Japan). Photographs by Hidenori Okada

RGAKFD: Russian State Archive of Film and Photo Documents (Krasnogorsk, Russia). Photographs by Peter Bagrov

SFI: Svenska Filminstitutet (Stockholm, Sweden). Photographs by Camille Blot-Wellens

VGIK: All-Russian State Institute of Cinematography (Moscow, Russia). Photographs by Peter Bagrov and Olga Chizhevskaia

In some cases, the illustrations have been provided by individuals:

LB: Luciano Berriatúa

BP: Brian Pritchard

DC/BP: David Cleveland and Brian Pritchard

Many thanks also to Brian Pritchard and David Cleveland for allowing us to reproduce some of the illustrations from their publication *How Films Were Made and Shown. Some Aspects of the Technical Side of Motion Picture Film (1895-2015)*, Manningtree, Essex: David Cleveland (2015).

Acknowledgements

I would like to warmly thank the contributors to this new, expanded edition of Harold Brown's *Physical Characteristics of Early Films as Aids to Identification*:

Peter Bagrov, for researching Soviet film stocks and gathering features to identify Russian and Soviet films;

Luciano Berriatúa, with whom all this started 20 years ago, for his crucial contribution on Agfa;

Christophe Dupin, for writing a remarkable biographical piece on Harold Brown, highlighting the richness of his knowledge and his pioneering work in the archive field;

Martin Koerber, for supporting the project from the beginning, sharing information, our enjoyable technical talks, and writing the Foreword;

James Layton, for his enthusiasm, and for spending so many evenings writing the most comprehensive and exhaustive study of Eastman Kodak's edge-printing system ever;

Pierrette Lemoigne, for enriching our knowledge on Éclair;

Eric Loné, who supported the work of Harold Brown, preparing a French version of the book as early as 1992[20], and contributing in many aspects to this new edition, providing both new texts and illustrations;

Jacques Malthête, who agreed to write an exhaustive paper on the films of Méliès, offering a complete and documented perspective on their identification;

Hidenori Okada, for his contribution on Fuji, including the latest emulsions, and Akira Tochigi, for his help and for translating the text into English;

Brian Pritchard, who has been so helpful from the beginning of the project, and was so generous in sharing knowledge.

And a host of others who helped to make this project come true:

Elaine Burrows and Sabine Lenk, for sharing with us the scans of Harold Brown's materials.

Bryony Dixon, Jane Fernandes, Kieron Webb, Carolyne Bevan, and Nathalie Morris, for their invaluable help with the collections of the British Film Institute (London and Berkhamsted).

Manfred Gill and Uwe Holz (Industrie- und Filmmuseum Wolfen), for their hospitality and all the information they shared with us.

Stéphanie Salmon and Anne Gourdet-Marès (Fondation Jérôme Seydoux–Pathé, Paris), for their hospitality, generosity, patience, and knowledge.

Marc Sutherland (Agfa, Mortsel), for his kindness and generosity.

Florian Wrobel, for his fundamental contribution to the Orwo study.

Jean-Pierre Martel, former president of Kodak Industrie and founder of the Cercle des Conservateurs de l'Image Latente (CECIL, Fragnes), and Jean-Pierre Thouvenot, former engineer of the Kodak Technical Department of Vincennes and later Head of the Departments of Finishing, Papers and Films, in Chalon, for their help and generosity.

20 *Les Caractéristiques physiques des premiers films comme aides à l'identification*. Traduit de l'anglais par Eric Loné et corrigé par Harold Brown. Centre National de la Cinématographie. Service des Archives du Film (1992).

Frank Böhme (FilmoTec, Wolfen), for his help and patience.

Jean-Marc Lamotte (Institut Lumière, Lyon) for his precious attention to the text about Lumière.

Adelheid Heftberger, Egbert Koppe (Bundesarchiv); Paolo Cherchi Usai (George Eastman Museum); the team of the Special Collections Reading Room ("Espace Chercheurs") of the Cinémathèque française (Paris); Bruno Mestdagh, Arianna Turci (Cinémathèque royale de Belgique); Caroline Fournier (Cinémathèque suisse); Marcello Seregni (Cineteca Italiana, Milan); Alice Rispoli (Cineteca del Friuli, Gemona); Laurent Bismuth (Direction du Patrimoine – CNC); Nikolaus Wostry (Filmarchiv Austria, Vienna); Shivendra Singh Dungarpur (Film Heritage Foundation, Mumbai); Josetxo Cerdán los Arcos and my former colleagues at the Filmoteca Española (Madrid), especially Alfonso del Amo and Encarni Rus; Elżbieta Wysocka, Monika Supruniuk, Michał Pieńkowski (Filmoteka narodowa – Instytut audiowizualny); Luca Giuliani; Peter Sansom (Harman Technology Ltd., Worcester); Jan-Erik Billinger, Mathias Rosengren, Ola Törjas, and Jon Wengström (Svenska Filminstitutet). And Nicola Mazzanti, without whom this new edition would probably not exist as it is.

The members of the FIAF Executive Committee and the Technical Commission of FIAF, for unreservedly supporting this project.

Special thanks to the unrivalled production team of this new edition:

The publisher, Christophe Dupin, for his crucial support and enthusiasm over the years.
The copy-editor, Catherine A. Surowiec, for her invaluable help, advice, understanding, and knowledge.
The graphic designer, Lara Denil, for her tireless patience and receptiveness.
Christine Maes and Barbara Robbrecht at the FIAF Secretariat, for their additional proofreading.

I write these lines 30 years after the first edition of the book that contributed to changing our approach to film, with the hope that Harold Brown might approve of this new edition, and that new generations of archivists will continue his clear-sighted work.

Camille Blot-Wellens
Stockholm, October 2020

PART I

Physical Characteristics of Early Films as Aids to Identification

Harold Brown

1. Preface to the 1990 Edition

1.1 The author worked in the National Film Archive in Britain from its inception in May 1935 until his retirement in August 1984. In the course of his handling of many films of the period prior to the First World War, he found that he was able to recognise the product of certain film-makers of that time by the appearance of the films themselves, as distinct from the subject matter as seen on the screen. Thus it was often possible to pick up a roll of film and look at a few feet, and say with confidence that it was made by, for example, Hepworth or Lumière; this by a sort of tangible* familiarity with a considerable number of known Hepworths and Lumières.

1.2 There came a time, probably around the end of the 1950s, when Ernest Lindgren, then Curator of the National Film Archive, encouraged the author to try to describe the tangible* features by which he recognised the product of the different makers, and to write it down and illustrate it for the benefit of other film archivists.

1.3 Eventually he produced a paper for the 1967 FIAF Congress in East Berlin, called "Notes on Film Identification by Examination of Copies". This ran to 30 pages of typescript, and included 50 illustrations from frames of film, reproduced actual size on two 10-inch x 8-inch (26cm x 20cm) photographic plates. A copy of this paper was given to each archive represented at that Congress. Subsequently, from time to time, the author was asked for further copies, and some were made by photocopying, and given to various enquirers.

1.4 In March 1968 the theme of film identification was further pursued by FIAF in a symposium at Gottwaldov in Czechoslovakia, at which the author showed slides illustrating characteristic features of some more of the early producers, and also distributed to the participants a list of 44 Gaumont films ranging in date from 1906 to 1914, and showing how edge marks, title style, form of trademark, and tinting of the titles of this producer varied systematically from time to time, and particularly how the production serial number which appeared on the main title and intertitles of the films was related to the date of the film.

1.5 In 1985, the author was invited to deliver the annual Ernest Lindgren Memorial Lecture at the National Film Theatre in London. This was on the subject of this aspect of identification of early films.

---

* Editor's Note: In his Preface to the 1990 edition, Harold Brown used the adjective "intangible" twice: (1) in the context of "a sort of intangible familiarity", and (2) "hitherto intangible features". After considerable reflection, the copy-editor and the editor decided that the adjective should be replaced in both instances by "tangible", more adapted to and accurately expressing Brown's hands-on approach of direct examination of film materials. "Tangible" derives from the Latin verb *tangere*, to touch. According to the *Oxford English Dictionary*, it means "1. Capable of being touched; affecting the sense of touch; touchable. Hence, material, externally real, objective. 2. That may be discerned or discriminated by the sense of touch." And as a noun, "1. physical and material assets which can be precisely valued or measured. 2. That can be laid hold of or grasped by the mind, or dealt with as a fact; that can be realized or shown to have substance; palpable." The Oxford definitions of "intangible" – "incapable of being touched; not cognizable by the sense of touch; impalpable; assets which cannot easily or precisely be measured" – thus seem to be in fundamental contradiction with Brown's demonstration, since he is referring to handling and examining actual films.

1.6 The present publication[1] is an assembly and re-editing of the content of these three works, together with some more material of the same kind, gathered during the years 1987-1990, by inspection of more films, mainly at the National Film Archive's Conservation Centre at Berkhamsted, and also at the National Film & Sound Archive in Canberra, and at the Československý Filmový Ústav: Filmový Archiv in Prague.

1.7 It is important for readers to realise that this publication is by no means exhaustive. Study of as many as possible of other producers' films would yield more information of the same kind. It is to be hoped that other workers who have access to relevant films will contribute to FIAF any information which they can find.

1.8 Grateful thanks are due to the National Film Archive, London, for the provision of vital facilities for the inspection of films, and production of the illustrations, at its Conservation Centre. I also wish to express my appreciation for the help and co-operation so willingly given by members of the staff of the Archive. They are many, and I dare not name any, for fear of omitting any. They know who they are, and that I greatly value all their varied contributions.

Harold Brown, 1990

1 Editor's Note: In the 1990 publication, Harold Brown often used the word "paper" to refer to his text, as it clearly had its origins in his 1967 paper "Film Identification by Examination of Film Copies" and subsequent papers delivered at archival congresses, conferences, symposia, and workshops. The word "publication" is actually more appropriate for the 1990 book published by FIAF, so it has been decided to replace the word "paper" with "publication" in all relevant instances, whenever Harold Brown refers to the 1990 *Physical Characteristics of Early Films as Aids to Identification*.

2. Introduction to the 1990 Edition

2.1 Now and for many years past, all the dimensions of cinema film, and the positions of everything that goes on it, has been standardised to within as close tolerances as it is universally and commercially practicable to achieve. In the first 20 years of cinema, this was not so. The industry was founded on the gauge of film used by Edison in his Kinetoscope, but various people, in a number of countries, made films only more-or-less the same as those. And it is the variations between different makers' products that makes possible the recognition of the products of some of the producers (persons and companies) of that period.

2.2 It is normal now for film processing and printing work to be carried out by specialist laboratories, who have a number of film-producing organisations among their customers; and those customers may sometimes employ one laboratory, and sometimes another.

2.3 During about the first 20 years of cinema history, however, it was at least almost universally the case, that the film-makers processed their own negatives, and made and processed their own prints. This led to each having consistent peculiarities which make the films of many of the producers of the period recognisable, quite apart from whether or not their name appeared on the film. Some of the peculiarities of the different film-makers are of no more than curiosity interest. Others are capable of enabling us to recognise which company made the film, and with some greater or lesser precision, when it was made; when this is not so easily discoverable by other means. That is: some of these characteristics, peculiar to certain early film-makers of this period, are useful aids to identification. I emphasise the word "aids". You cannot fully identify a mysterious film by use of physical characteristics alone. Reference to written sources of information is vital for this purpose. What the physical characteristics can do is to materially narrow down the area within which it is necessary to search the literature. See a list of some relevant publications at the end of this section.

2.4 Note that some events occurred at a precisely determined date (e.g., the introduction of "positive" perforations in 1924). Other changes were spread over a period of time which cannot be precisely stated.

2.5 The various features dealt with in this publication are noted on the Contents page. Information relating to each producer is generally arranged under the headings of the several sections. Other odd items about a particular producer are given in a little "Essay" on that producer. Some of the essays include a Film List.

2.6 This publication deals primarily with features which are observed on the film itself, as opposed to what may be seen at a screening of the film. However, it deals with some features which, although they are actually visible on the screen, are so inconspicuous or of such short duration, that they cannot be studied or compared at a normal screening.

2.7. Some relevant publications[2]:

- *The Bioscope*: A British film trade paper which commenced publication in 1908.
- *Variety*: A theatrical paper which embraced films.
- *Kinematograph and Lantern Monthly*: devoted to still and moving projected pictures around the turn of the 20th century.
- Library of Congress, *Catalog of Copyright Entries.*
- *American Film-Index* by Einar Lauritzen and Gunnar Lundquist.
- The early film catalogues of Georges Méliès, Gaumont, Cecil Hepworth, Warwick Trading Co., Pathé Frères, and probably others.

2 Other references will be suggested.

- The "Film Index" of the British Film Institute.
- The "Producer Lists" of the National Film Archive, London.

Notes on the illustrations

Almost all the illustrations are reproduced the actual size of the original film.

Some of the relevant features are very small and may need to be viewed through a magnifying glass (e.g., the trademarks in scenes).

In some of the illustrations the quality of the picture itself is very poor. This is because, in these cases, I have processed the illustration in a way to enhance the relevant feature (e.g., perforation images) and have sacrificed picture quality in order to achieve this.

Some illustrations include the perforations, while some do not. When perforations appear, they are the perforations of original contemporary positives or (rarely) negatives. When an illustration has been taken from a duplicate copy (when, to show the perforations of the duplicate is both irrelevant to the purpose of the illustration and could be misleading), they are not included.[3]

3 It will be indicated when an illustration was not made from an original.

3. Perforation Shapes

3.1 Until about 1905 the size of perforations was smaller than the present standard "negative" or "positive" perforations. These were of different precise shapes, as shown by these illustrations.[4]

**Un homme de têtes / The Four Troublesome Heads*
(Méliès, 1898) – FE

From that time a change to larger perforations, similar in shape to the present negative perforation, gradually took place.

Bell & Howell

This is also known as the "Bell & Howell" perforation (B.H.), after the engineers who devised its precise dimensions.

3.2 Perforations of this shape and size were then used for all films, both negative and positive, until 1924. In 1924 Kodak introduced the familiar large rectangular perforation for positive projection prints, commonly called the "positive" perforation. This perforation is also known as the "Kodak Standard" perforation (K.S.). This perforation is 0.003 inches (0.076mm) higher than the negative perforation.

Kodak Standard

3.3 After that time, negatives continued to have the "negative" (B.H.) perforations, but almost immediately thereafter virtually all projection positives had the new positive perforation, except for the use by Pathé of their own peculiar perforation, which continued until toward the end of the 1920s.[5]

4 More examples of early perforations can be found in Section 7, "Frame Characteristics", pp.75-87.
5 For further information, see the chapter on Pathé, p.205.

3.4 Throughout the period from about 1905 until the end of the silent era, the Pathé organisation used a perforation of distinctive shape. This was of similar shape to the perforation "Bell & Howell", but of the same height as the "positive" perforation, and the corners were rounded.

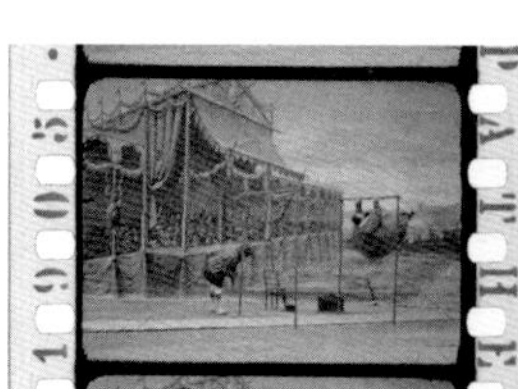

**Les Levanni, barristes comiques / The Levanni* (1903, print dated 1905) – FE

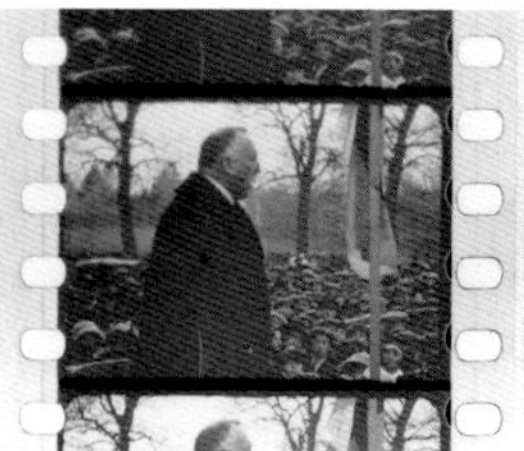

Unidentified
(Probably post-1910)

3.5 During the earliest years of cinema, some film-makers had other variations of perforation which were peculiar to themselves.

Among these were:

- The brothers Louis and Auguste Lumière, of France.

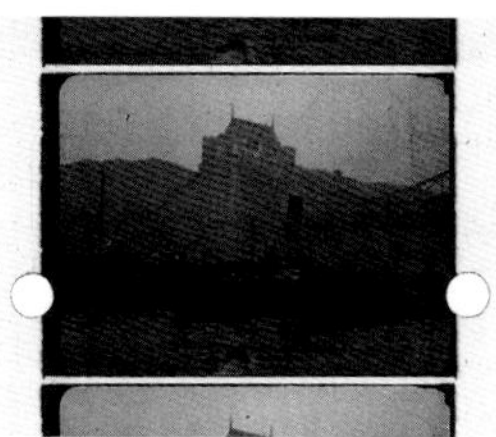

Note: One pair of circular perforations per frame.

- Max Skladanowsky, of Germany.

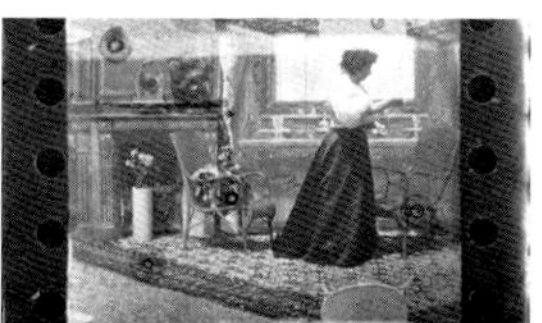

Note: Four pairs of very small circular perforations per frame.

- The firm of Prestwich, of Britain.

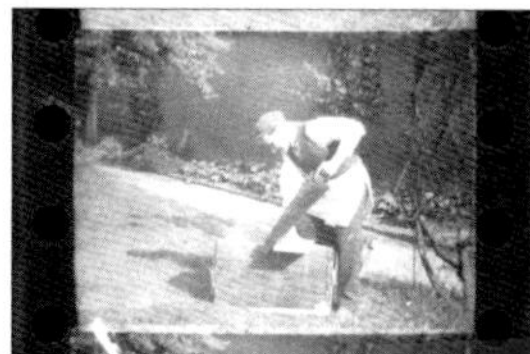

Note: Three pairs of circular perforations per frame, of a size similar to the Lumière perforations.

4. Embossed and Punched Marks

4.1.1 Some of the earlier producers embossed their names or trademarks at the beginning of their films. These are not apparent on the screen but may be seen on the film in the hand, if the beginning of the film has not been lost. These embossed marks also are usually not apparent on a copy when the film is duplicated. I have produced some specimens by taking pencil rubbings on tissue paper. Some of these are, unfortunately, not very clear because the embossings are not very deep, but they can be discerned with care.[6]

4.1.2 Some examples are:

- Signature "G. Méliès"

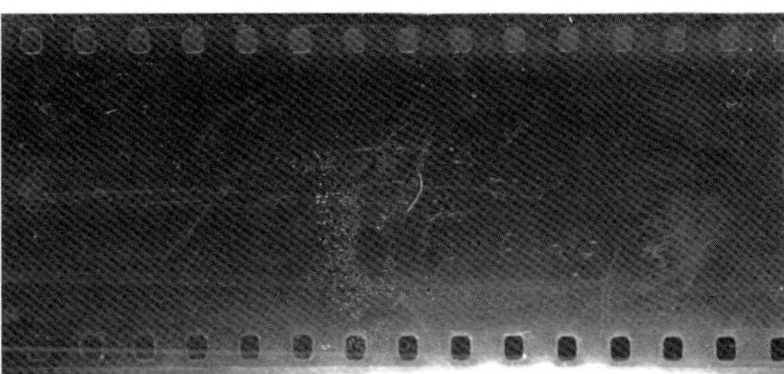

**Un homme de têtes / The Four Troublesome Heads* (Méliès, 1898) – FE

- *"Méliès Star", in rectangle

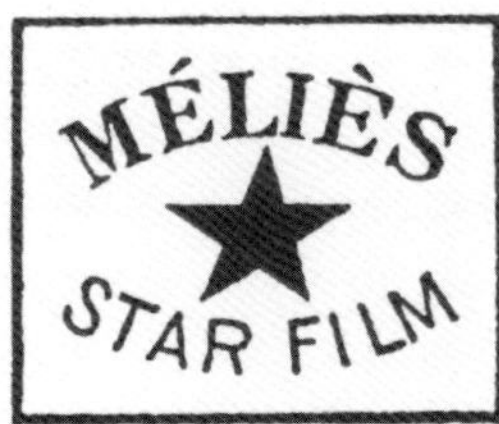

- Gaumont Company

- ELGE

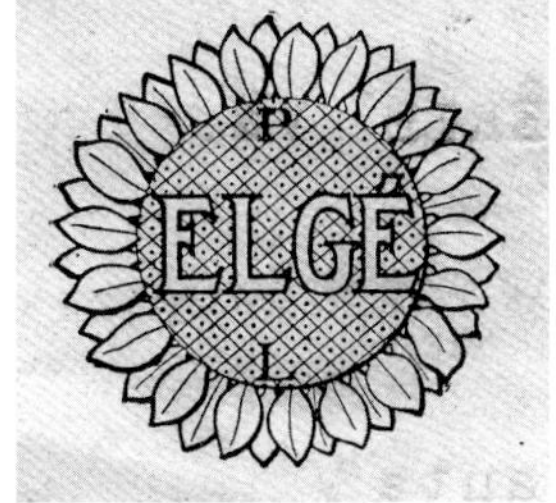

*Letter sent by Gaumont to Eduardo Gimeno (1905) – FE

6 When the original specimens provided by Harold Brown were not legible enough, the Editor has tried to provide other examples. These new examples are indicated with an asterisk.

- Charles Urban Trading Co.

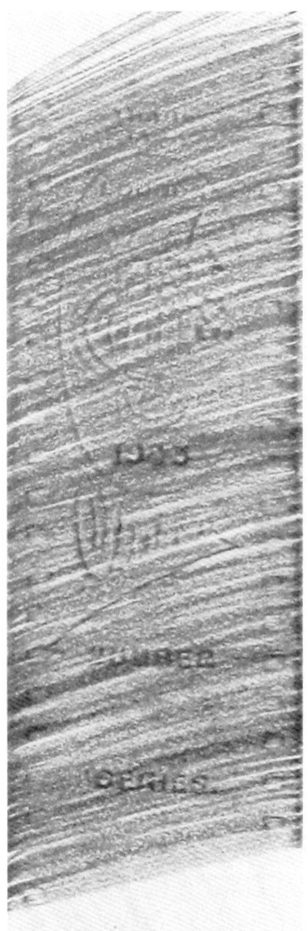

Charles Urban Trading Company

IMPORTANT.

All genuine Urban Films bear a facsimile of this Trade Mark with counter signature of C. Urban embossed on the beginning of each Film.

All Films of our Subjects not so marked are either pirated duplicated copies or rejected misprints.

You accept all so-called Urban Films without this Trade Mark at your own risk.

*Urban: *We Put the World Before You by Means of the Bioscope and Urban films* (1903)

- Warwick Trading Co.

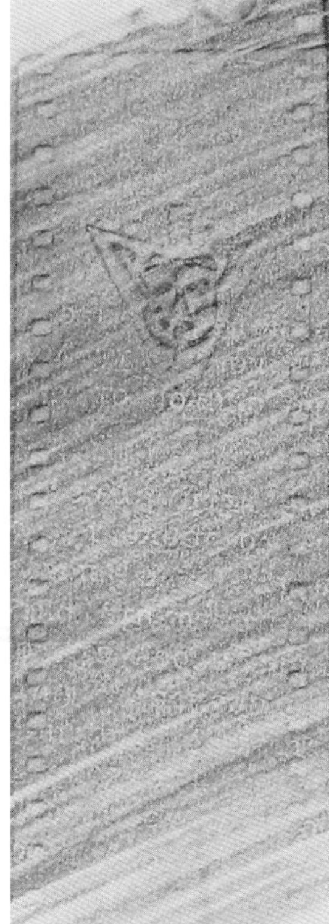

Warwick Trading Company

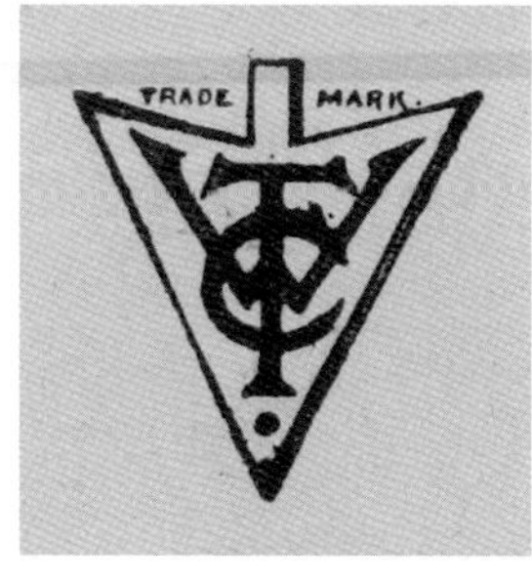

*Hopwood: *Living Pictures* (1899)

**The Kiddies and the Rabbits* (1904) – FE

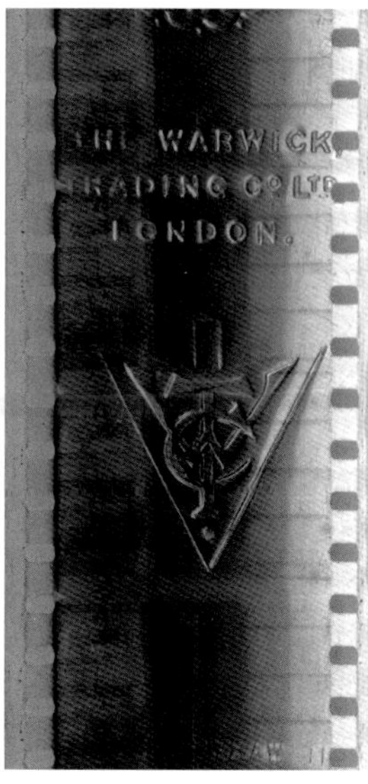

**Fox Hunting* (1906) – FE

- *Robert William Paul

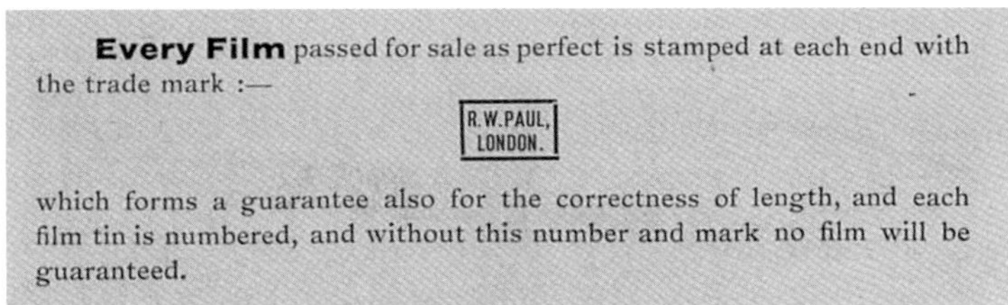

Every Film passed for sale as perfect is stamped at each end with the trade mark :—

R.W.PAUL, LONDON.

which forms a guarantee also for the correctness of length, and each film tin is numbered, and without this number and mark no film will be guaranteed.

* Paul: *Animatograph Films* (1903)

4.1.3 Beware that not all the names embossed at the beginnings of films are those of their makers. Some are just the names of traders who bought and sold the copy. Such a case is PHILIPP WOLFF.

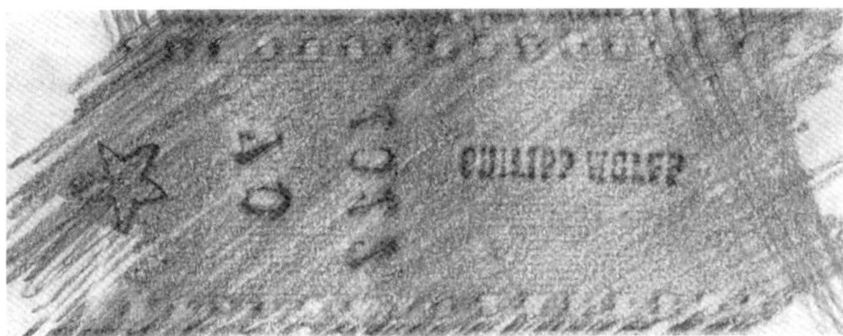

Philipp Wolff

PHILIPP WOLFF

*Hopwood: *Living Pictures* (1899)

4.2.1 Méliès had another useful practice. Starting during 1896, he embossed his catalogue number at the beginning of his films.

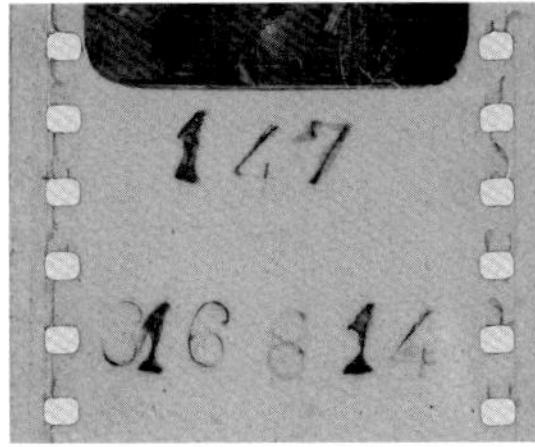

Visite sous-marine du "Maine" / Divers at Work on the Wreck of the "Maine" (1898)

In this case, No. 147 (*Divers at Work on the Wreck of the "Maine"*). At that time, film stock was manufactured in lengths of about 20 metres (65 feet). Méliès gave a separate number to each 20-metre length, so that a film 40 metres long had two consecutive numbers.

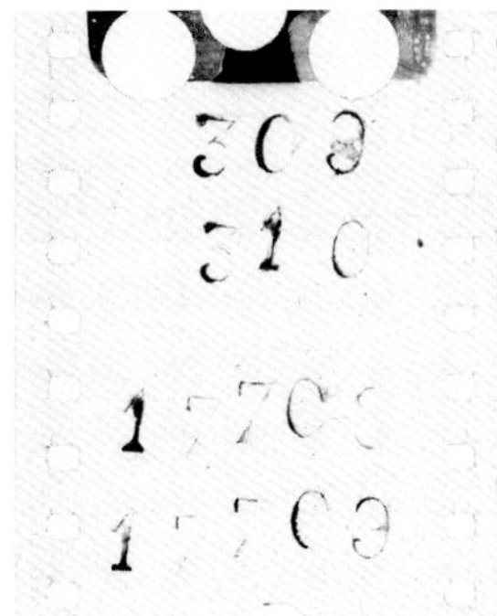

Nouvelles luttes extravagantes / Fat and Lean Wrestling Match (1900)

In this case, Nos. 309/310 (*Fat and Lean Wrestling Match*). He continued this practice even when he was making films which ran to 300 metres or more. Reference from these numbers to Méliès' own catalogues, or to *Sight and Sound Index Series,* No. 11, August 1947, "An Index to the Creative

Work of Georges Méliès 1896–1912", compiled by Georges Sadoul, or to "Essai de reconstitution du catalogue français de la Star-Film", published in 1981 by the Service des Archives du Film du Centre national de la cinématographie, Bois d'Arcy, France, will help to provide identification of the film.[7]

4.2.2 In one early specimen which the N.F.A. has, the numbers are scratched on, not embossed.

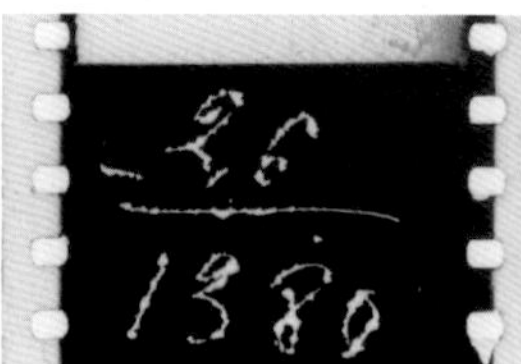

Scratched number 26: *Une nuit terrible / A Terrible Night* (1896)

Whether this represents his early practice before he had the means of embossing (which may be so, although some other films of about the same time are left blank), or whether it was just an odd exception, I feel it would be rash to assume without the evidence of more cases.

4.2.3 There is also a second set of numbers, on both the scratched and embossed specimens, the significance of which I do not know. In the cases where there are two or more catalogue numbers, there is also the same number of consecutive unknown numbers. I do feel confident that these numbers were put on this film by Méliès, and not by someone else later, because of the presence of the second numbers.

4.2.4 Another practice of Méliès, which can identify films as his, was to punch a hole the shape of a five-pointed star (his trademark) at the start of the negative. This printed as a black star on positive prints. His catalogue states that this was in the first frame.

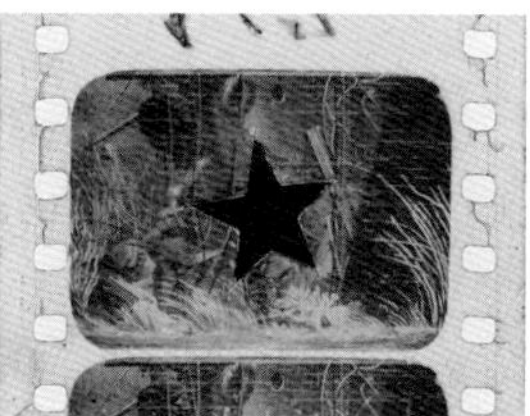

Visite sous-marine du "Maine" /
Divers at Work on the Wreck of the "Maine" (1898)

Actually, it was sometimes in the second or third frame, and, in at least one case, it was on the blank spacing just before the first frame.

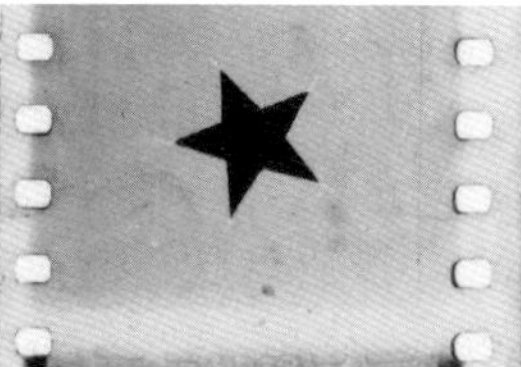

4.3 Hepworth also put his name at the start of some of his early films. Although these were put on photographically, it is convenient to refer to them here.

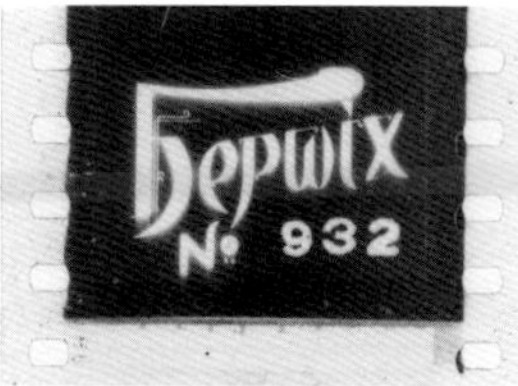

The serial number 932, by reference to Hepworth's catalogue, identifies the film as *The International Exchange*.

7 See also "Identifying a Georges Méliès Film", pp.169-176.

5. Producers' Edge Marks

5.1 The practice by some producers of printing their name on the margins of their films goes back to very near the beginnings of cinema, and virtually ceased at the time of the First World War, although there was some slight use thereafter.

Note that edge marks are sometimes very faint, and need to be looked for. The illustrations all show good clear examples.

5.2 **Lumière**

The Lumière Brothers started making films in 1895. Their early prints had the mark "LUMIERE" printed on one margin in purple ink near the beginning and end. From 1896 the mark read "LUMIERE LYON DEPOSE".

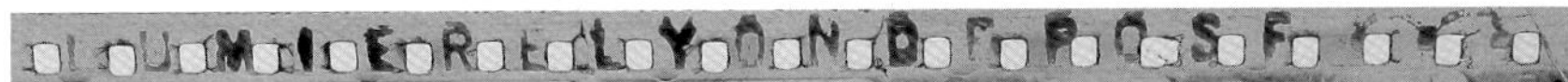

**Défilé de cuirassiers* (1896-1897). Print ca. 1898 – SFI

5.3 Note that all the edge marks subsequently referred to here are printed photographically. It may be seen that while many read from the emulsion side, some read from the base side. I know no significance to this.

5.4 Pathé

In 1905 Pathé started to print "PATHE FRERES PARIS 1905" in both margins.

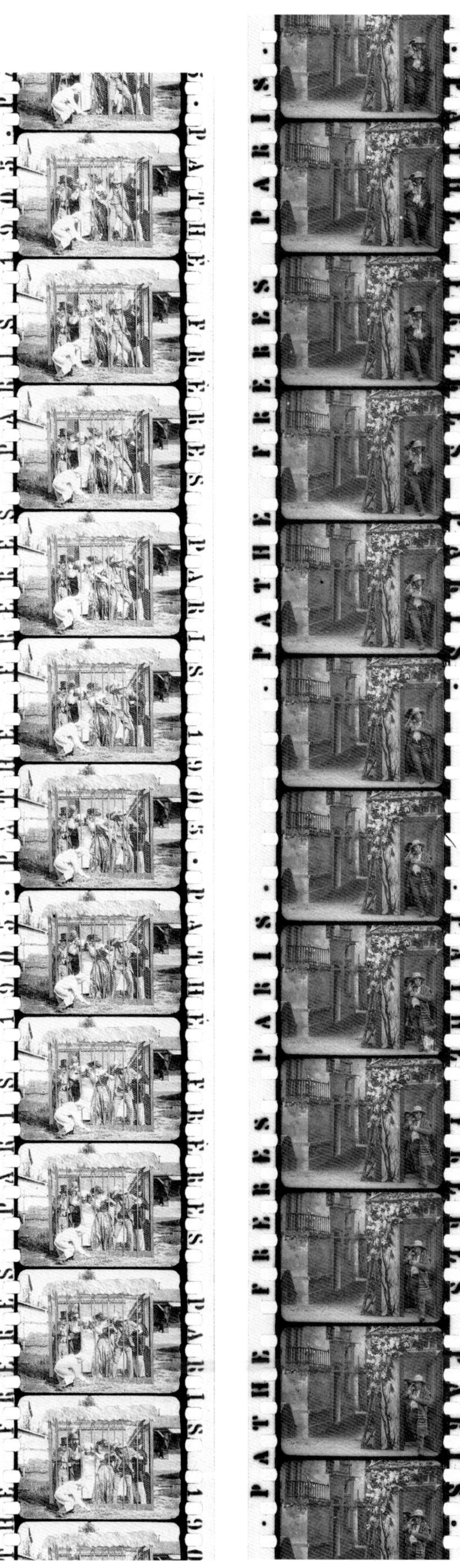

1905

1906

The date 1905 was subsequently omitted, presumably at the end of that year, and the mark continued in the form "PATHE FRERES PARIS" on both edges until about April 1907. Note that on some films of 1906 there is a gap between each printing of "PATHE FRERES PARIS", in the position where the year "1905" had been.

On other films there is no such gap. It seems likely that the ones with the gap are from the beginning of 1906, and that later the gap was closed.

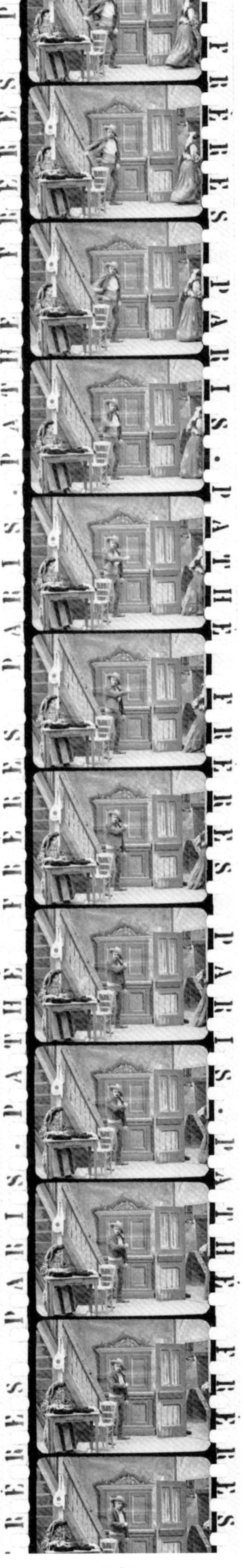

1906

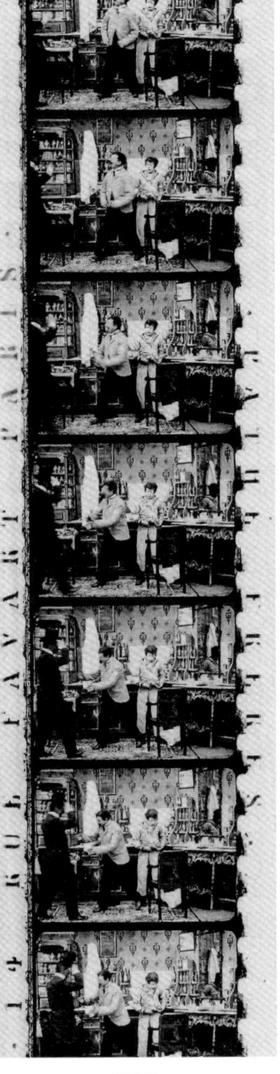

1907

From April 1907 until 1909 the mark was "PATHE FRERES" on one margin and the address "14 RUE FAVART PARIS" on the other.
See also the chapter on Pathé, p.199.

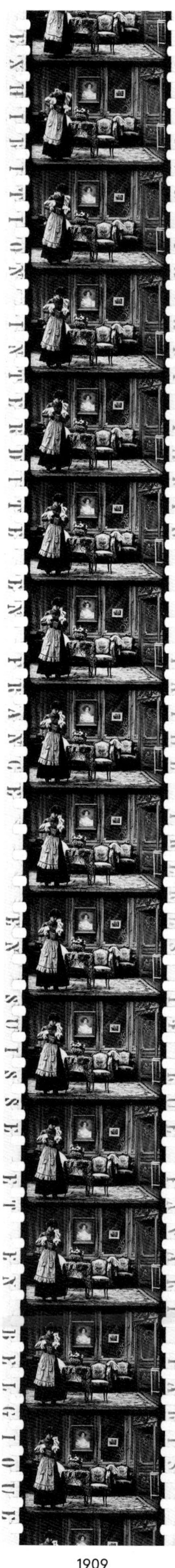

1909

From the beginning of 1909 onward the mark was, on one edge, "PATHE FRERES 14 RUE FAVART PARIS", and on the other, "EXHIBITION INTERDITE EN FRANCE EN SUISSE ET EN BELGIQUE".

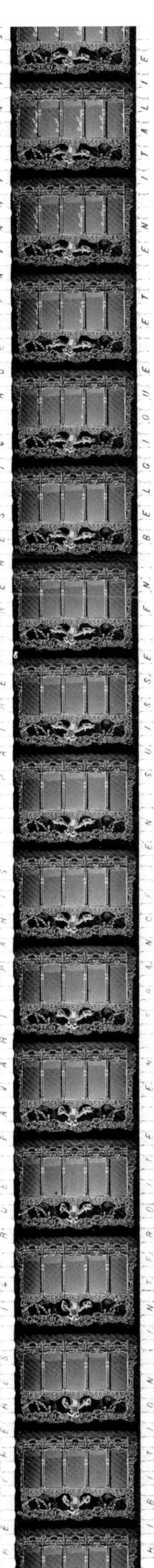

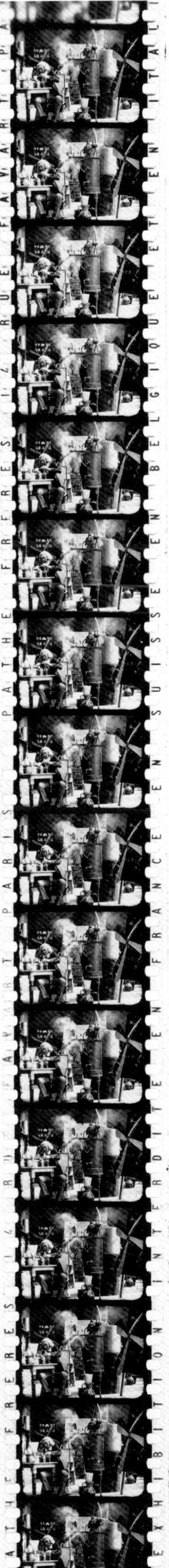

In about 1911 the words "EN ITALIE" were added, and about the same time the style of lettering was changed. The earlier films have an upright "stencilled" style of lettering with pronounced serifs. The ones of 1911 onward have thin "solid" letters. Some of these have sloping letters, and others have upright letters
See also the chapter on Pathé, p.199.

5.5 Gaumont

The earliest Gaumont films – before 1907 – had no edge mark. The edge mark started to be included at about the beginning of 1907. If one takes a close look at the edge mark, one can see that between each print of the name "GAUMONT" there is a letter.

In this example, the letter is orientated at 90 degrees to the name "GAUMONT". It may be regarded as lying on its back. This interim letter was placed thus sideways throughout 1907 and probably during 1908.[8]

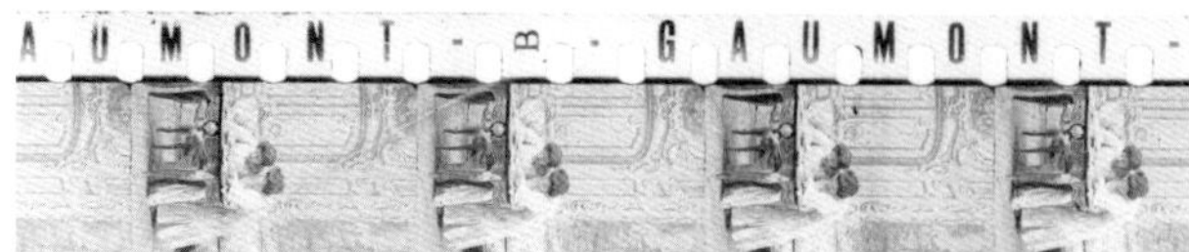

1907

Here, the interim character is similar to the letter Y. The three films which the N.F.A. has with the "Y" were all made during 1909 or at the very beginning of 1910.[9]

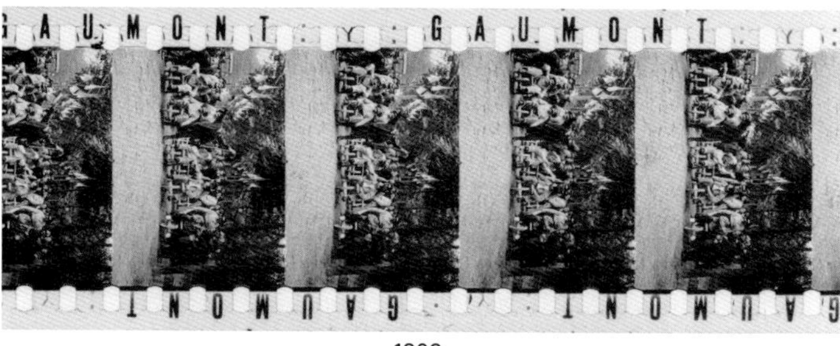

1909

After that time, until 1914, there was a letter standing the same way as the "GAUMONT" name.[10]

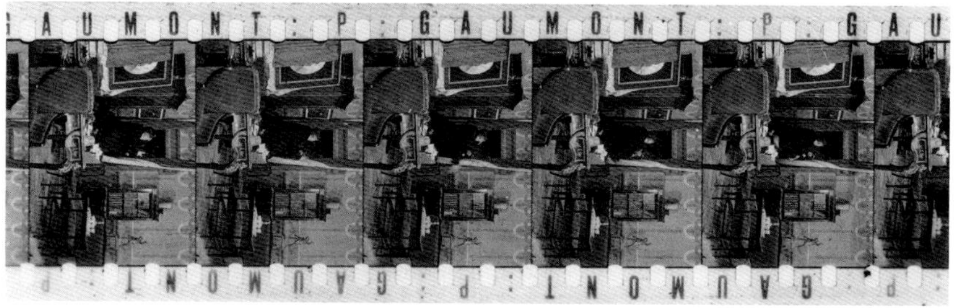

1910

---

8 The letter can change according to the films. Examples of letters observed on prints of films released in 1907: A (*Un Monsieur aimanté*, N° 1646), B, C (*La Course à la saucisse*, N° 1587; *La Terroriste*, N° 1590), D (*Une héroïne de 4 ans*, N° 1664; *Le Piano irrésistible*, N° 1670), E (*Le Lit à roulettes*, N° 1665), F (*Le Noël de Madame*, N° 1812), G (*Le Noël de Madame*, N° 1812; *Les Chansons ont leur destin*, N° 1831).
On prints of films released in 1908: G (*L'Enfance charitable*, N° 1834), H (*Le Récit du colonel*, N° 1824; *L'Homme de marbre*, N° 1832; *Le Violon*, N° 1833; *Le Buffet anthropophage*, N° 1837 and *L'Équilibriste*, N° 1857) and K on the print of an unidentified title. (Eric Loné)

9 There are more examples of films released in 1909. The inscription follows the same system as 1907 and 1908, but the letter is not always orientated at 90 degrees; it's even straight in most of the cases. The following letters are ordered according to the date of release:
R (90°), on *Les Deux devoirs* (N° 2197), released 15.02.1909.
G on *Un Monsieur qui a mangé du taureau* (N° 2276), released 3.05.1909.
T (90°) on *Le Gui porte-bonheur* (N° 2252), released 12.04.1909, and on *La Bouilloire magique* (N° 2318), released 21.06.1909.
Y on *La Fée des grêves* (N° 2414) and *Le Violon brisé* (N° 2415), released 13.09.1909, as well as on *Sur la Côte d'Émeraude* (N° 2436), released 20.09.1909.
Z on *Les 1000 francs de Grenouillard* (N° 2408), released 4.10.1909.
A on *50 degrés à l'ombre* (N° 2427), *La Femme doit suivre son mari* (N° 2434) and *La Coiffure du commissaire* (N° 2451), all released 11.10.1909.
B on *L'Épave* (N° 2507), released 13.12.1909.
C on *Douloureux cambriolage* (N° 2500), released 6.12.1909 and *Calino au théâtre* (N° 2509), released 13.12.1909.
It appears that there was possibly a logical order in the system of edge marks, but the fact that sometimes the system seems to be illogical (for instance, the G of May 1909, between R and T) could mean that the system did not follow a logic, or that this print was made later. (Eric Loné)

10 Between 1910 and 1915 (and even later), there is always a letter placed in the same alignment as Gaumont:
D on *Léocadie*, released in 1910.
E on *Le Fil de la vierge*, released in 1910.
J and K on *La Côte d'Azur pittoresque*, released in July 1910.
L and M on *Le Tri-porteur* (N° 2823, August 1910).
M on *Monsieur veut se marier* (N° 2908, September 1910).
T on *Le Portrait de Mireille* (N° 2978), released in November 1910.
O and P on *Calino déjeune en ville* (N° 3291) released in 1911.
B on *Calino fait l'omelette* (N° 3732), released in 1911.
H on *La Hantise* (N° 3985), released on 18.10.1912.
I on *Le Pont sur l'abîme* (N° 3956), released in September 1912.
R on *Les Chasseurs de lion*, released in 1913.
It is common that the same letter appears throughout the print, but in some cases the letter changes according to the shots. (Eric Loné)

In each copy there is the same letter throughout the copy; but different copies of the same film may have different letters. On some films the mark appears on the titles, but not on the picture; on others it is on the picture but not on the titles; while on some it is on both picture and titles, and with some it is not on either. I have looked for some significance in these variations, and in the use of different letters, but have found none; except that the films which the N.F.A. has without an edge mark (apart from the earliest films before 1907, and newsreel material which did not normally have the edge mark), all occur in the last quarter of 1911. All the Gaumont edge marks are the same on both edges of the film.

See also the details set out in the Gaumont film list in Appendix 1, pp.122-125.

5.6 Cines

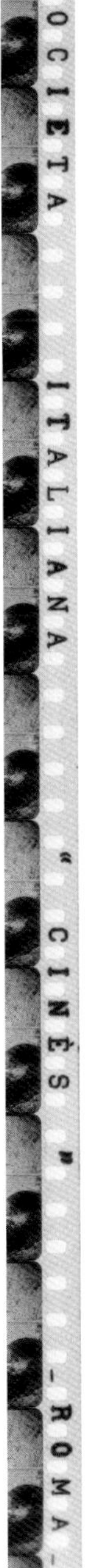

1909-1912

Almost all the films of the years 1909 through 1912 carry the edge mark 'SOCIETA ITALIANA «CINES» – ROMA –'. The films of 1913 onward do not have this edge mark. The N.F.A. has a copy of *Caius Julius Caesar,* made in 1914. This has the mark "INGLESE" on each edge at 14-frame intervals. A section on the very end of the film has the word "SPANA" in similar form. This seems to be merely to indicate the language of the copy, English and Spanish respectively. All the Cines films have the same mark on both edges.

5.7 Vitagraph

Vitagraph films printed in the United States of America from 1909 onward had the edge mark "THE VITAGRAPH Co OF AMERICA".

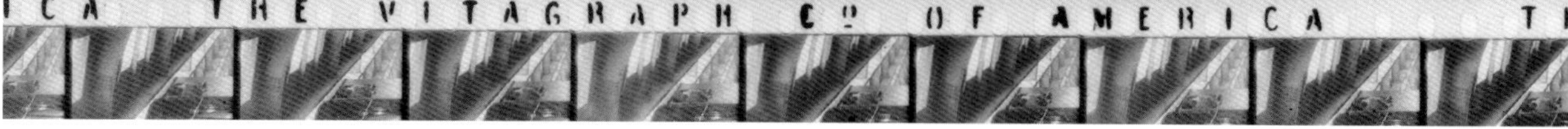

1909-1914

Vitagraph films for circulation in Europe were printed in Paris, and from the same time had the edge mark "THE VITAGRAPH Co PARIS".

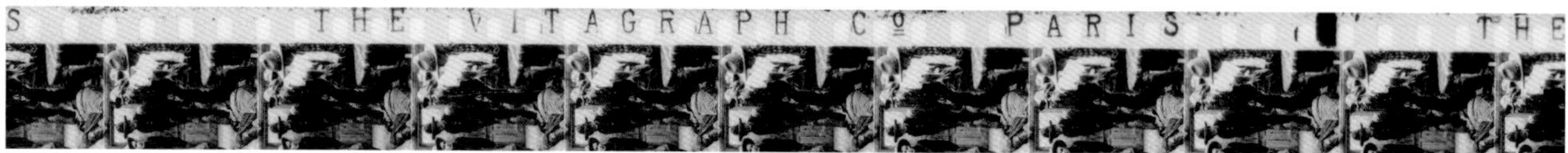

1909-1914

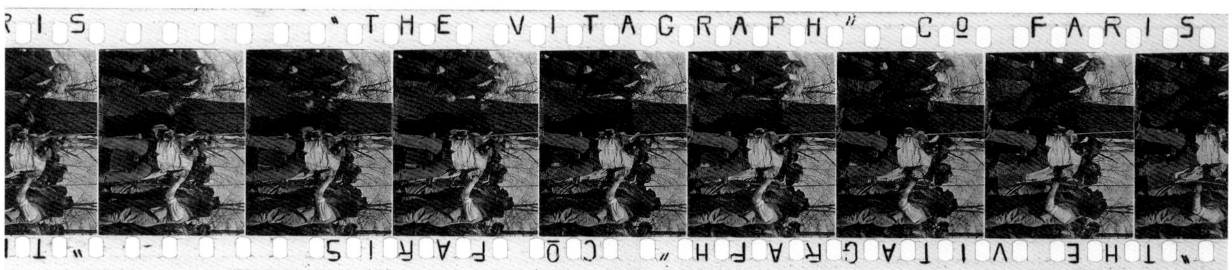

1909-1914

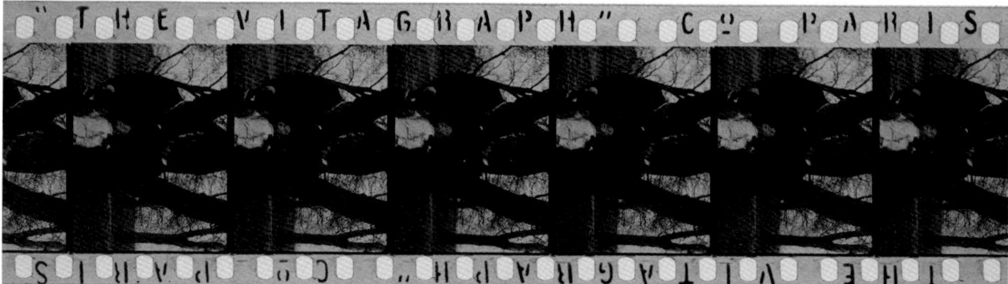

*1913 – AFA

This was discontinued on Paris prints from 1914. Note the different letter styles. The same mark appears on both edges.

5.8 Selig

The Selig Polyscope Company printed their name on one margin. In 1909 and 1910, it was as in the example below: PROPERTY OF SELIG POLYSCOPE COMPANY, CHICAGO, ILL. U.S.A.

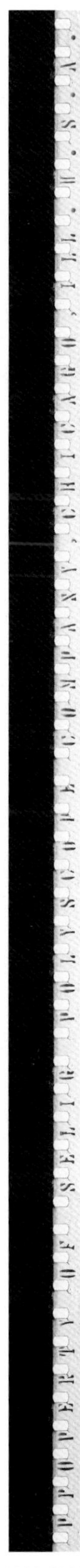

1909-1910

After 1910 the letters "U.S.A." were omitted. After 1911 the edge mark was discontinued altogether. The Selig name reads on the bottom edge. The upper edge bore the message of the Motion Picture Patents Co. (See Paragraph 5.17, p.67).

5.9 **Pasquali**

The Italian company Pasquali seems to have had no consistent practice, but an edge mark may be found on some films.

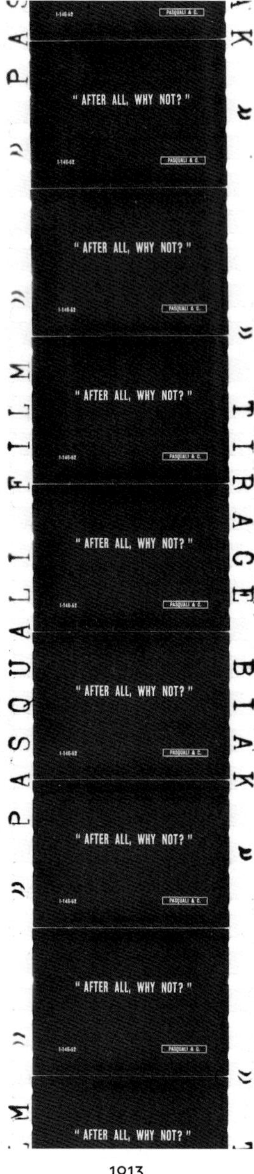

1913

"BIAK" is apparently the name of the laboratory which made the prints.

5.10 **Warwick**

The Warwick Trading Company, of Britain[11], printed its name in the margins of some films over a period which I have not got precisely defined, but which was approximately 1910-14.

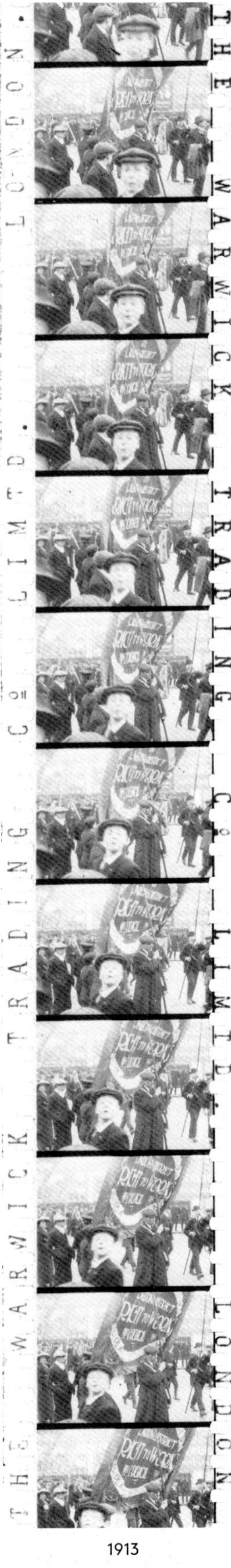

1913

11 Originally called Maguire & Baucus, established in 1894 near Liverpool Street Station, London, its manager, Charles Urban, changed the name to Warwick when the firm relocated to Warwick Court in London in September 1897. Urban left in 1903 to form the Charles Urban Trading Company. (Brian Pritchard)

5.11 Ambrosio

The Ambrosio Company of Torino (Turin)[12] used the following edge mark:

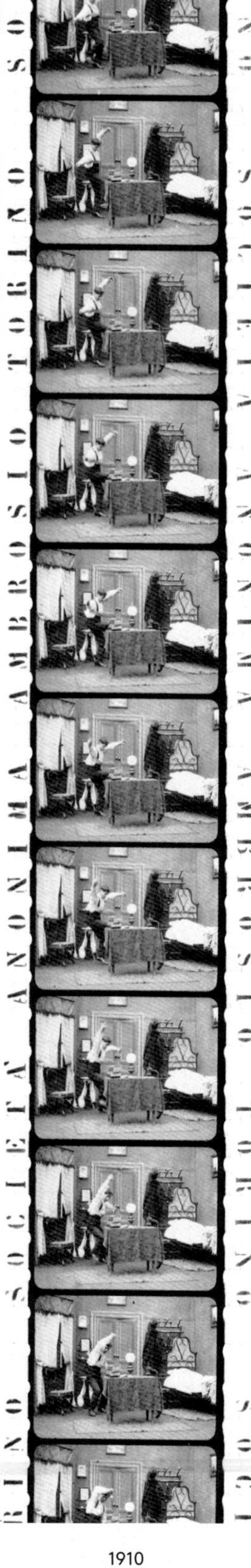

1910

12 Formed in 1904 and originally called Arturo Ambrosio & C., it became Società Anonima Ambrosio Film, Torino, on 16 April 1907; the company was declared insolvent 4 December 1924. (Brian Pritchard)

5.12 **Sascha**

Sascha of Austria used this mark:

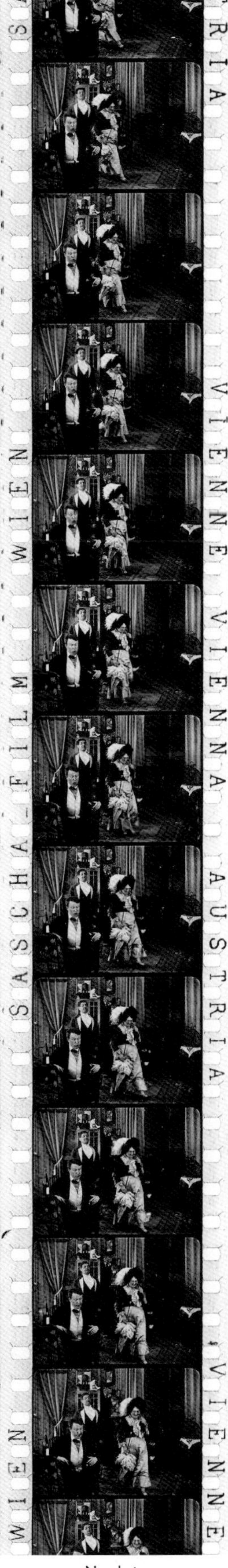

No date

5.13 **Messster**

Messter[13] of Germany used:

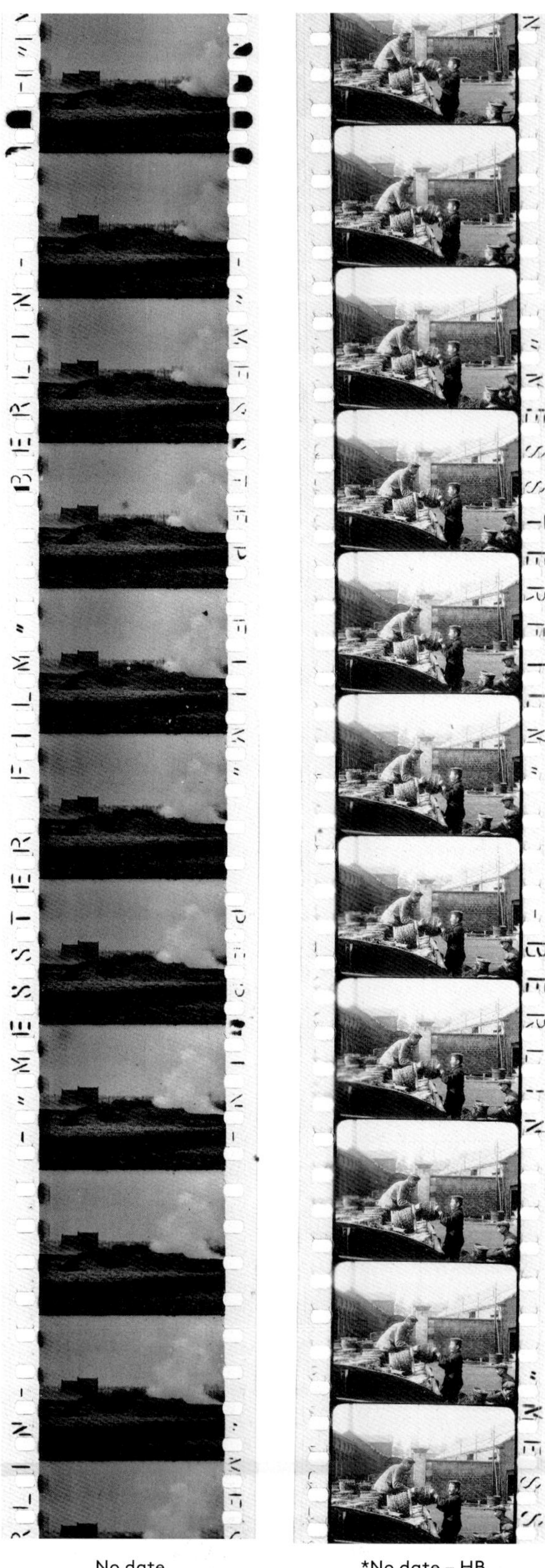

No date

*No date – HB

13 Oskar Messter, known as "the Father of the German Film Industry", had an optical company. In early 1896 he started designing film devices, and at the end of 1896 he started producing films. Soon after, he built a film studio in Berlin. In 1901, his production was so important that he decided to reorganise his companies into specialised firms for production, distribution, and manufacture. In 1914, he started a newsreel (*Messter-Woche*). He sold the firm to Ufa in April 1918. Deac Rossell, in Stephen Herbert, Luke McKernan (eds.): *Who's Who of Victorian Cinema*, London: British Film Institute (1996), pp.96-97. See also *Special-Catalog No. 32 über Projections- und Aufnahme-Apparate für lebende Photographie, Films, Graphohons, Nebelbilder-Apparate, Scheinwerfer, etc. der Fabrik für optisch-mechanische Präcisions-Instrumente von Ed. Messter, Berlin 1898*, in *Kintop 3* (1995). More information can be found in Martin Koerber's article, "Oskar Messter, Film Pioneer: Early Cinema between Science, Spectacle and Commerce", in Thomas Elsaesser, Michael Wedel (eds.): *A Second Life. German Cinema's First Decades*, Amsterdam: Amsterdam University Press (1996), pp.51-61. (Camille Blot-Wellens)

5.14 Éclair

Éclair of France actually printed the year in the margins during 1910 to 1913 with a mark which appeared in several forms.

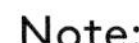

Note:

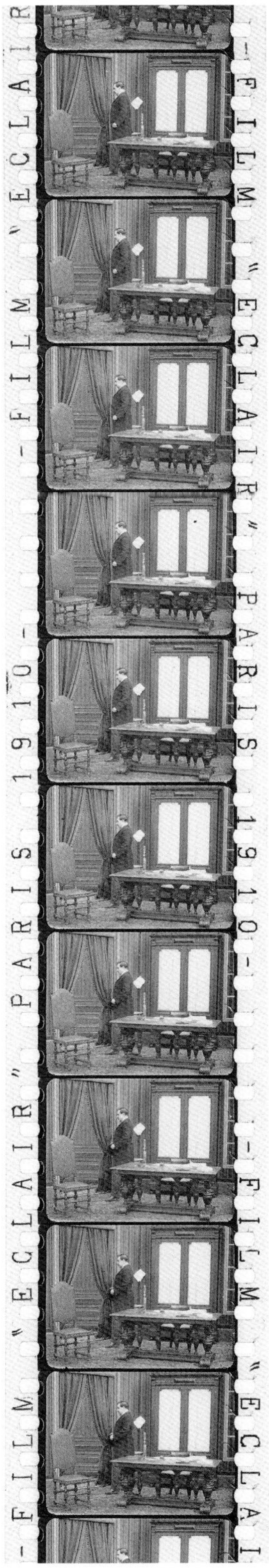

1910 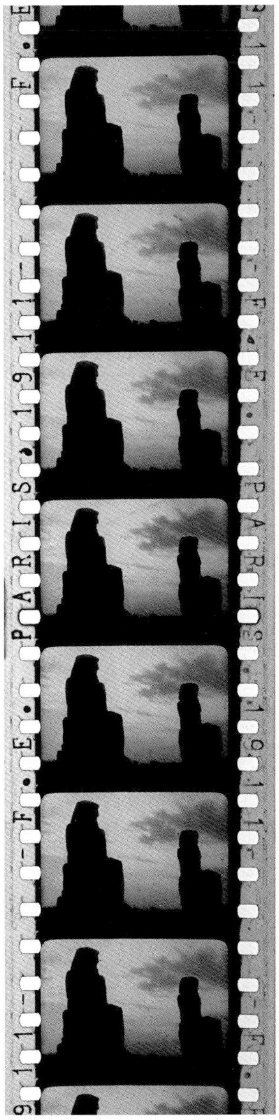 | *1911 – SFI 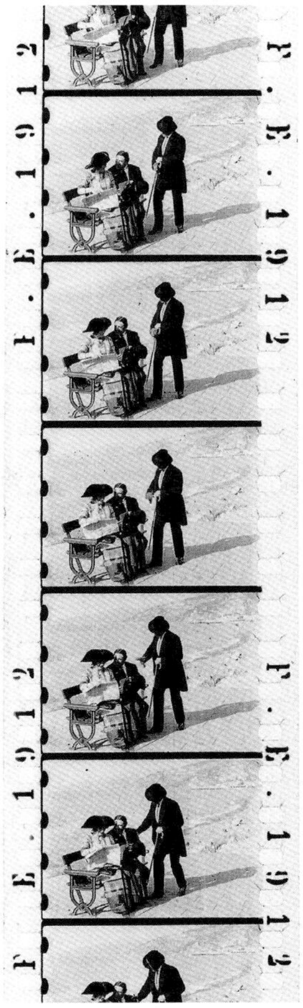 | 1912 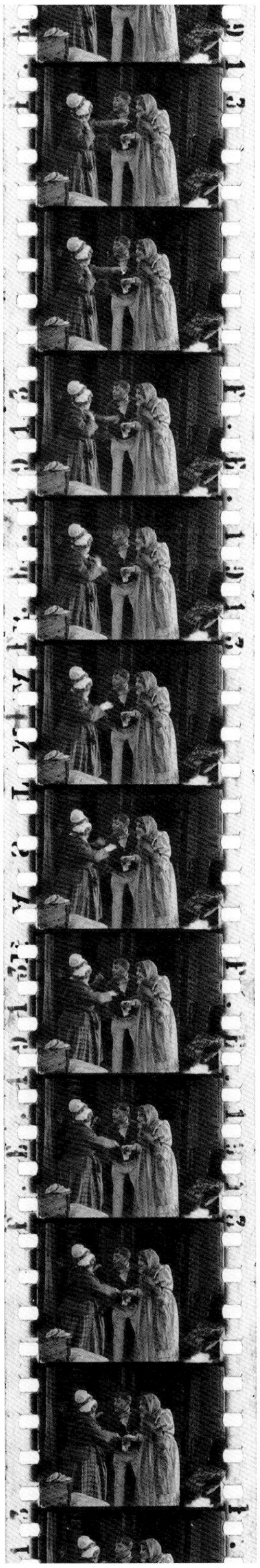 | 1913 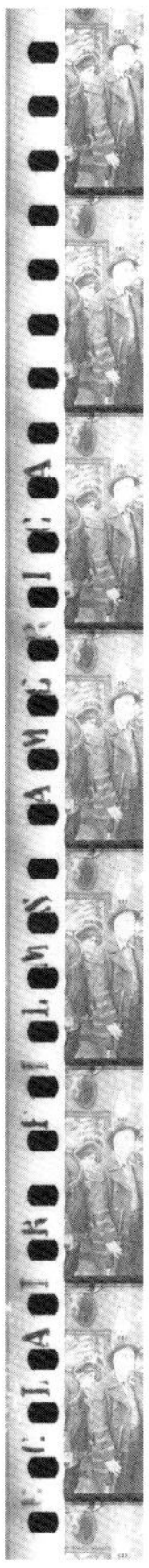 | No date

"F.E." = "Films Eclair"

In this example the name KODAK has clearly been put there by the producing company, and not by Kodak, supposedly as some claim to the quality of the print.

One wonders whether such an action may have had an influence on the Eastman Kodak Company themselves placing their name on the film three years later.

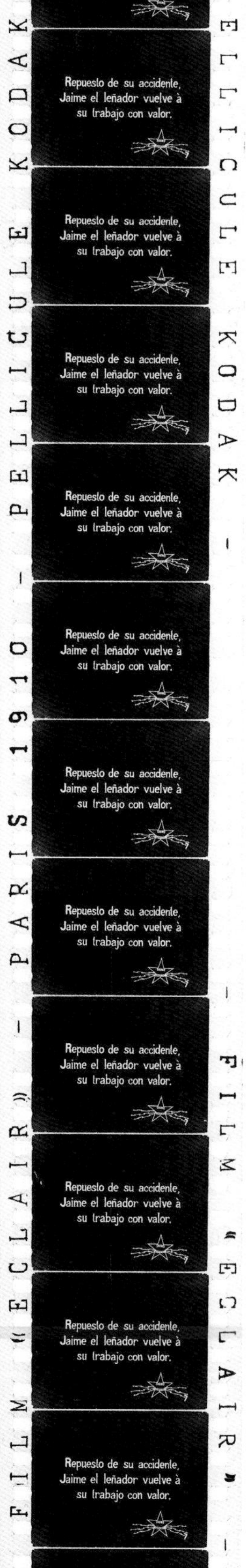

1910

1910

This example does not actually have the name "Eclair".

This one has an x before each fourth repetition of the name and year on one side.

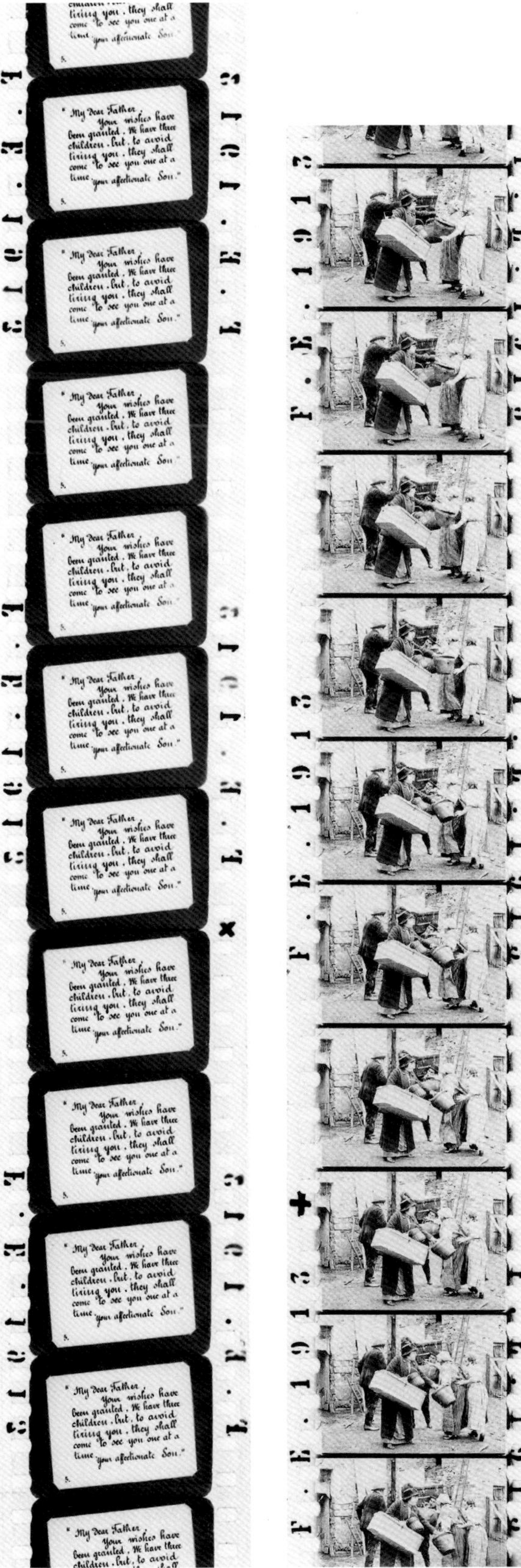

1913 with x

1913 with +

This similarly has a + after each fourth name and year, also on one side only.

5.15 Urban

The Charles Urban Trading Company[14] marked some of their films with the letters "C.U." repeated along the margin.

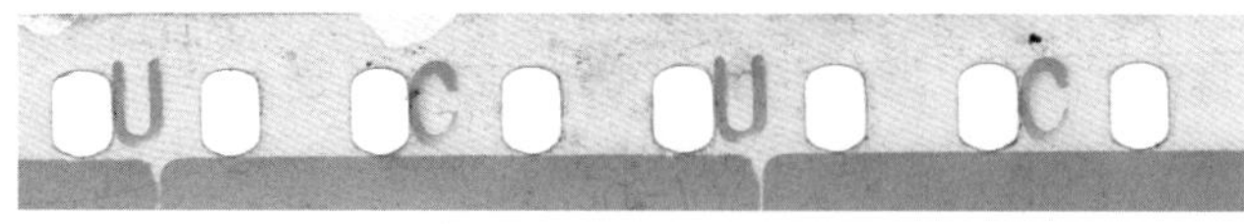

*BP

5.16 Lux[15]

An example of the edge mark on a film by Lux of France:

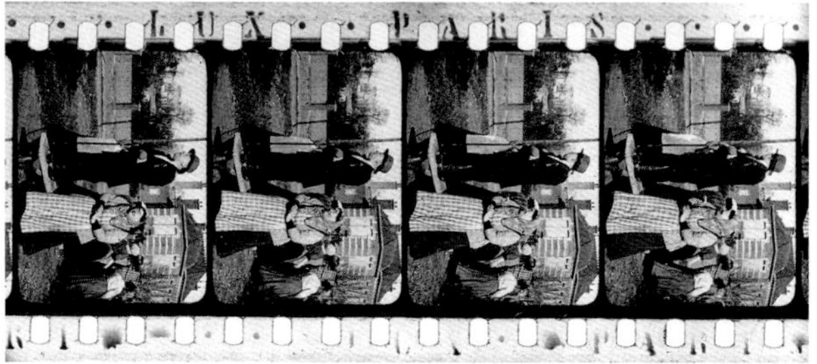

No date

14 Formed in 1903 when Charles Urban left the Warwick Trading Company. (Brian Pritchard) Urban would be one of the most important figures in the British film industry until World War I, notably thanks to his documentaries and natural colour films (Natural Colour Kinematograph Company). He stopped producing films around 1923. Luke McKernan, in Stephen Herbert, Luke McKernan (eds.), *Who's Who of Victorian Cinema*, London: British Film Institute (1996), pp.144-145. See also Luke McKernan, *Charles Urban. Pioneering the Non-Fiction Film in Britain and America, 1897-1925,* Exeter: University of Exeter Press (2013).

15 See also the chapter on Lux, pp.189-193.

5.17 **Motion Picture Patents Company**

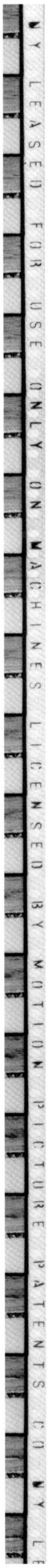

This company was not actually a producer, but it is convenient to place it here. The company was formed to protect and exploit Thomas Edison's patent claims. On some films, of the companies which acknowledged these, may be seen the edge mark.

This ties the film to the years 1909-1914. That is, to the years of what has been called "The Patents War", in the U.S.A.

1909-1914

5.18 Post-War Edge Marks

During at least part of the First World War and afterward, until about 1921, Pathé printed the initials "P G" (for *Pathé Gazette*) and a number, in the margins of at least some of their newsreels:

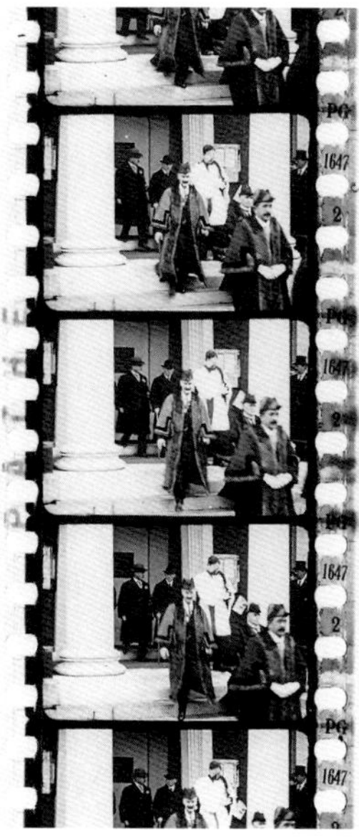

1914-1921

They also printed, in similar style, the initials "P W" in the margins of their weekly magazine (*Pathé Weekly*) during about the same period.

5.18.1 The only other case of such edge marks of which I am aware was also by Pathé, but rather as distributor than as producer. This was in the U.S.A. by Pathé Exchange, Inc., who printed the words "Property of Pathé Exchange Inc." in the margins of films during the early and mid-1920s.

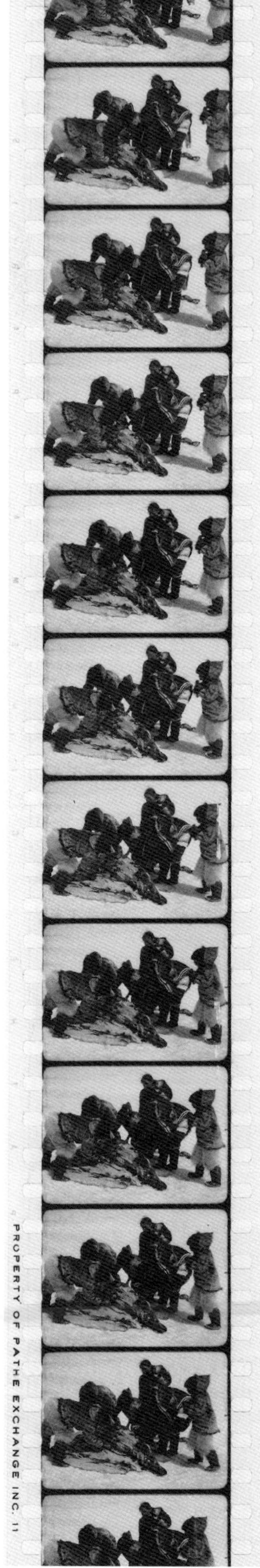

No date

6. Stock Manufacturers' Edge Marks

6.1 It has for many years been the custom of film stock manufacturers to photographically print their names and sometimes other marks in the margins of their films. Note that these marks are sometimes very faint, and need to be looked for. The illustrations all show good clear examples.

6.2 **Eastman Kodak**

The practice appears to have commenced in 1913 with the Eastman Kodak Company, who then printed the word "EASTMAN" in large "stencilled" letters on one margin.

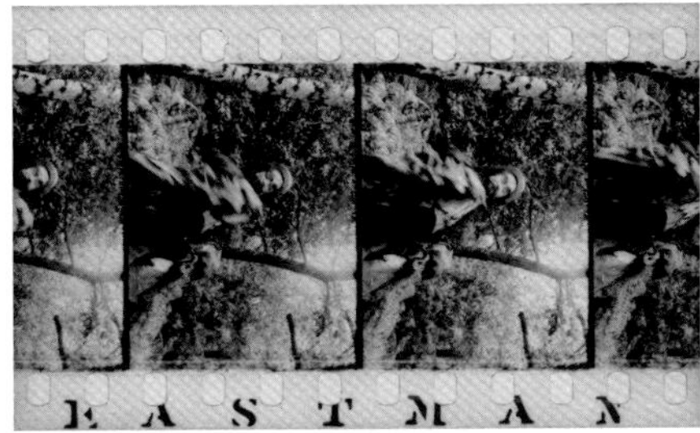

1913/1914

This style was continued until about the middle of 1914, when the lettering was changed to a smaller style, and a dash was included 2 or 3 frames from the name.

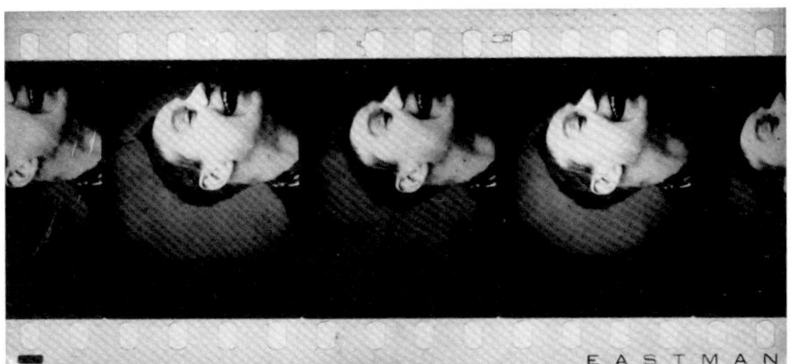

1914/1915

This continued throughout 1915. In the early part of 1916 the films had two small dots in place of the dash.

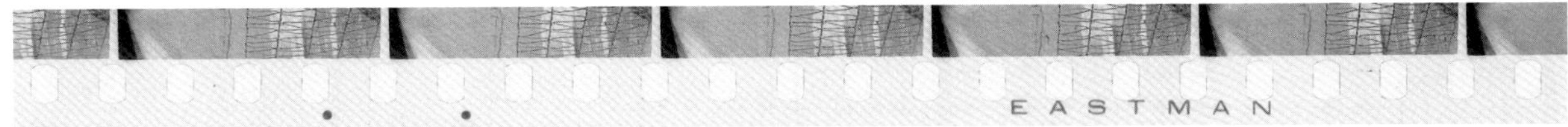

Early 1916

6.2.1 During 1916 the Eastman Kodak Company began a systematic series of year symbols on their stocks made in Rochester, New York, U.S.A.[16] Although this system continues far past the period broadly covered here, I feel that it will be useful to include the information here. A comparable series was begun in 1917 on Kodak stock made at Harrow in the United Kingdom. Stock made by Kodak in Canada from 1925 onward used another similar series of marks. This information was conveyed by Kodak to the National Film Archive in the late 1940s[17], originally confidentially, but was subsequently released. The whole scheme is set out in Paragraph 6.2.4. No such marks were put on the stocks made by Kodak in France and Germany during the 1930s. French stock was just marked "Kodak France" and German stock "Kodak A.G.".

16 See also the chapter on Eastman Kodak, pp.229-256.
17 According to documents held in the Harold Brown Collection at the British Film Institute, Kodak revealed the codes in July 1953.

6.2.2 In 1927 Kodak took control of the Pathé film stock factory in France. Thereafter, the Pathé stock was still marked with the name "Pathé", but the Kodak U.K. symbols may be found on it.

6.2.3 It may be noted that the U.S.A. system of symbols repeats every 20 years; the U.K. system repeats every 19 years, and the Canadian every 11 years.

6.2.4 I find it convenient to call the symbols by names. Thus:

U.S.A. by the names of the geometric shapes:
● : Circle ■ : Square ▲ : Triangle + : Cross

U.K. by "Letter" names:
▼ : U ▙ : L ▬ : Dash

Canada has an additional symbol: ◆ Diamond

Year	U.S.A.	U.K.	Canada	Year	U.S.A.	U.K.	Canada	Year	U.S.A. only
1916	●			1947	■ ▲	▙ ▼	● ▙	1978	▲
1917	■	▼		1948	● ● ●	+	● ▬	1979	● ●
1918	▲	▙		1949	+	+ ▼	● ▼	1980	■ ■
1919	● ●	▬		1950	▲ +	+ ▙	▙ ●	1981	▲ ▲
1920	■ ■	▼ ▼		1951	● +			1982	○ □ ×
1921	▲ ▲	▙ ▙		1952	■ +		NB 1982 PATHE (France only) ○ □ △	1983	× △ ×
1922	● ■	▬ ▬		1953	+ ▲			1984	△ □ △
1923	● ▲	▼ ▙		1954	+ ●			1985	□ ○ △
1924	▲ ■	▬ ▙		1955	+ ■			1986	△ ○ △
1925	■ ●	▼ ▬	● ▙	1956	●			1987	□ △ △
1926	▲ ●	▬ ▼	● ▬	1957	■			1988	+ + △
1927	■ ▲	▙ ▬	● ▼	1958	▲			1989	× + △
1928	● ● ●	▙ ▼	▙ ●	1959	● ●			1990	△ + △
1929	+	+	▬ ●	1960	■ ■			1991	× + ×
1930	▲ +	+ ▼	▼ ●	1961	▲ ▲			1992	□ + △
1931	● +	+ ▙	◆	1962	● ■			1993	+ △ △
1932	■ +	+ ▬	◆ ●	1963	● ▲			1994	+ ○ △
1933	+ ▲	▼ +	◆ ▬	1964	▲ ■			1995	+ □ △
1934	+ ●	▙ +	◆ ▙	1965	■ ●			1996	× ○ △
1935	+ ■	▬ +	◆ ▼	1966	▲ ●			1997	× □ △
1936	●	▼	● ▙	1967	■ ▲			1998	× △ △
1937	■	▙	● ▬	1968	+ +			1999	○ × △
1938	▲	▬	● ▼	1969	+			2000	□ □ △
1939	● ●	▼ ▼	▙ ●	1970	▲ +			2001	△ △ ○
1940	■ ■	▙ ▙	▬ ●	1971	● +			2002	○ □ ○
1941	▲ ▲	▬ ▬	▼ ●	1972	■ +			2003	○ △ ○
1942	● ■	▼ ▙	◆	1973	+ ▲			2004	△ □ ○
1943	● ▲	▬ ▙	◆ ●	1974	+ ●			2005	□ ○ ○
1944	▲ ■	▼ ▬	◆ ▬	1975	+ ■			2006	△ ○ ○
1945	■ ●	▬ ▼	◆ ▙	1976	●			2007	□ △ ○
1946	▲ ●	▙ ▬	◆ ▼	1977	■			2008	+ + ○

6.2.5 In the earlier years, stock manufactured in the first half of the year had the symbols closely following the word "Kodak". On stock manufactured in the second half of the year the symbol is about three-eighths of an inch (1 cm.) from the word "Kodak". This distinction was not continued after the general introduction of triacetate film. From that time all the stock, wherever manufactured, was marked with the U.S.A. symbols, which followed the word "film" in the edge mark. The country of manufacture is indicated by the position of a dot inserted in the word "SAFETY", thus:

S•AFETY : U.S.A. SA•FETY : Canada
SAF•ETY : U.K. SAFE•TY : France

6.2.6 In reading the symbols, take caution in connection with that for the "Cross".
Some are quite clear, thus: ✚
Some appear rather as: ✦
Others have thick arms, thus: ⯃
which can be confused with "Circles".
The U and the Dash when faintly printed are also easily confusable.

6.2.7 Note that the symbols indicate the year of manufacture of the stock, which is not necessarily the year in which it was exposed and developed. Thus it is possible for stock to pre-date or post-date an event recorded on it.

6.2.8 The symbols are photographically printed onto the stock, and on negative and positive stocks appear as "black" characters on a "clear" background on the film on which they are originally printed. They can then, of course, be copied onto another film where they will appear as "clear" on a "black" background, and can be copied yet again onto a third film with a further reversing of the "black" and "clear" to the manner of the original. One needs to be aware of the possibility of being deceived by this, since one may thus find the symbol of an earlier year copied onto stock of a later year.

6.2.9 The "black" character on "clear" background rule does not always apply to reversal stocks. On these there are some cases where the original mark appears as "clear" on a "black" margin, and some in which there is a "clear" panel in which a "black" symbol appears.

6.2.10 In 1981 it was observed that the nature and position of the symbols had been changed slightly. On some colour print film the symbol, instead of being solid, has been like "computer" characters, thus: ▲ = Triangle. On black & white release positive, squares have been open instead of solid, thus: □, and placed between the words "Eastman" and "safety" instead of after the word "film".

6.2.11 From 1982 Kodak introduced a new system of symbols consisting of three characters for each year and incorporating a new character "X". This scheme is planned up to the year 2021. In this new scheme the characters are sometimes "solid" and sometimes "open".

6.2.12 In 1989 Kodak introduced a "barcode" on the margins of their stocks. Kodak have published the data on this.

6.3 Agfa

The name "Agfa" first appears on stock in the early 1920s.[18] Through to 1923 the letters had thick strokes and A's with flat tops.

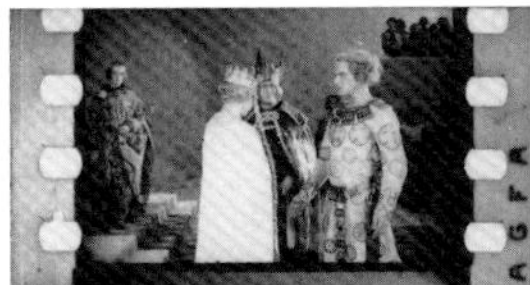

Up to 1923

From 1924 the letters are thinner and the A's have pointed tops.

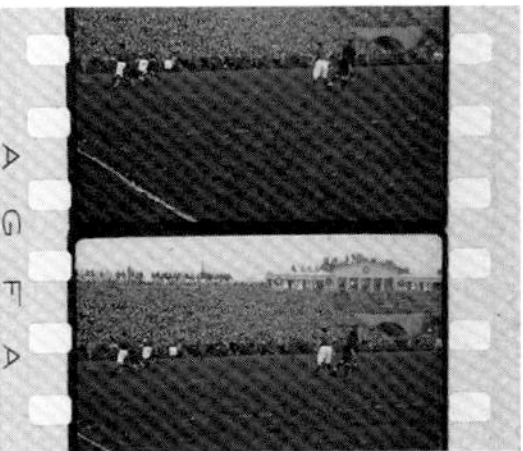

*After 1924 – CRB

18 See also the chapter on Agfa, pp.257-266.

6.4 Selo

Film marked "SELO" in letters with pronounced serifs appeared about 1928 and continued until about 1933. This stock was manufactured in Britain, and almost entirely used there.[19]

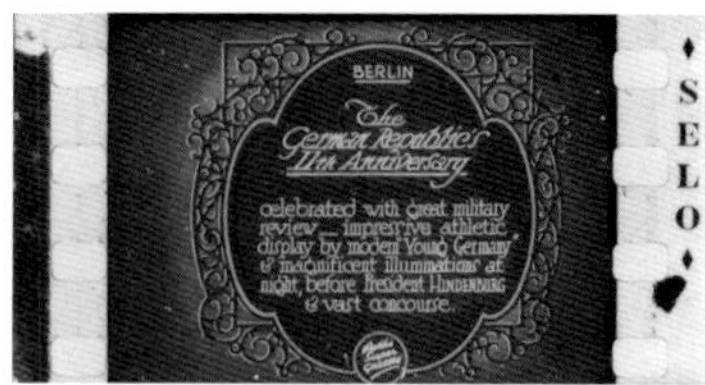

1928-1933

6.5 Brifco

The mark "Brifco" appears on film made in Britain and used there in the years 1920 to 1925. It usually appears only very faintly.

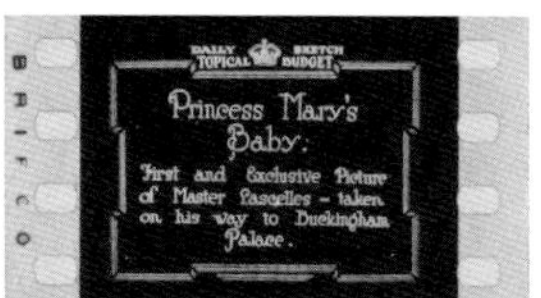

1920-1923

6.6 Gevaert

We illustrate three forms of "Gevaert" edge marks:

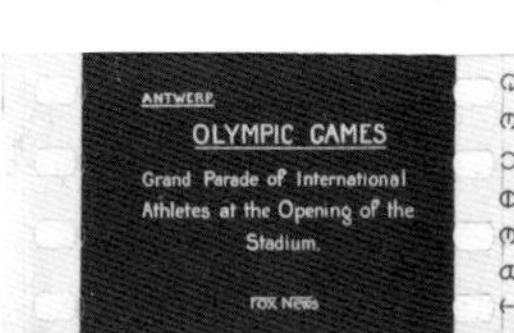

From a film of 1920:
1920

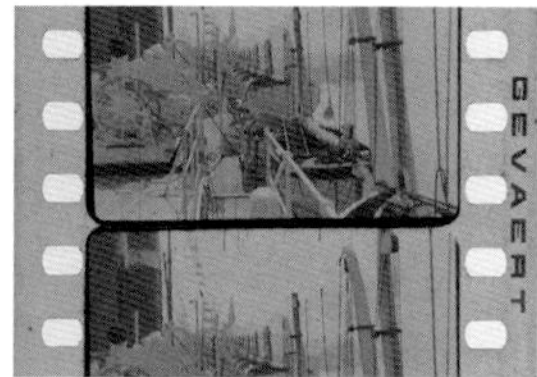

From a film of 1921:
*1921 – CRB

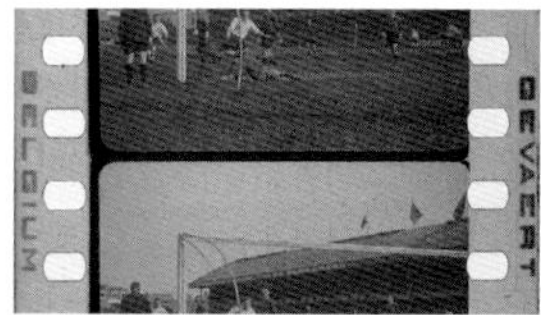

From a film of 1923:
*1923 onward – CRB

We have no specimen of film on Gevaert stock which is known to be from 1922. The same mark as in 1923 continued in use through the 1930s.[20]

It may be observed that many of these specimens are on titles. These titles all come from newsreels of events whose dates were easily ascertained quite precisely, so that we can accurately date the film, and relate the edge marks to the dates.

We used contemporary prints of these newsreel items. It would be unlikely for newsreels to be using old film stock, so the relation between date of event and the manufacture of the stock is reliably close.

---

19 Selo was formed by Ilford, Imperial, Gem, and a consortium of seven other UK manufacturers, eventually under the umbrella name of APM (Amalgamated Photographic Manufacturers Ltd., London). Selo films were sold by these companies under their own labels. The Selo company was situated in Woodman Road, Warley, Brentwood, Essex. Although the name Selo was officially dropped in 1946, the Brentwood factory continued to be known as "The Selo Factory" until it was sold and the site was subsequently levelled in the early 1980s. (Brian Pritchard) See also Ilford, p.309.

20 See also the chapter on Gevaert, pp.267-271.

6.7 Pathé

The same method has been applied to dating Pathé stock of the 1920s according to certain figures printed in the margin. The words "PATHE CINEMA FRANCE" or "PATHE CINEMA PARIS" are followed by a 4-digit number and then by one or two smaller groups of numbers.

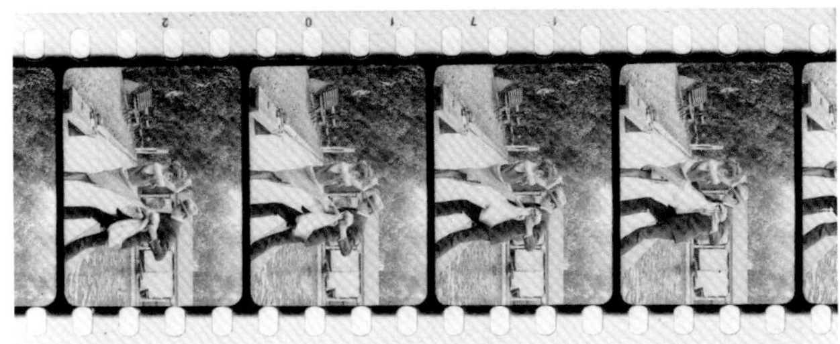

1921-1927

The following example has this edge mark faintly and reading from the base side.

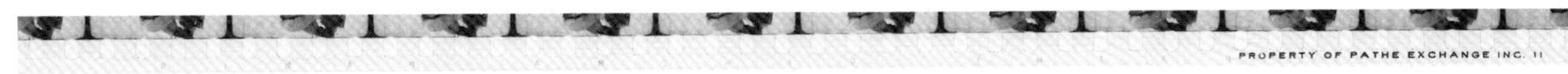

No date

See the first 2 digits of the 4-digit number:

1921: 16 / 17
1922: 18 / 19 / 20 / 21
1923: 22 / 23 / 24 / 25
1924: 26 / 27
1925: 28 / 29
1926: 30 / 31
1927: 32 / 33 / 34 / 35 / 36

The above numbers are applicable to positives. The numbers on Pathé negatives follow a different series which so far we have not been able to relate to dates.[21] The N.F.A. has very few negatives on Pathé stock of the 1920s.

It may be noted that this series of marks terminates at 1927, the year in which the raw stock manufacturing facilities of Pathé were acquired by Kodak.[22] Thereafter the name Pathé still appeared on the film, but not the same number series.

It may be observed that some of the marks are read from the emulsion side of the film, and some from the base side. I am not aware of any significance to this.

6.7.1 On some of the film stock of this period, manufactured by Pathé but used by Gaumont, the name "Pathé" does not appear, but the film has three little marks which look like windmills.

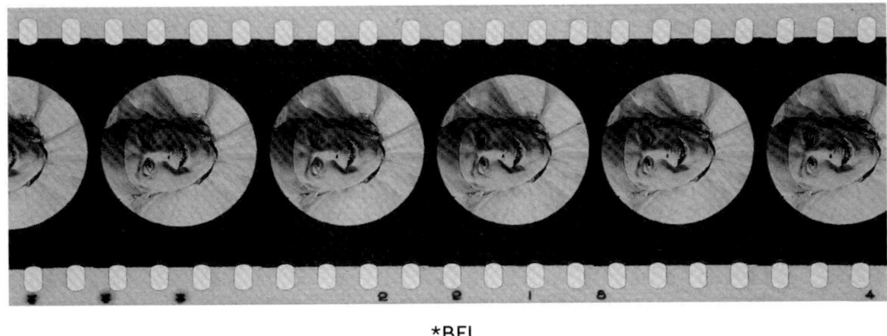

*BFI

The same number series applies.

21 It is now possible to date Pathé negative film stocks as well. See the chapter on Pathé, pp.222-228.
22 Pathé still manufactured film stock after 1927, at least until the early 1930s, following the same system. See also the chapter on Pathé, pp.206-207.

6.8 Lignose

This name has been seen on films made in the second half of the 1920s, but not at any other time. The films we have seen are all clearly of German production.[23]

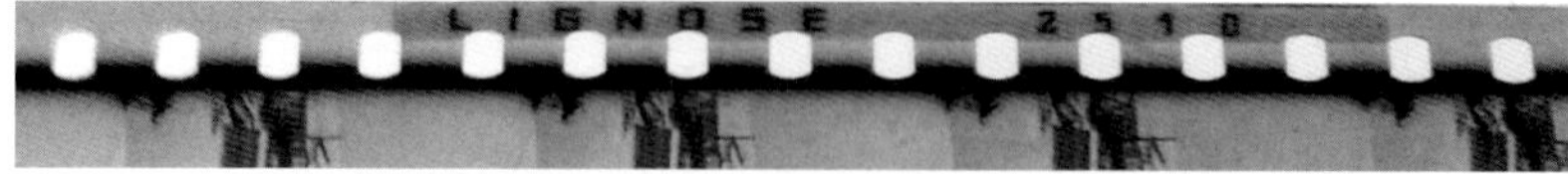

*NFA

6.9 Goerz Tenax

The same observation applies as to "Lignose" above (6.8).[24]

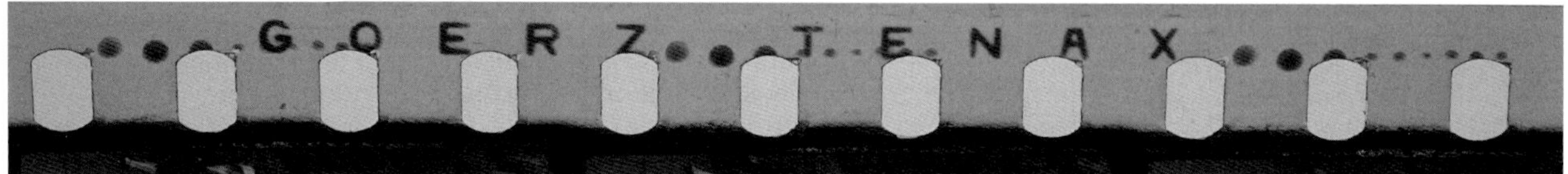

*Pre 1926 – BFI

23 Lignose-Hörfilm GmbH linked with British Phototone in 1928. (Brian Pritchard)

24 Goerz was founded in 1886 and started making photographic products in 1908. One of the cameras the firm made was called Tenax. In 1926 the German company combined with ICA, Contessa-Nettel, and Ernemann to become Zeiss Ikon. (Brian Pritchard)

7. Frame Characteristics and Features of Margins

7 .1 The principle here is that each frame of film is photographed in a camera through an aperture more or less the size of one picture frame. Today there are very precisely formulated international standards for the dimensions of the camera aperture, so that virtually all cameras and the film exposed in them (except for special formats) are identical in this respect. But in the early years of the industry there was no precise standardisation, and the exact size and shape of aperture varied from maker to maker and from user to user. Some producers adhered to camera and printer aperture characteristics of their own, by which they may be recognised.

Thus:

7.1.1 If the camera aperture is of somewhat less height than the length of film used for each frame, there will be a line of unexposed film between each frame of negative. This will appear as a dark line on a positive print (a wide frame-line). If the aperture height is exactly equal to the length of film used for each frame, then there will be no unexposed "frame-line", and each frame of picture will exactly meet the adjacent ones. It is also possible for the aperture height to be greater than the length of film used, and so each frame will overlap the next, and between each frame of negative there will be a strip exposed twice. This is found only rarely, however. In this case the "frame-line" area will appear lighter than the adjacent picture on the positive print, being composed of overlapping portions of the two picture frames.

7.1.2 The width of the camera aperture may be exactly the width of film between the two rows of perforations. It may be less, in which case the picture will not extend the full width between the two rows of perforations and there will be a blank border on each side, which will be transparent in negative and could be printed black in positive. The width of the aperture may be greater than the width of film between the two rows of perforations, so that the picture spreads onto the perforation area.

7.1.3 The rectangular shape of the aperture may be a perfect sharp-cornered rectangle; or the corners of the aperture may be rounded to greater or lesser degree.

7.1.4 The negatives made in the cameras were printed.

Here it is necessary to explain that when printing copies of films from a negative which is made in the camera, there are two different methods.
(a) Frame by frame (or step) printing.
(b) Continuous printing.

In a step printer, the negative and the unexposed positive copy film are passed through the gate of the printing machine in the same way as film passes through the gate of a camera, or of an ordinary projector, i.e., one complete frame at a time. The printer gate has an aperture approximately the size of one frame, and complete frames are printed one after another. In a print made by this kind of printer the printing aperture also images its shape on each frame of the film, in a manner similar to the camera aperture. Thus, on a print made on a step printer we may be able to see the image of both camera aperture and printer aperture. During the first 20 years of cinema, most prints were made on step printers.

In a continuous printer the negative, and the positive print film, are drawn smoothly (not frame-by-frame) past an illuminated slit, and the positive is thus exposed. In this case there is no frame-size aperture to image its shape on the print, as in a step printer.

7.1.5 By the differences in camera and printer aperture, which can be seen by looking at the film itself (but which do not appear on the screen), we can recognise in some cases the films of different producers. There is a further feature to look for in this connection. This is the position of the frame-line in relation to the perforations. For many years past, the position of the frame-line has been standardised midway between two perforations, but during about the first 20 years of the cinema, while some makers had the frame-line lying level with somewhere in the space between the perforations, some had it across the middle of the perforations; indeed, it might fall on any intermediate position. This is a further characteristic which points to a film being made by certain particular producers.

A number of illustrations are provided. For purposes of description of the position of the frame-line in relation to the perforations, I distinguish: mid-space; top space; bottom space; mid-perf; top perf; bottom perf.

One will not necessarily always be able to say, from comparison with the illustrations, that a specimen is from a certain maker; but one may well be able to determine certain makers whose product it is NOT, which still helps towards identification.

7.2 Méliès[25]

7.2.1 This was made on a continuous printer.[26] It can be seen that there is no black line between the frames; the corners of the picture are sharp square; the printing slit has extended over the perforations, so that there is a black band on the left and the right of the picture. The black band on the left is wider than the one on the right. The frame-line is at the bottom of the space between the perforations. These features are characteristic of all the seven original Méliès prints in the National Film Archive, produced up to about the middle of 1897.

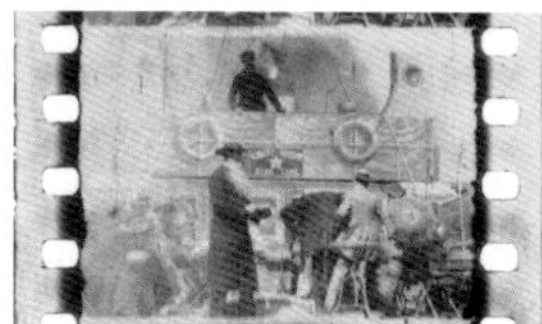

Entre Calais et Douvres / Between Calais and Dover (1897)

7.2.2 This is another Méliès print, but in this case made on a step printer.[27] In this case, the height of the printer aperture was small, and has thus left a clear space between the frames. It was narrow, and has thus created a clear space to the left and right of the picture. The corners were deeply rounded, which has also left clear round corners. This appearance is found on Méliès films from about mid-1897 through 1901.

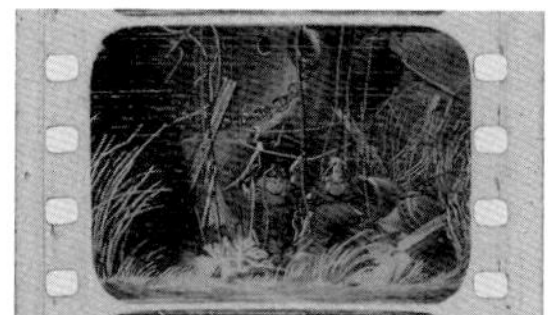

Visite sous-marine du "Maine" / Divers at Work on the Wreck of the "Maine" (1898) – BFI

7.2.3 Another Méliès example, printed on a step printer.[28] Observe that in this one there is also a clear space between frames, but the corners of the picture are not deeply rounded, but are nearly sharp square. Also, the picture extends the whole way between the two rows of perforations. The original Méliès prints which the N.F.A. has of 1902 through 1904 are all like this.

The N.F.A. does not have original prints of any of Méliès' later films (except one of 1912 which was printed by Pathé, and thus has the characteristics of Pathé), and when these films are duped it is usual that some or all of these features are obscured or completely obliterated.

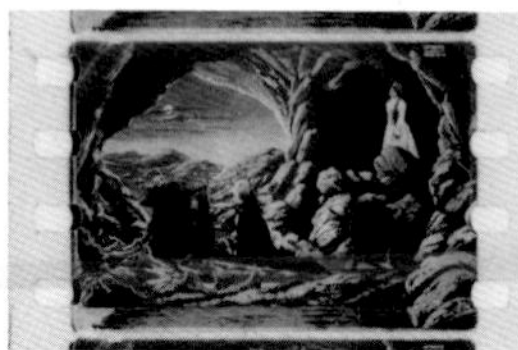

Royaume des fées / Fairyland; or, the Kingdom of the Fairies (1903)

25 See also "Identifying a Georges Méliès Film", pp.169-176.
26 According to Jacques Malthête, these physical characteristics don't necessarily imply the use of a continuous printer.
27 See Footnote 25.
28 See Footnote 25.

7.3 Paul

7.3.1 These frames are quite different from any of the Méliès films in a number of ways. The first, and most noticeable, is that the whole of the perforated margins is black. Also, there is only a thin line between the frames or none at all. The picture has sharp square corners and occupies the whole of the width between the two rows of perforations. The frame-line of all of them is through the perforation. All of the N.F.A.'s original prints of the films of Robert Paul from his beginnings in 1896 up to 1904 have this appearance.

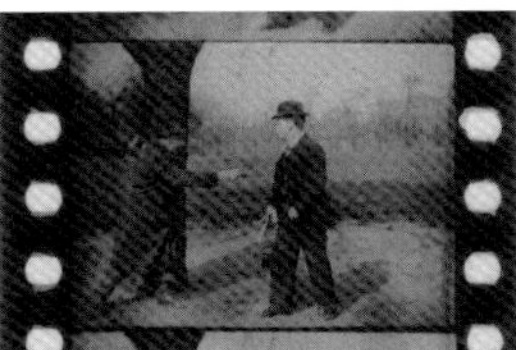

A Wayfarer Compelled to Disrobe Partially (1897)

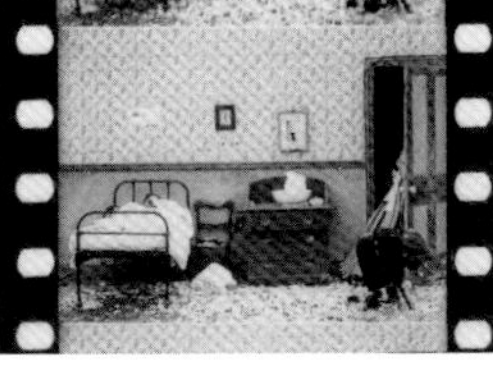

Children in the Nursery (1898)

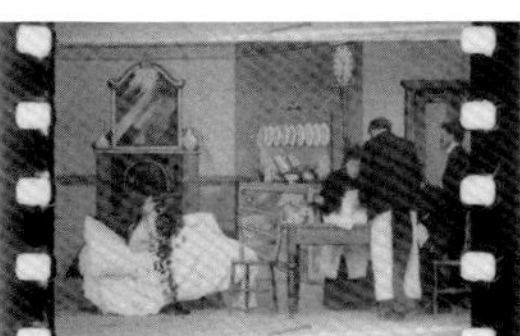

The Waif and the Wizard (1900)

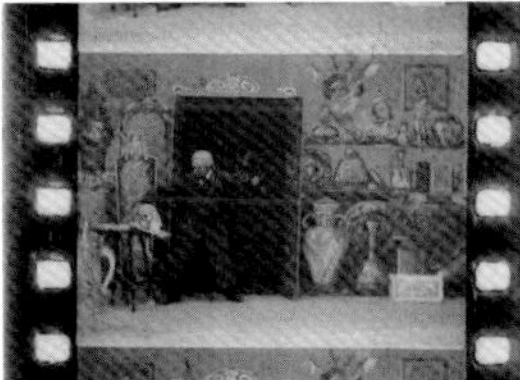

The Haunted Curiosity Shop (1901)

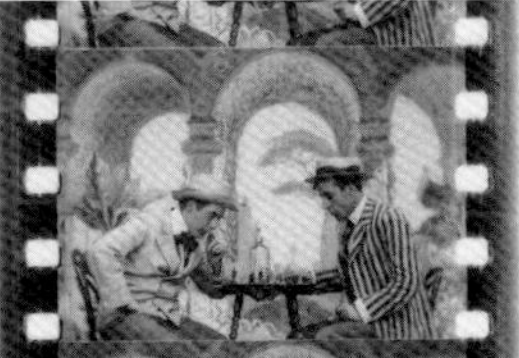

A Chess Dispute (1903)

7.3.2 From 1904 onward, Paul's films no longer have black margins. There is still only a thin frame-line between frames, and the picture still occupies the whole of the film between the rows of perforations. Observe that in all these prints of Paul, the frame-line is across the perforations. (Compare this with Edison prints and some early Pathé.)

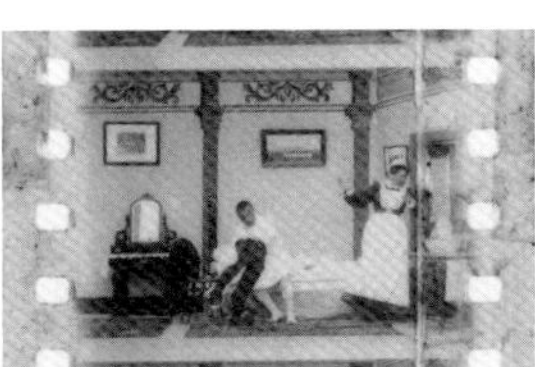

Mr. Pecksniff Fetches the Doctor (1904)

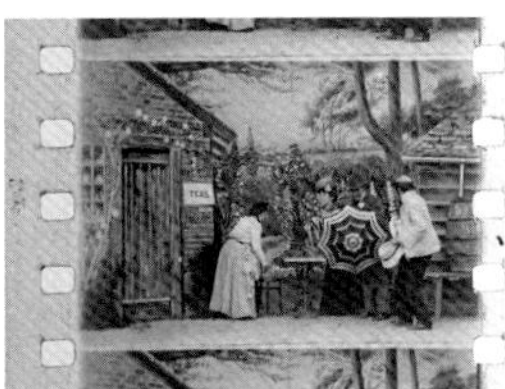

* *Short-Sighted Sammy, or the Stolen Spectacles* (1905) – SFI

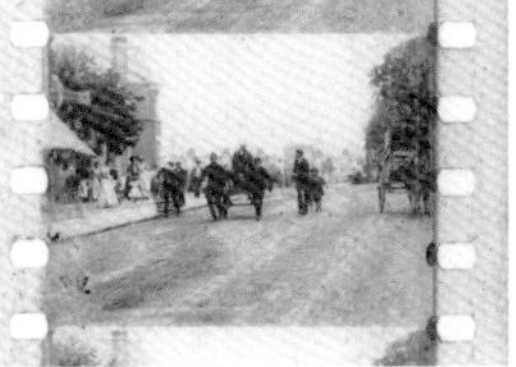

Is Spiritualism a Fraud? (1906)

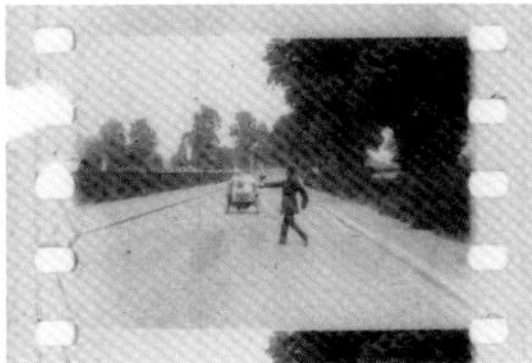

The ? Motorist (1906)

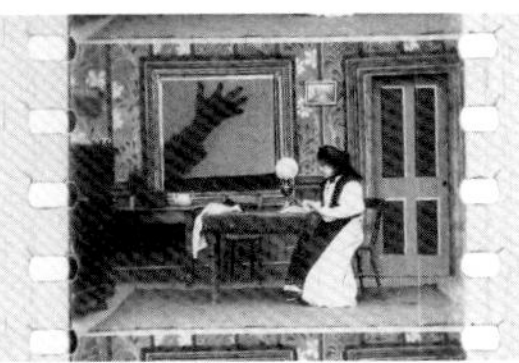

**The Fatal Hand* (1907) – SFI

7.4 Edison

This refers to Edison's early small-perforation films, not to the Edison Company's films of ca.1906-1915.[29]

7.4.1 This also has black margins, but has a quite different appearance from the prints of Paul. See that between the perforations of the positive, there is the image of the perforations of the negative. In making the print, the perforations of the negative and of the print film do not coincide. This is so in all of the few early Edison prints I have seen.

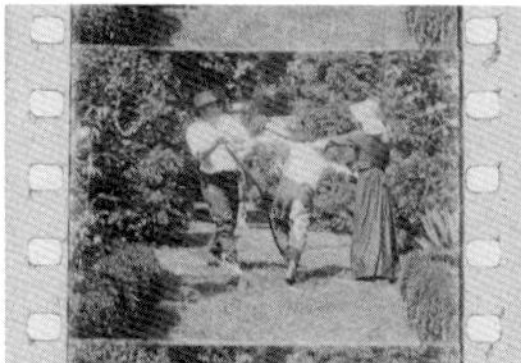

**Bad Boy and the Gardener* (1896) - SFI

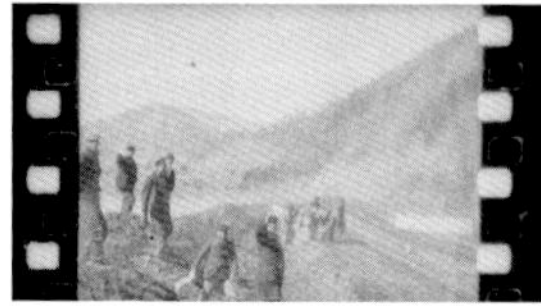

New Black Diamond Express (1900)

7.4.2 Compare with the prints of Paul, where the image of the perforations of the negative only shows very slightly, if at all. Here you can see a little bit of negative perforation (left-hand row); but not in any Paul film do images of the perforations of the negative lie completely between the perforations of the positive as in the Edison films.

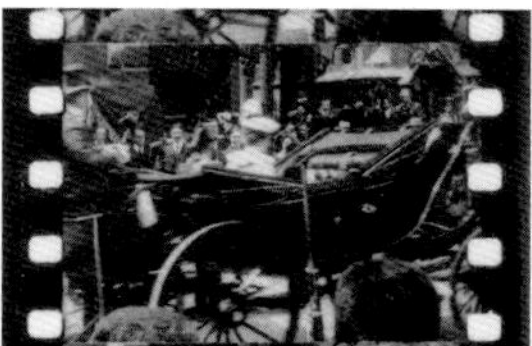

President Loubet's Visit (1903)

*7.4.3 Edison Company Films 1909-1912

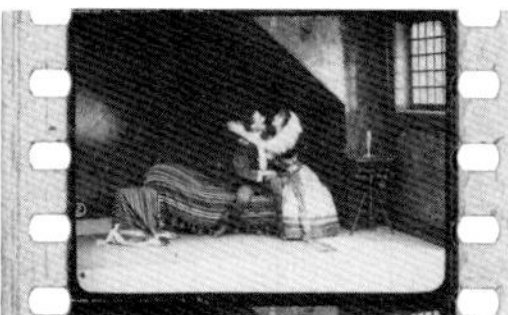

Fenton of the 42nd (1909) - GEM

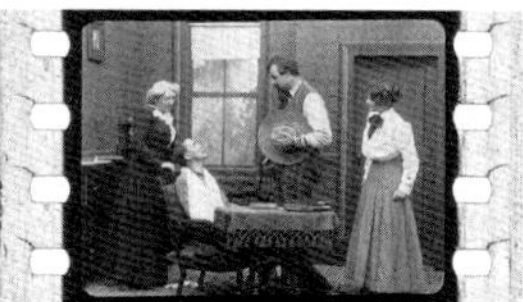

Arms and the Woman (1910) - GEM

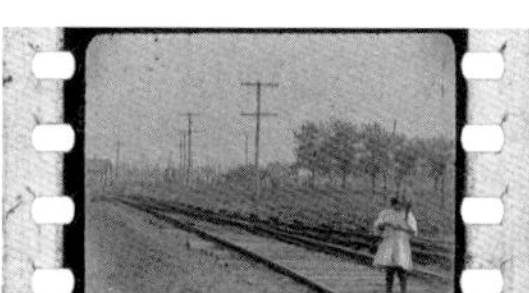

The Switchman's Tower (1911) - GEM

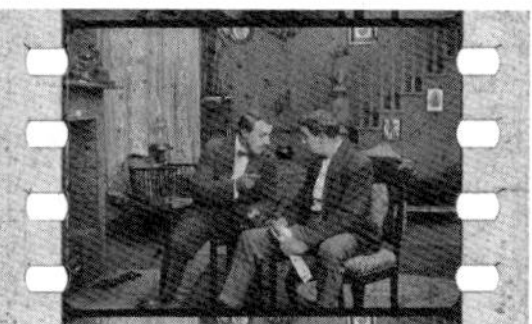

A Baby's Shoe (1912) - GEM

29 Examples of films produced by Edison between 1909 and 1912 are reproduced in 7.4.3. More information on Thomas Alva Edison can be found on the Library of Congress webpage: www.loc.gov/collections/edison-company-motion-pictures-and-sound-recordings/. See also Charles Musser, *Edison Motion Pictures (1890-1900): An Annotated Filmography*, Gemona: Le Giornate del Cinema muto / Washington D.C.: Smithsonian Institution Press (1997).

7.5 **Pathé** (Early Films with Small Perforations)

Some of the early Pathé films also had black margins, but note:

1. A black line between frames.
2. Black round corners to the picture.
3. The image of one circular perforation of the negative on each side (not very clear). Look at the point of the arrow.

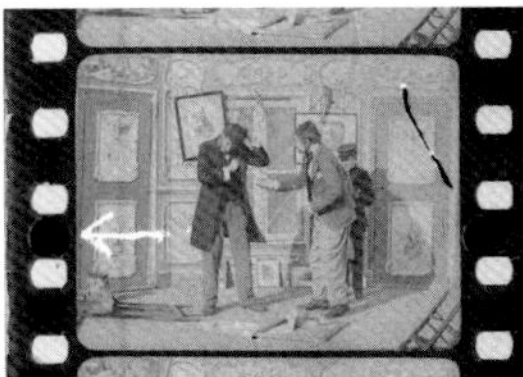

The Artist (1900)

Pathé clearly used some negative with Lumière-type perforations, but made prints on film with Edison-type perforations.

Some other early Pathé films had clear margins and had the image of a circular perforation near the top of the frame.[30]

Excentricités américaines / American Eccentricities (1901)

7.6 **Lumière**

The Lumières themselves, as well as supplying prints with their own type of perforations, also supplied prints with Edison-type perforations.

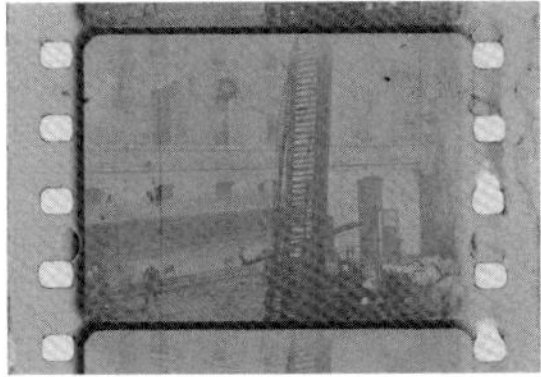

Paris: un incendie / Fire in a Court (1896-1897)

Here the image of part of the perforation of the negative can be seen on the left between the bottom two perforations of the frame.

Note also:

1. The picture itself, by reason of the camera aperture, has round corners.
2. The printer aperture also had round corners.
3. There is a black line between frames.

30 See also the chapter on Pathé, pp.201-205.

Most of the Lumières' prints were like this, but some had a clear space between frames.[31]

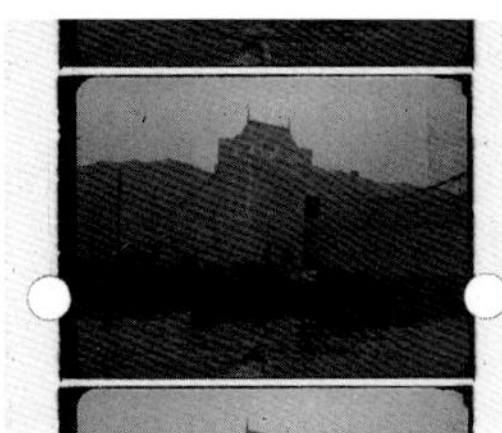

Pont de Westminster / Westminster Bridge (1896)

Another feature of all the Lumière prints was that they have become discoloured to a deep amber shade; this should not be confused with amber tinting.

Some prints of some other producers are discoloured, but not all. It may be that other producers sometimes used film stock manufactured by Lumière, who were already photographic manufacturers before they took up cinematography.

7.7 Warwick

Early small-perforation films:

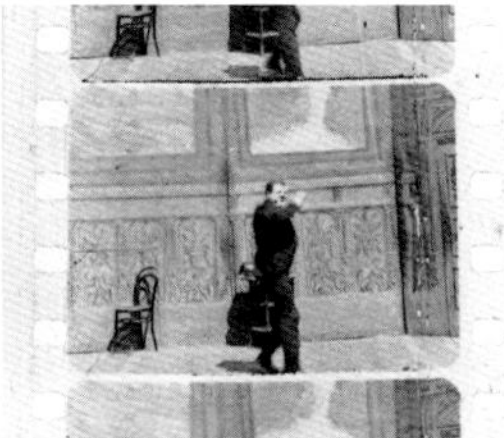

Will Evans, the Musical Eccentric (1899)

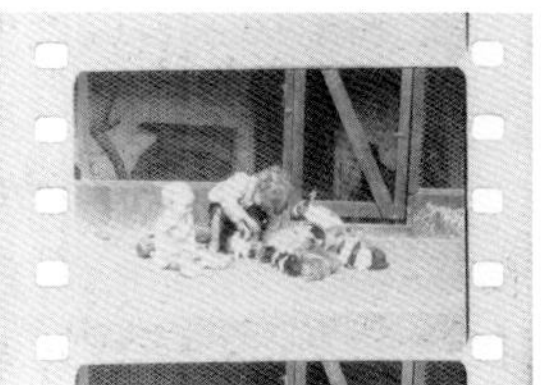

**Kiddies and Rabbits* (1904) – FE

Note:

1. Thin clear frame-line.
2. Round clear corners, due to the shape of the printer aperture. (Compare with Hepworth.)
3. The frame-line lies between perforations.

Later, the perforations change.

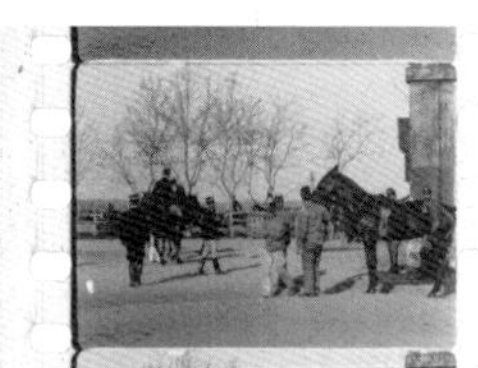

**Tor di Quinto* (ca. 1906) – FE

31 See also the chapter on Lumière, pp.165-168.

7.8 Hepworth

Early small-perforation films:

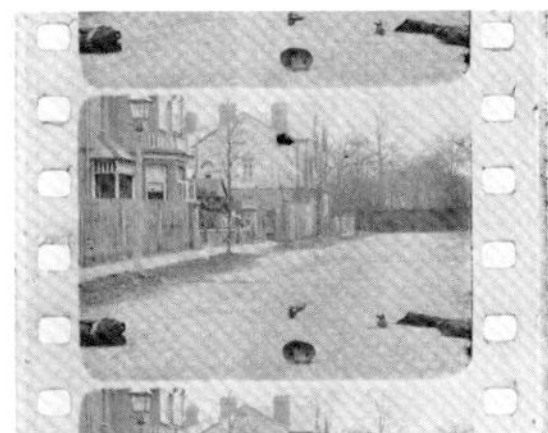
How to Stop a Motor Car (1902)

Note the similarity to Warwick.

Cecil Hepworth worked for the Warwick Trading Company before setting up on his own, so the similarity is not surprising. However, if one looks at the emulsion surface of Warwick prints, they have a fairly glossy appearance. Hepworth's similar prints have a dull matte appearance. Hepworth prints also generally exhibit a somewhat lower contrast than Warwick.

7.9 Vitagraph

From 1906 to 1916 Vitagraph had all their films with the frame-line across the perforations. The actual printer-aperture shape and the nature of the frame-line varied. This variation was random, as far as I have been able to see.

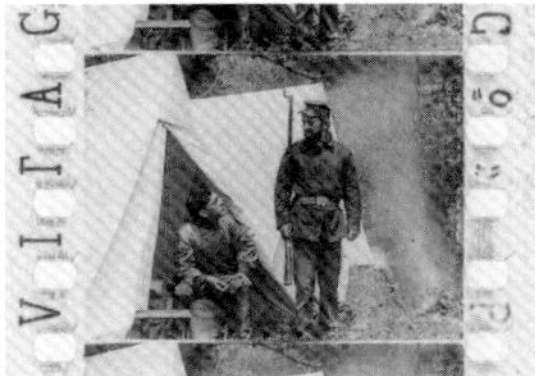
Betty's Choice (1909) – BFI

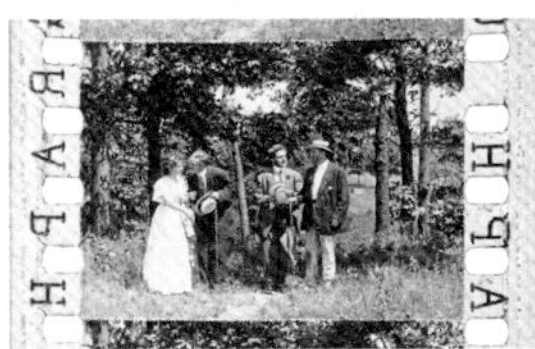
**Ransomed; or, A Prisoner of War* (1910) - HB

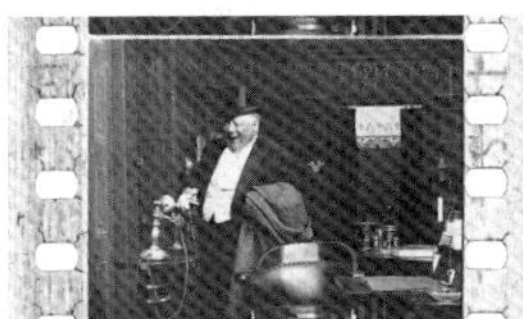
**The Tired, Absent-Minded Man* (1911) – GEM

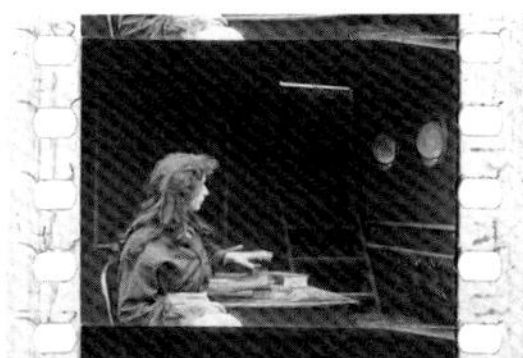
**His Last Fight* (1913) – BFI

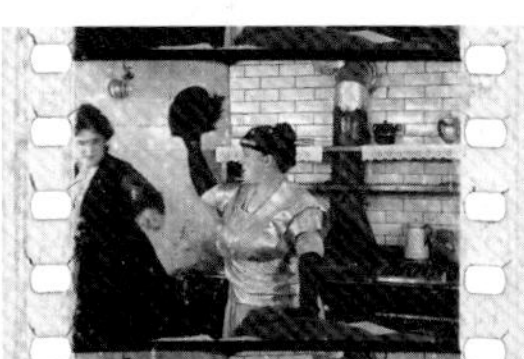
**His Wife Knew About It* (1916) – BFI

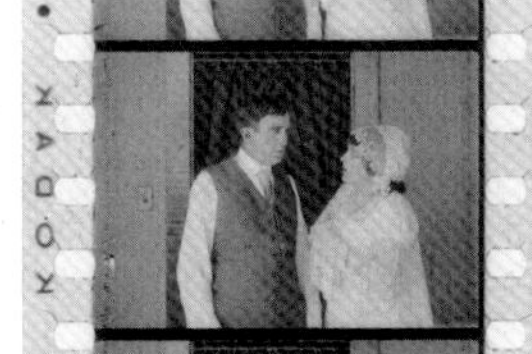
His Lesson (1917)

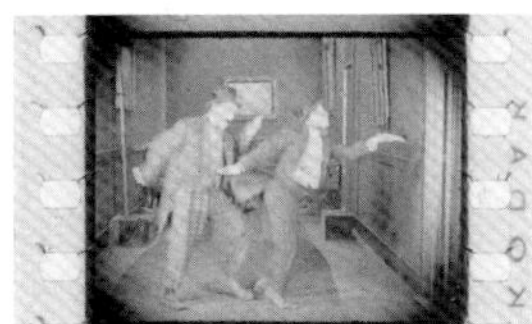
Damsels and Dandies (1919)

7.10 Selig

The N.F.A. has Selig Polyscope films ranging in date from 1909 through 1915. I examined some 50 of these, and have found just two consistent characteristics throughout that time: namely, the frame-line is always across the perforations, and there is always a black frame-line. The two specimens reproduced show the range of variation in frame-line thickness, and also show that in some prints the picture extends right to the perforations, and some have a black border to the left and right.

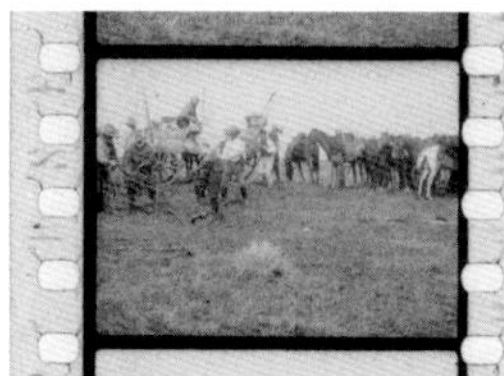

Ranch Life in the Great South-West (1910)

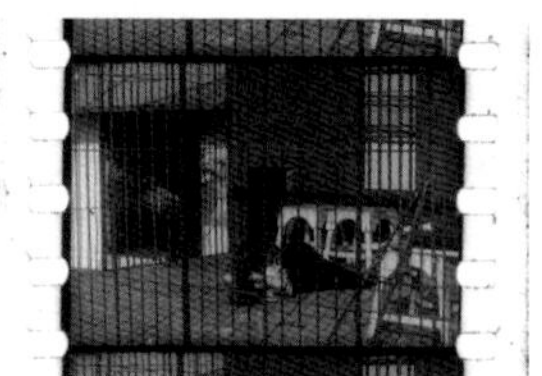

The Artist and the Brute (1913)

7.11 Pasquali

Pasquali had no one single consistent character. I show seven specimens:

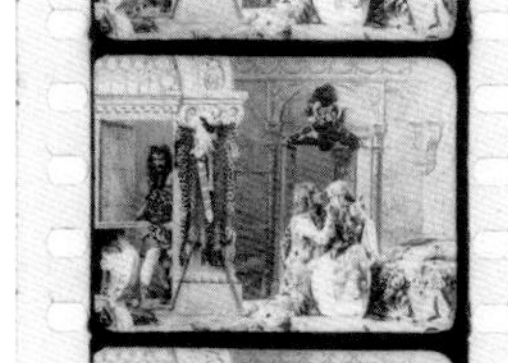

Alboino e Rosmunda / Alboino and Rosmund (1909)

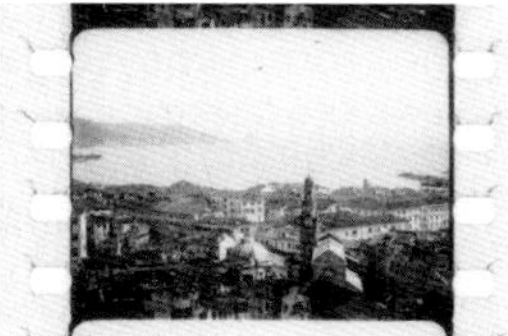

Primavera a Sanremo / Spring in San Remo (1911)

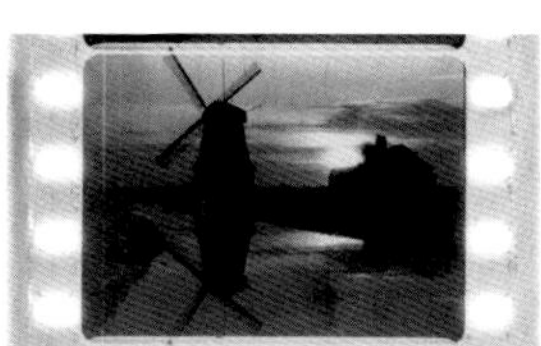

L'Olanda pittoresca / Picturesque Holland (1911)

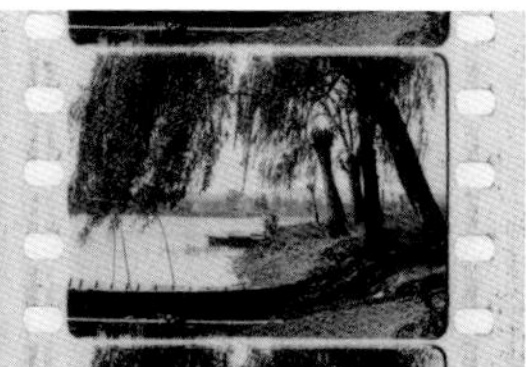

Madrid, la città del sole / The Town of the Sun - Madrid (1912)

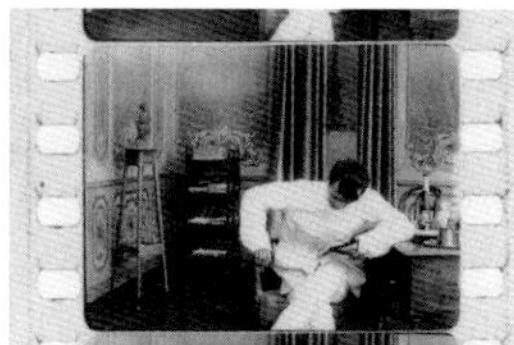

Polidor al club della morte / Polidor, a Member of the Death Club (1912)

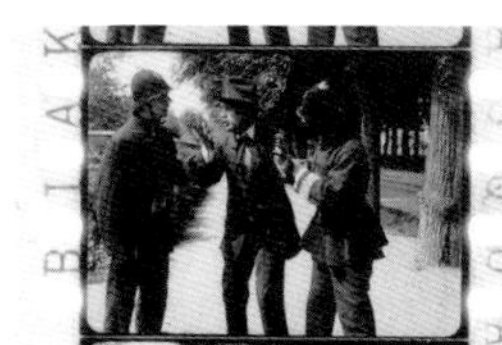

Il fascino dell'innocenza / The Fascination of Innocence (1913)

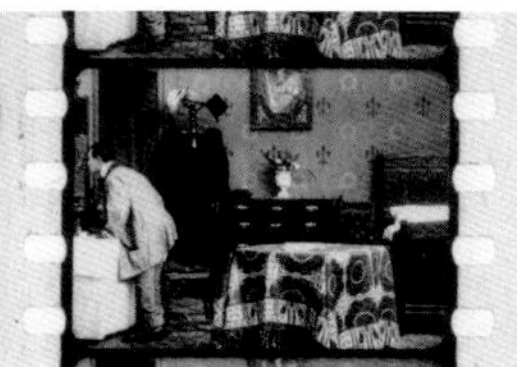

Polidor coi baffi / Polidor's Moustache (1914)

Picturesque Holland, The Town of the Sun – Madrid, and *Polidor, a Member of the Death Club* are somewhat similar to Cines prints, but without the distinctive crescents at the corners.

The Fascination of Innocence is similar to Itala prints. It may be that Pasquali, being a small producer, sometimes used the laboratory which made the Cines prints, and sometimes the laboratory which made the Itala prints. Note that the print of *The Fascination of Innocence* has the edge mark "BIAK".

7.12 Gaumont

Gaumont prints from 1906 to 1914 have a distinctive appearance.[32]

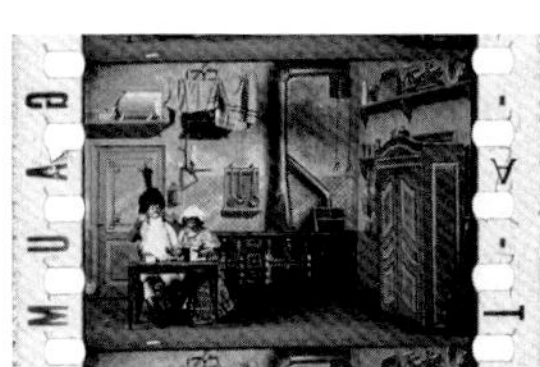
* *Le Bonnet à poils / The Fur Hat* (1908) - FE

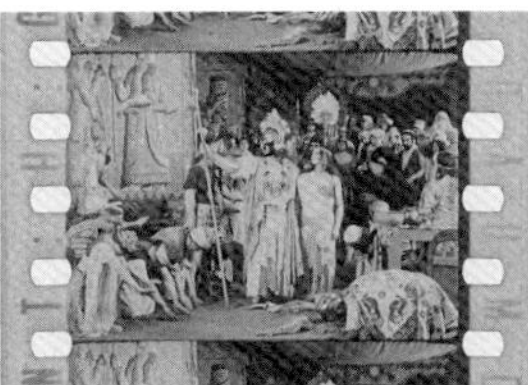
* *Esther* (1910) - BFI

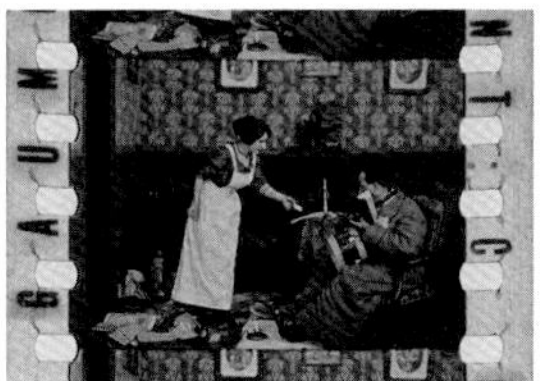
* *Calino courtier en paratonnerres / Calino's New Invention* (1912) - BFI

Note:

1. The frame-line is a very thin line.
2. The frame-line lies mid-perf.
3. The picture itself spreads onto the perforation area.

7.13 Cines

Cines film of the years 1909 to 1914 is quite recognisable.

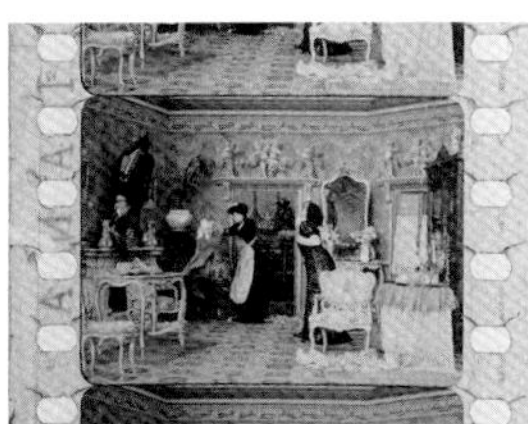
* *Il vezzo di perle perduto / The Pearl Necklace Lost* (1910) - CdB

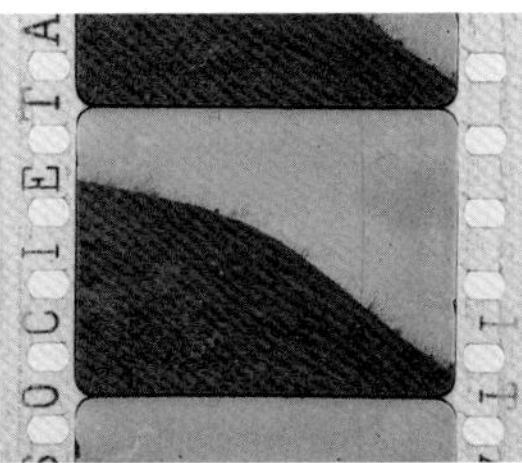
* *Il sogno di gloria di Tontolini / Tontolini's Glorious Dream* (1911) - CdB

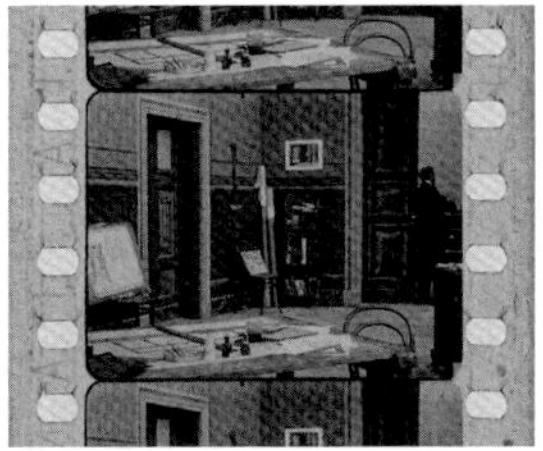
* *Salvata / The Danger Line* (1912) - CdB

Note:

1. The boundary of the actual picture has deeply rounded corners.
2. Outside the picture is a black border which also has rounded corners, but not so deeply rounded. This leaves crescent-shaped black corners which are quite distinctive. The thickness of the black border varies from film to film but the crescents remain distinctive.

32 This particular appearance, recognized as a characteristic of Gaumont films, is also observable in American productions (Vitagraph, Essanay, Biograph, etc.), but at this period in France, it's essentially a characteristic specific to Gaumont. (Eric Loné)

7.14 Thanhouser

The character of Thanhouser films varies considerably, but the characteristics do group into certain periods. See the Film List (pp.141-143).

1910

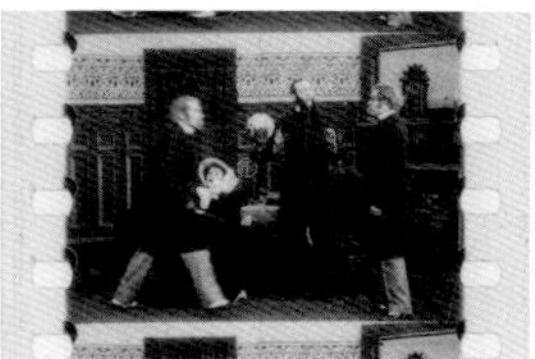

Daddy's Double (1910)

1912

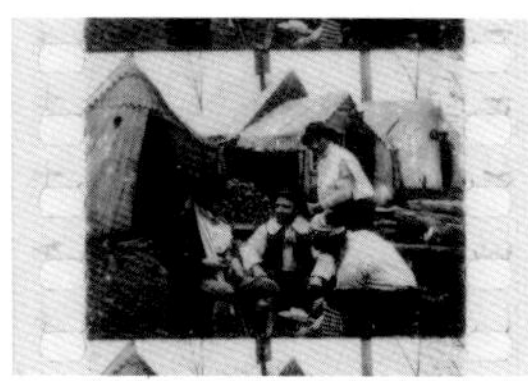

The Old Curiosity Shop (1912)

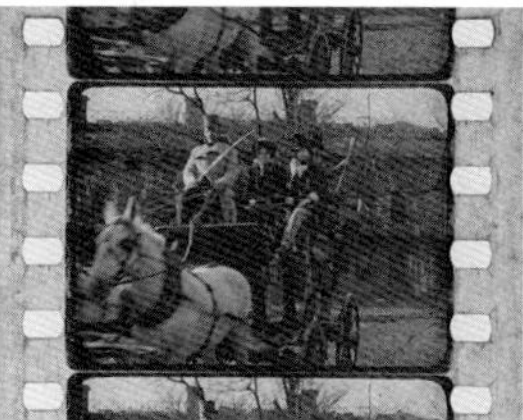

Nicholas Nickleby (1912) - BFI

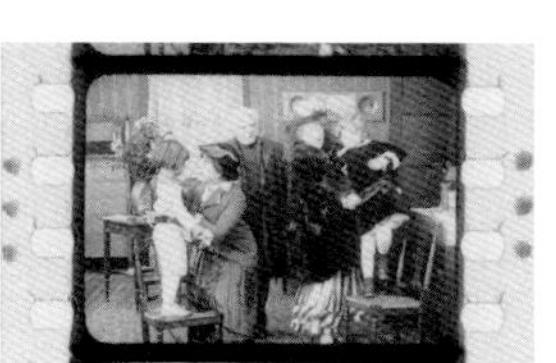

Treasure Trove (1912)

1913

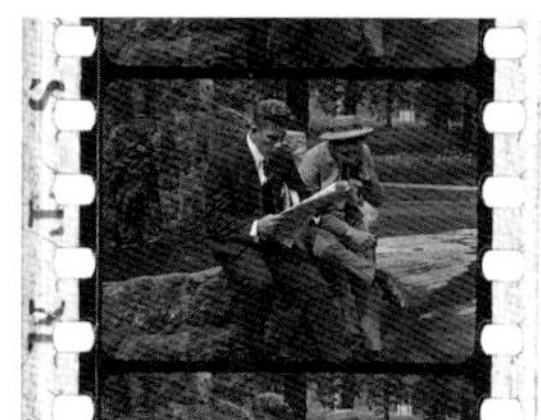

The Farmer's Daughters (1913) - BFI

1914

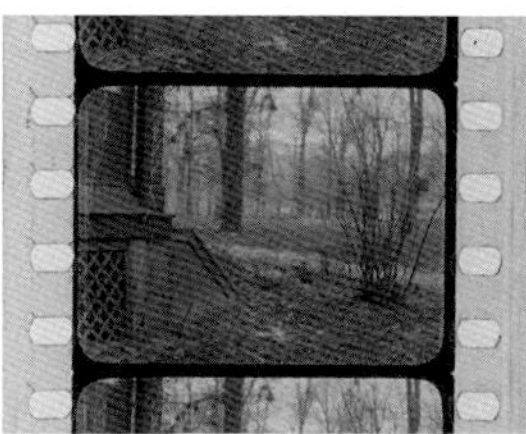

An Elusive Diamond (1914) - BFI

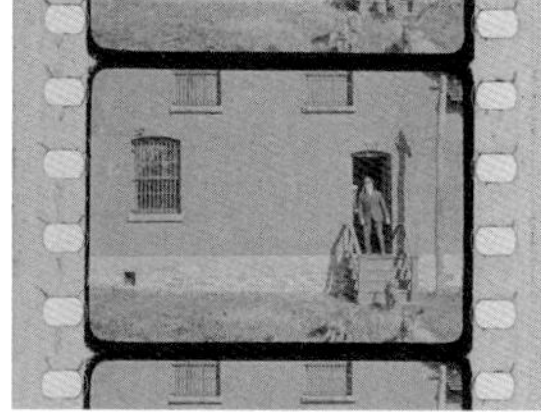

The Center of the Web (1914) - BFI

1915

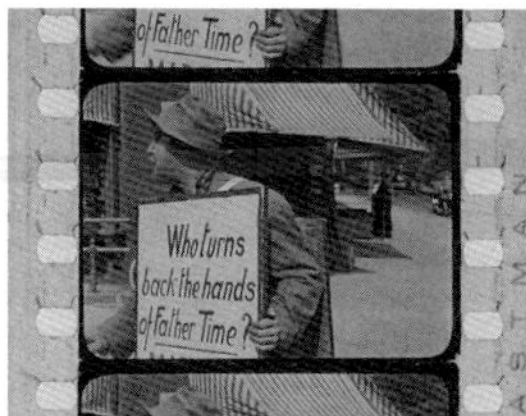

Madame Blanche, Beauty Doctor (1915) - BFI

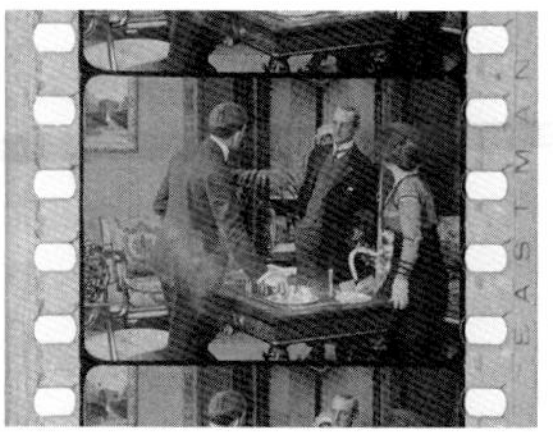

A Telephone Tragedy (1915) - BFI

7.15 **Pathé** (Films of 1905 Onward[33])

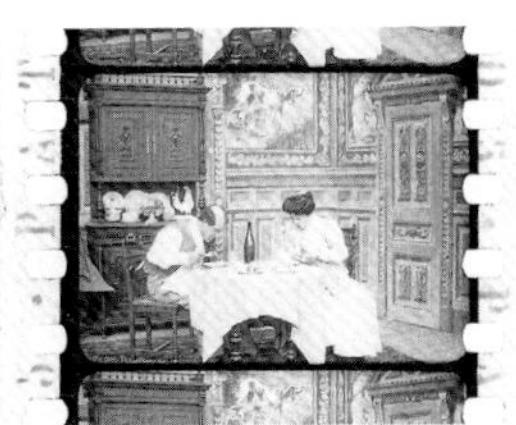

Voyage irréalisable / An Impracticable Journey (1905)

7.16 **Lubin**

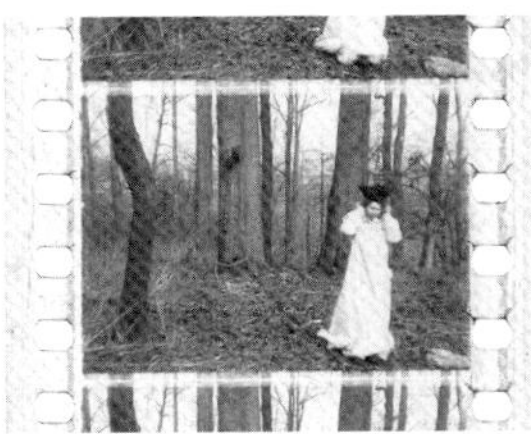

* *The Wreckers of the Limited Express* (1906) - GEM

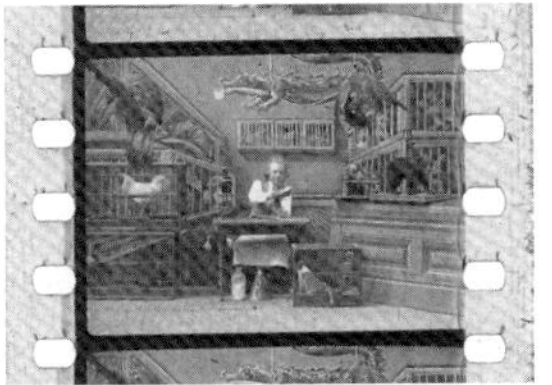

* *Professor Wise's Brain Serum Injector* (1909) - GEM

From 1910 through 1914, Lubin films had a camera aperture with sharp square corners. The printer aperture had slightly rounded corners, which creates a distinctive appearance.

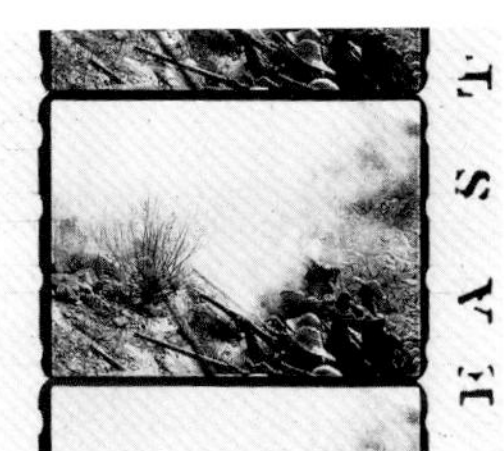

[Unidentified] (1910-1914)

7.17 **Kalem**

Many Kalem films had thick black frame-lines between the frames, and a thick black border on each side of the picture. The picture itself had fairly deeply rounded corners.

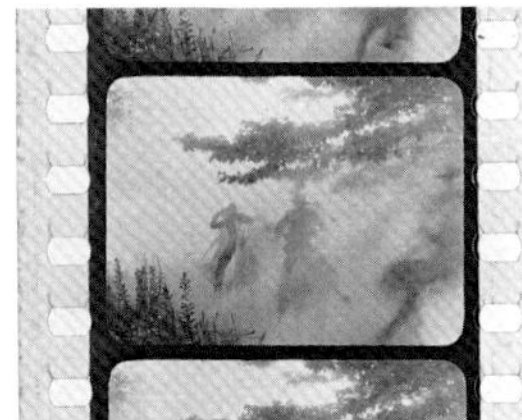

Unidentified

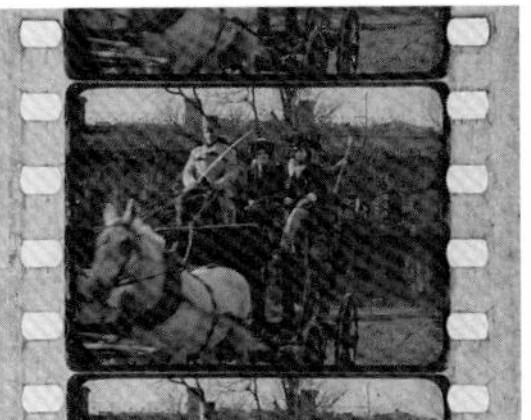

* *A Lad from Ireland* (1910) - GEM

33 See also the chapter on Pathé, pp.195-205.

Not all Kalem films were so. The example below has a very thin frame-line placed at the bottom of the perforation.

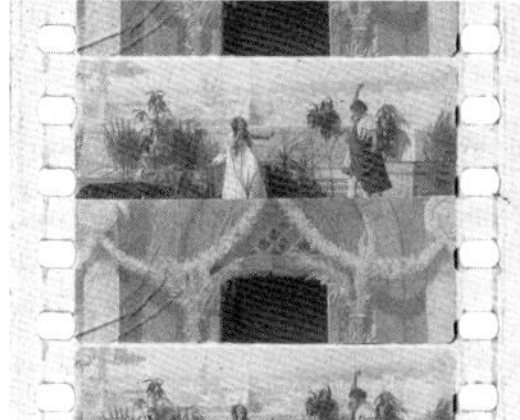

* *Ben-Hur* (1907) - GEM

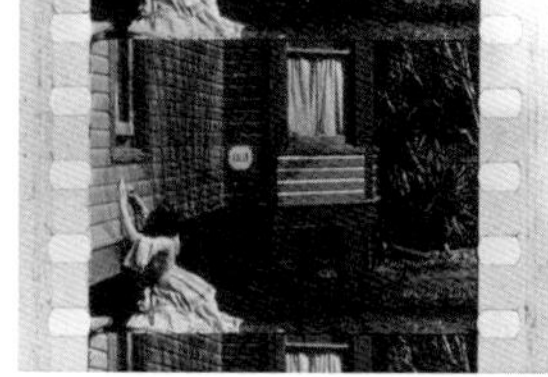

Adventures of a Spy Girl (1912)

7.18 Essanay

Many Essanay prints had the thick black frame-line and black borders, identical to the Kalem films described above in Paragraph 7.17.

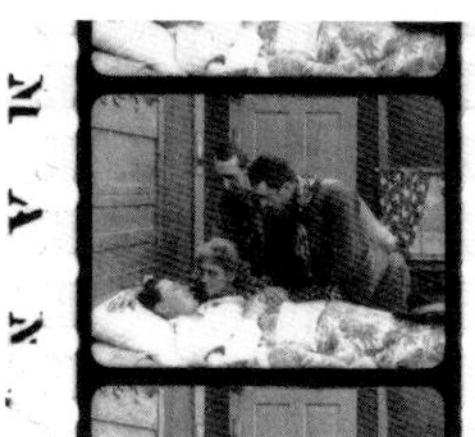

Unidentified

7.19 Ambrosio[34]

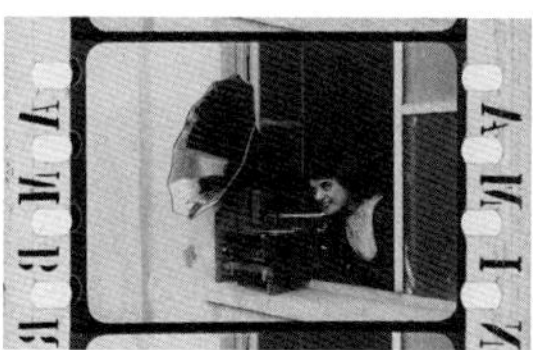

Robinet ha un tic per il ballo / Tweedledum's Dancing Fits (1910) - BFI

7.20 Itala

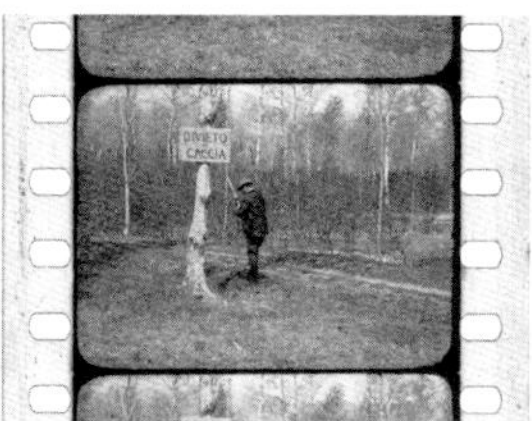

* *Il cuore più forte del dovere* (Rossi & Itala, 1907) - GEM

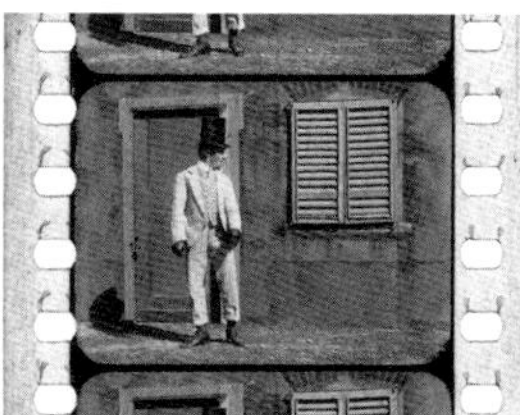

* *Cretinetti ficcanaso / Mr. Nosey Parker* (1909) - GEM

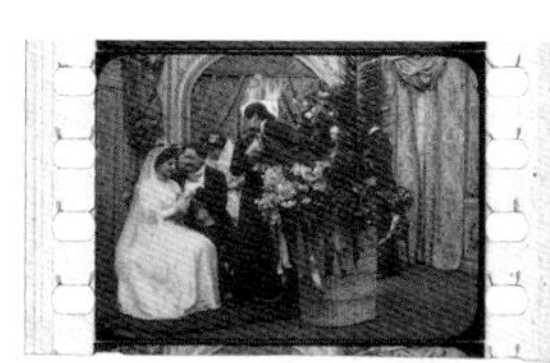

* *Cretinetti re dei reporters / Foolshead, Chief of the Reporters* (1910) - GEM

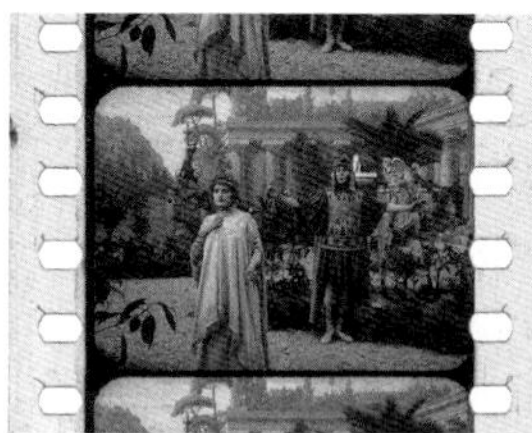

* *La caduta di Troia / The Fall of Troy* (1911) - GEM

34 No major difference was observed in the frame characteristics of productions between 1909 and 1913. (Camille Blot-Wellens)

7.21 American Biograph Company

All the original prints of the American Biograph Company had the distinctive characteristic that the black image of part of one perforation of the negative can be seen on each side of each frame of the print.

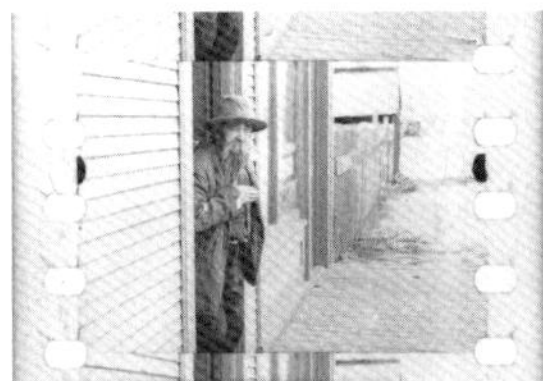

The Old Actor (1912)

The Biograph camera punched one perforation on each side of each frame, after that frame was exposed. These perforations were not beside that last exposed frame, but a little way from it. The mechanism which pulled each frame of stock through the camera gate did not pull a precise amount. Thus the position of those negative perforations is not constant in relation to the frames beside which they lie. If you project one of these films in a manner so that the perforations can be seen on the screen, the picture and the perforations of the print will be steady, but the images of the perforations of the negative will jump up and down.

7.22 Lux

Lux of Paris[35]

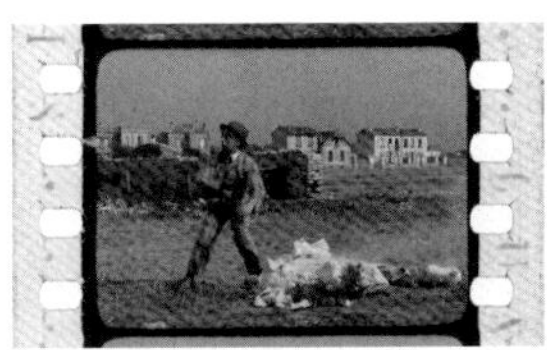

* *Patouillard crieur de journaux / Bill as a Newsboy* (1912) – FE

35 See also the chapter on Lux, pp.189-190.

8. Title Styles

8.1 Some producers used a consistent style of main title and intertitles, and also changed that style from time to time. Thus one can, in those cases, deduce a period within which a film was produced by reference to the style of its main title or an intertitle used in it.[36]

8.2 In studying any particular film, one has to face the possibility that the titles on the film are not the original ones put there by the film's producer, but are replacements, inserted either for change of language, or to deliberately conceal the origin of the film.

8.3 **Hepworth (1907-1913)**

Cecil Hepworth's titles are:

1907–1908

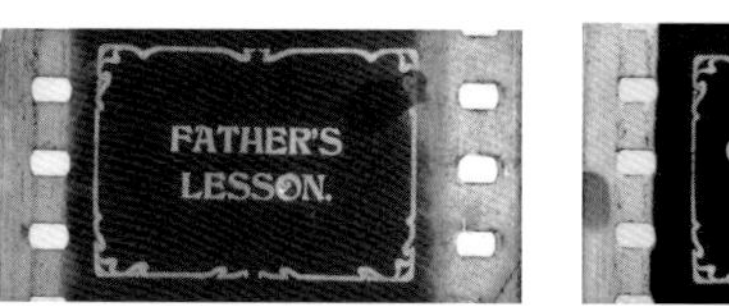

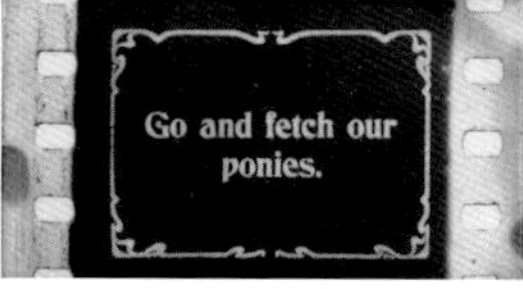

Father's Lesson (1908)

Main title, and intertitle. These are tinted green.

1909

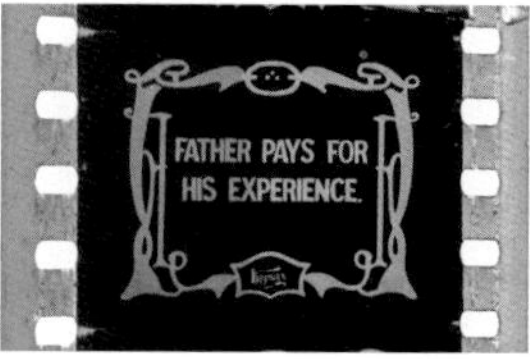

A Cheap Removal (1909) - BFI

Note that this is an intertitle and uses all capital letters. This is tinted pink.

1910–1912

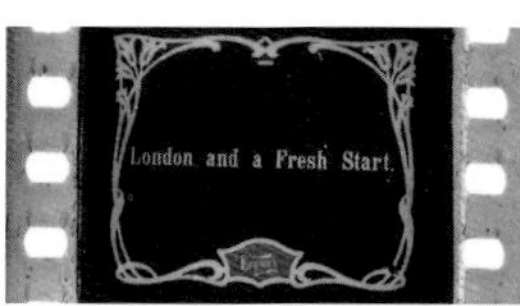

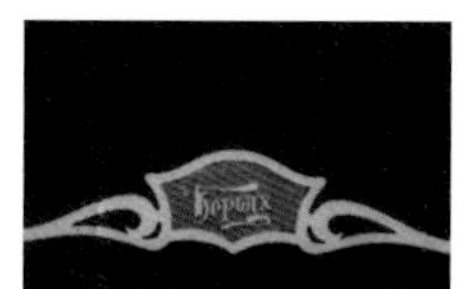

A Burglar for One Night (1911)

Note that the name in the bottom of the decorative border is "Hepwix". This is tinted amber.

36 See examples in the Colour Section, pp.145-160. Examples from other film companies not present in the 1990 edition are also reproduced in the Colour Section. These are indicated with an asterisk.

1913

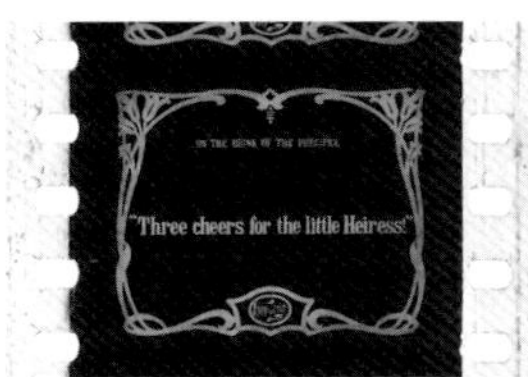

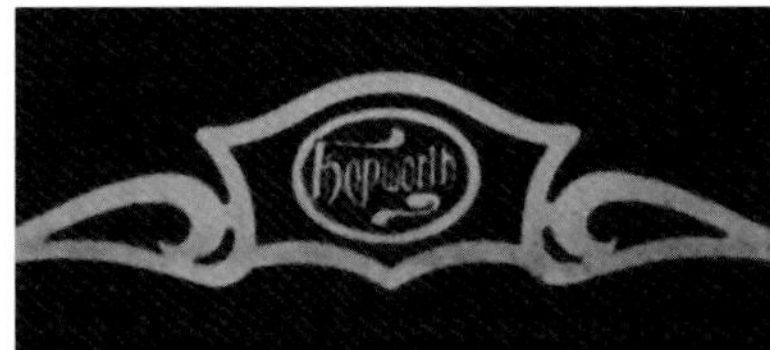

On the Brink of the Precipice (1913) – BFI

This is similar to 1910–1912, also tinted amber, but the name is "Hepworth".

8.4 Gaumont (1906-1914)

8.4.1 Gaumont film is usually easily recognisable as Gaumont. On most copies, other than newsfilms, the name is printed along both edges. (See detail in the section "Producers' Edge Marks".) However, in any Gaumont film on which the name is not printed, there are characteristics which distinguish it. (See Paragraph 7.12, p.83).

8.4.2 There is some evidence of date to be derived from the form of Gaumont's original titles. On the main titles of their films up to and including 1907, there appeared the name "ELGE", derived from the initial letters of the name "Léon Gaumont". Before 1907 this trademark appeared within a circular floral border.

The Traitorous Guest (Pre-1907). Tinted Pink

During 1907 there was no floral border.

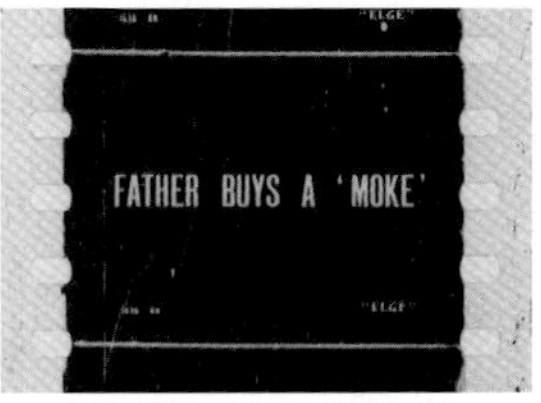

L'Âne récalcitrant /
Father Buys a 'Moke' (1906). Tinted Blue

In the all-too-frequent event that the main title is missing, the name may also be found on intertitles.

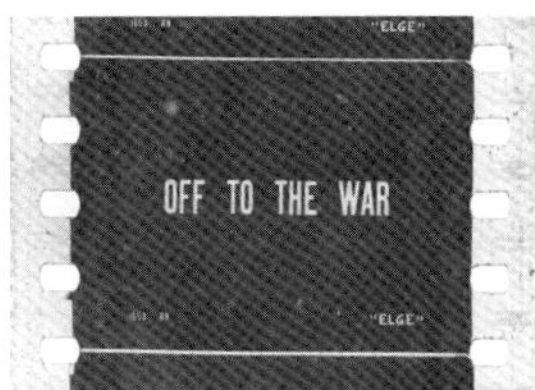

La Fiancée du volontaire /
The Hand of the Enemy (1907). Tinted Blue

8.4.3 From 1909 through to the middle of 1910 the form of name on the titles was "Gaumont".

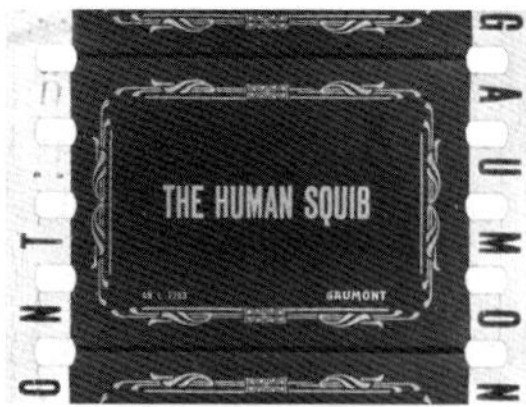

Le Poivrot incendiaire /
The Human Squib (1909). Tinted Blue

8.4.4 Later in 1910 the name "Gaumont" on the main titles was enclosed in the floral border; this style continued through 1913.

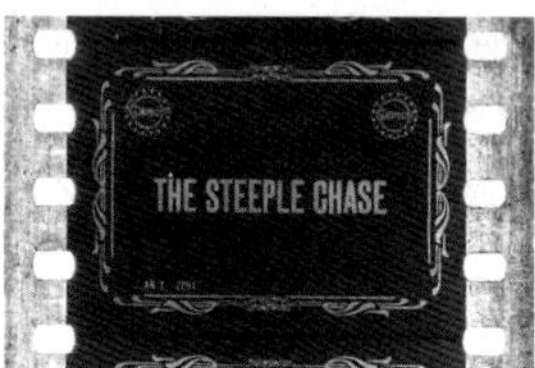

[*Le Grand Steeple-chase*] or [*La Course du steeple*] /
The Steeplechase (1910). Tinted Amber

8.4.5 Until 1910, with few exceptions, the titles of the films were tinted blue. Throughout 1910 the titles were tinted amber.

8.4.6 From the latter part of 1910 the form of trademark on the intertitles was the letter "G" in a floral border.

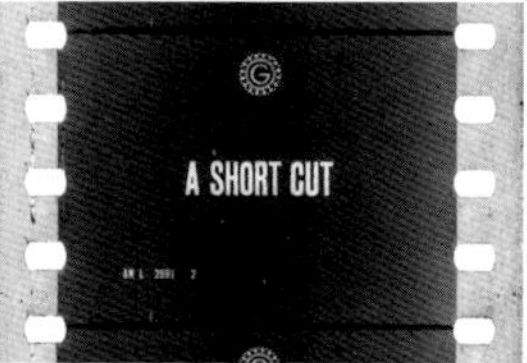

[*Le Grand Steeple-chase*] or [La Course du steeple] /
The Steeplechase (1910)

8.4.7 From 1911 onwards the form of trademark on the titles remained as in 1910, but they were tinted green.

8.4.8 From the start of 1912 through to 1914, the trademark on the intertitles, as well as on the main titles, was "Gaumont", in the floral border.

8.4.9 If one looks more closely, there is some small writing consisting of a four-figure number followed by the letters "AN". A number in this form seems to have been normal throughout 1907. I have not seen enough films of 1908 to make a statement about that year.

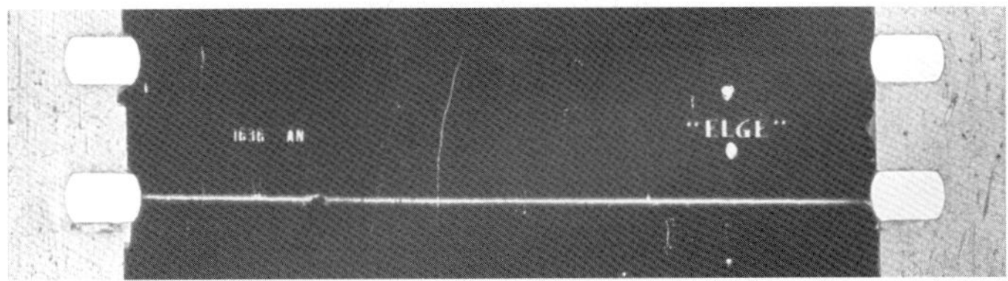

L'Âne récalcitrant / Father Buys a 'Moke' (1906)

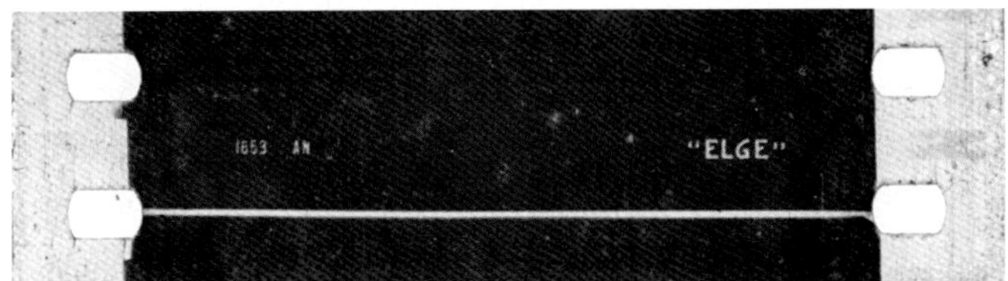

La Fiancée du volontaire / The Hand of the Enemy (1907)

8.4.10 At the beginning of 1909 the form changed to the letters "AN L", followed by a 4-figure number.

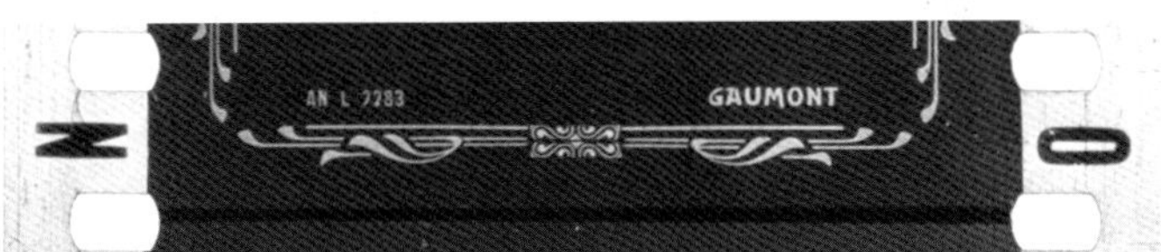

Le Poivrot incendiaire / The Human Squib (1909)

This form continued throughout 1910 and 1911. Concerning the meaning of the letters: I take "AN" to be the first letters of the French word for "English", i.e., "ANglais". The significance of the "L" I do not understand, but it may be for "Londres" (London), to distinguish from a differently spelled version for other English-speaking areas (e.g., the U.S.A.). On the German-language copies which I have seen, the letters are "AL", presumably for "ALlemand". Some have also the letter "B", perhaps for "Berlin". On Dutch-language copies it is "NL".[37]

8.4.11 Early in 1911 there was a change in the titles. In the margin at the bottom left of the title appeared a black patch.

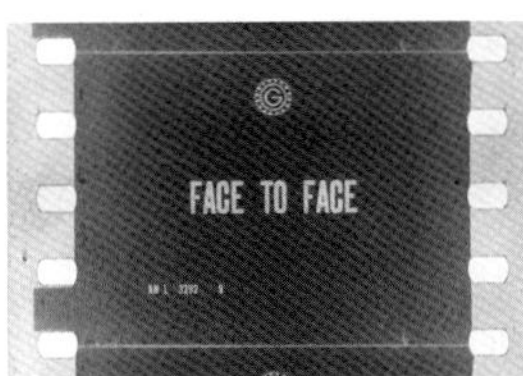

[*Le Fils de Locuste*] / [*In the Days of Nero*] (1911). Tinted Green

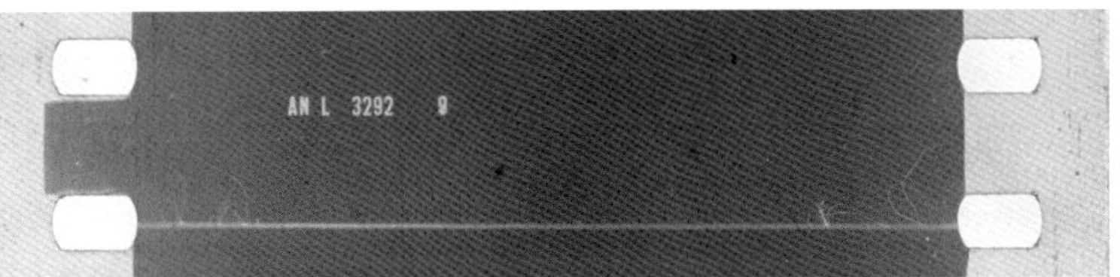

Detail

37 Spanish titles have the letters "ES B", the B possibly standing for Barcelona. (Camille Blot-Wellens)

Near the end of 1911 a large white number appeared in that black patch.

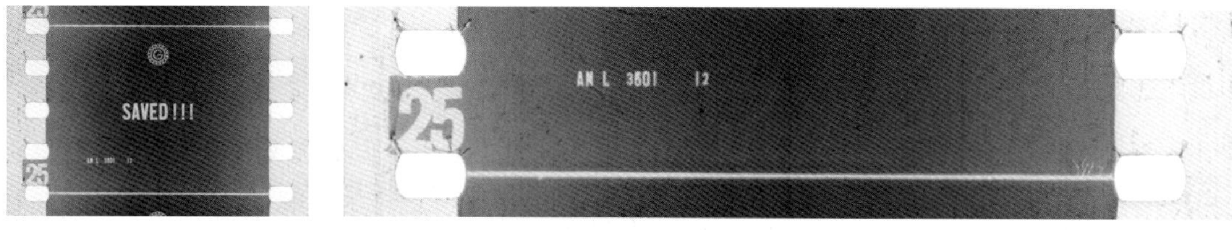

Island Maiden (1911). Tinted Green

Detail

The explanation of that number is as follows:

After the main title, one counts all the scenes, intertitles, and letters (inserts) in a single series through the film. The intertitles have their numbers in large white printed figures in the black patch. Letters (inserts) have their numbers written in black figures in the same position in the clear margin. Picture scenes are not marked.

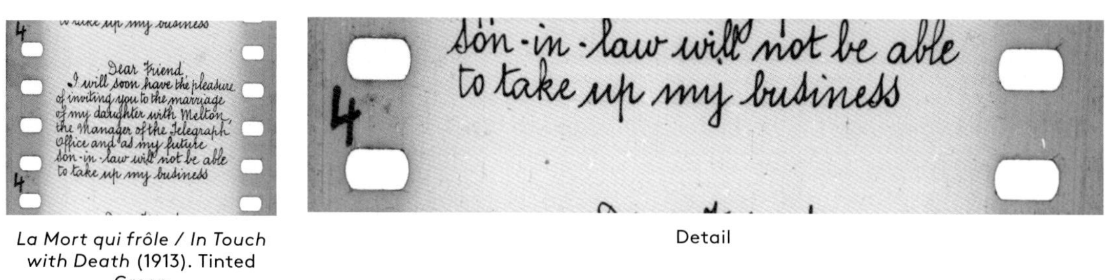

La Mort qui frôle / In Touch with Death (1913). Tinted Green

Detail

Thus, a typical sequence might be:

MAIN TITLE

Scene	1	- not marked
Scene	2	- not marked
Intertitle	3	- white number on black patch
Scene	4	- not marked
Letter	5	- black written figure in clear margin
Scene	6	- not marked
Scene	7	- not marked
Scene	8	- not marked
Intertitle	9	- white number on black patch
Scene	10	- not marked
Scene	11	- not marked
Intertitle	12	- white number on black patch, etc.

8.4.12 Early in 1912 the small four-figure number and its associated letters disappear from the bottom left-hand corner of the intertitles and appear on the black margin patch, but twisted round sideways.

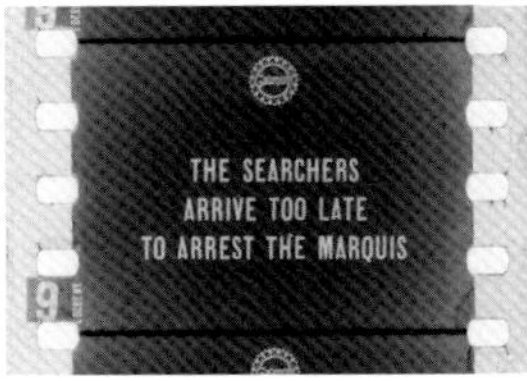

La Cassette de l'émigré / The Refugees' Casket (1912). Tinted Green

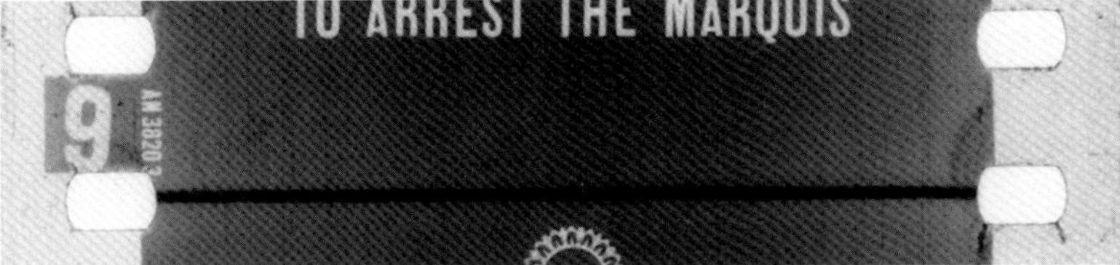

Detail

This figure has the form "AN 3820", then a small space and a figure "3". This last figure indicates that this is the third intertitle in the film. These numbers are not always as clear as in this example, and the single figure is sometimes cut off by the perforations, so it is worth looking at all the intertitles.

8.4.13 The 4-figure number, whether on the screen area or on the black margin patch, is a production serial number. The list of "Gaumont Films 1906–1914" is arranged in order of this serial number. It will be seen that this also proves to be the date order of the films. (There is one exception: *The Inundations in France*, Number 2656, appears earlier than its number suggests. I attribute this to its being a subject of topical interest; so that it was sent out more quickly than normal.) Thus, by noting the serial number on the title of any unknown Gaumont film of the period, and referring to the list, it is possible to establish its date with a fair degree of accuracy.

8.4.14 The Gaumont film list (reproduced in Appendix 1, pp.122-125), with the explanation of the abbreviations used, sets out the features described above in respect of the films examined.

8.5 **Cines** (1909-1914)

8.5.1 I have examined 57 films of Cines of the years 1909 to 1914.[38]

8.5.2 Cines film of these years is quite recognisable. (See Section 7.13, p.83).

8.5.3 Cines changed the style of their main title and intertitles from time to time, and in the absence of other clues this may help with some indication of the date of the films.

8.5.3.1 In 1909 they used a form of trademark on the main titles consisting of a wolf with two babies (Romulus and Remus) and the name CINES beneath it.

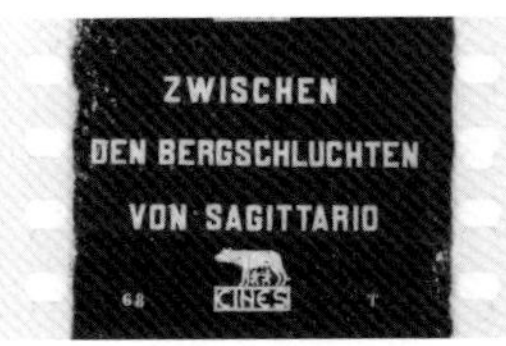

Gole del Sagittario / Sagittario (1909). Tinted Red

38 More examples of title styles are in the Colour Section.

The associated intertitles were of this style:

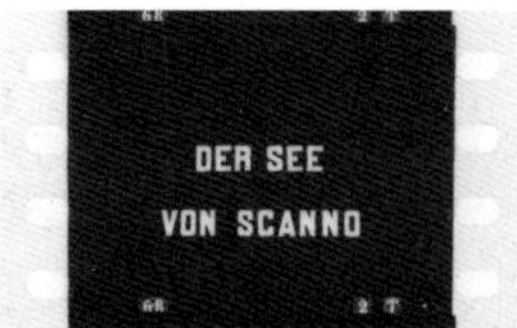

Gole del Sagittario / Sagittario (1909)

8.5.3.2 In 1910 and throughout 1911 the main titles were in this style, with intertitles in this style:

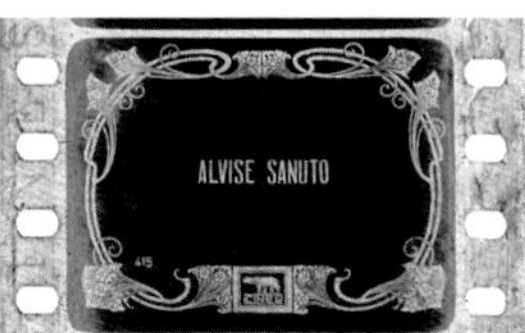

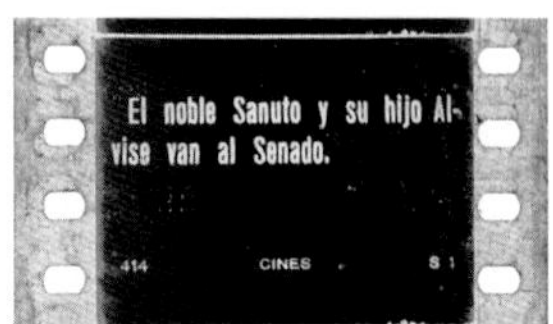

Alvise Sanuto / Venetian Chivalry (1910). Tinted Amber – BFI

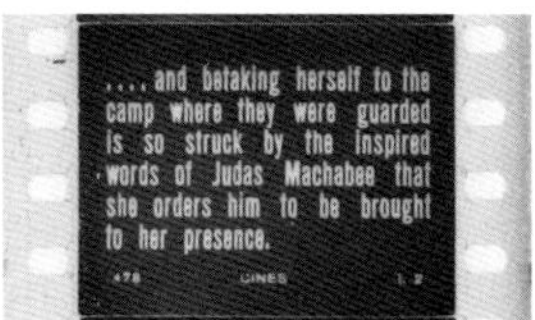

I Maccabei / Judas Maccabaeus (1911). Tinted Amber

8.5.3.3 None of the Cines films of 1912 in the N.F.A. has a main title; they are all missing, so I have no evidence as to their style.

8.5.3.4 The N.F.A. has three Cines films of 1913 with their original main titles, and these three all incorporate the name "Cines" into a 5-petalled flower motif.

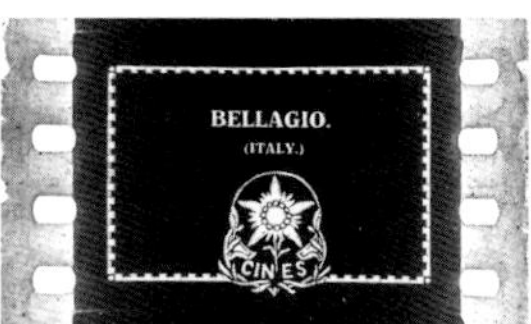

Bellagio (1913). Tinted Green – BFI

Among the films of 1913 the floral motif appears in two different sizes. I know of no significance to this. Most of the intertitles of 1913 are as in *By Unseen Hands*, but two from early 1913 are different.

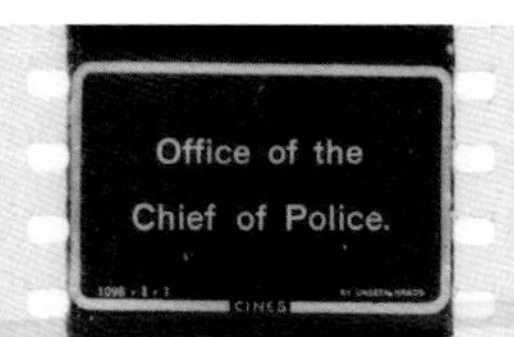

Le mani ignote / By Unseen Hands (1913). Tinted Red

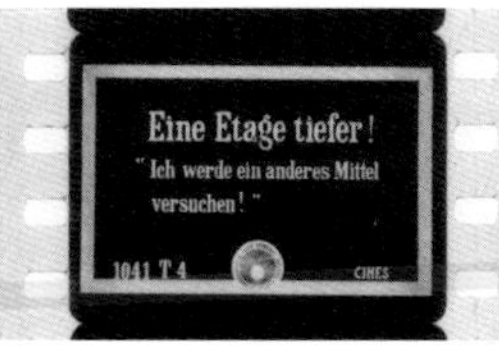

Kri Kri e la suocera / Bloomer's Mother-in-Law (1913). Tinted Pink – BFI

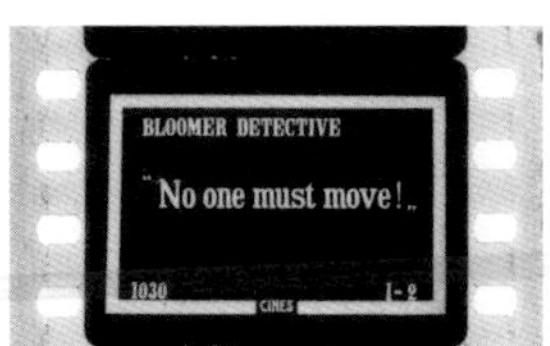

Kri Kri Detective / Bloomer Detective (1913). Tinted Pink

8.5.3.5 Some of the titles of Cines films are printed with sharp square corners to the frame and some with rounded corners. I have found no useful significance to this variation.

8.5.3.6 In some of the titles the wording is all in capital letters, while others use both capital and small letters.

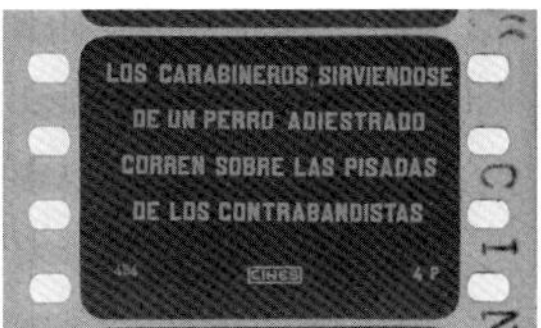

Dramma alla frontiera / Frontier Drama (1911). Tinted Amber – BFI

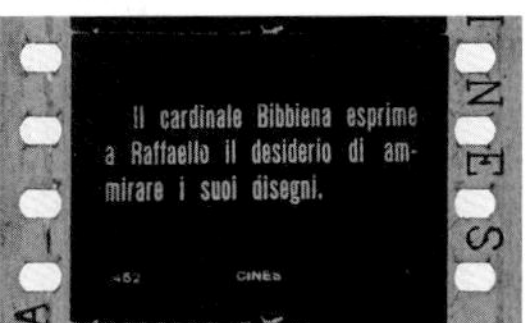

* *Raffaello e la Fornarina / Raphael and His Model* (1911) – CdB

I have found no consistency in the use of the two practices which can help with dating.

8.5.3.7 The titles in languages other than Italian have a letter indicating the language. Those in Italian have no letter. (See explanations in the Cines Film List, Appendix 2, pp.126-131).

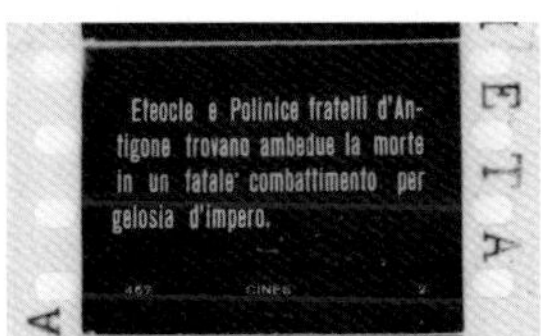

Antigone / Antigone (1911). Tinted Amber

8.5.4 The main titles and intertitles of the films have a production serial number. These numbers have been related to the dates of mention in the British trade press, so that the approximate date of any Cines film can be determined by reference to this serial number and the list of 57 Cines films in Appendix 2.

8.5.4.1 The list is arranged in order of serial number. The serial numbers and the dates of review do not follow in exactly the same order, but do so approximately. The one exception to this is Number 314, which should be about October 1910. No review at this period has been found, but it does appear in *The Bioscope* in March 1915. I guess that what happened is that in 1910 the film was not considered of sufficient interest to British audiences to distribute it in Britain, but in the circumstances of the war during 1915 there was felt to be a market for it.

8.5.4.2 On some titles the serial number appears in the bottom left-hand corner and in others in the right-hand corner. Sometimes this variation exists within a single film. I find no informative significance to this.

8.5.4.3 Some details of some films on the list are left blank. This is because we do not have the particular piece of information: e.g., if we have only a duplicate copy of the film, this will have no information about the tinting, and any edge mark which was on the original may not have been copied onto the duplicate; some short comic films had no intertitles, so there can be no information about title style; where the main title is missing there can be no information about its style or tinting.

8.6 **Vitagraph** (1906-1919)

8.6.1 In the N.F.A. there are over 150 Vitagraph films of this period, and I have not examined all of them. What follows is based on the examination of about 50 of them.

8.6.2 Vitagraph did not place any production serial number on their films, so we do not have that as a guide to date them; however, the style and tinting of the intertitles was changed from time to time, and this can provide some indication of the date of the films.

Thus:

8.6.3 1907. A decorative style of lettering was used on a plain background, in black & white (i.e., with no tinting).

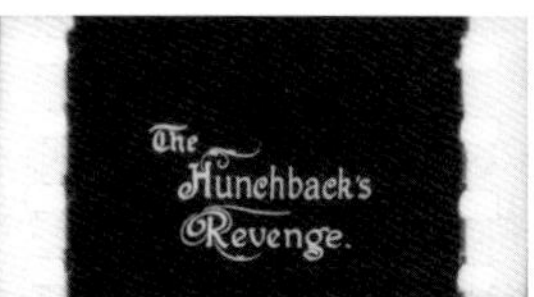

Francesca da Rimini (1907)

8.6.4 1908. The lettering is less ornate and sometimes the Vitagraph trademark of an eagle and "V" appears in black & white.

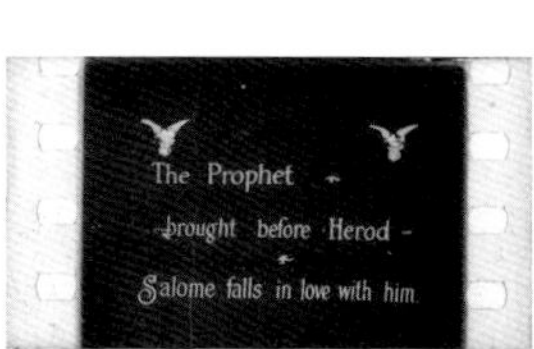

Salome (1908)

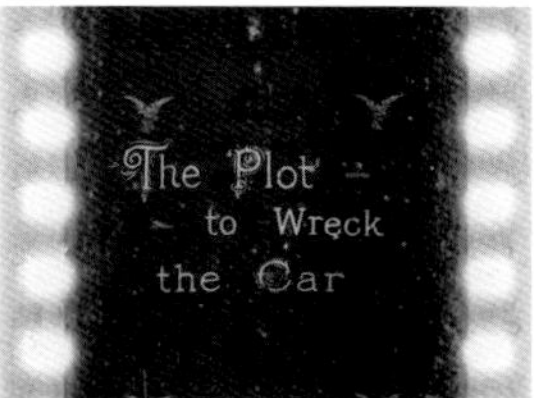

An Auto Heroine (1908)

8.6.5 1909. The intertitles used through 1909 were tinted amber.[39] This required that the titles were all spliced into the copies, so that there is a join at the beginning and end of every title. In this year also the edge mark "VITAGRAPH CO. OF AMERICA" or "VITAGRAPH CO. PARIS" was introduced.

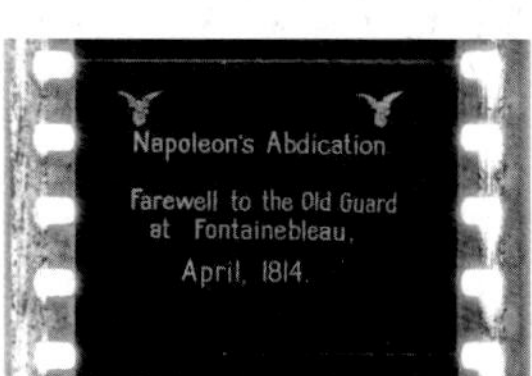

Napoleon (1909)

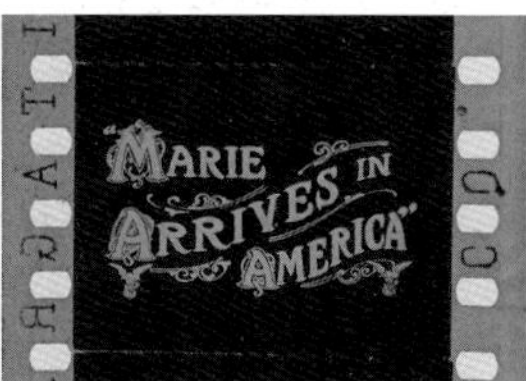

An Alpine Echo (1909) – BFI

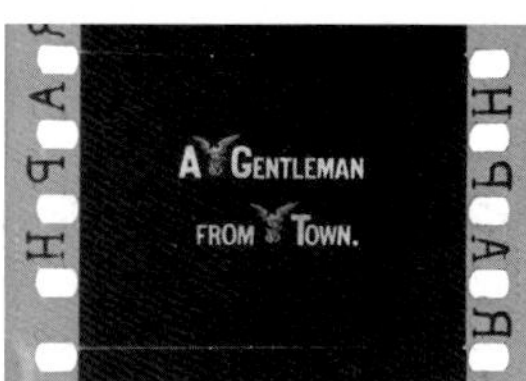

Betty's Choice (1909) – BFI

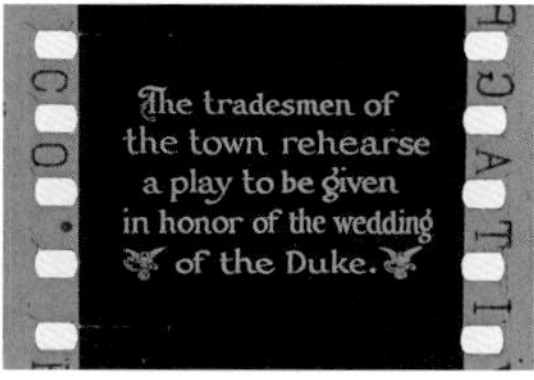

A Midsummer Night's Dream (1909) – BFI

Copies for distribution in Europe were printed in Paris.

39 See colour examples of 1909, 1910, 1911, 1912, 1913, 1916, 1917, 1918, and 1919 in the Colour Section, pp.149-151.

8.6.6 1910. The 1909 style seems to have continued into 1910, but in this year was introduced a new style, with the wording placed in a decorative border. These titles were nearly always tinted blue, otherwise amber.

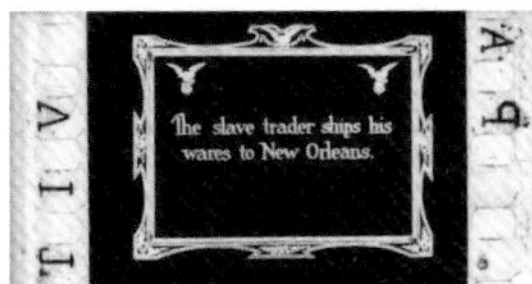

Uncle Tom's Cabin (1910). Tinted Blue

Daisies appears to be have a special title style designed to suit the particular subject, while the title in *Ransomed* has the same design as in *Uncle Tom's Cabin*.

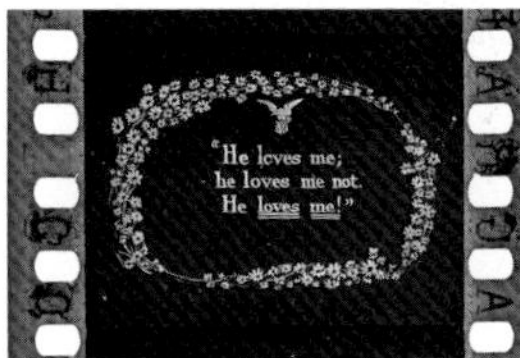

Daisies (1910). Tinted Blue – BFI

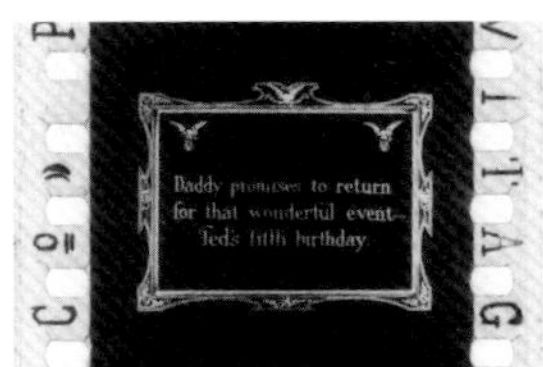

Ransomed; or, A Prisoner of War (1910). Tinted Blue

8.6.7 1911. I considered whether the presence of the trademark on the intertitles was significant, but it seems to be included only when there is not much wording, though it does not appear at all from 1915 onward. The two specimens from 1911 have the same decorative border as the 1910 ones.

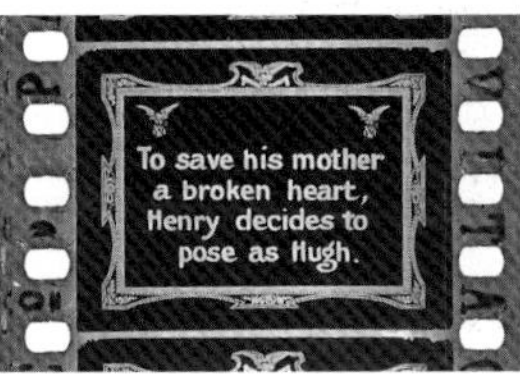

A Dead Man's Honor (1911). Tinted Blue – BFI

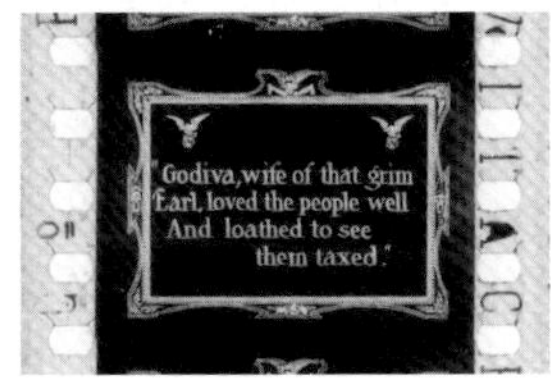

Lady Godiva (1911). Tinted Blue

Notice that *A Dead Man's Honor* has a plain block style of lettering with no serifs, unlike the other 1910-1911 ones.

8.6.8 1912. There is a similar, but not identical, decorative border, but the title of the film appears on the intertitles.

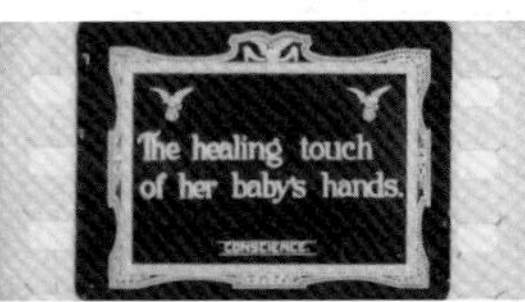

Conscience (1912). Tinted Blue

8.6.9 1913. *Coke Industry* has the same decorative border as *Conscience*, but, perhaps because it is a factual film, it does not have the film title.

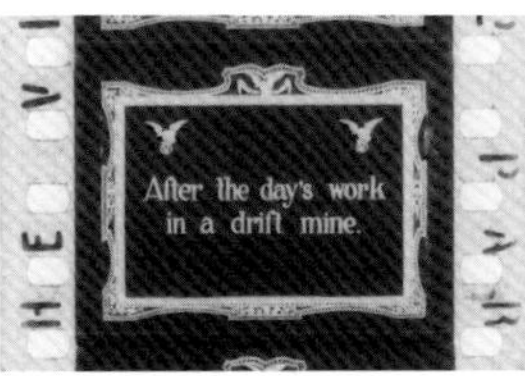

Coke Industry (1913). Tinted Blue

His Last Fight does have the film title. Also note that while the border is very similar, the name "Vitagraph" is incorporated into the bottom of the border.

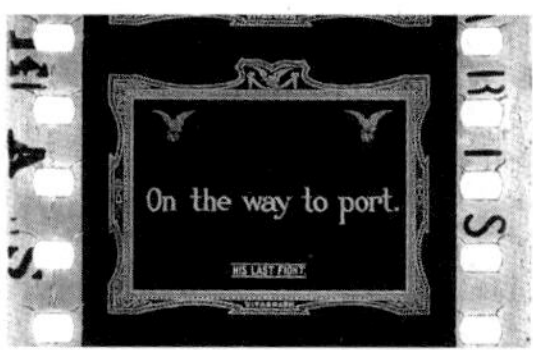

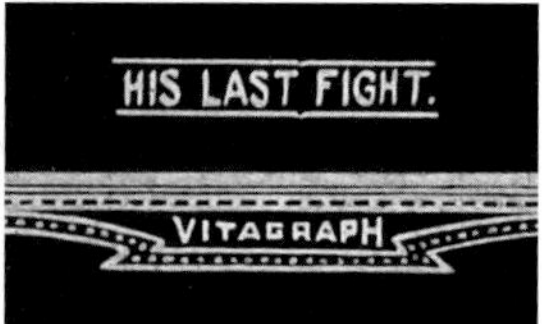

His Last Fight (1913). Tinted Blue – BFI

8.6.10 1914. The same style continued in use into 1914. From this year also, most of the copies have no edge mark, though there are some later copies which do.

Also in this year another change of practice began. Up to this time all the titles were spliced into the copies. Now some of the films had their titles spliced into the negative and printed in one continuous piece with the adjacent picture, so that (apart from joins/splices due to damage) there are no joins/splices at the beginning and end of titles, and each title takes the tint of the section of film in which it is placed, instead of all the titles in the film having the same tint.

8.6.11 1915.

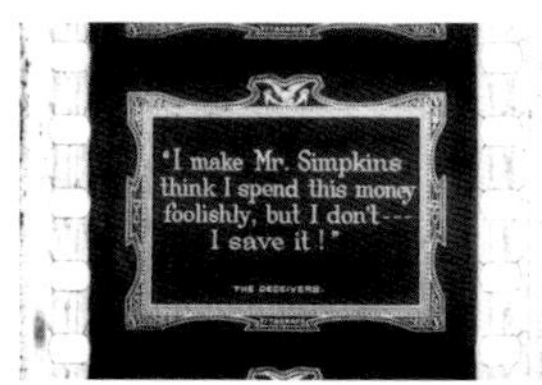

The Deceivers (1915)

The Deceivers exhibits the same decorative border as *His Last Fight*, but observe that the film title does not have the line above and below it. When titles were inserted into the prints, their tint is blue. There also seems to have been less use of tinting of the picture.

8.6.12 1916.

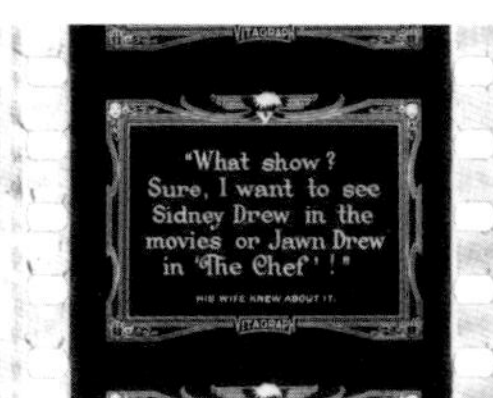

His Wife Knew About It (1916). Tinted Amber

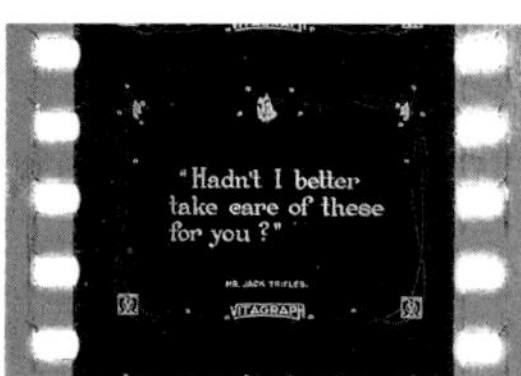

Mr. Jack Trifles (1916). Tinted Blue – BFI

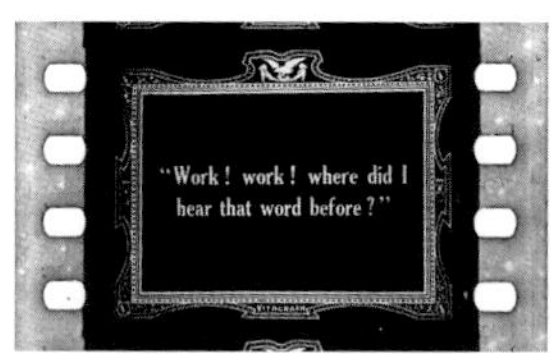

The New Porter (1916). Tinted Blue – BFI

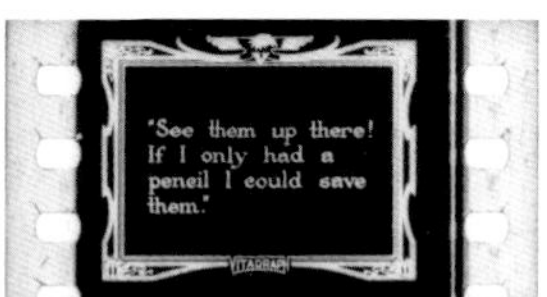

[Unidentified] (1916)

The name "Vitagraph" continues to appear in the decorative border, but different styles of border are used and the film title is not always included.

8.6.13 1917-1919. There seem to have been very few Vitagraph films in these years. Those in the N.F.A. are all comedies.

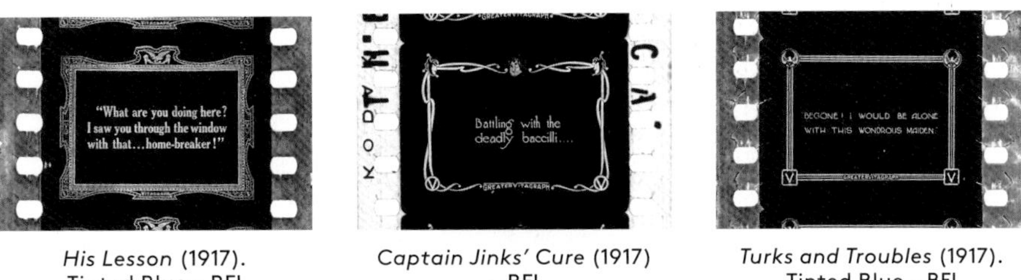

His Lesson (1917). Tinted Blue – BFI

Captain Jinks' Cure (1917) – BFI

Turks and Troubles (1917). Tinted Blue – BFI

These are all from films of 1917, but note that the style of *His Lesson* was also used in 1916. The title of *Captain Jinks' Cure* is similar to *Mr. Jack Trifles* of 1916, but the name in the border is "Greater Vitagraph". The title style of *Turks and Troubles* of 1917 is similar to *Skids and Skalawags* of 1918.

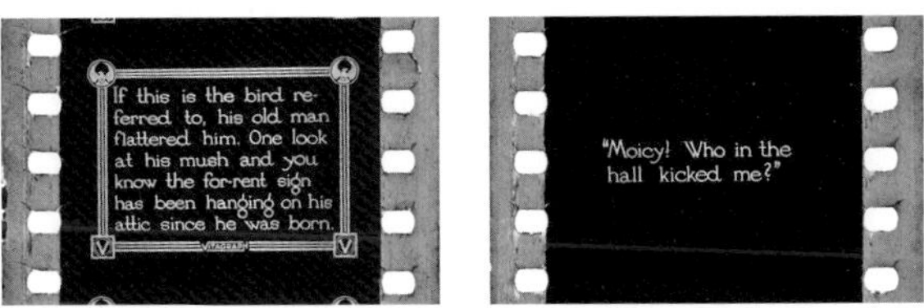

Skids and Skalawags (1918). Tinted Blue – BFI

(Note "Vitagraph" and "Greater Vitagraph".)

This is from the same film. This represents the convention, which became widely adopted, that narrative titles are put in a border, and speech titles have a plain background.

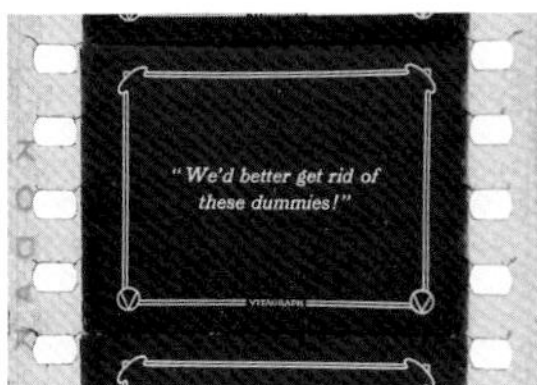

Damsels and Dandies (1919). Tinted Blue – BFI

The title above is from a film of 1919, and the other "Vitagraph" film of 1919 in the N.F.A. has the same style of intertitle.

8.6.14

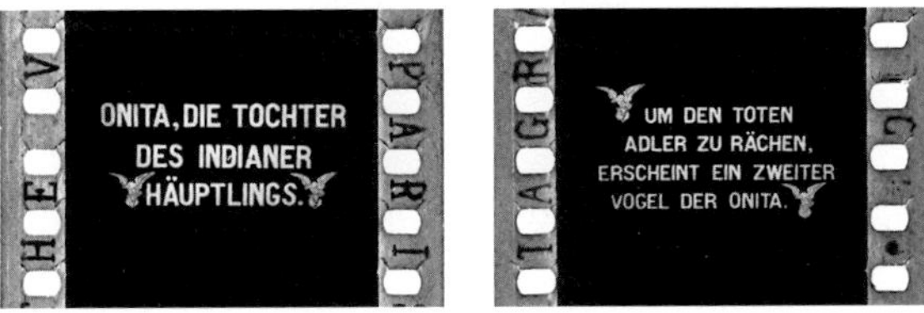

Red Eagle (1911). Tinted Mauve – BFI

Both titles are from a film of 1911. These titles are in German. It seems that Vitagraph used different styles of title in different languages. I have seen on the screen, but not held in my hand, films with other styles in other languages. Perhaps archives which have these films would provide specimens and information about their dates to FIAF.

8.7 Thanhouser (1910-1915)

The intertitle styles of Thanhouser films do provide some indication of date. The titles of the films of 1910 and 1911 are as in *Cinderella*.

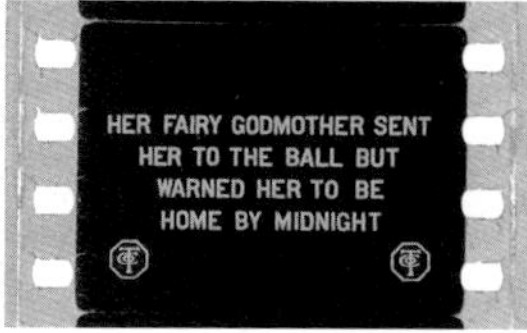

Cinderella (1910-1911). Tinted Amber – BFI

8.7.1 Those of 1912 and 1913 are similar, except that the film title appears on the intertitles. In the earliest film of 1912 which we have at the N.F.A., it is at the top of the frame (as in *Nicholas Nickleby*), while in the others it is at the bottom (e.g, *The Farmer's Daughters*).

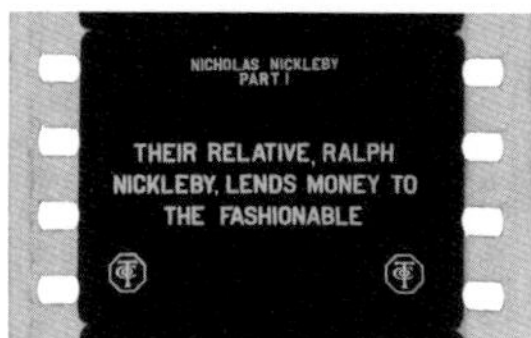

Nicholas Nickleby (1912). Tinted Amber – BFI

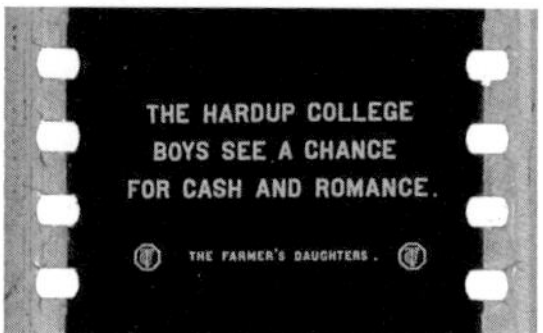

The Farmer's Daughters (1912-1913). Tinted Amber – BFI

Further, the intertitles of four of the films of this period have the film title enclosed in curved brackets [parentheses]. These are all from the second half of 1912. This feature may help further to determine the dates of otherwise undated films.

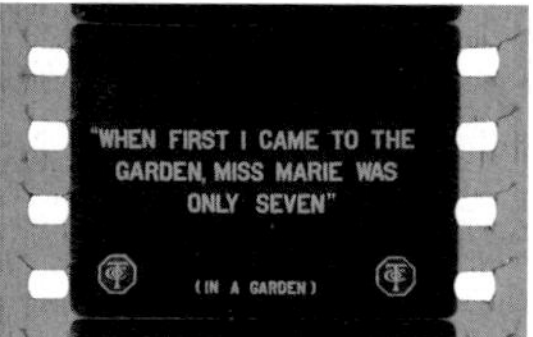

In A Garden (1912). Tinted Amber – BFI

8.7.2 Until at least the end of 1913 the titles were tinted amber[40] (with the exception of two Thanhouser films which were printed by Pathé[41]), but at some time in 1914 the tinting ceased and they were black & white, while the style of the titles otherwise remained the same.

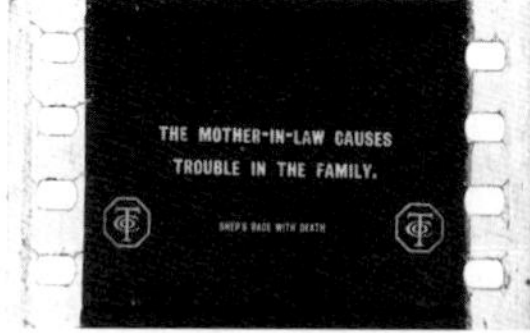

Shep's Race With Death (1914) – BFI

40 See colour examples of titles from 1910-1911, 1912 and 1913 in the Colour Section, pp.151-152.
41 See Essay Thanhouser (pp.117-118) and Film List (pp.141-143).

8.7.3 Of the four films which we have of 1915, two have main and intertitles in the style of *A Telephone Tragedy* and *John T. Rocks and the Flivver*.

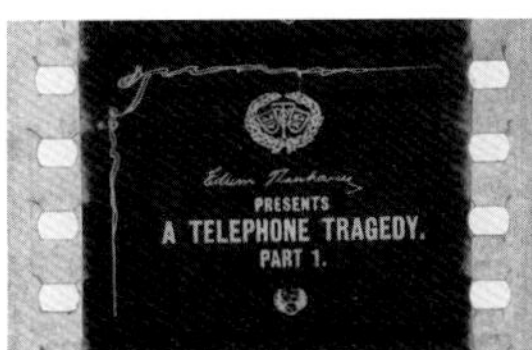

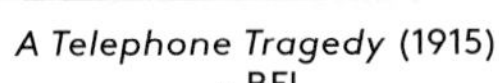
A Telephone Tragedy (1915) - BFI

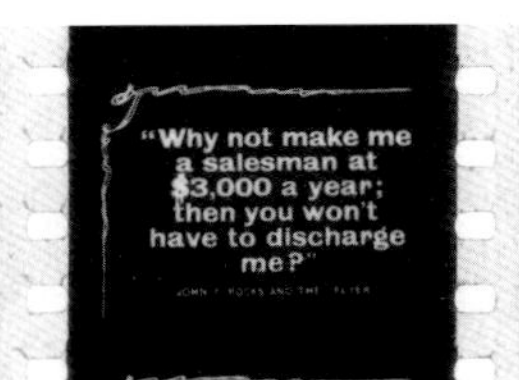

John T. Rocks and the Flivver (1915)

One has no intertitles, and one has a unique style, having the name "Falstaff" in it (*Madame Blanche, Beauty Doctor*). It seems clear that there was some association between Thanhouser and Falstaff. This is a case where narrative titles have a decorative border, and speech titles are on a plain ground.

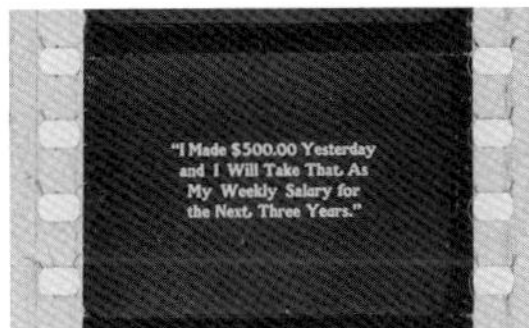

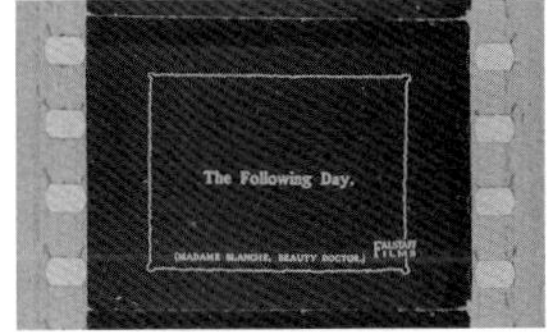

Madame Blanche, Beauty Doctor (1915) - BFI

8.7.4 The earliest of our Thanhouser films, *Daddy's Double* (1910), has titling on the front stating that it is controlled, outside the United States, by the Gaumont Company.

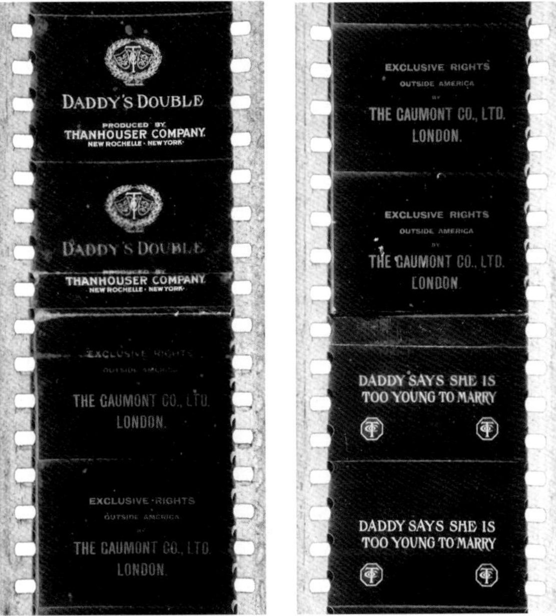

Daddy's Double (1910). Tinted Red - BFI

This titling is printed all on one piece of film with the main title and first intertitle, which have the Thanhouser trademark, which also appears in some scenes. This is thus clearly NOT some subsequent replacement title. It seems clear, therefore, that Gaumont distributed some Thanhouser productions at about this time. This film does NOT have the edge mark which was customary on Gaumont's own productions.

8.8 Selig (1908-1915)

I examined 50 Selig Polyscope films. The intertitles of the Selig films are of four main kinds, and there are variations within each kind. The earliest ones, on the films of 1908 and 1909, have a plain background, tinted red, and have the words in all capital letters.[42] The size of the letters varies from film to film. These are just not distinctive at all.

8.8.1 The form of intertitle used, roughly through 1911, has the wording within a large diamond-shaped border. There is a letter "S" in the top and the bottom corners of the diamond. These were tinted amber.

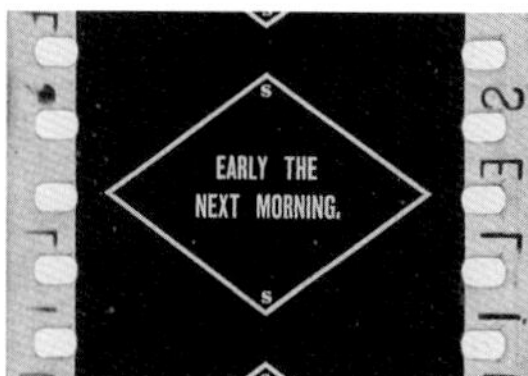

One Hundred Years After (1911). Tinted Amber - BFI

8.8.2 From early 1912 through mid-1914, variations on that style of title appeared. First, the title of the film appeared in the top of the diamond; then what seems to be a production number appears in the bottom left-hand corner outside the diamond; then a small white square appears in the left- and right-hand corners of the diamond; then the name "Selig Polyscope Co." appears in the bottom right-hand corner. These were tinted or black & white, as shown on the Selig film list.[43]

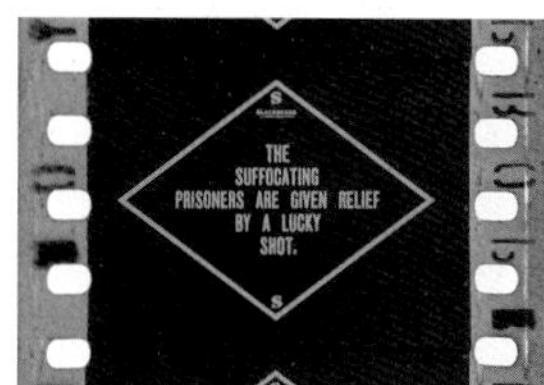

Blackbeard (1911).
Tinted Amber - BFI

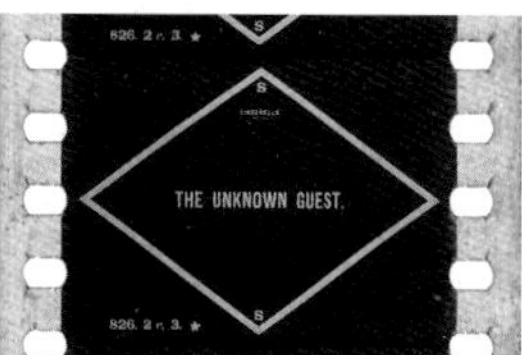

Cinderella (1911).
Tinted Amber - BFI

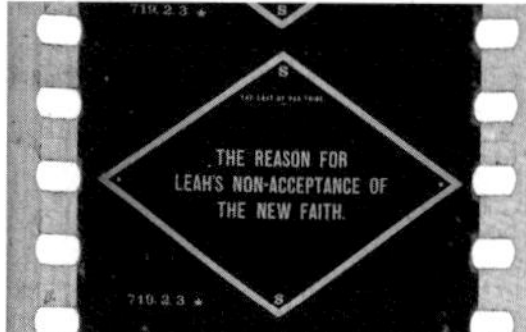

The Last of Her Tribe (1912).
Tinted Amber - BFI

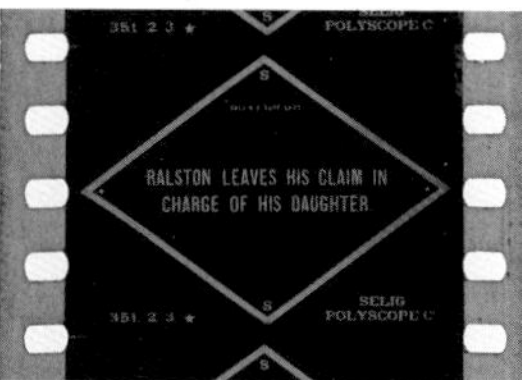

Sallie's Sure Shot (1913).
Tinted Pink - BFI

8.8.3 About the middle of 1914 another style began to be used. This consisted of a rectangle with a little diamond in each corner, and the letter "S" in each diamond. These all have the title of the film within the top of the border. Some have the name of the Selig Polyscope Co. within the bottom of the border, and some also have a copyright date, but I have not been able to relate these variations more closely to date.

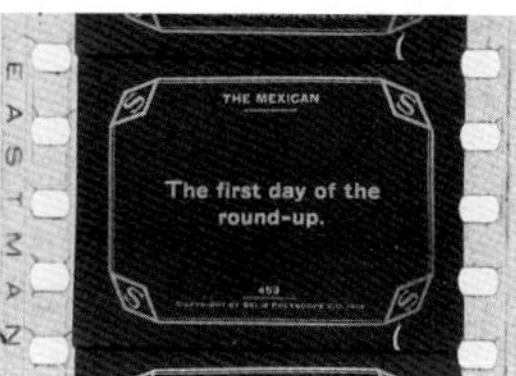

The Mexican (1914) - BFI

8.8.4 Within the date range stated in Paragraph 8.8.2, in September 1913, we have two films with a different title style.

Details of these titles are shown in the film list.

42 See colour titles of 1911, 1912 and 1913 in the Colour Section, pp.152-153.
43 See Appendix 4, pp.137-140.

8.8.5 There are what appear to be some inconsistencies in the scheme indicated in Paragraphs 8.8.2 to 8.8.4. This may be explained in connection with the sources from which I had to obtain the dates. These came from three sources: the Library of Congress copyright registration; the *American Film-Index 1908-1915* by Einar Lauritzen and Gunnar Lundquist; and from *The Bioscope*. There can be a difference amounting to months in the date of copyright registration, of American release, and of the appearance of the film in Britain. I was not able to find references to all the films in any one of these sources. It could be that in some cases the date which I have found is of a reissue.

8.8.6 All the Selig films of 1915 use the convention that narrative titles have a border and speech titles have plain grounds.

8.9 **Éclair (1908-1915)**

8.9.1 Éclair allotted serial numbers to their productions. These numbers can be used, in conjunction with the film list, to provide an approximate date for any films on which the number appears. Unfortunately, they did not usually put the number on the intertitles, but only on the main title; and this only from sometime in 1911. The number does appear on some letters (inserts) and intertitles. The only ones I have seen are on films of 1910 or 1911. Of approximately 40 films which I have been able to examine, only eight still have their main titles.

8.9.2 Éclair used different styles of main titles and intertitles at different periods.[44]

Main titles

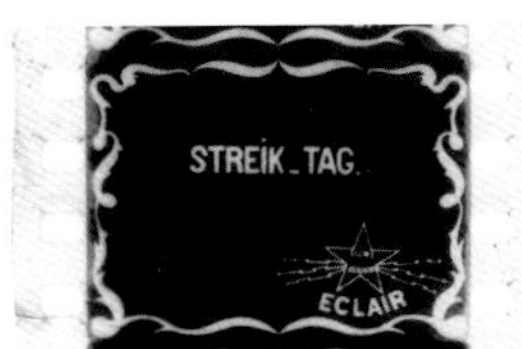

Journée de grève (1909). Tinted Green

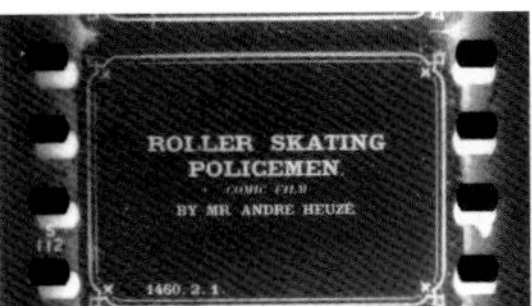

Les Agents à roulettes / The Roller-Skating Policemen (1911). Tinted Green. Note: Photograph of a duplicate negative.

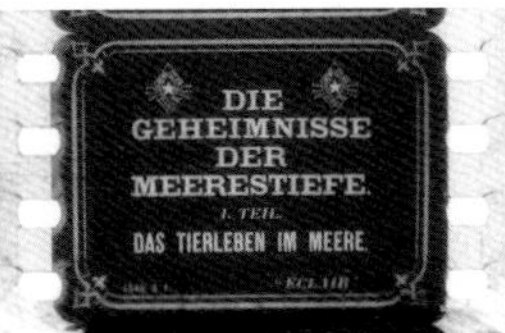

La Vie au fond des mers / Life at the Bottom of the Sea (1911). Tinted Green

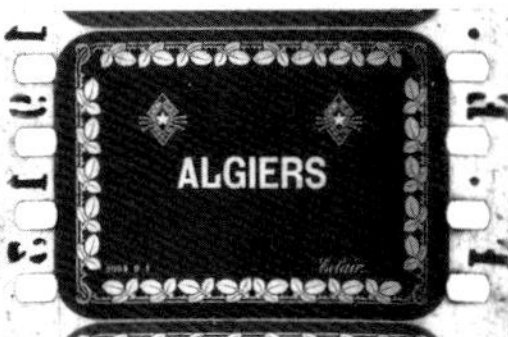

Alger la blanche / Algiers (1913). Tinted Green – BFI

Intertitles

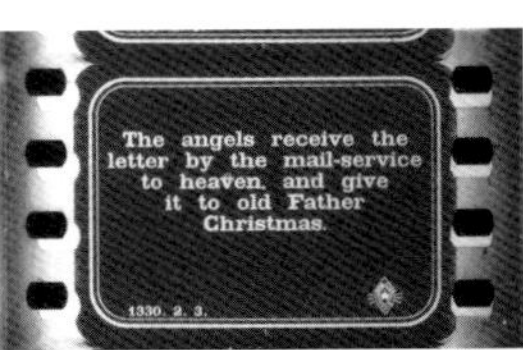

La Lettre au Père Noël / Letter to Father Christmas (1910). Tinted Green. Note: Photograph of a duplicate negative.

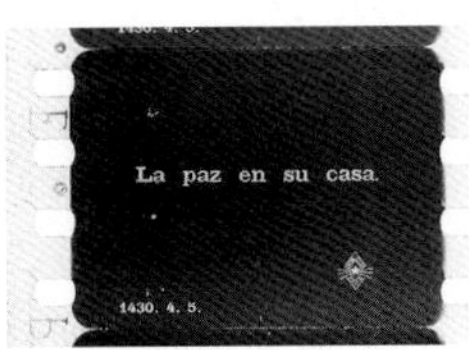

Le Gilet à pointes / The Waistcoat with Points (1911). Tinted Green

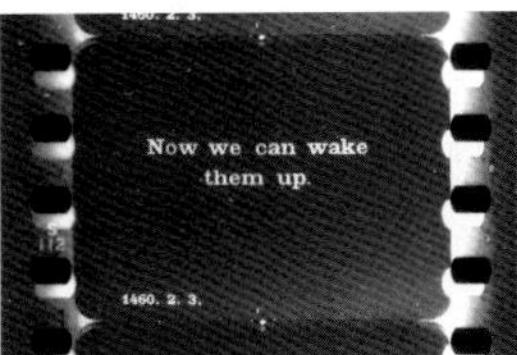

Les Agents à roulettes / The Roller-Skating Policemen (1911). Tinted Green. Note: Photograph of a duplicate negative.

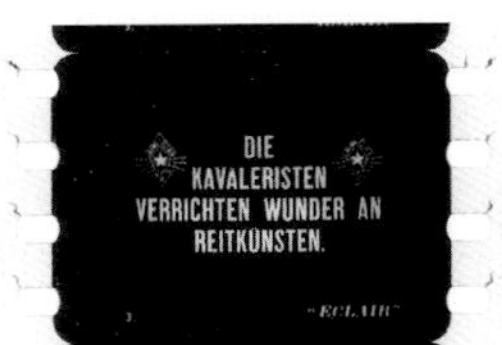

Les Centaures portugais / The Portuguese Centaurs (1911). Tinted Green

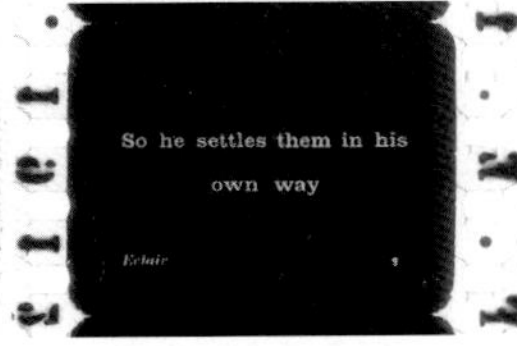

Rival de son maître / His Master's Rival (1912). Tinted Green

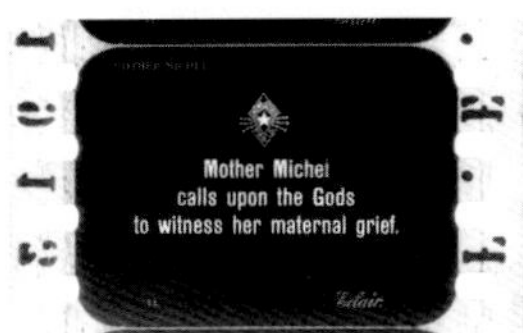

C'est la mère Michel / Mother Michel (1913). Tinted Green

44 The styles of title used on each of the films which have them is shown in the Eclair film list in Appendix 3, pp.132-136. See examples of colour titles from 1911, 1912 and 1914 in the Colour Section, pp.153-155.

Note:

8.9.2.1 On all the intertitles there is a number indicating the order of the intertitle in one of the bottom corners.

8.9.2.2 On most, the diamond-shaped Trademark has a dark background.

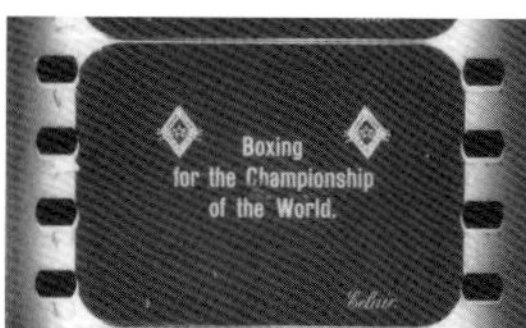

Willy contre le bombardier Wells / Little Willie v. Bombardier Wells (1913). Tinted Green.
Note: Photograph of a duplicate negative.

Here the diamond is white.

8.9.2.3 In *Gontran and His Accomplice*, there is a circular emblem with some German wording in it. This probably appeared only on copies circulated in Germany.

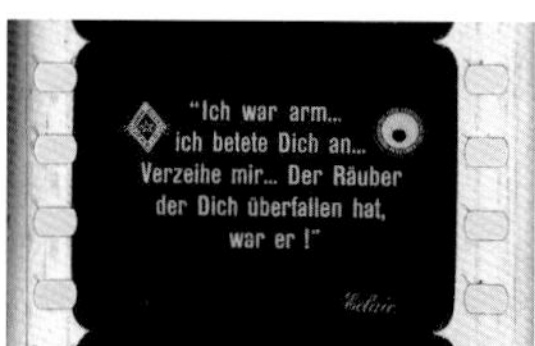

Gontran et son complice / Gontran and His Accomplice (1913).
Tinted Green – BFI

8.9.2.4 In *First Love* there is the letter "E" in the middle of the diamond. In the others there is a five-pointed star.

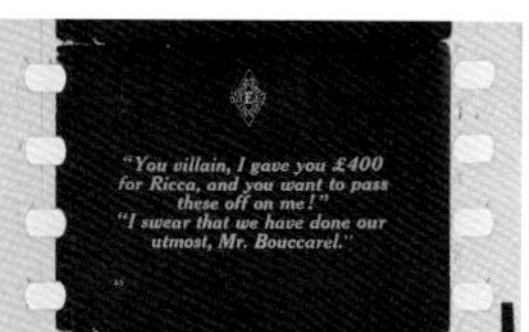

Premier amour / First Love (1912).
Tinted Green – BFI

8.9.2.5 The title of *Silent Jim* comes from a copy circulated in America. Note that the same film has the following edge printing:

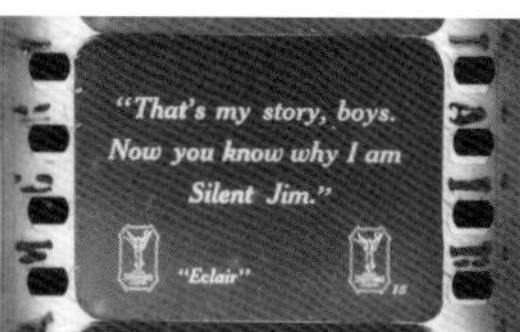

Silent Jim (1913). Tinted Green.
Note: Photograph of a duplicate negative.

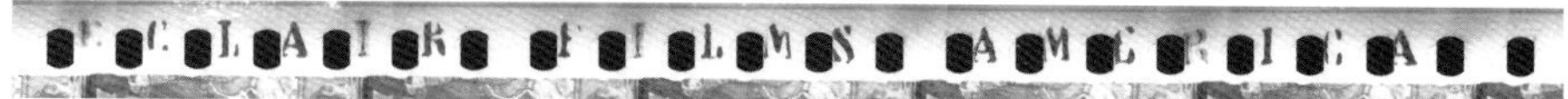

No date

8.9.2.6 The two films which we have of 1915 use the convention of a plain ground for speech titles.

8.9.3 Éclair made a series of films with a mischievous small boy. His name was variously rendered, sometimes within a single film, spelled either "Willie" or "Willy". He was also sometimes called "Little" Willie or Willy, and sometimes not. In the film list I have used the form on the main title where there is one, or the form used in the *Bioscope* reference in other cases. Further to this confusion of names, some other producers had films about characters called "Willy" or "Willie". It is wise to consider any cases carefully, and study the edge marks, title styles, and frame characteristics of any film being examined.

8.9.4 Some Éclair films, made in 1911, and perhaps in other years, were reissued in 1915, and sometimes under different titles. Thus *The Portuguese Centaurs* and *The Portuguese Cavalry* are the same film. The other case which we have is *The Waistcoat with Points*. In this case, note that our copy has the intertitle style associated with 1915.[45]

45 See the Eclair Film List in Appendix 3, pp.132-136.

9. Production Serial Numbers

9.1 It was the practice of some producers, in the period up to sometime during the 1914-1918 war, to place a number somewhere within the main title and/or intertitles. This was some sort of serial number of the production. With some producers this number ran through the whole of their productions of whatever kind of subject. Others had separate series for different kinds of film. Some also included a number indicating the order of the intertitles. Some also included letters which constitute the initial letters of the title of the film, presumably in the language of origin rather than the language of the titles. Others included a letter or letters identifying the language of the titles. All this was apparently to assist in assembling the various scenes and titles into complete copies; but to us they can constitute clues to the dates and identities of the films.

Such details as I have on this aspect of the subject are included in the essays on individual producers.[46]

9.2 **Pasquali (1909–1915)**

9.2.1 The titles of some but not all the Pasquali films have a production serial number which, by relating these to the known dates of some films, can aid in dating others. However, in the case of Pasquali, the numbers do not run straight through all the productions of all kinds; but there are separate series of numbers for different groups of films.

9.2.2 Thus, the comedy films of the actor Ferdinand Guillaume, in his role of Polidor, have their own number series.

The N.F.A. has just 5 Polidor films. Their titles, the serial numbers given on their titles, and the dates of review in *The Bioscope* are given in the list below.

Comparing the intervals between the serial numbers, and the intervals between the dates of the reviews, it appears that these Polidor films were issued at the rate of approximately one each week. It will be seen that Number 72 in the list below is recognised as a Polidor film, but it has no title. We can discover its approximate date by reference to its serial number and the dates of review of the adjacent numbered films. Working at weekly intervals forward from Number 49 would make Number 72 come on 22 August 1912. Working back from Number 85 would make it come on 1 August 1912. So we can confidently expect to find the issue of Number 72 between these two dates. However, I have not found any Polidor film mentioned in *The Bioscope* in that period. Minor films were not always included in the reviews or lists. Does anyone else have access to any other publication which may refer to this?

In this list, the original Italian title is stated first, followed by the English title.

Title	Serial	Bioscope Review
Uno scandalo in Casa Polidor *Scandal at Polidor's*	49	14.03.12
(Polidor)	72	
Polidor statua *Polidor a Statue*	85	31.10.12
Polidor al club della morte *Polidor, a Member of the Death Club*	96	02.01.13
Polidor coi baffi *Polidor's Moustache*	198	07.05.14

46 For Production Serial Numbers: Gaumont, see pp.54, 91-93; Cines, see p.95; Éclair, see pp.103, 183. Complete the information with the Film Lists for Gaumont, Cines, Éclair, Selig, and Thanhouser, see Appendices 1-5, pp.122-143.

9.2.3 Pasquali's travel films, of which the N.F.A. has seven, also have their own series of numbers:

In the list below, the Italian title, when known, is given first, followed by the English title. The approximate dates of other numbered members of the series may be discovered by interpolation, and possibly by extrapolation.

Title	Serial	Bioscope Review
Fra i ghiacciai del Görnergrat *Gornegrat, the Highest Railway in the World*	07	02.02.11
On the Sea Shore of Italy	09	09.03.11
La città eterna *The Eternal City*	018	15.06.11
Primavera a Sanremo *Spring in San Remo*	020	13.07.11
L'Olanda pittoresca *Picturesque Holland*	027	09.11.11
L'isola di Helgoland *The Island of Hellgoland*	033	28.12.11
Madrid, la città del sole *The Town of the Sun – Madrid*	040	06.06.12

9.2.4 Of the few other Pasquali films in the N.F.A. only three have serial numbers, and we have a review date for only one of these. This is far too small a number of films to be of much value. This may be an opportunity for anyone who has any numbered Pasquali films, of which they know the date, to convey the information to FIAF.

Title	Serial	Bioscope Review
Alboino e Rosmunda *Alboino and Rosmund*	None	1909
Il fascino dell'innocenza *The Fascination of Innocence*	146	18.12.13
[La maschera di ferro] *The Man in the Iron Mask*	None	1914
Man Proposes but -!	222	1915
Gli ultimi giorni di Pompei *The Last Days of Pompeii*	None	?
La maschera che sanguina *The Swell Mobsmen*	241	?

9.3 Selig (1908-1915)

9.3.1 Some of the Selig films have a number on their titles which seems to be a production number. However, there seems to be no logical order or chronology to these. They are included in the Selig film list in Appendix 4 (pp.137-140), in case any system should appear.

10. Letters, etc.

10.1 Letters and other documents commonly formed part of the story of films of this period.

10.2 One sometimes finds the same style of letter in different films of the same producer. For example, see *The Detective's Dog*.

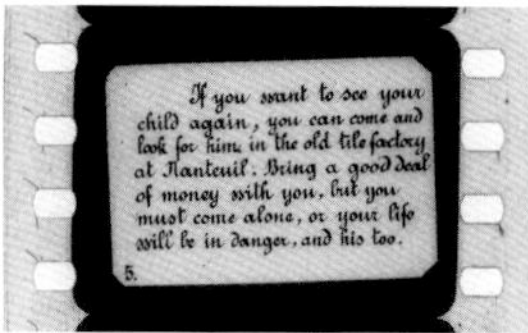

Le Chien du détective / The Detective's Dog (1912)

Letters in this style appear in four Éclair films within a date range from early 1912 to mid-1913. Note the clipped corners of the white paper. An identical style of writing appears on the letters in all four films.

10.3 I have also observed the same background news page used in different films of the same producer, with, of course, different main items. See examples from the Pasquali films *The Fascination of Innocence* (reviewed in *The Bioscope*, 18.12.1913) and *Polidor's Moustache* (reviewed in *The Bioscope*, 07.05.1914).

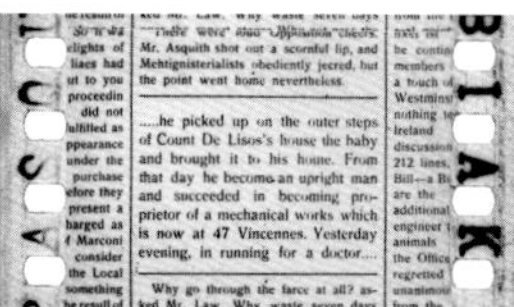

Il fascino dell'innocenza / The Fascination of Innocence (1913)

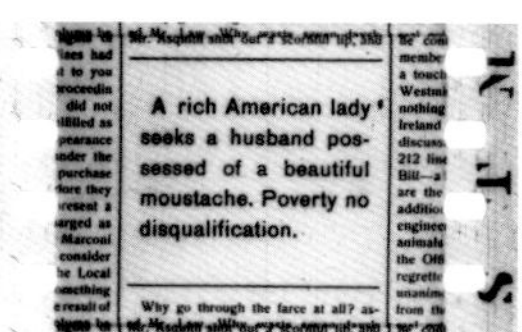

Polidor coi baffi / Polidor's Moustache (1914)

10.4 Some European producers sometimes put their letters on squared paper. This is not seen in British or American films.

10.5 It is noticeable and valuable that when letters, cheques, telegrams, calendars, and the like had dates on them, these were universally current dates. So such dates may reliably be taken as the time of production of the film, except where the film is clearly representing some other historical period.

As examples, see:

1.

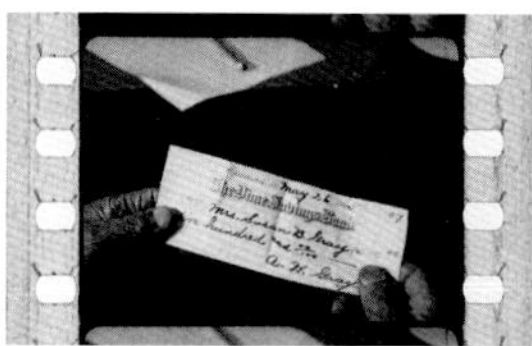

Madame Blanche, Beauty Doctor (1915) – BFI

The date on the cheque is "May 26, 1915". This is from the Thanhouser film *Madame Blanche, Beauty Doctor* (reviewed in *The Bioscope*, 10.07.1915).

2.

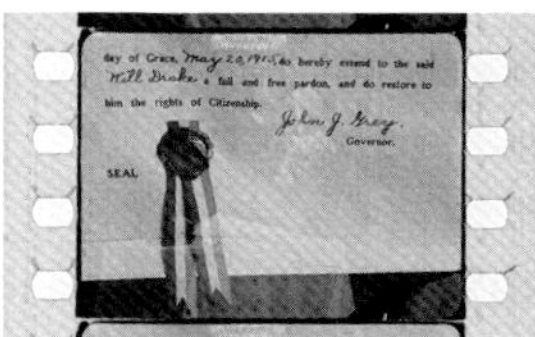

A Telephone Tragedy (1915) – BFI

The date of the pardon document is "May 20, 1915". This is from the Thanhouser film *A Telephone Tragedy* (reviewed in *The Bioscope*, 07.10.1915).

3.

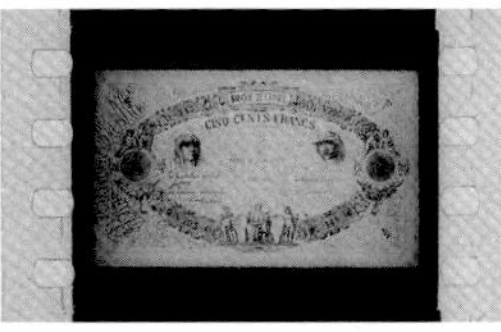

Un médecin distrait / An Absent-Minded Doctor (1910)

The date in the middle of the banknote, "PARIS le 3 avril 1909", conforms with the date of the Éclair film in which it appears, *An Absent-Minded Doctor* (reviewed in *The Bioscope*, 08.09.1910). Look very carefully; the date is very difficult to read.

10.6 Not all these things are direct indications of producer or date, but can constitute helpful clues.

11. Trademarks in Scenes

11.1 In the early years of cinema there was a significant trade in illicitly duplicated copies of films. To copy someone else's film was apparently not an offence in law. It is an offence to copy someone else's trademark. In order to frustrate illicit copying, some producers incorporated their trademark into some of the scenes of their films. This practice serves not only to identify the producer, but can also give some clue to date, since producers used this device during certain years, and, in some cases, varied the form of the trademark from time to time.

11.2 **Méliès**[47]

Méliès sometimes incorporated his own name or his trademark "Star Film" into some scene or scenes in his films. The specimen given is from *Between Calais and Dover*, and the name is disguised as the name of the shipping line and appears in the middle of the frame.

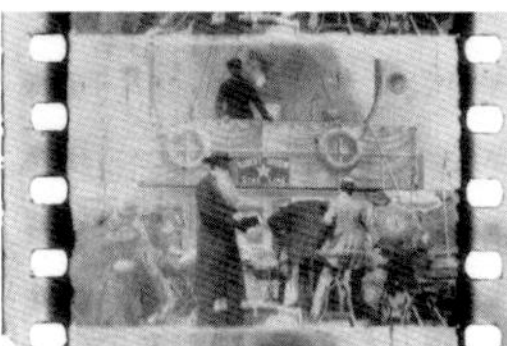

Entre Calais et Douvres / Between Calais and Dover (1897)

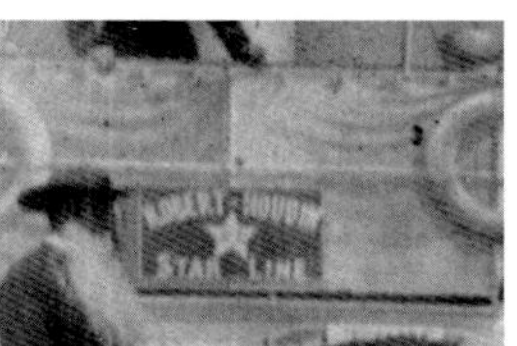

Detail

11.3 **Hepworth**

Cecil Hepworth used the trademark "Hepwix". This name appears in scenes in a number of his films.

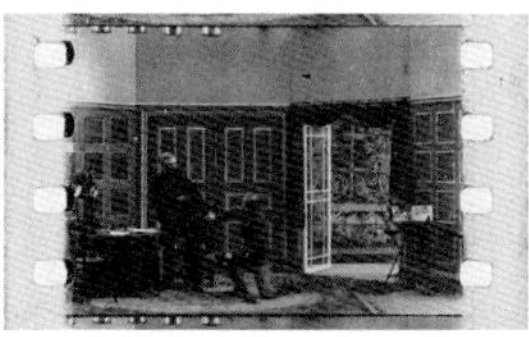

Falsely Accused (1905) – BFI

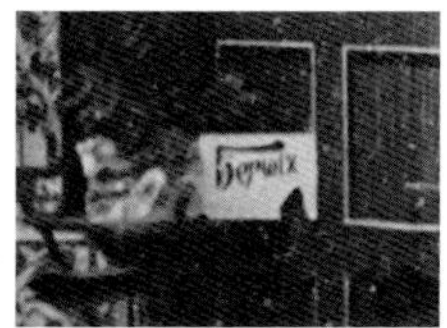

Detail

Where the trademark "Hepwix" appears, the film must be before 1913, when Cecil Hepworth separated from his business partner Monty Wicks.

11.4 **Pathé**

In their early films around 1900, Pathé often included a board with the letters "P.F.", for "Pathé Frères".[48] In the specimen given, this was, more ingeniously than usual, placed on the "Stars & Stripes" board which the "American" carries.

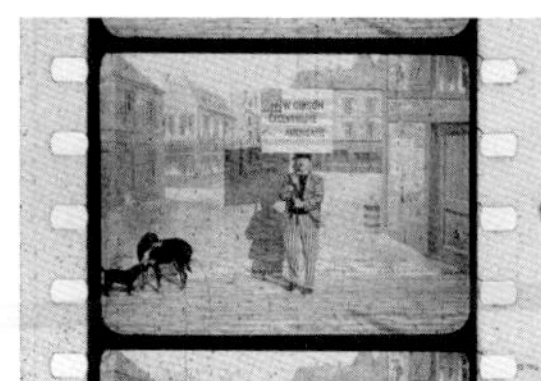

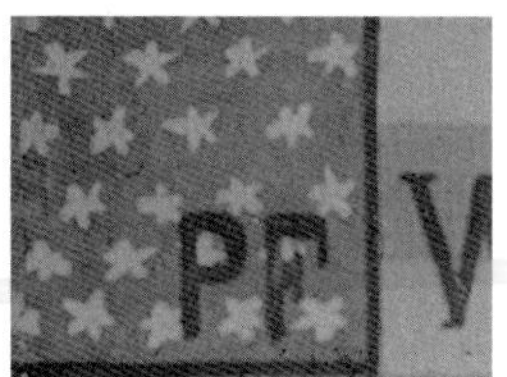

Excentricités américaines / American Eccentricities (1901)

(Incidentally, I just cannot imagine any United States citizens representing their national flag with only 9 stripes!!)

47 See also the chapter "Identifying a Georges Méliès Film", pp.169-173.
48 See also the chapter on Pathé, pp.195-197.

In their later films Pathé often placed a white cockerel in their scenes. I understand that they abandoned this practice after 1909.

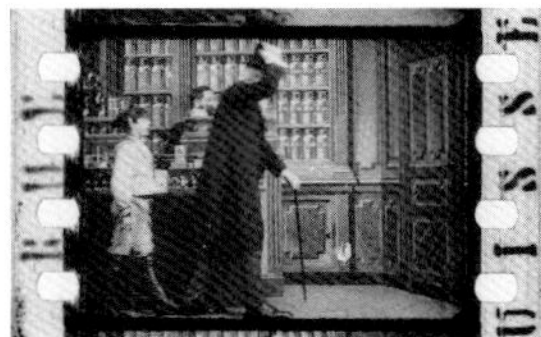
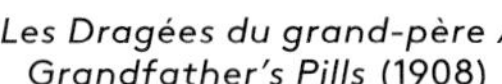
Les Dragées du grand-père / Grandfather's Pills (1908)

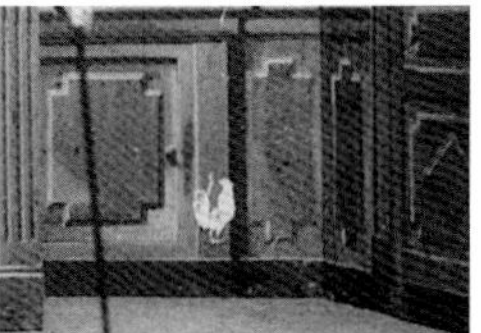
Detail

11.5 Gaumont

Gaumont included their name in some form in some of their films from 1907 right through 1913. The form of the trademark was changed from time to time.

Thus:

- *The Floor Polisher* (11.07.1907) has "ELGE", derived from the initial letters of "Leon Gaumont":

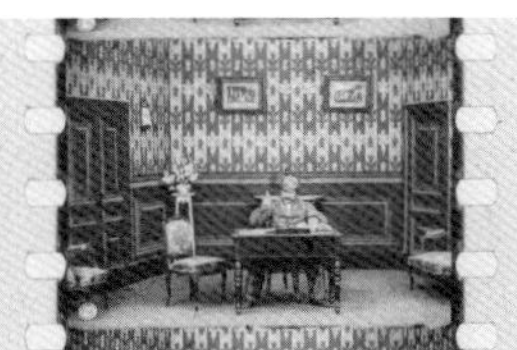
Le Frotteur / The Floor Polisher (1907)

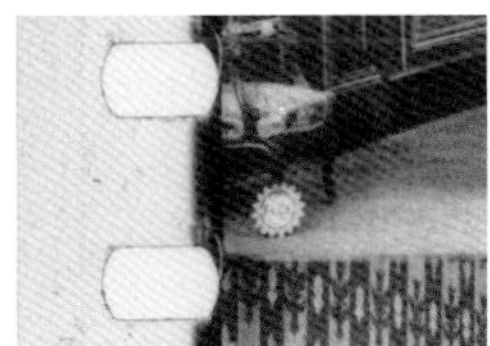
Detail

- *The Electric Policeman* (11.02.1909) has "Gaumont":

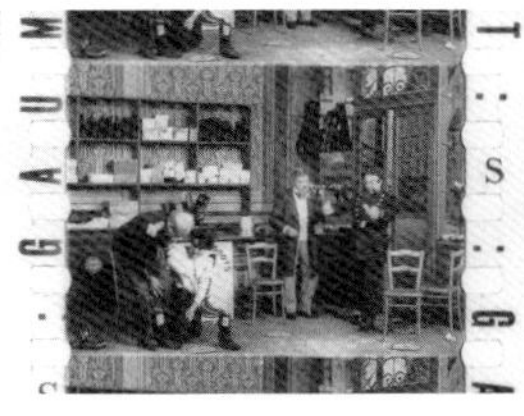
The Electric Policeman (1909)

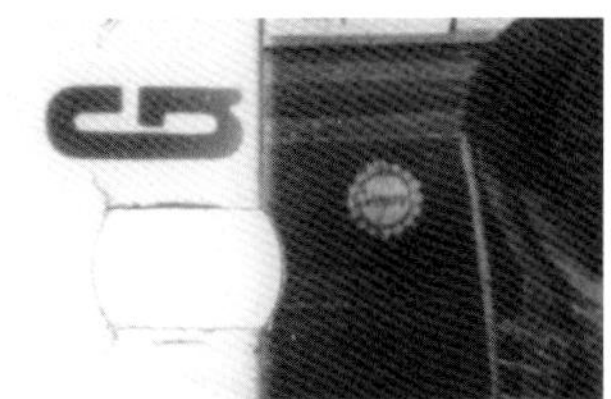
Detail

- *Esther* (02.05.1910) has "Gaumont":

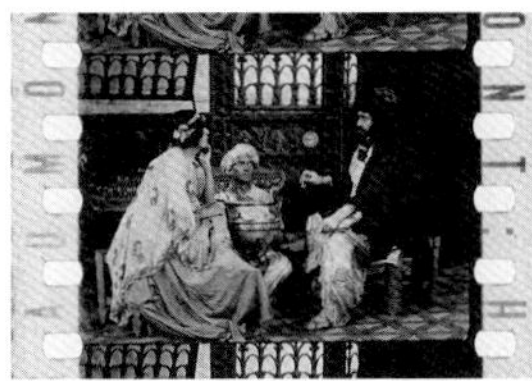
Esther (1910) – BFI

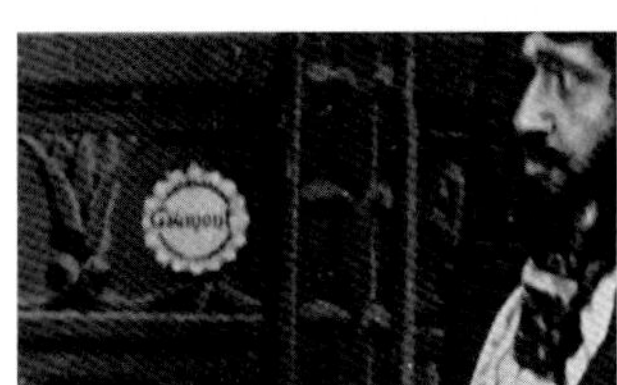
Detail

- *The Police of the Future* (09.06.1910) has "G":

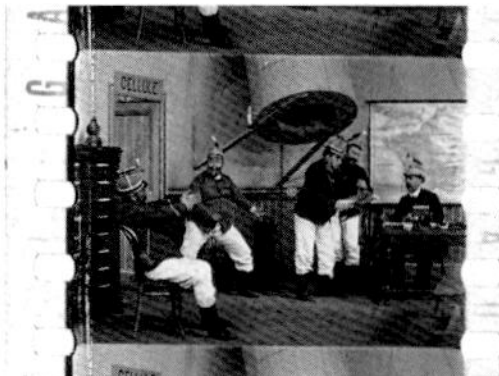

La Police de l'an 2000 / The Police of the Future (1910) – BFI

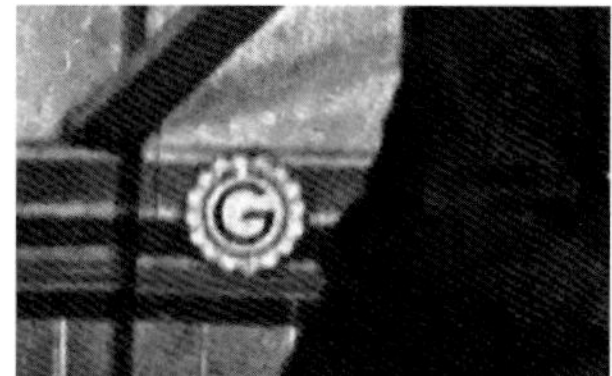

Detail

- *Mind the Wasps!* (03.11.1910) has "G":

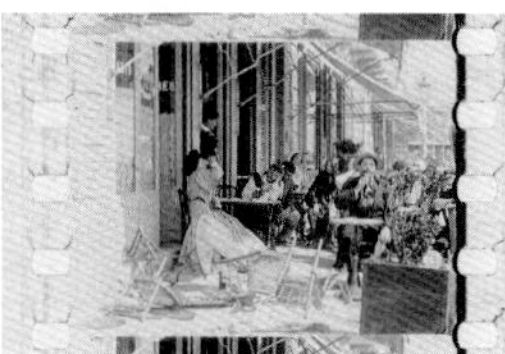

La Guêpe / Mind the Wasps! (1910)

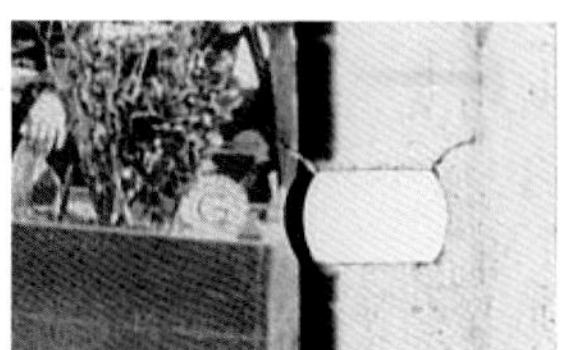

Detail

- *Bobby Raises the Wind* (25.07.1912) has "G":

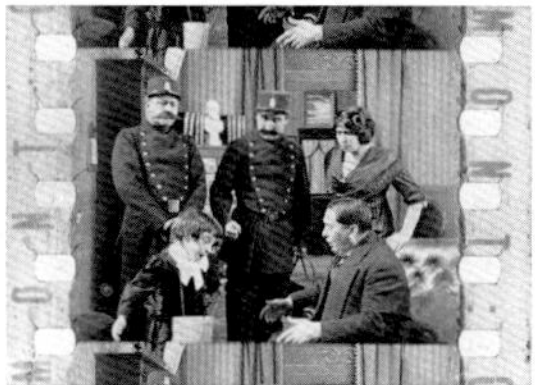

Bébé veut payer ses dettes / Bobby Raises the Wind (1912)

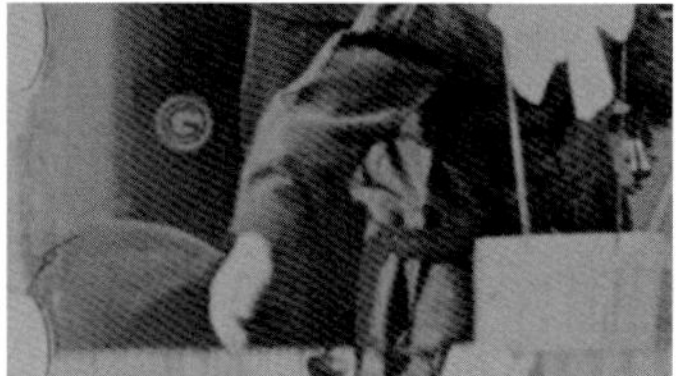

Detail

- *Simple Simon's Mother-in-Law* (24.07.1913) has "G":

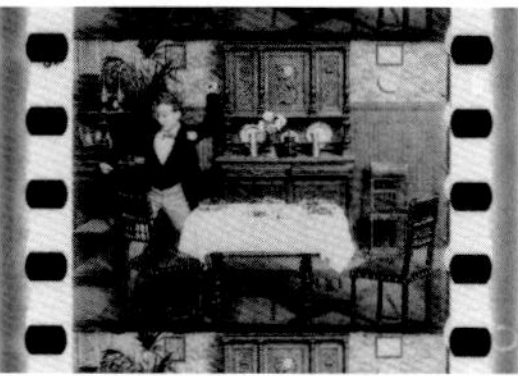

[*Onésime aime trop sa belle-mère*] / *Simple Simon's Mother-in-Law* (1913). Note: Photograph of a duplicate negative.

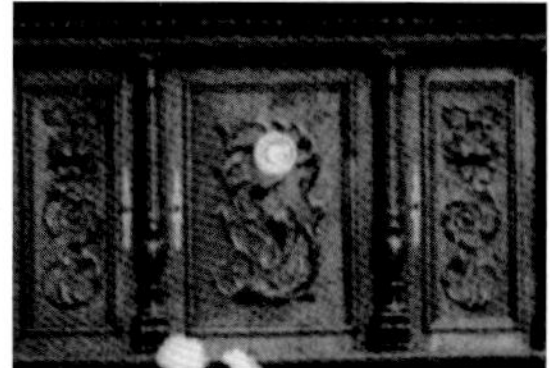

Detail

I have not found a specimen from 1908. It is clear that Gaumont used the same form of trademark in scenes during fairly definable periods. Thus, the form of trademark gives some indication of when a film was made. "ELGE" was used through 1907, and probably 1908. "Gaumont" was used from the beginning of 1909 to mid-1910, and "G" thereafter through 1913. In *Calino's New Invention* (22.02.1912), the name "Gaumont" is painted on a large basket forming part of a scene.[49]

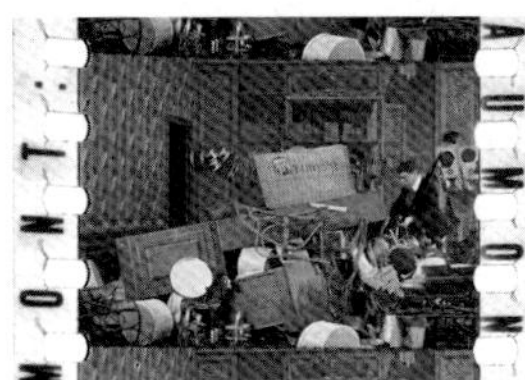

Calino courtier en paratonnerres / Calino's New Invention (1912)

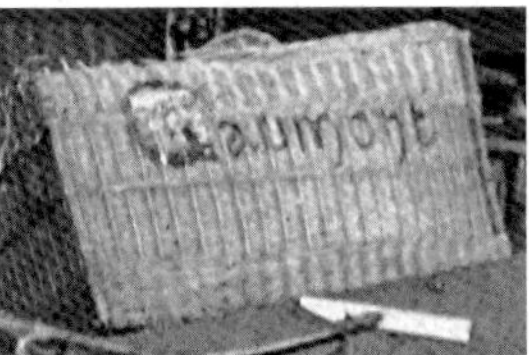

Detail

11.6 Vitagraph

Vitagraph placed their "Eagle and 'V'" trademark in many films through the years 1907 to 1910 inclusive. I have found no Vitagraph film later than 1910 having the trademark in any scene. The style of the trademark varies.

- *Liquid Electricity* (1907)

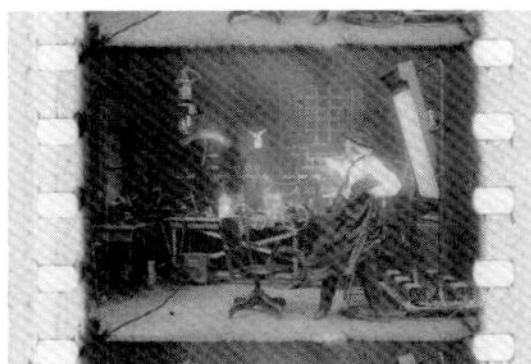

Liquid Electricity (1907)

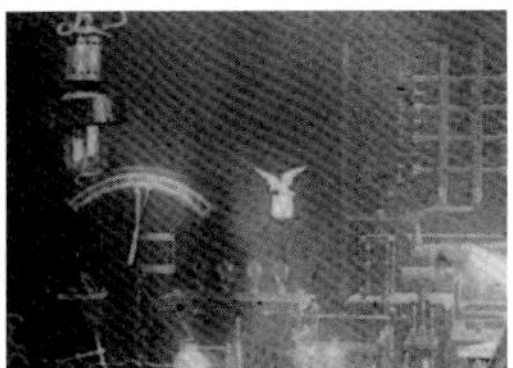

Detail

Here it is solid white.

- *An Auto Heroine* (1908)

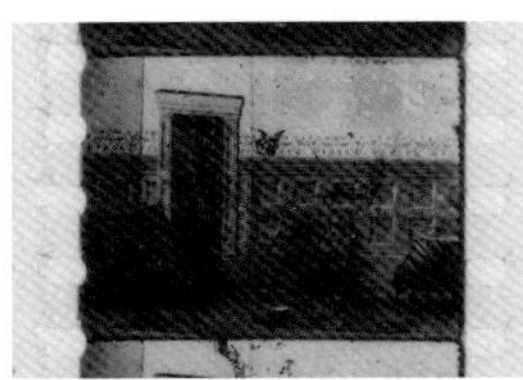

An Auto Heroine (1908)

Detail

Dark in contrast to the white wall.

- *Napoleon* (1909)

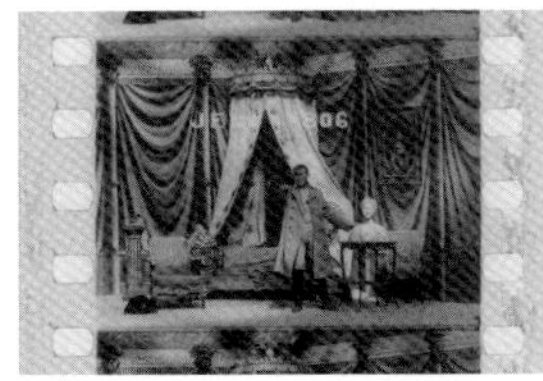

Napoleon (1909)

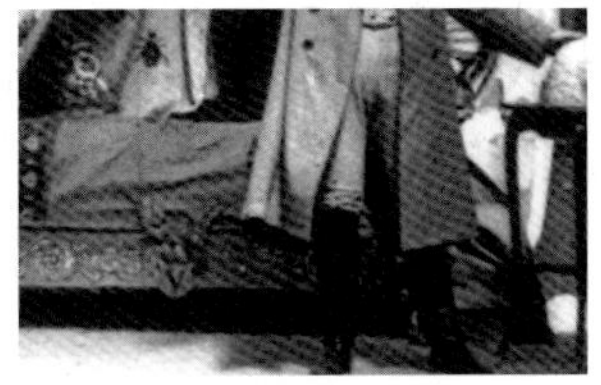

Detail

Incorporated into the woodwork near the tail of Napoleon's coat.

49 Nevertheless, the indication "ELGÉ" was observed on an accessory in a film released in 1903. Indeed, production companies seem to have stopped inserting their trademark in the settings from 1913 and onwards, but it seems that the company went on using accessories that had the trademark on them even after 1913. For instance, in an episode of Feuillade's *Les Vampires* (1915), a "G" appears on a wicker trunk. The trademark, usually observable in fiction films, was also used in some documentaries or newsreels. (Eric Loné)

- *Daisies* (1910)

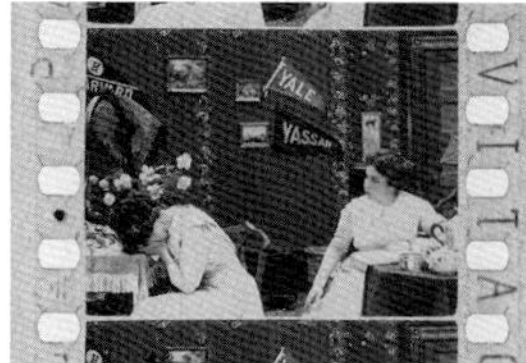

Daisies (1910) – BFI

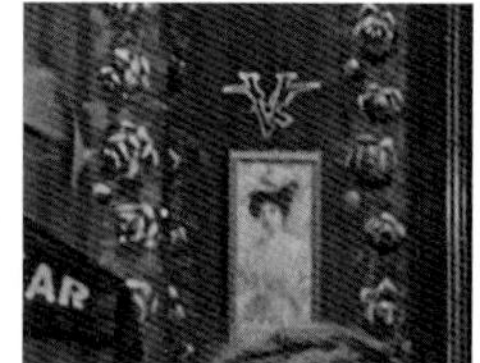

Detail

Among several other pictures, etc., adorning the wall.

- **The Tired, Absent-Minded Man* (1911)

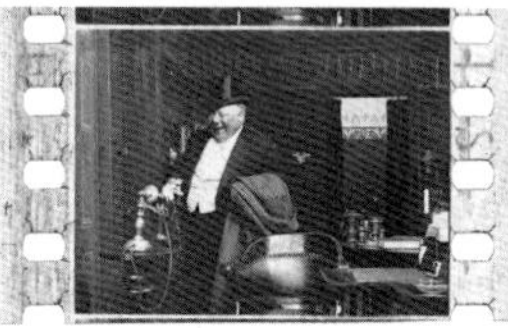

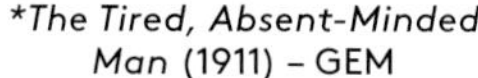

**The Tired, Absent-Minded Man* (1911) – GEM

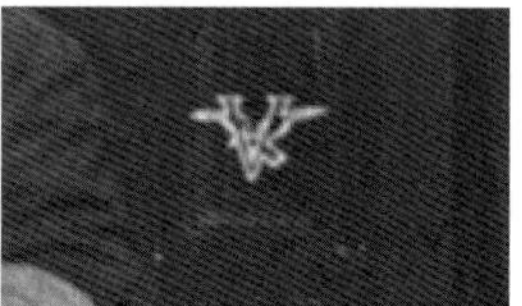

Detail

In the case of Vitagraph, it is not suggested that the form of trademark is related to date. It appears to be selected just to show against the background.

11.7 Thanhouser

The N.F.A. has five Thanhouser films which have the trademark appearing in scenes. These are all in the date range 1909 through 1912. None have been found with a trademark after 1912. I have chosen to use these two examples because they show peculiarities.

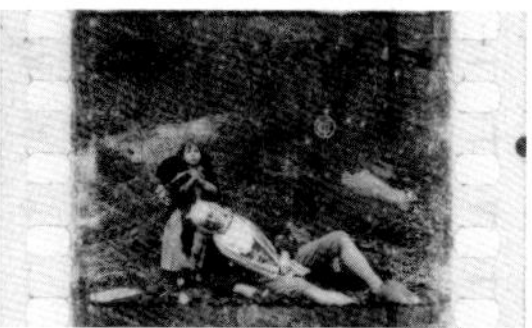

The Old Curiosity Shop (1912)

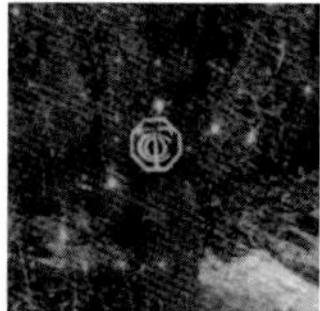

Detail

Above is a comparatively rare case of a trademark appearing in an exterior scene. It may be that it is a studio shot depicting an exterior. This is from *The Old Curiosity Shop*, based upon the novel of Charles Dickens. This is reviewed in *The Bioscope* of 22.02.1912. This film was clearly printed by Pathé. It has Pathé edge mark, Pathé perforations, and the whole tangible feel and appearance of Pathé film – <u>except</u> for the frame characteristics, and the Thanhouser trademark.

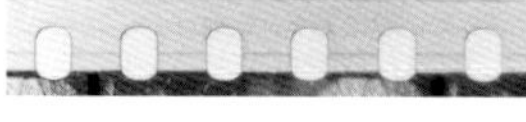

Pathé Perforations

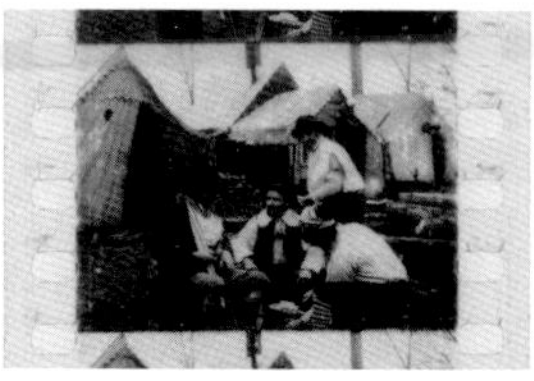

The Old Curiosity Shop (1912)

This other film is similar in its relation to Pathé. The film came to the N.F.A. with replacement titles which are neither Thanhouser nor Pathé, but are of Belgian creation. The Belgian title is *Le Laitier millionnaire*, and it was reviewed in *The Bioscope* on 19.12.1912 under the title *The Millionaire Milkman* as a Pathé film.

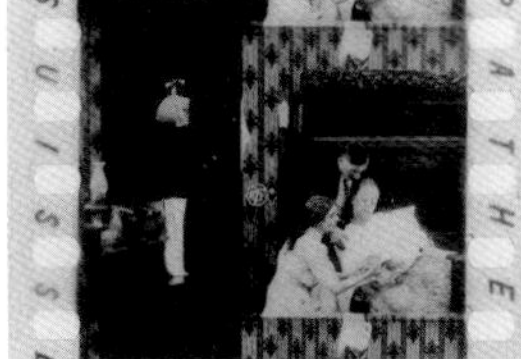

The Millionaire Milkman (1912)

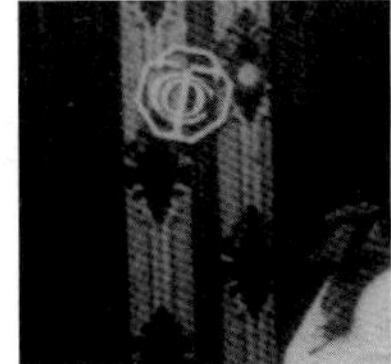

Detail

Pathé printed and distributed both Thanhouser and Méliès films in the period around 1911/1912. It may be that they also distributed the product of some other makers.

11.8 Kalem

Kalem sometimes displayed their trademark within scenes. But I have no more information than that.

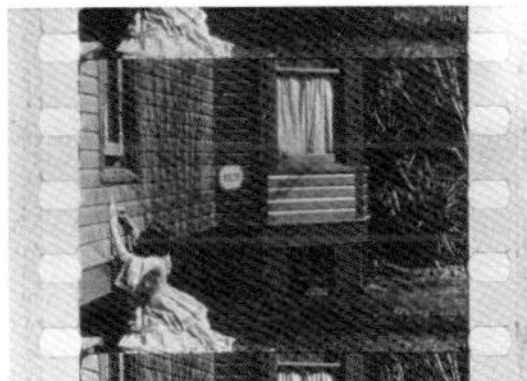

Adventures of a Spy Girl (1912)

Detail

11.9 Éclair

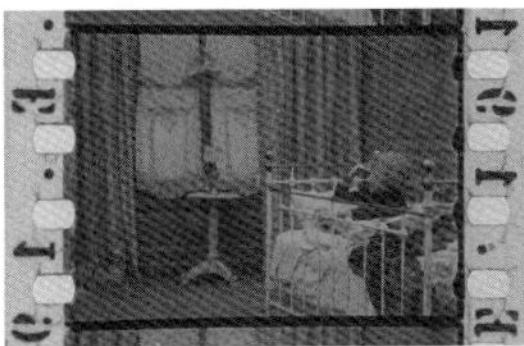

**Willy et le prestidigitateur / Willy and the Conjuror* (1912) – SFI

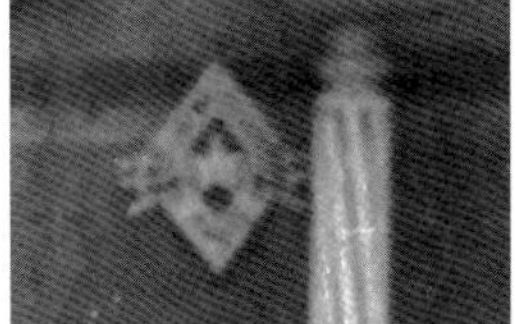

Detail

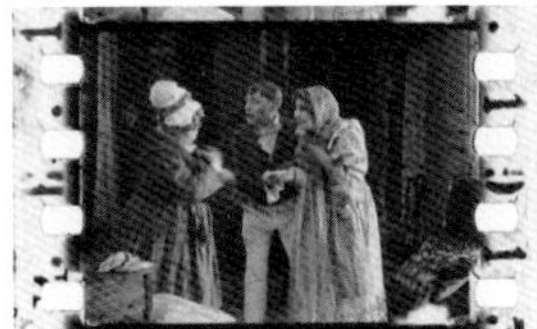

C'est la mère Michel / Mother Michel (1913)

The specimen comes from *Mother Michel*, of 1913, which is after all the other producers which I know had ceased to put trademarks in scenes.

11.10 *Logos from other film companies

- **Saturn-Films**

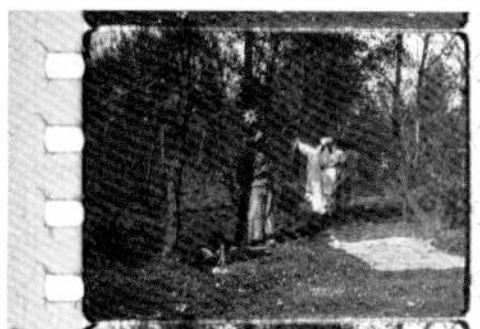
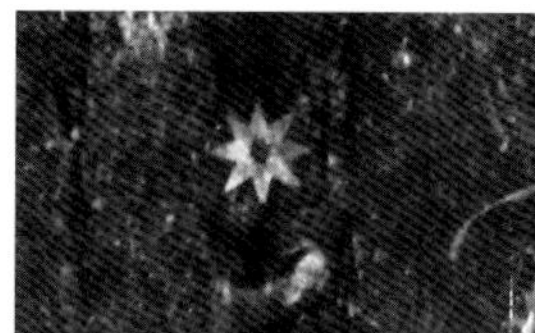

Sklavenschicksal (1906) - FINA

- **Edison**

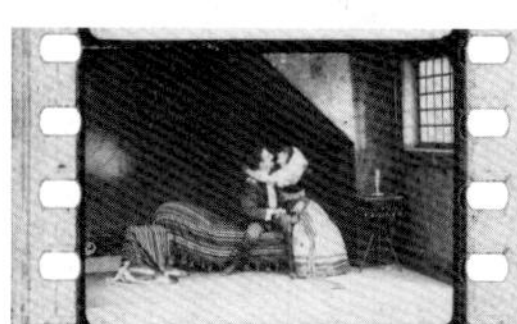

Fenton of the 42nd (1909) - GEM

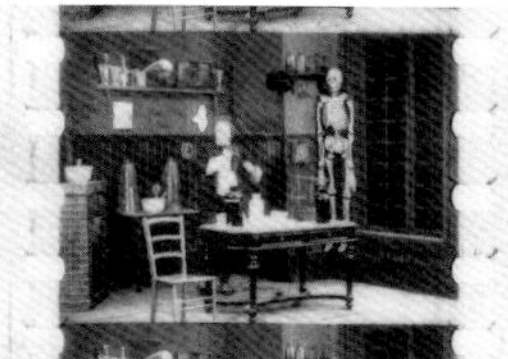

A Trip to Mars (1910) - HB

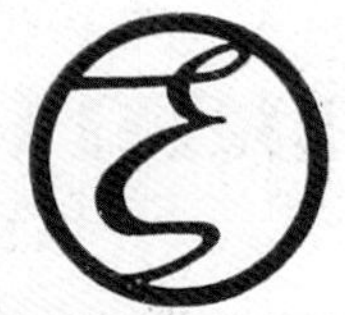

Moving Picture World (May 6, 1911)

- **Selig**

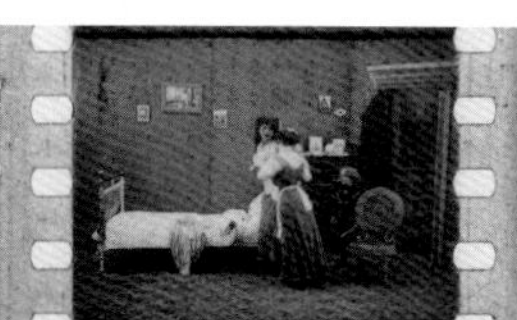

The Infant Terrible (1909) - HB

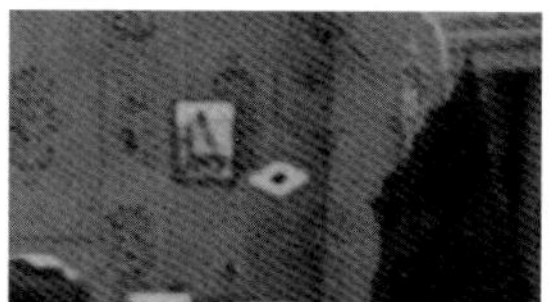

Detail

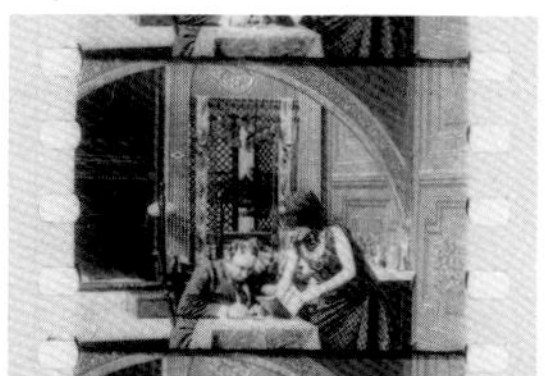

Up San Juan Hill (1909) - HB

12. Essay Thanhouser

Films of 1910–1915

12.1 There are less-consistent clear characteristics with Thanhouser than with some other producers.

12.2 In almost all the films which I have inspected the frame-line lies in mid-space.

12.3 The frame characteristics vary considerably.

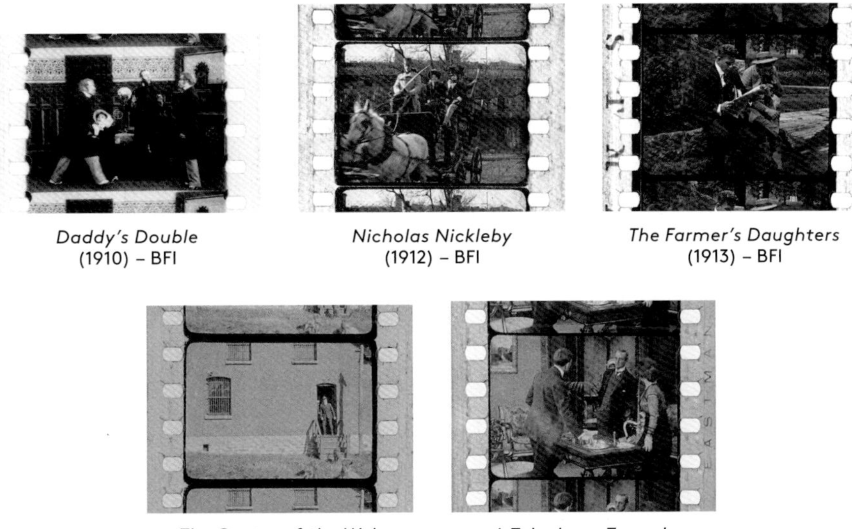

Daddy's Double (1910) - BFI

Nicholas Nickleby (1912) - BFI

The Farmer's Daughters (1913) - BFI

The Center of the Web (1914) - BFI

A Telephone Tragedy (1915) - BFI

12.4 Note that two of the films, *The Millionaire Milkman* and *The Old Curiosity Shop*, both of 1912, bear the edge marks of Pathé.

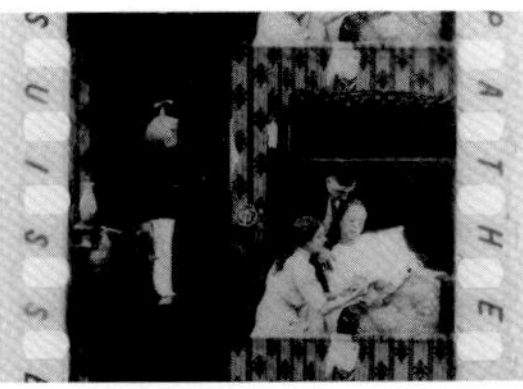

The Millionaire Milkman (1912)

These films were clearly printed by Pathé. They have the appearance of Pathé as regards the emulsion surface, and also the Pathé shape of perforation. There can be no doubt, however, that these are Thanhouser productions, since the Thanhouser trademark appears in some of the scenes. Also note that these films do NOT have the frame characteristics of Pathé's own productions. It appears, therefore, that Pathé distributed at least some Thanhouser productions in about 1912. It is known that Pathé printed and distributed some films of Georges Méliès in 1912, so it is quite credible that they distributed others in about the same period.

12.5 Observe that after 1912 none of the films has the trademark in scenes.

12.6 The titles in *Under Two Flags* are in German, and they may be original titles produced by Thanhouser, or they may be replacements by some user. They are on a plain background, so this remains uncertain.

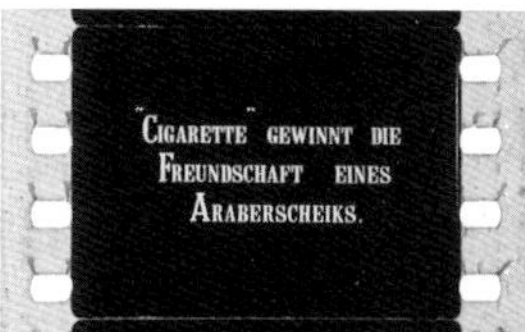

Under Two Flags (1912) – BFI

The titles on *The Millionaire Milkman* are replacements and not original Thanhouser titles, so they can provide no guide to the dates of other films.

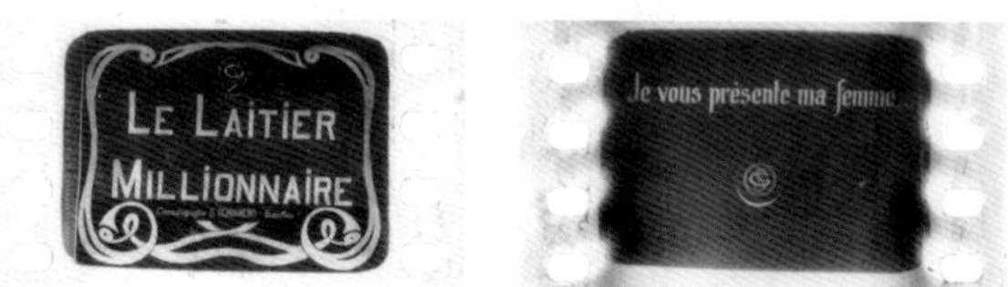

**The Millionaire Milkman / Le Laitier millionnaire* (1912)

12.7 Apart from the two films which have Pathé's edge mark, none of the films has a producer's edge mark.

12.8 ALL of the films from mid-1913 onward are printed on Eastman stock and have the edge mark corresponding to their time of production. (See Paragraph 6.2, p.69). So one may be fairly safe to assume that any film of that period which is not on Eastman stock is not a Thanhouser film.

12.9 It is clear that in the case of *The Decoy* there was association between Thanhouser and the distribution company Princess. Similarly in the case of *Madame Blanche, Beauty Doctor* with the distributor Falstaff.

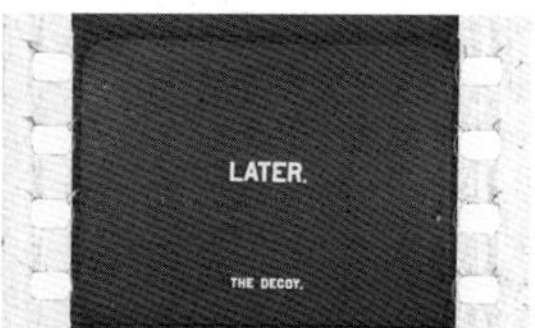

The Decoy (1914) – BFI

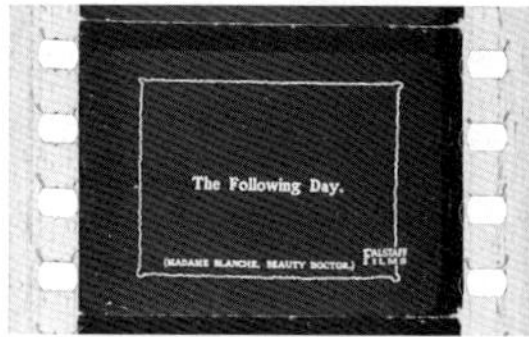

Madame Blanche, Beauty Doctor (1915) – BFI

See also the Thanhouser film list (Appendix 5, pp.141-143).

13. Essay Hepworth

Films of 1896–1913

13.1 When he commenced business, Cecil Hepworth was in partnership with his cousin, Monty Wicks, hence the trademark "HEPWIX", which appeared in scenes, and on the intertitles of some of his films. Hepworth parted from his cousin in 1913, and then put the name "HEPWORTH" on his titles.

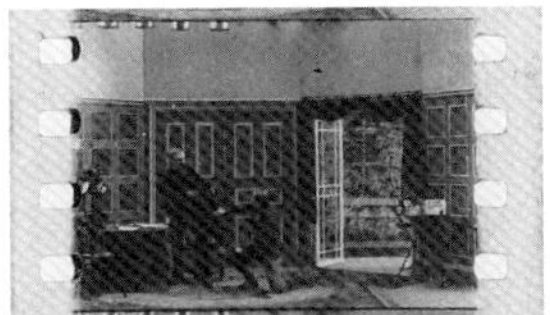

Falsely Accused (1905) – BFI

13.2 Hepworth had a number of practices which can help recognition of his films:

13.2.1 In his earliest films he cut the ends of his negatives in a concave curve. The image of this can be seen in complete prints.

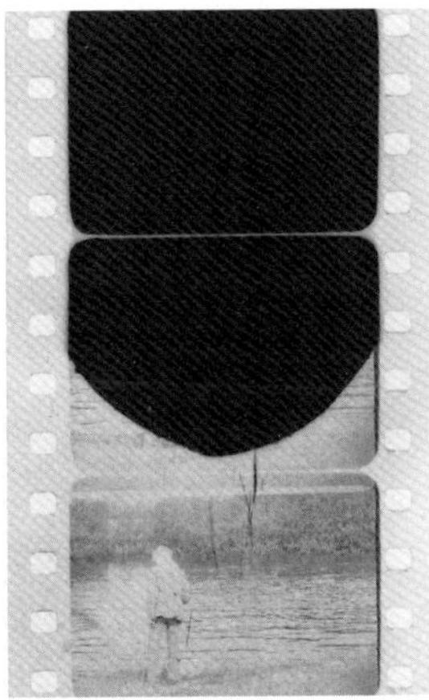

13.2.2 When he began to make films which were not just all one continuous shot, he sometimes placed a few black frames between shots.

13.2.3 Later, about 1907, some of his films have a black spot over the join/splice between scenes. This was made by punching a hole in the negative.

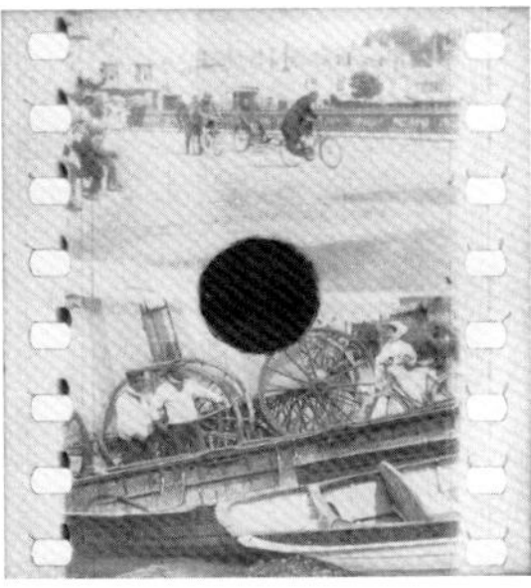

13.2.4 Later still, from about 1910, he used a great many fades out and in between scenes. These fades were very short, being only 6 to 8 frames long.

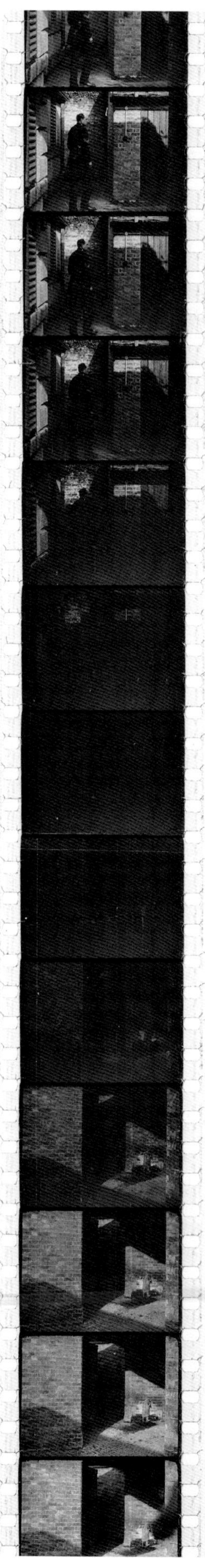

13.2.5 Several producers printed numbers on their intertitles indicating the order of the titles. Hepworth had his own peculiar numbering system of a mark, somewhat in the manner of Roman numerals, incorporated into the top of the decorative border of the title. This consisted of dots, dashes, crosses, and loops; a dot being 1, a dash representing 5, a cross 15, and a loop counting 20.

Thus:

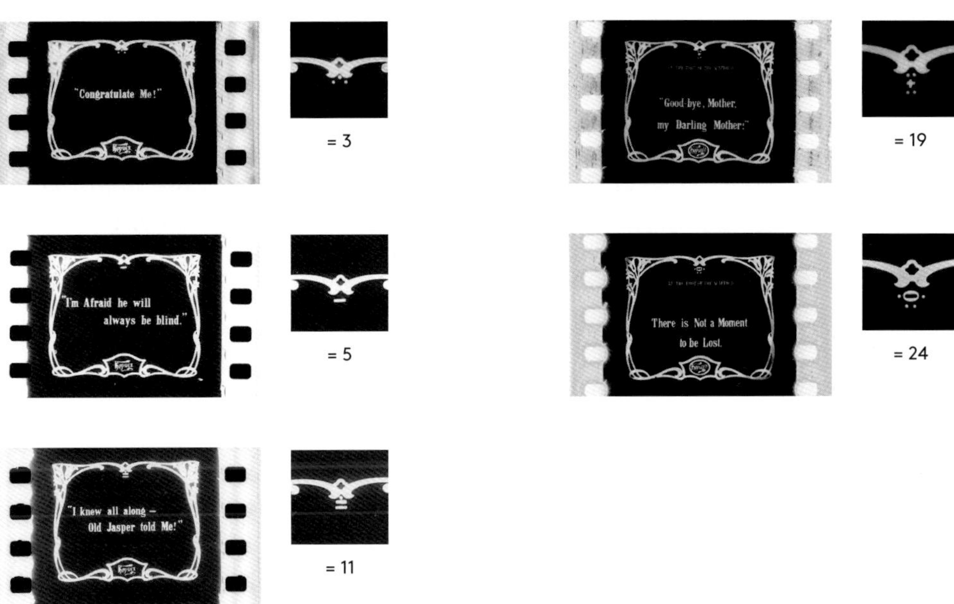

13.3 Hepworth's early prints were very similar to those of Warwick of the same time.

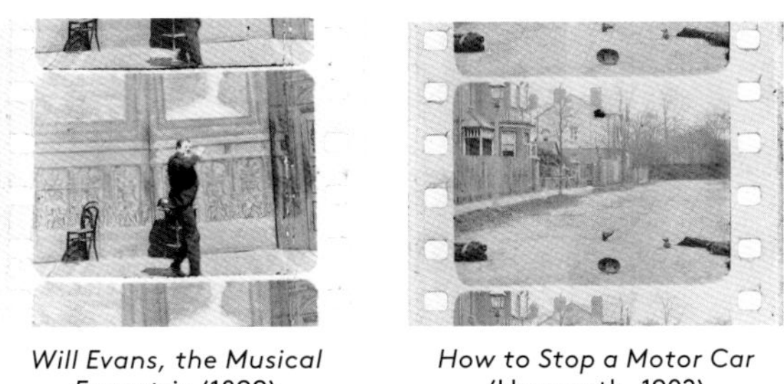

Will Evans, the Musical Eccentric (1899)

How to Stop a Motor Car (Hepworth, 1902)

Here are some examples of his later prints:

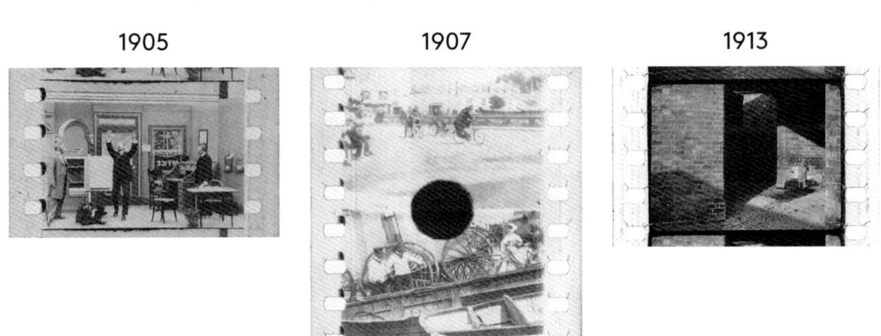

Appendices: Film Lists

1. Gaumont (1906-1914)

1.1 Explanation of abbreviations used in the film list

Title	The original French title is given first when known; then the English title when known; then any other language title.
Review Date	This is the date of the first known mention in the British trade press, in the order Day / Month / Year. Most are from *The Bioscope*. Those which do not give a day, are from a monthly publication.
Serial Number & Letter	Serial number and letters.
Trademark on Main Title	When the main title is missing from the copy, the title is enclosed within parentheses "()".
Trademark on Intertitle	On both Main Titles and Intertitles GAU : The name "Gaumont" appears in full. G : Only the initial letter "G" appears. F : The trademark has the floral border. No IT : No Intertitle. The film has no intertitles, or no original Gaumont ones. No TM : No Trademark. There is no Trademark on the intertitles.
Tint of Titles	A = Amber / B = Blue / G = Green / P = Pink / W = Black & White.
Serial Number	An asterisk * is in this column when the serial number appears in the projected area of the titles.
Margin Patch	An asterisk * is in this column when the titles have the black margin patch.
Number in Margin Patch	A letter "L" is in this column when a large white number appears in the black margin patch. The letters "Ls" are in this column when the small serial number appears in the black margin patch.
Edge Mark	N when there is none in the film. S when the interim letter is placed sideways. Y when the interim character is similar to the letter Y. P when the mark is on the edge of picture. T when the mark is on the edge of titles. / when only part of the copy has an edge mark.
?	Indicates this detail is not known. In these cases, I may have seen only a duplicate on which these details were obscured; or otherwise did not have sufficient access to determine them.
[...]	This is believed to be not the true title or the original title is uncertain.

1.2 Film List

Title	Review Date	Serial Number & Letter	Trademark on Main Title	Trademark on Intertitle	Tint of Titles	Serial Number	Margin Patch	Number in Margin Patch	Edge Mark
The Traitorous Guest, Or, The Great Silver Robbery		None	ELGE F	No IT	P				N
(La Naissance, la vie et la mort de Notre-Seigneur Jésus-Christ) *The Birth, the Life and the Death of Christ*	05.1906	None		ELGE F	W				N
La Marâtre *(The Stepmother)*		None		ELGE	B				N
L'Âne récalcitrant *Father Buys a 'Moke'*	04.07.1907	1636 AN	ELGE	No IT	B	*			S
Le Frotteur *The Floor Polisher*	11.07.1907	1649 AN	ELGE	No IT	B	*			N
La Fiancée du volontaire *(The Hand of the Enemy)*	22.08.1907	1653 AN		ELGE	B	*			S
La Course des potirons *[Pumpkin Race]*	05.03.1908	?		No IT					S
The Electric Policeman	11.02.1909	ANL 2189	GAU	No IT	W	*			S
L'Aveugle de Jérusalem *(The Blind Man of Jerusalem)*	08.04.1909	?		No IT					S
How Percy Won the Beauty Competition	27.05.1909	None	GAU	GAU	B				N
Dickes Fell		ALB 2224	?	?	?	*			?
Le Poivrot incendiaire *The Human Squib*		ALN 2283	GAU	No IT	B	*			S
Monsieur a le hoquet		ALB 2326			B	*			S
La Course de brouettes *(A Barrow Race)*	09.09.1909			No IT					Y
[Idylle Corinthienne] *(The White Slave)*	23.12.1909	ANL 2515		No IT	?	*			P
The Fatal Wedding	23.12.1909	ALB 2516	?	?	?	*			?
Lysistrata ou La Grève des baisers *Lysistrata, or, The Cessation of Kisses*	10.03.1910	ALB 2642	?	?	?	*			?
The Inundations in France	21.01.1910	ANL 2656	GAU	No IT	B	*			Y
Esther *(Esther) (Part 1)* *(Esther) (Part 2)*	 02.05.1910 02.05.1910	 ANL 2707 ANL 2714		 No TM No TM	 A A	 * *			 PT PT
Music Gone Mad	05.05.1910	ALB 2718	?	?		*			?
(Smuggling Up To Date)	01.06.1910	?							P
La Police de l'an 2000 *(The Police of the Future)*	09.06.1910	?							P
[Le Grand Steeple-chase] or [Une course de steeple] *The Steeplechase*	06.10.1910	ANL 2891	GAU F	G F	A	*			P
La Guêpe *(Mind the Wasps!)*	03.11.1910	ANL 2940		GF	A	*			P

Title	Review Date	Serial Number & Letter	Trademark on Main Title	Trademark on Intertitle	Tint of Titles	Serial Number	Margin Patch	Number in Margin Patch	Edge Mark
Eugène amoureux *(Eugene in Love)*	03.03.1911	?		No IT	?				P
Monsieur Prudhomme donne la comédie *Mr. Longwhisker's Private Theatricals*	09.03.1911	ANL 3264	GAU F	G F	G	*			PT
The Siamese Sisters	19.01.1911	ANL 2371	?	?	?	*	?		?
Bébé roi de Rome *(Napoleon and His Son)*	30.03.1911	ANL 3290		G F	G	*	*		PT/
Le Fils de Locuste *[La Locuste]*	13.04.1911	ANL 3292		G F	G	*	*		P
Le Rembrandt de la rue Lepic *(How to Fake a Rembrandt)*	13.04.1911	ANL 3307		G F	G	*	*		P
(Opening Flowers)	13.04.1911	ANL 3308		G F	G	*	*		P/
Fabrication du papier *(Manufacture of Paper)*	28.09.1911	ANL 3523		G F	G	*	*		PT/
(Island Maiden)	16.11.1911	ANL 3601		G F	G	*	*	L	N
Les Chardons *(The Thistles)*	07.12.1911	ANL 3618		G F	G	*	*	L	N
Calino et ses pensionnaires *(Calino and His Boarders)*	21.12.1911	ANL 3633		G F	G *	*	*	L	N
Calino courtier en paratonnerres *(Calino's New Invention)*	22.02.1912	ANL 3753		GAU F	G	*	*	L	PT
La Cassette de l'émigré *(The Refugee's Casket)*	02.05.1912	AN 3820		GAU F	G		*	Ls	
Bébé veut payer ses dettes *(Bobby Raises the Wind)*	25.07.1912	AN 3884		GAU F	G		*	Ls	PT
Sur les rails *The 4:40 Express*	12.09.1912	AN 3936	GAU F	GAU F	G		*	Ls	T
Onésime et le chien bienfaisant *(Simple Simon's Canine Friend)*	10.10.1912	AN 3958		GAU F	G		*	Ls	P
Calino chef de gare *(Calino as a Station Master)*	17.10.1912	AN 3966		GAU F	G		*	Ls	PT/
Crabes de mer *(Sea Crabs)*	19.12.1912	AN 4080		GAU F	G		*	Ls	P
La Dentellière *(The Lacemaker's Romance)*	02.1913	AN 4118		GAU F	G		*	Ls	P
Le Rêve du cocher *The Cabby's Nightmare*	27.02.1913	AN 4173	GAU F	GAU F	G		*	Ls	P
La Mort qui frôle *In Touch with Death*	01.05.1913	AN 4223	GAU F	GAU F	G		*	Ls	P
Islands of New Zembla	08.05.1913	AN 4256	GAU F	GAU F	G		*	Ls	T
Great North American Timber Trade	29.05.1913	AN 4279	GAU F	GAU F	G		*	Ls	T
Good for Evil	26.06.1913	AN 4309	GAU F	GAU F	G		*	Ls	PT

Title	Review Date	Serial Number & Letter	Trademark on Main Title	Trademark on Intertitle	Tint of Titles	Serial Number	Margin Patch	Number in Margin Patch	Edge Mark
[Onésime et l'affaire du Toquard-Palace] (Simple Simon Stays at the Royal Hotel)	03.07.1913	AN 4319		GAU F	G		*	Ls	T
La Disparition d'Onésime (Simple Simon's Disappearance)	12.07.1913	AN 4337		GAU F	G		*	Ls	PT
[Onésime aime trop sa belle-mère] (Simple Simon's Mother-in-Law)	24.07.1913	AN 4353		GAU F	G		*	Ls	PT
Léonce cinématographiste (Leonce as a Cinematographer)	07.08.1913	AN 4370		GAU F	G		*	Ls	PT
La Voix qui accuse (The Accusing Voice)	30.10.1913	AN 4465		GAU F	G		*	Ls	PT
Bout-de-Zan et le lion (Tiny Tim and the Lion)	12.1913	AN 4487		No Trade Mark			*	Ls	PT
Les Pâques rouges (At the Hour of Dawn)	02.02.1914	AN 4624		GAU F	G		*	Ls	PT

2. Cines (1909-1914)

2.1 Explanation of abbreviations used in the film list

Title	The original Italian title is given first when known. This is followed by the English title when known, which is followed by the title in the language of the copy in the N.F.A., when this is not English.
Review Date	This is the date of the first known mention in the British trade press, in the order Day / Month / Year.
Main Title Style	The entries in this column refer to the illustrative figures.
Tint of Main Title	There is probably no significant difference between the colours described as "Pink" and "Red". This probably represents only a random variation in the dyes at different times.
Intertitle Style	The entries in this column refer to the illustrations.
Tint of Intertitles	Colour letter as Main Titles.
Edge Mark	There is an asterisk * in this column when the film has the edge mark 'SOCIETA ITALIANA "CINES" ROMA'.
Language	The entry in this column indicates the Language of the Titles in the copy examined. The letter is the initial letter of the Italian name of the language and appears on the titles.

I	=	Inglese	=	English
T	=	Tedesco	=	German
S	=	Spagna	=	Spanish
/	=	The titles are in Italian and have no language letter.		

- - -	This feature is not on the film.
?	We do not have this information. This may be because we only have a duplicate which does not show this feature; or because the film is not in our archive and we are thus not able readily to inspect it.
Spec. 1.	*King of Italy's Bodyguard* has its own special design of main title.
Spec. 2.	*Cajus Julius Caesar / Caius Julius Caesar* has main and intertitles of its own special design which do not appear on any of the other film in the list. There is an edge mark on this film consisting of the word "INGLESE" on both edges, at a frequency of 14 frames.
[...]	The original title or date of release is uncertain.

2.2 Film List

Main Title Style

A

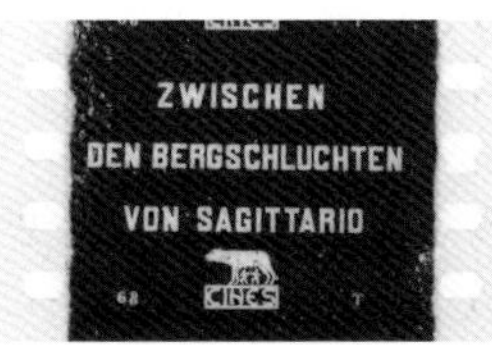

Gole del Sagittario / Sagittario (1909)

B

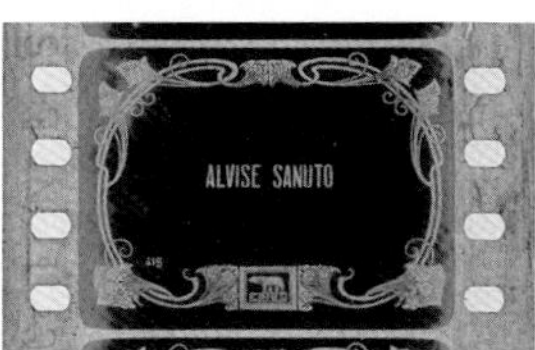

Alvise Sanuto / Venetian Chivalry (1910) – BFI

C

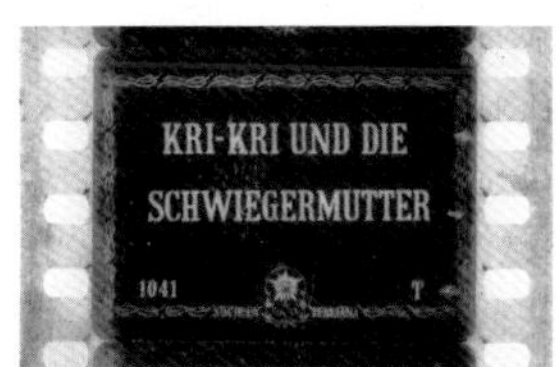

Kri Kri e la suocera / Bloomer's Mother-in-Law (1913)

D

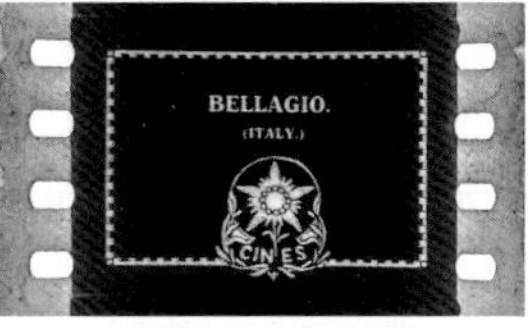

Bellagio (1913) – BFI

E

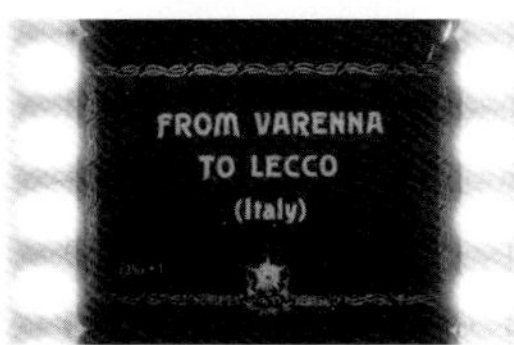

Da Varenna a Lecco / From Varenna to Lecco (1913)

Intertitle Style

1

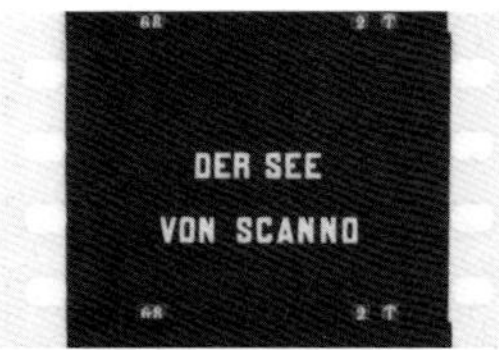

Gole del Sagittario / Sagittario (1909)

2

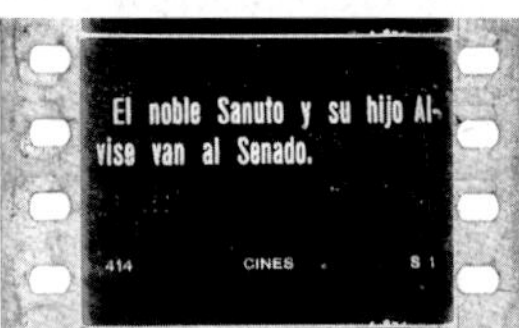

Alvise Sanuto /
Venetian Chivalry (1910) – BFI

3

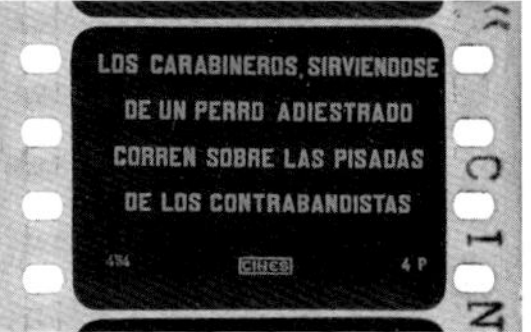

Dramma alla frontiera /
Frontier Drama (1911) – BFI

4

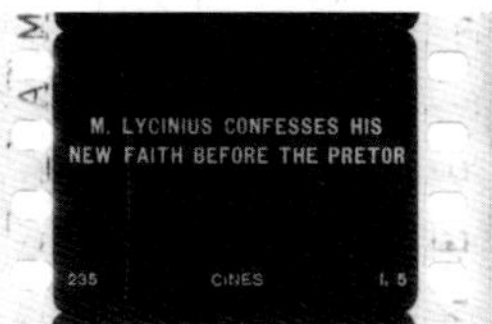

Dall'amore al martirio /
Love and Martyrdom (1910)

5

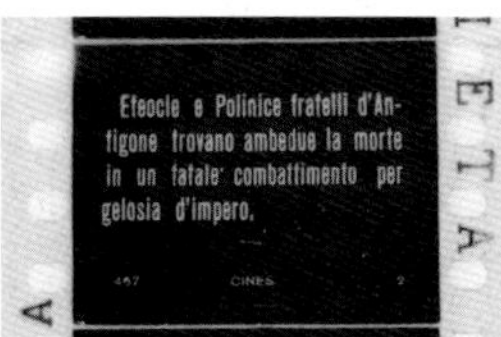

Antigone / Antigone (1911)

6

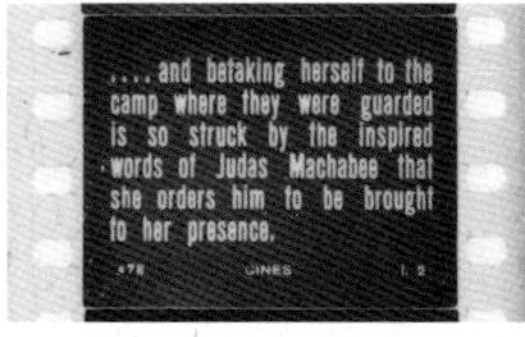

I Maccabei /
Judas Maccabaeus (1911)

7

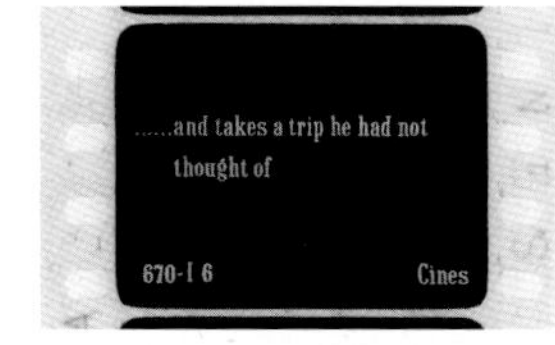

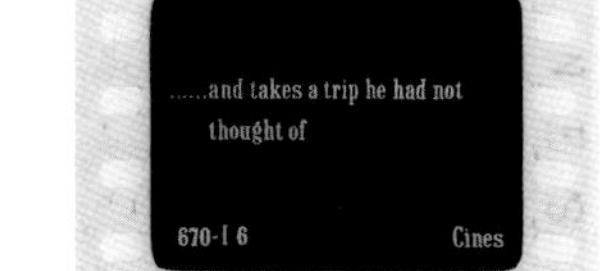

Tontolini e l'asino / Tontolini and the Donkey (1911) – BFI

8

I Calzoni del colonnello / The Colonel's Bet (1912)

9

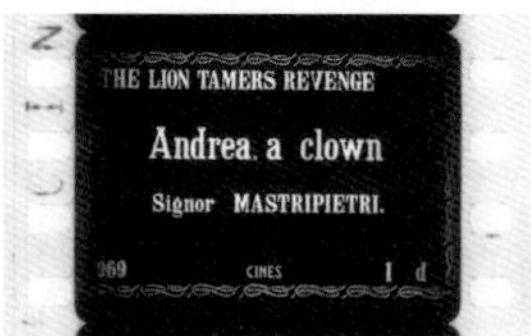

In pasto ai leoni /
The Lion Tamer's Revenge (1912)

10

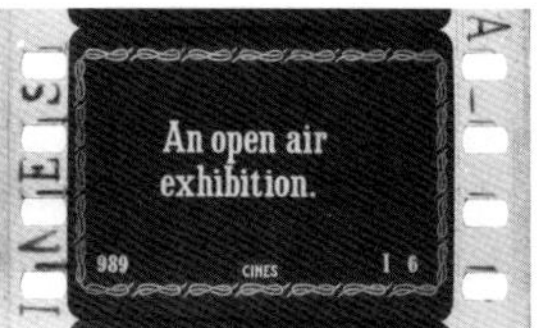

Checco e Cocò domatori /
Stout and Thynne as Lion Tamers (1913)

11

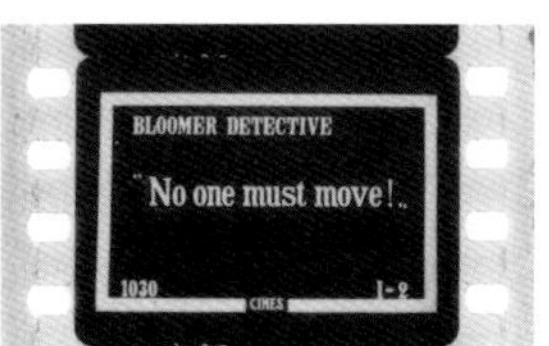

Kri Kri Detective /
Bloomer Detective (1913)

12

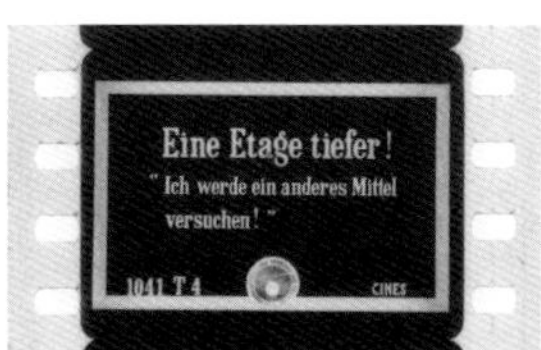

Kri Kri e la suocera /
Bloomer's Mother-in-Law (1913) – BFI

13

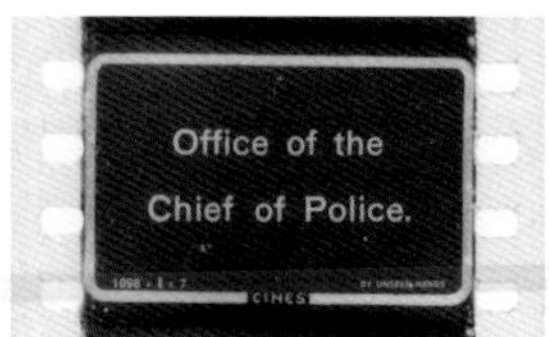

Le mani ignote / By Unseen Hands (1913

Title	Review Date	Serial Number	Main Title Style	Tint of Main Title	Intertitle Style	Tint of Intertitle	Edge Mark	Language
Il Poliziotto *The Policeman*	22.07.1909	44	A	Red	1	Red		I
La Bella andalusa *The Fair Dolores*	19.08.1909	67	---	?	1	Red	*	I
Gole del Sagittario *Sagittario*	09.09.1909	68	A	Red	1	Red	*	T
La Campana *The Bell*	04.09.1909	95	A	Red	1	Red	*	T
Bianca Capello *Bianca Capello*	21.10.1909	100	---	-	1	Amber	*	T
Angelo di pace *Angel of Peace*	04.11.1909	103	A	Amber	1	Amber	*	T
Il Cid *The Triumphant Hero*	10.02.1910	116	---	-	1	Amber	*	I
L'insorto della Vandea *The Vandean Rebel*	09.12.1909	123	B	Mauve	2	M	*	T
Livorno *Leghorn, Western Italy*		133	---	-	3	Blue	*	I
La congiura di Piacenza *The Piacenza Plot*	24.02.1910	169	B	Amber	2	Amber	*	T
Casa paterna *Home*	12.05.1910	210	---	-	4		*	I
Dall'amore al martirio *Love and Martyrdom*	02.06.1910	235	---	?	4	Amber	*	I
Faust *Faust*	02.06.1910	248	B	?	2	?	*	I
Era destino! *Destiny*	04.08.1910	275	B	Amber	---	?		I
Una scommessa di Tontolini *Tontolini's Bet*	11.08.1910	294	---	?	---	?	*	I
Tontolini ruba una bicicletta *Tontolini Steals a Bicycle*	01.09.1910	308	B	Amber	---	?	*	I
Squadrone Guardie del Re *[King of Italy's Bodyguard]*	[25.03.1915]	314	Spec 1		2	Amber	*	I
Le Sorelle Bartels *[Bartels Sisters in Their Acrobatic Repertory]*	03.11.1910	320	B	Amber	---	?		I
Alvise Sanuto *Venetian Chivalry*	03.11.1910	414	B	Amber	2	Amber	*	S
Acrobatica monociclista *Monocycle Display by Miss Lily*	19.01.1911	424	?	?	?	?	?	T
Santa Cecilia (La martire cristiana) *St. Cecilia*	02.02.1911	429	---	?	2	Amber	*	I
Amore e libertà *For Love and Country*	16.02.1911	440	---		2	Amber	*	I
Dramma alla frontiera *A Frontier Drama*	16.03.1911	454	B	Amber	3	Amber	*	S

Title	Review Date	Serial Number	Main Title Style	Tint of Main Title	Intertitle Style	Tint of Intertitle	Edge Mark	Language
Antigone *Antigone*	16.03.1911	467	---	?	5	Amber	*	/
Angelo tutelare *His Guardian Angel*	16.03.1911	468	---	?	3	Amber	*	S
Tontolini accalappiacani *Tontolini's Dogs*	16.03.1911	469	B	Amber	---	?	*	S
I Maccabei *Judas Maccabaeus*	16.03.1911	478	---	?	6	Amber	*	I
L'agente Tontolini e il suo commissario *Policeman Tontolini's Inspector*	01.06.1911	537	---	?	3	Amber	*	I
La sposa del Nilo *The Bride of the Nile*	08.06.1911	539	---	?	3	Amber	*	I
Salomone e C. *Solomon and Co.*	06.07.1911	566	---	?	3	Amber	*	T
L'anello della Regina Elisabetta *Queen Elizabeth's Ring*	21.09.1911	636	B	Amber	3	Amber	*	I
Il violino di Tontolini *Tontolini's Violin*	12.10.1911	645	B	Amber	3	Amber	*	I
Bruto *Brutus*	12.10.1911	658	---	?	?	Amber	*	I
L'Etna *Among the Etna Fire*		662	B	Amber	3	Amber	*	I
Tontolini e l'asino *Tontolini and the Donkey*	09.11.1911	670 670	--- B	? Amber	7 ?	Amber Amber	* *	I T
Equilibristi meravigliosi *Wonderful Equilibrium*	07.12.1911	710	?	?	?	?	?	T
Debito pagato *The Debt Paid*	30.05.1912	864		?	5	Pink	*	I
I calzoni del colonnello *The Colonel's Bet*	04.07.1912	888	---	?	8	Pink	*	I
Un furto misterioso *The Mysterious Theft*	29.08.1912	919	?	?	?	?	?	S
Il Goloso *The Glutton*	17.10.1912	966	---	?	9	Pink	*	I
In pasto ai leoni *The Lion Tamer's Revenge*	12.09.1912	969	---	?	9	Pink	*	I
Checco e Cocò domatori *Stout and Thynne as Lion Tamers*	07.11.1912	989	---	?	10	Pink	*	I
Kri Kri in prova *Bloomer Tries Acting*	20.11.1912	1006	---	?	11	Pink	*	I
Kri Kri Detective *Bloomer Detective*	02.01.1913	1030	---	?	11	Pink		I
Kri Kri e la suocera *Bloomer's Mother-in-Law*	16.01.1913	1041	C	Pink	12	Pink		T
Le mani ignote *By Unseen Hands*	20.03.1913	1098	---	?	13	Red		I

Title	Review Date	Serial Number	Main Title Style	Tint of Main Title	Intertitle Style	Tint of Intertitle	Edge Mark	Language
Costumi abruzzesi *Traditional Customs of the Abruzzi*	04.09.1913	1191	---	?	13	Red		I
Un sogno di Kri Kri *Bloomer's Dream*	04.09.1913	1208	---	?	13	Red		I
Bellagio *Bellagio*	06.11.1913	1245	D	?	13	Green		I
Kri Kri ama la tintura *Bloomer and the Dyer* (Review title) *Bloomer and the Paper-Hanger* (on intertitle)	20.11.1913	1246	---	?	13	Pink		I
Da Varenna a Lecco *From Varenna to Lecco*	13.11.1913	1250	E	Red	13	Red		I
Kri Kri gladiatore *Bloomer Gladiator*	13.11.1913	1256	?	?	?	?	?	T
Kri Kri domestico *Bloomer Manservant*	06.11.1913	1259	---	?	13	Pink		I
Kri Kri è miope *Bloomer Short-Sighted*	22.01.1914	1293	---	?	13	Pink		I
Vendemmia sicula *Sicilian Vintage*	30.04.1914	1363	?		?	Red		I
La rinunzia *The Gambler*	21.05.1914	1376	---	?	13	Pink		I
Cajus Julius Caesar *Caius Julius Caesar* (produced 1914)	29.04.1915	1405	Spec 2	Red		Red	I	I

3. Éclair (1908-1915)

3.1 Explanation of abbreviations used in the film list

Title	The original French title is given first when known; then the English title when known (or in any other language title known when the English title is unknown).
Review	This is the date of review or other mention in *The Bioscope*, in the order Day / Month / Year.
Serial Number	The production serial number when it appears on the film copy examined.
Main Title	The entries in this column refer to the illustrations.
Intertitle	The entries in this column refer to the illustrations.
Edge Mark	The film has the edge mark shown in the given illustration number. "No" means that there is no edge mark on the film. There is no illustration of the 1911 mark; this reads "F.E. PARIS 1911".
Language	The language of the titles on the copy examined.
/	The feature does not exist on the film, or is missing from the copy examined, or is represented only by replacements.
?	Means that we are not able to determine this detail.
[...]	Means that the original title is uncertain.

3.2 Film List

Main Title Style

A

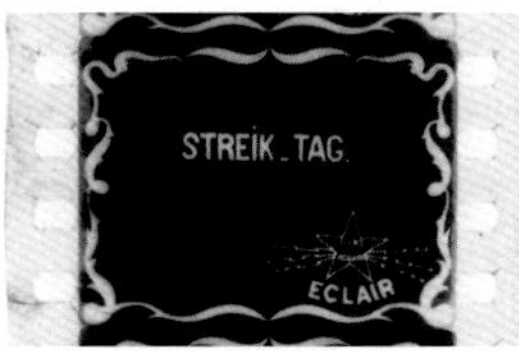

Journée de grève (1909)

B

**Louqsor et Thèbes / Luxor and Thebes, Egypt* (1911) – SFI

C

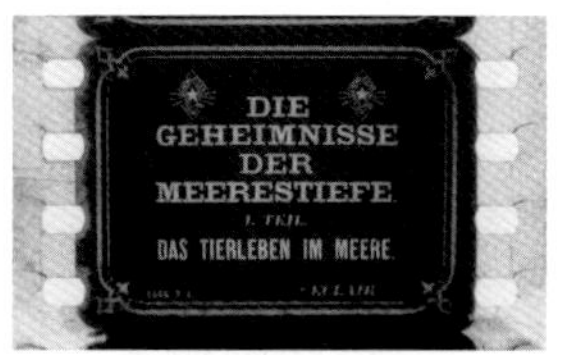

La Vie au fond des mers / Life at the Bottom of the Sea (1911)

D

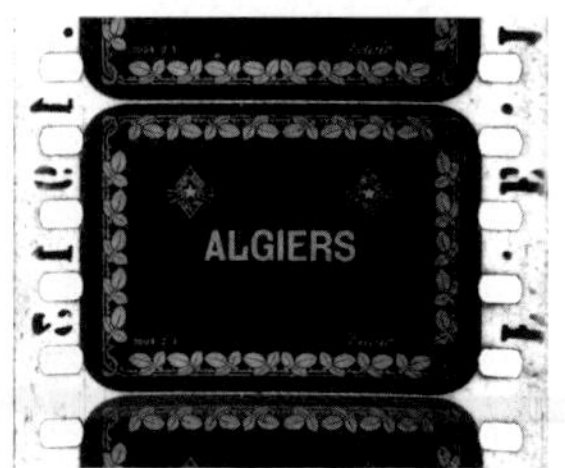

Alger la blanche / Algiers (1913) – BFI

Intertitle Style

1

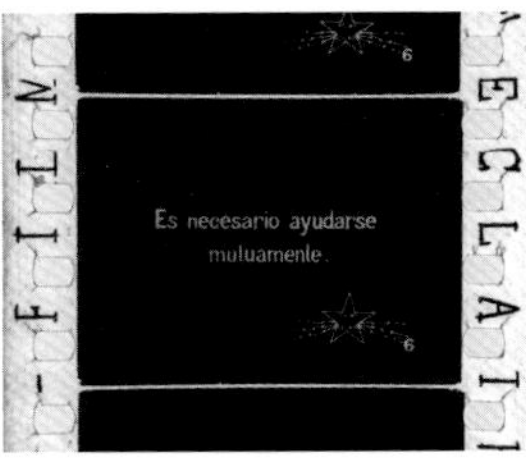

Une chasse fructueuse /
[Cazador furtivo] (1910) – BFI

2

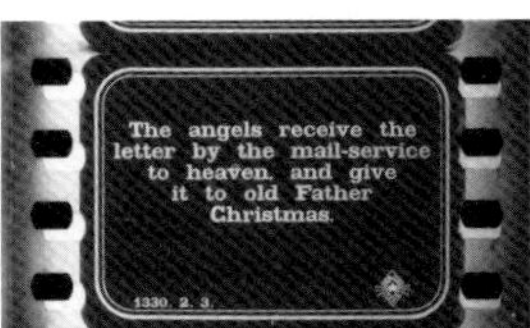

La Lettre au Père Noël /
Letter to Father Christmas (1910).
Note: Photograph of a duplicate negative.

3

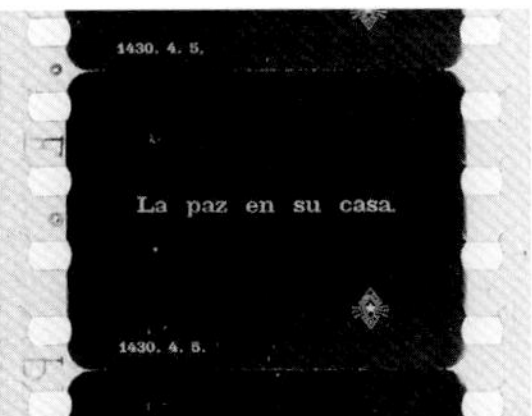

Le Gilet à pointes /
The Waistcoat with Points (1911)

4

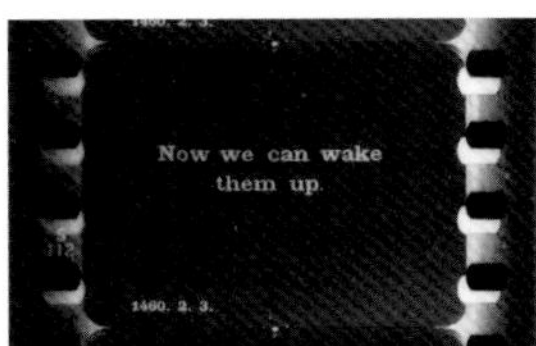

Les Agents à roulettes /
The Roller-Skating Policemen (1911).
Note: Photograph of a duplicate negative.

5

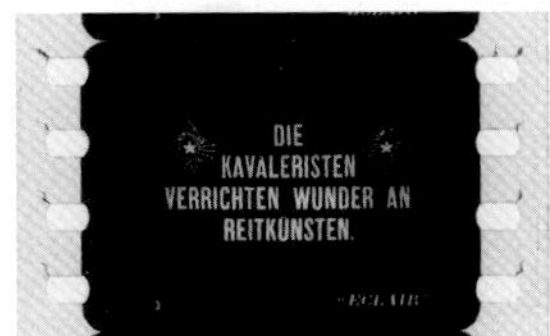

Les Centaures portugais /
The Portuguese Centaurs (1911)

6

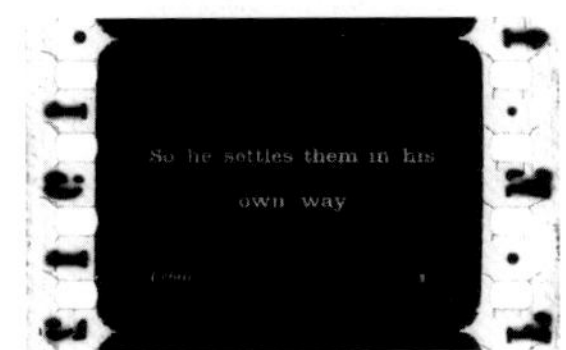

Rival de son maître /
Servant's Holiday (1912)

7

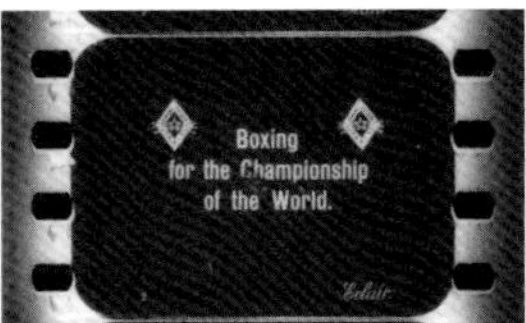

Willy contre le bombardier Wells /
Little Willie v. Bombardier Wells (1913).
Note: Photograph of a duplicate negative.

8

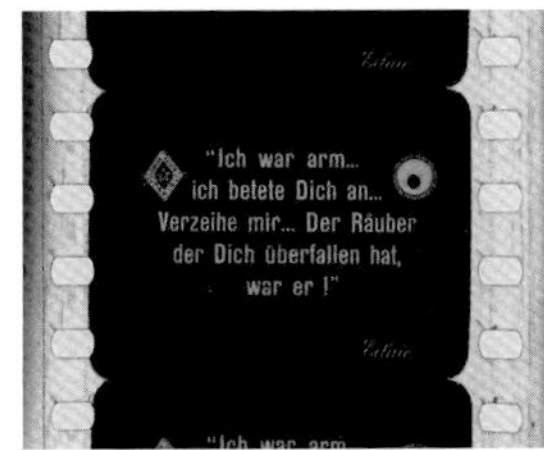

Gontran et son complice /
Gontran and His Accomplice (1913) – BFI

9

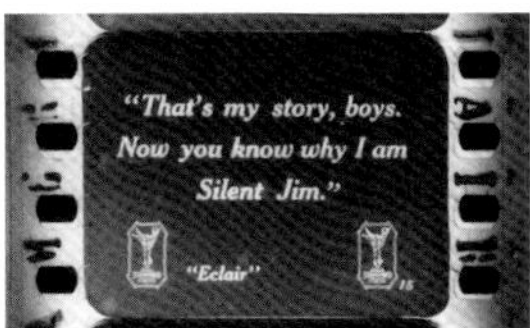

Silent Jim (1913).
Note: Photograph of a duplicate negative.

10

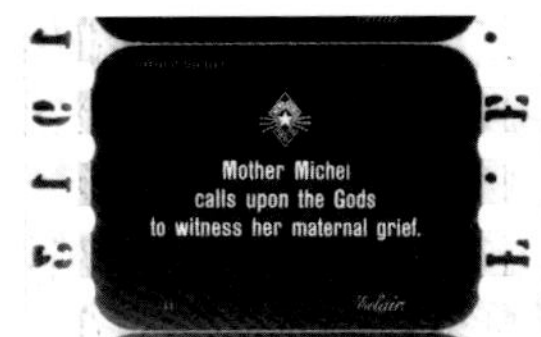

C'est la mère Michel / Mother Michel (1913)

11

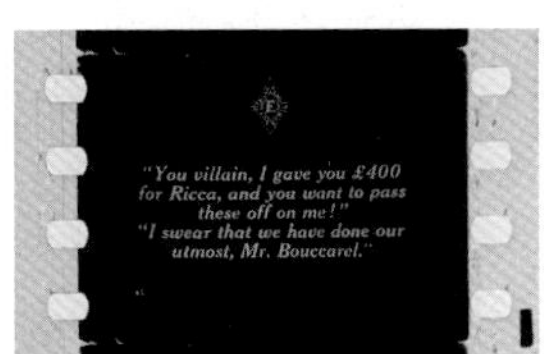

Premier amour / First Love (1912) – BFI

Producer's Edge Mark

I

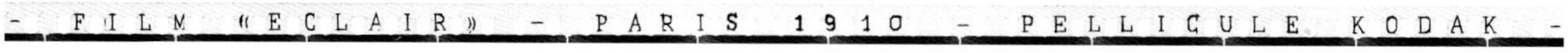

1910

II

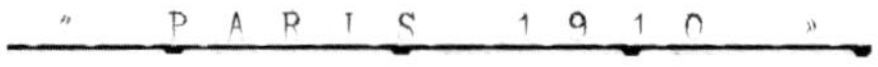

1910

III

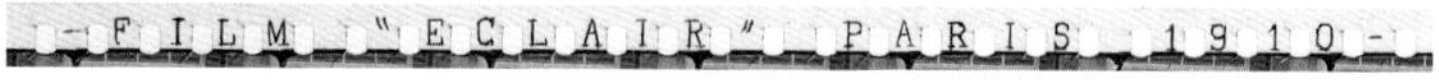

1910

IV

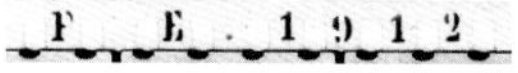

1912

V

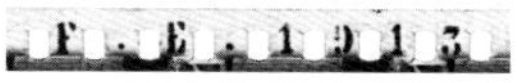

1913

VI

1913 with +

VII

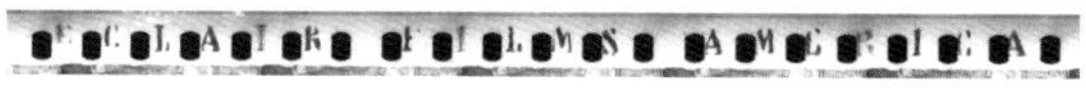

No date. Note: Photograph of a duplicate negative.

VIII

1913 with x

Title	Review	Serial Number	Main Title	Intertitle	Edge Mark	Language
Nick Carter: Le Roi des Détectives. Nr.1: Le Guet-apens	1908		/	/	No	/
Nick Carter – Le Club des suicidés *The Suicide Club*	1909		/	1	No	German
Une Pouponnière à Paris *Baby Rearing in Paris*	16.09.1909		A	1	No	German
Journée de grève *Streik Tag*	1910		A	1	No	German
Une chasse fructueuse *Un cazador furtivo*			/	1	I	Spanish
Un médecin distrait *An Absent-Minded Doctor*	08.09.1910		/	/	II	English
His Last Experiment	1910		/	1	III	English
La Lettre au Père Noël *Letter to Father Christmas*	22.12.1910	1330	/	2	II	English
Le Gilet à pointes *The Waistcoat with Points*	13.04.1911 11.02.1915	1430	/	3	1911	Spanish
Les Agents à roulettes *The Roller-Skating Policemen*	1911	1460	B	4	1911	English
La Parabole de l'Enfant prodigue *Parable of the Prodigal Son*	25.05.1911		/	/	1911	/
La Vie au fond des mers *Life at the Bottom of the Sea*	21.09.1911	1545	C	5	No	German
Les Centaures portugais *Portuguese Centaurs*	28.09.1911	1550 1550	C	5 5	No No	English German
Le Premier cigare de Willy *Little Willie's First Cigar*	02.11.1911	1576	C	5	No	German
Copenhague à vol d'oiseau *Copenhagen*	09.11.1911	1584	/ C	5 5	No	English German
Willy Professeur de Skating *Little Willie Professor of Skating*	23.11.1911		/	5	No	
Un Cri dans la nuit *A Cry in the Night*	18.01.1912	1632	?	?	?	S
Le Chien du détective *The Detective's Dog*	08.02.1912		/	5	No	English
La Guerre de l'Indépendance américaine *War of American Independence*	18.04.1912		/	Plain	No	German
Brousse (Turquie d'Asie) *Brusia (Asiatic Turkey)*	06.06.1912	1725	C	5	No	German
Willy étrenne son costume de marin *Willy's Sailor Suit*	27.06.1912		/	5	No	German
Tachkent *Tachkent, Asiatic Russia*	04.07.1912		/	5	No	English
Le Bonnet blanc *The White Cap*	11.07.1912	1755	?	?	?	Spanish
Infants Progress	1912		/	5	No	English
Rival de son maître *Servant's Holiday*	1912		/	6	IV	English

Title	Review	Serial Number	Main Title	Intertitle	Edge Mark	Language
Les Rivières indochinoises[50] *Rivers of Indo-China*	1912		/	5	IV	English
Willy contre le bombardier Wells *Little Willie v. Bombardier Wells*	30.01.1913		/	7	V	
Zigomar peau d'anguille *Zigomar: The Eel's Skin*	06.02.1913		/		V	English
La Ruse de Willy *Willy's Ruse*	20.02.1913		/	7	VI	English
Gontran et son complice *Gontran and His Accomplice*	27.02.1913		/	8	No	German
Silent Jim (Éclair-Universal)	22.05.1913		/	9	VII	English
An Ingenious Lover	19.06.1913		/	5	VI	English
Willy, le tambour et les lunettes *Willie Deceives His Grandma*	19.06.1913		/	5	V	English
[Willy diplomate] *Willy Plays a Part*	03.07.1913		/	5	VIII	English
Alger la blanche *Algiers*	14.08.1913	2094	D	5	V	English
En Malaisie *[In the Malay Islands]*	20.11.1913		/	5	VI	English
Willy and the Parisians *Willy and the Parisians*			/	5	VI	English
C'est la mère Michel *Mother Michel*	04.12.1913		/	10	VI	English
[Celles qui restent au logis] *Woman Who Stayed at Home*	c.1914/15		/	10/11	No	English
Premier amour *First Love*	03.06.1915		/	10/11	No	English
Humorous War Review	11.11.1915			/	No	/

50 This film seems to have been released in 1913, not in 1912.

4. Selig Polyscope (1908-1915)

4.1 Explanation of abbreviations used in the film list

Production Number — See Paragraph 9.1, p.106.

Reference date — See Paragraph 8.8.5, p.137.

Edge Mark — USA = Has the edge mark of 1909-1911 including the letters "U.S.A.".

1909-1911

EM = Has the edge mark without the letters "U.S.A.".
See Paragraph 5.8, p.57.

Intertitle Style — The film has intertitles of the style shown in the illustration indicated. The letter states the tint of the title.[51]
There are a few titles with a pinkish amber colour. It may be that these represent a period of change from Amber to Pink.
The earliest of the Selig films had a plain background. However, *Ranch Life in the Great South-West* has a production number in the bottom right-hand corner. *The Still Alarm* has a production number within the diamond, below the wording.

--- — This feature is not on the film.

? — I have not been able to find the information on this detail.

51 The tint defined as Amber by Harold Brown can differ considerably on the film materials inspected in April 2019 (Camille Blot-Wellens). See also the Colour Section, pp.152-153.

4.2 Film List

Intertitle Style

1

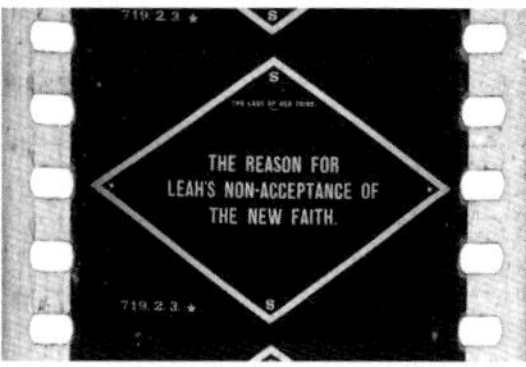

The Last of Her Tribe (1912) – BFI

2

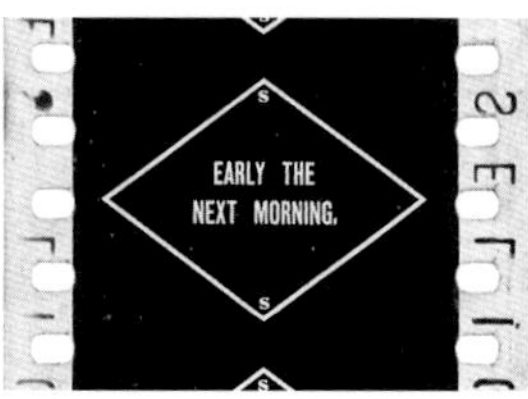

One Hundred Years After (1911) – BFI

3

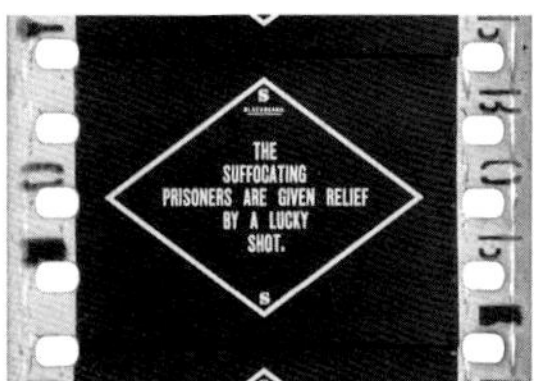

Blackbeard (1911) – BFI

4

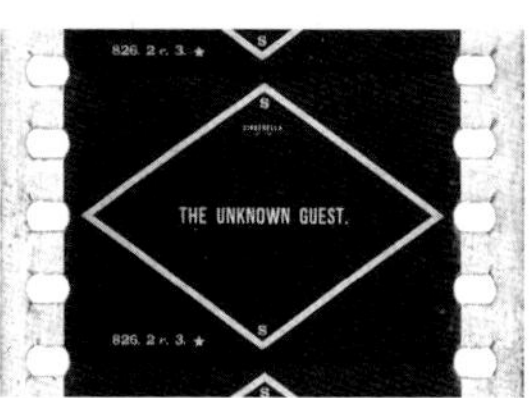

Cinderella (1911) – BFI

5

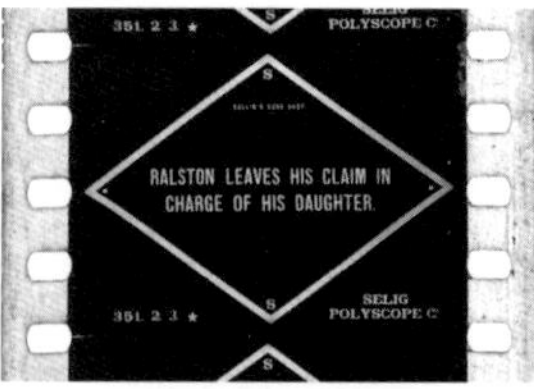

Sallie's Sure Shot (1913) – BFI

6

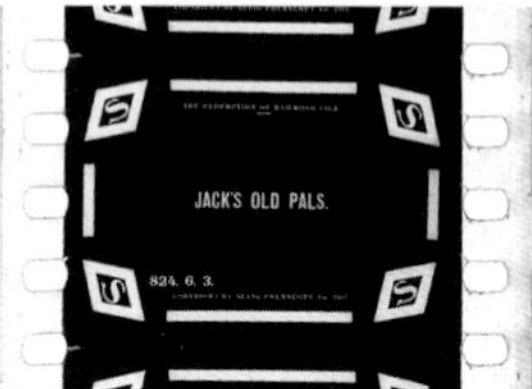

The Redemption of Railroad Jack (1913) – BFI

7

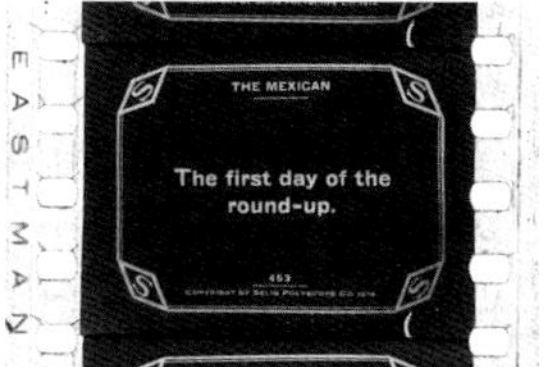

The Mexican (1914) – BFI

Title	Production Number	Reference Date	Edge Mark	Intertitle Style	Tint
The Cattle Rustlers	---	12.09.1908	?	Plain	Red
The Infant Terrible	---	08.07.1909	---	Plain	Red
Up San Juan Hill	354	27.11.1909	---	Plain	Red
Buried Alive	---	31.12.1909	USA	Plain	Red
The Devil, the Servant and the Man	842	05.02.1910	---	1	Amber
Ranch Life in the Great South-West	---	22.12.1910	USA	Plain	Pink
The Curse of the Red Man	---	11.02.1911	?	2	Amber
The Seminole's Sacrifice	---	04.03.1911	?	2	Amber
The Outbreak	---	11.03.1911	EM	2	Amber
Back to the Primitive	---	13.05.1911	EM	2	Amber
The Still Alarm	731	21.05.1911	EM	2	Amber
One Hundred Years After	---	06.1911	EM	2	Amber
The Totem Mark	---	02.09.1911	EM	2	Amber
An Indian Vestal	---	07.10.1911	EM	2	Amber
Blackbeard	---	12.11.1911	EM	3	Pinkish Amber
Cinderella	826	30.12.1911	---	4	Amber
The Bandit's Mask	---	27.01.1912	EM	3	Pinkish Amber
Two Old Pals	---	27.01.1912	EM	3	Pinkish Amber
The Trade Gun Bullet	986	07.09.1912	?	4	Amber
A Man Among Men	54	06.11.1912	---	1	Amber
The Ranger and His Horse	312	07.12.1912	---	1	Amber
The Last of Her Tribe	719	14.12.1912	---	1	Amber
Opitsah (Apache for "Sweetheart")	715	14.12.1912	---	1	Pink
The Artist and the Brute	751	31.01.1913	---	1	Pink
A Prisoner of Cabanas	764	03.04.1913	---	5	Pink
Arabia Takes the Health Cure	230	17.04.1913	---	5	Pink
Buck Richards' Bride	789	10.05.1913	---	5	Pink
Indian Summer	799	17.05.1913	?	5	
Sallie's Sure Shot	351	28.06.1913	---	5	Pink
A Western Romance	806	28.06.1913	---	5	Pink
The Trail of Cards	808	11.07.1913	---	5	
The Wild Ride	810	12.07.1913	---	5	
The Redemption of Railroad Jack	824	11.09.1913	?	6	
The Cattle Thief's Escape	338	25.09.1913	---	6	
In the Midst of the Jungle	763	13.11.1913	---	5	Pink
The Tide of Destiny	833	22.11.1913	---	5	Pink
Little Lillian Turns the Tide	912	28.02.1914	---	5	Pink

Title	Production Number	Reference Date	Edge Mark	Intertitle Style	Tint
The Midnight Call	?	26.03.1914	?	1	Pink
Little Miss Bountiful	124	14.04.1914	---	5	Pink
The Reporter on the Case	?	15.07.1914	?	7	
The Family Record	---	22.07.1914	?	7	
The Water Rat	?	29.08.1914	?	1	Pink
The Mexican	453	24.09.1914	---	7	
The Tragedy that Lived	?	30.09.1914	?	7	
The Broken X	989	24.10.1914	---	7	
Tiger Bait	---	22.02.1915	---	7	
A Night in the Jungle	---	23.03.1915	?	7	
In the Amazon Jungle	---	13.05.1915	---	7	

5. Thanhouser (1910-1915)

5.1 Explanation of abbreviations used in the film list

Title	The title of the film, which may be on the film or may have been discovered by research.
Review Date	The date of the earliest reference which I have found, in the order Day / Month / Year.
Title Style	The intertitles are in the style illustrated in the Title Style illustration number.
Tinted Amber	When the title is tinted, it's indicated with a check mark "√". Otherwise, the title is in black & white.
Frame Type	The frame characteristics are as the illustration indicated. A question mark (?) indicates that I do not know; the character of the original is obscured on the copy seen.
Trademark	There is a check mark "√" in this column when the trademark appears in any scene. It may be noted that no trademarks appear in scenes after 1912.
Curves	Of the films in which the film title appears on the intertitles, the title is enclosed in curved brackets (parentheses). Note that all the films whose titles have the curved brackets are in the second half of 1912.
Edge Mark	Two films have the Pathé edge mark from 1911. These are *The Old Curiosity Shop* and *The Millionaire Milkman*. The last word of the full title of the film *John T. Rocks* is rendered in different references as *"Flyer"* and as *"Flivver"*.

5.2 Film List

Title Style

1

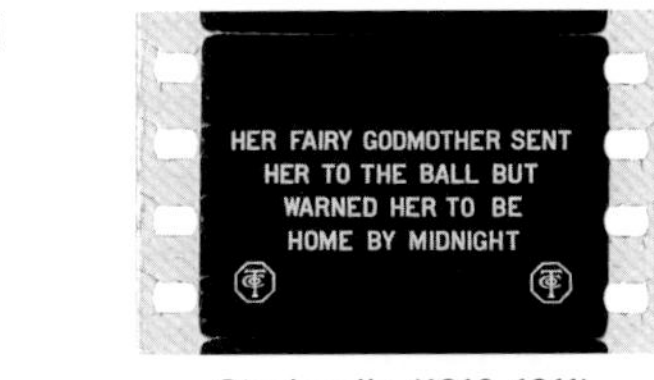

Cinderella (1910-1911)
- BFI

4

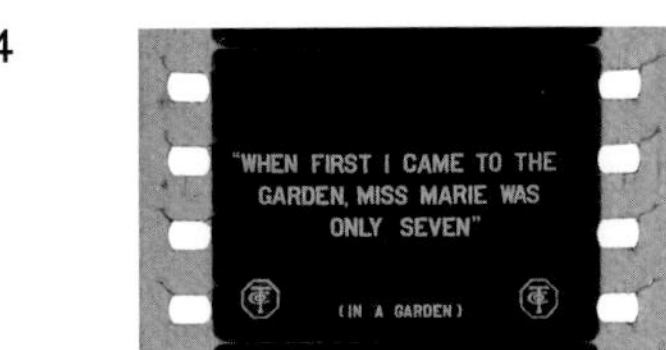

In A Garden (1912)
- BFI

7

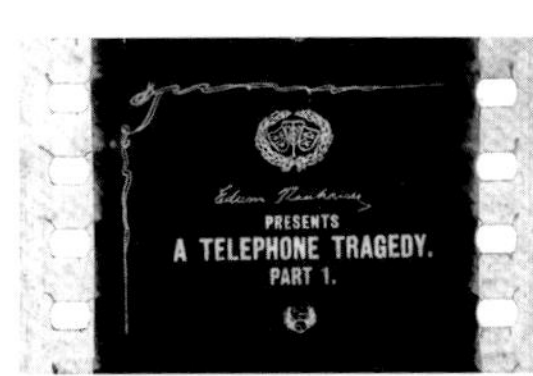

A Telephone Tragedy (1915)
- BFI

2

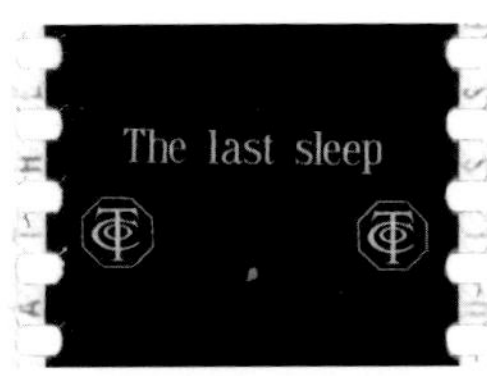

The Old Curiosity Shop (1912)

5

The Farmer's Daughters (1913)
- BFI

8

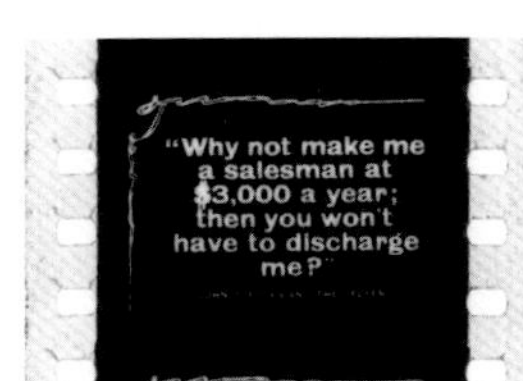

John T. Rocks and the Flivver
(1915)

3

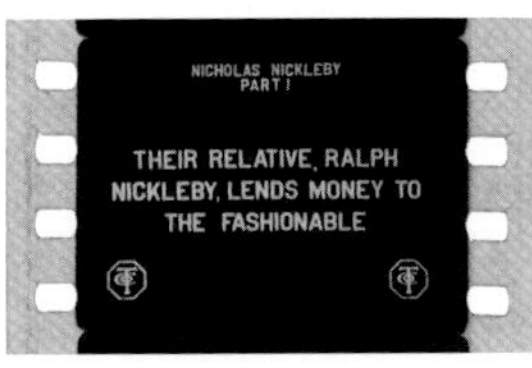

Nicholas Nickleby (1912)
- BFI

6

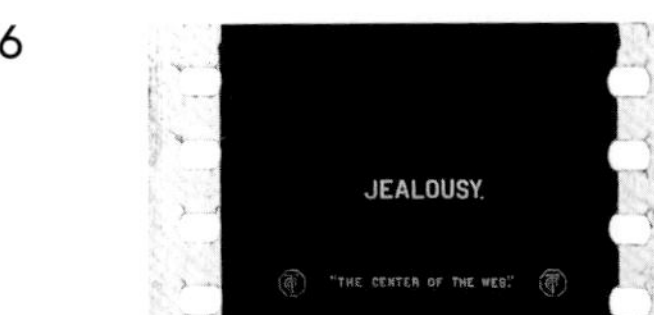

The Center of the Web (1914)
- BFI

Frame Type

A

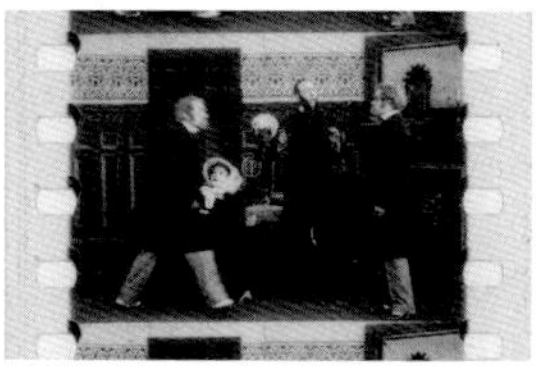

Daddy's Double
(1910)

B

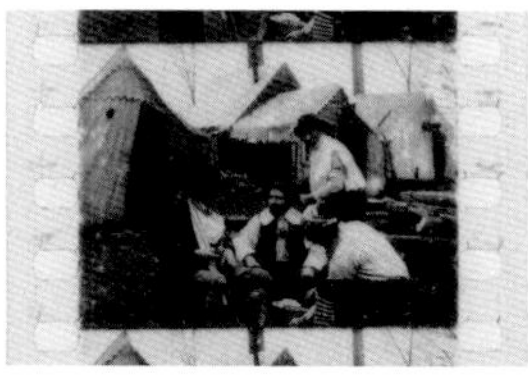

The Old Curiosity Shop (1912)

C

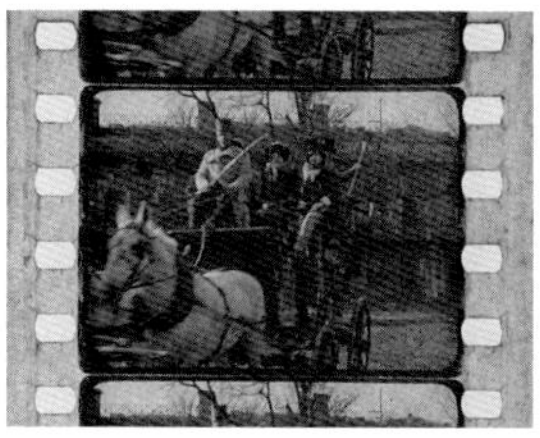

Nicholas Nickleby (1912) – BFI

D

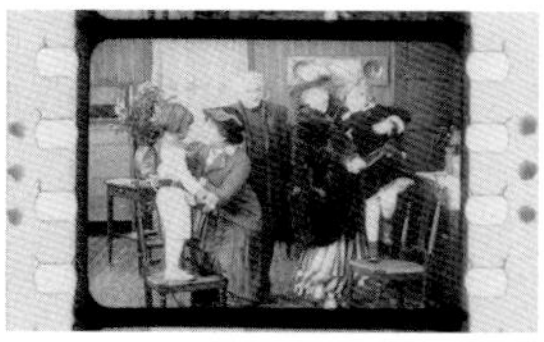

Treasure Trove
(1912)

E

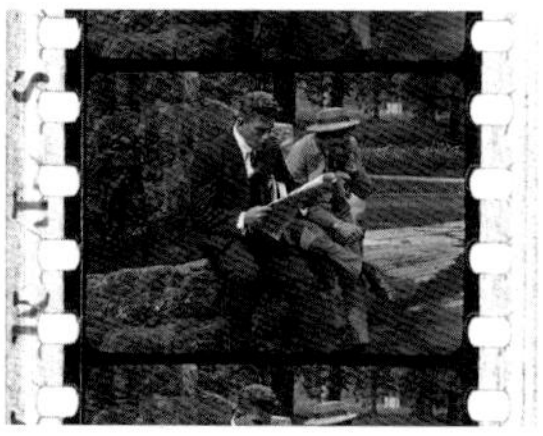

The Farmer's Daughters
(1913) – BFI

F

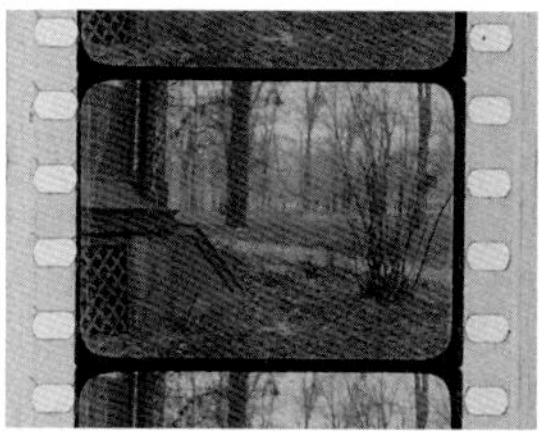

An Elusive Diamond (1914)
– BFI

G

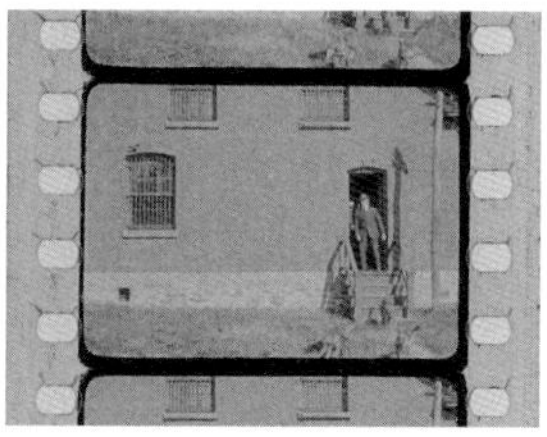

The Center of the Web (1914)
– BFI

H

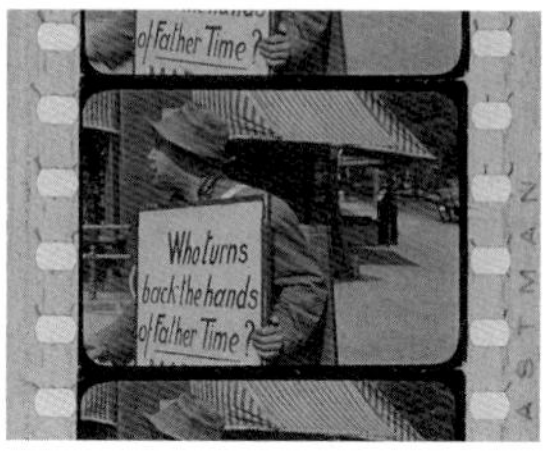

Madame Blanche, Beauty Doctor (1915) – BFI

Title	Review Date	Title style	Tinted Amber	Frame Type	Trademark	Curves
Daddy's Double	02.04.10	1	√	A	√	
The Pillars of Society	06.05.11	1	√	?	√	
Cinderella	06.12.11	1	√	?		
The Old Curiosity Shop	22.02.12	2		B	√	
Nicholas Nickleby	06.03.12	3	√	C		
Under Two Flags	09.07.12		√	C		
Treasure Trove	27.07.12	4	√	D	√	√
Undine	21.09.12	4	√	?		√
In a Garden	09.10.12	4	√	?		√
The Millionaire Milkman *Le Laitier millionnaire*	19.12.12	Belg.		?	√	
The Star of Bethlehem	24.12.12	4	√	?		√
Just a Shabby Doll	15.03.13	5	√	?		
Tannhäuser	19.07.13	5	√	?		
The Girl of the Cabaret	16.08.13	5	√	?		
The Farmer's Daughters	04.10.13	5	√	E		
Uncle's Namesakes	13.12.13	5	√	?		
An Elusive Diamond[52]	24.01.14	5	√	F		
Their Best Friend	28.02.14	5	√	F		
The Decoy (Princess Film Company)	11.07.14			F		
In Danger's Hour	19.09.14	6		?		
Shep's Race with Death	03.10.14	6		G		
The Center of the Web	12.12.14	6		G		
Their One Love	08.05.15	None		?		
Madame Blanche, Beauty Doctor (Falstaff)	10.07.15			H		
A Telephone Tragedy	07.10.15	7		?		
John T. Rocks and the Flivver	23.10.15	8 7		?		

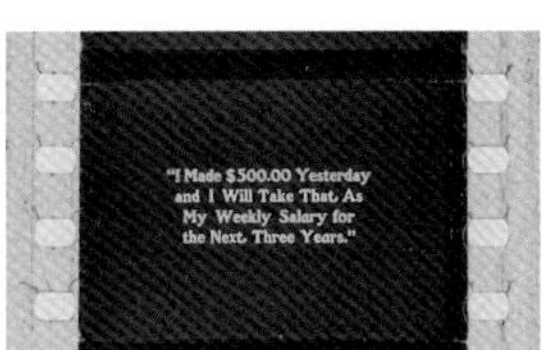

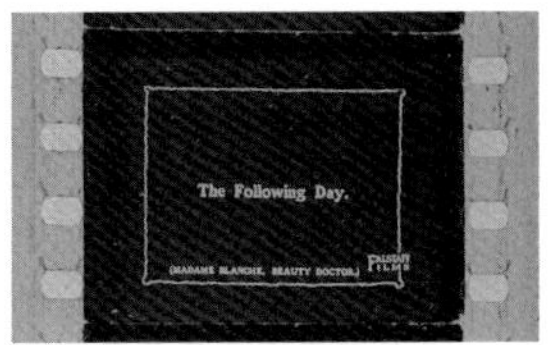

Madame Blanche, Beauty Doctor (1915) – BFI

52 This title also appears as *The Elusive Diamond* in the trade press of the period.

Title Styles – Illustrations in Colour

The colours reproduced in the illustrated sample frames cannot be regarded as exact equivalents, but can only give an idea of the original elements, since the pictures were taken by different persons, using different devices, at different times.

Hepworth

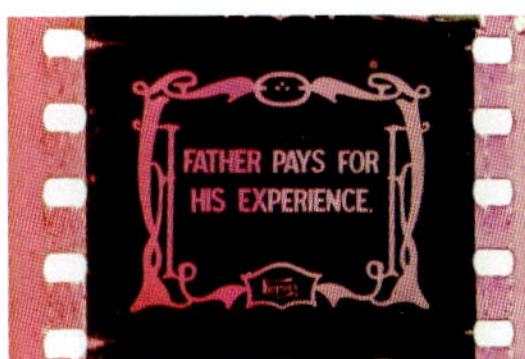

Intertitle (English).
A Cheap Removal (1909) – BFI

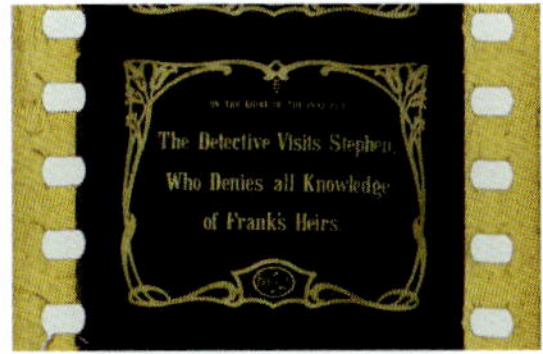

Intertitle (English).
On the Brink of the Precipice (1913) – BFI

Gaumont

1906

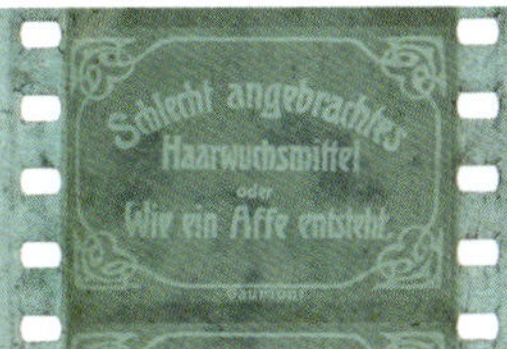

*Main title (German).
La Vérité sur l'homme singe – CNC

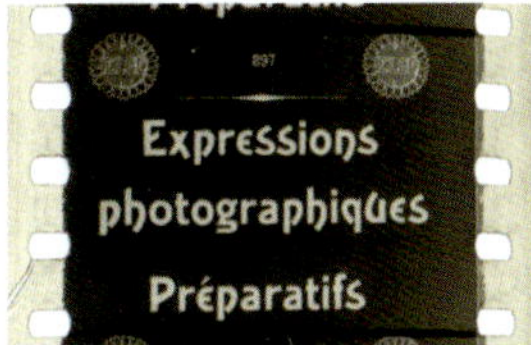

*Intertitle (French).
Expressions photographiques – CNC

1909

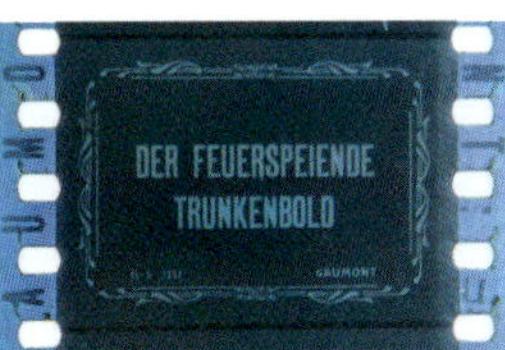

*Main title (German).
Le Poivrot incendiaire – SFI

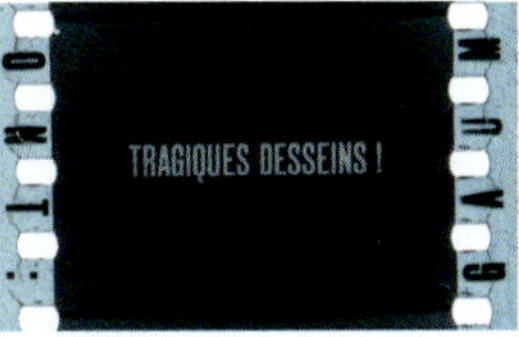

*Intertitle (French).
Les Mille francs de Grenouillard – CNC

1910

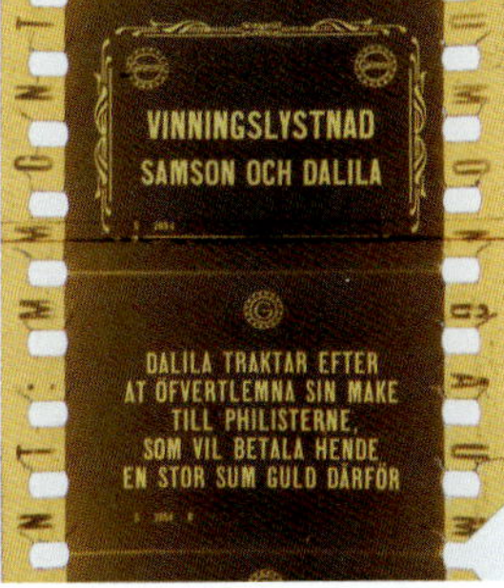

*Main title and intertitle (Swedish). *Les Sept péchés capitaux. L'Avarice. Samson et Dalila* – SFI

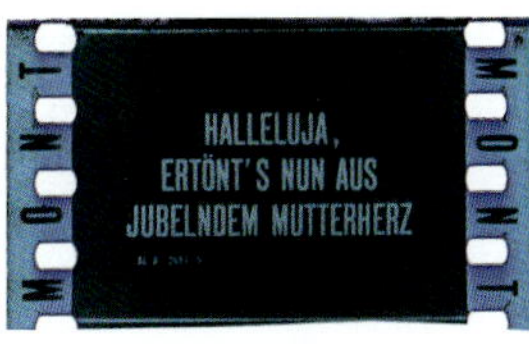

*Intertitle (German).
Pâques florentines – GEM

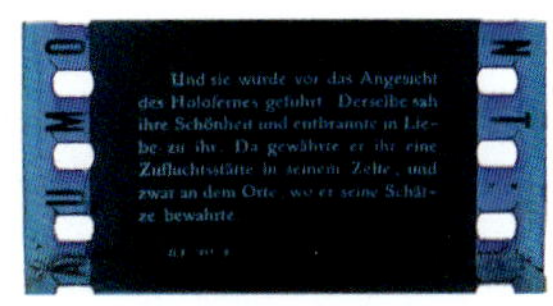

*Intertitle (German).
Judith et Holopherne – GEM

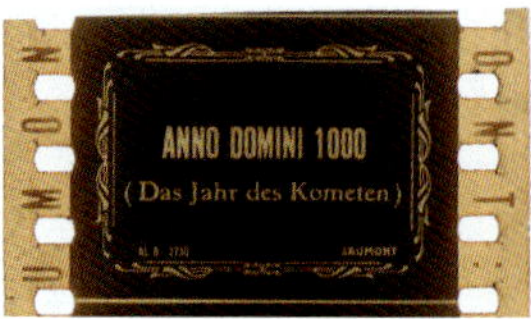

*Main title and intertitle (German). *L'An Mille (Anno Domini 1000)* – GEM

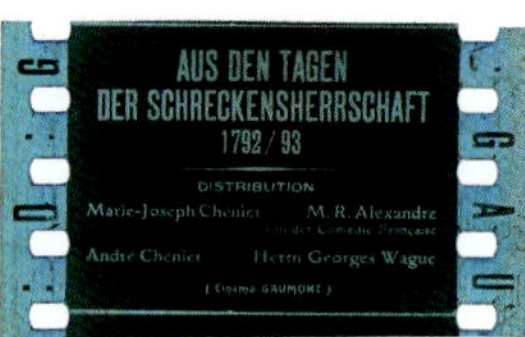

*Main title (German). *André Chénier* – GEM

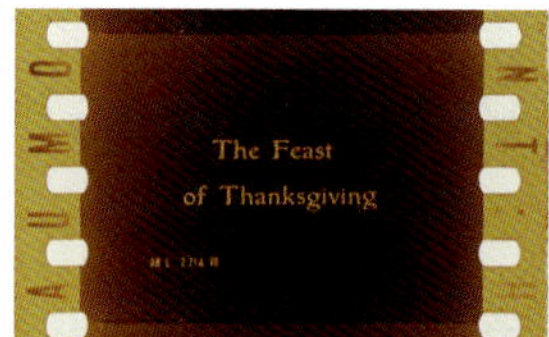

Intertitle (English). *Esther* – BFI

1911

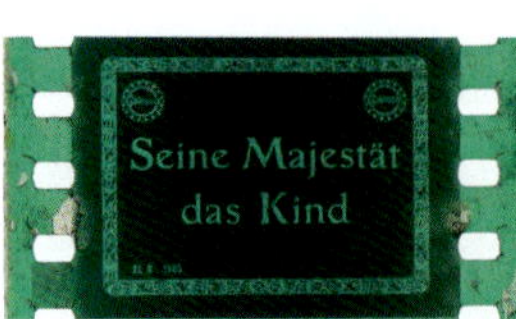

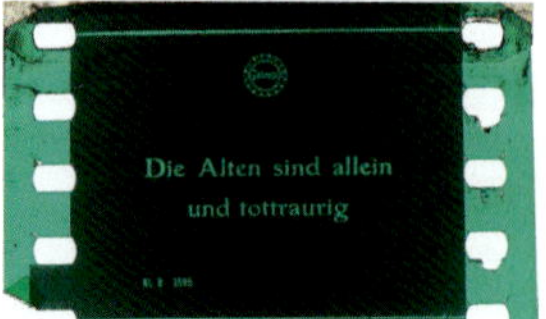

*Main title and intertitle (German). *Le Moïse du moulin* – GEM

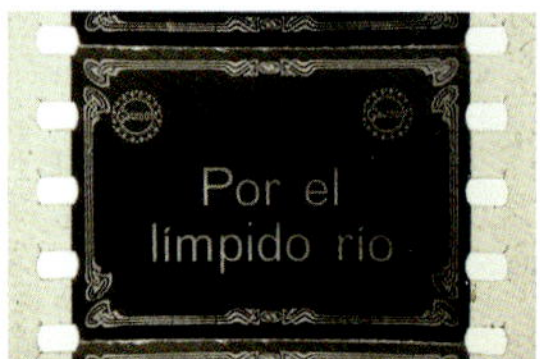

*Main title and intertitle (Spanish). *Sur la jolie rivière* – FE

1912

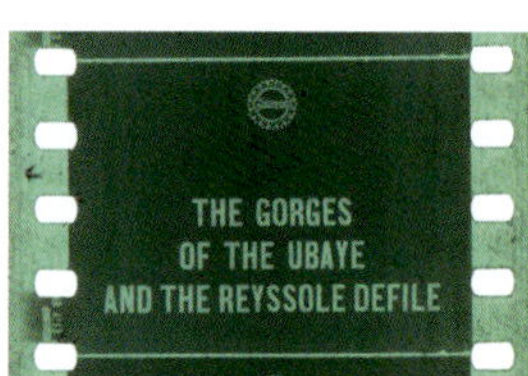

*Intertitle (English). *De Saint Paul à Entrevaux* – CNC

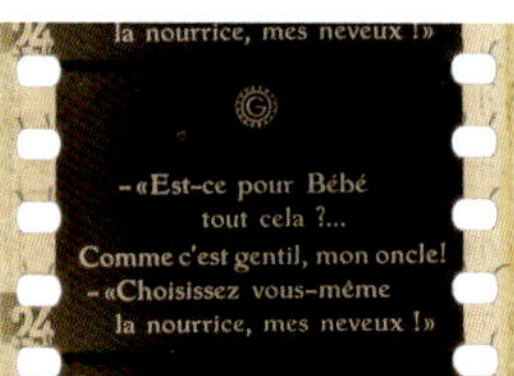

*Intertitle (French). *Oncle Thomas tombe en enfance* – CNC

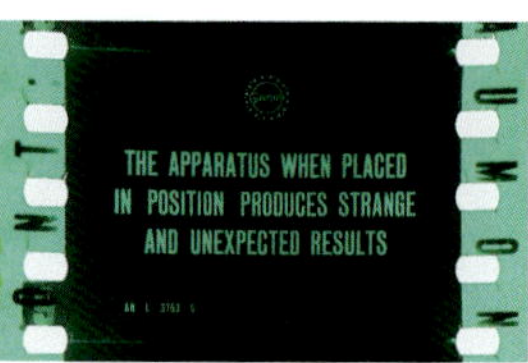

Intertitle (English). *Calino courtier en paratonnerres* – BFI

1913

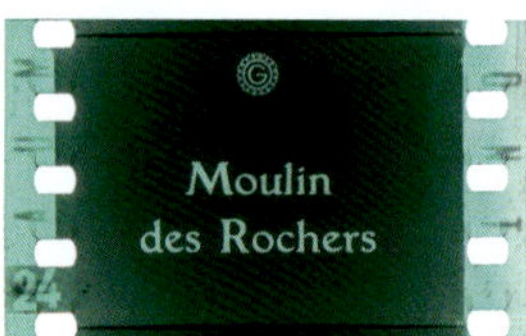

*Intertitle (French). *La Vallée de Chevreuse* – CNC

Cines

1908

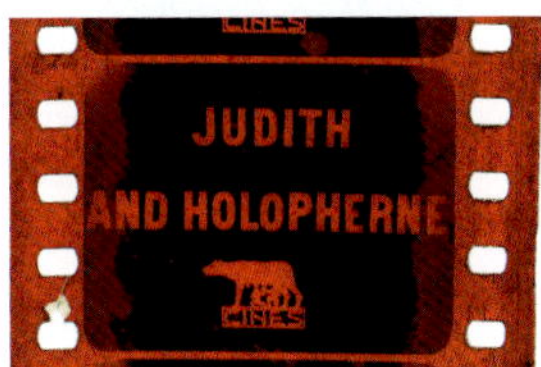

*Main title (English).
Giuditta e Oloferne – HB

1909

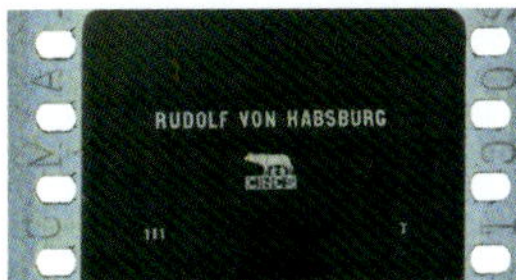

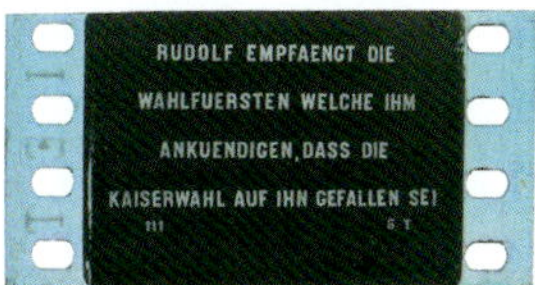

*Main title and intertitle (German). *Rodolfo d'Asburgo* – GEM

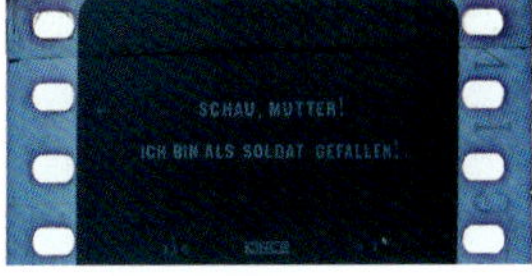

*Intertitle (German).
Il piccolo garibaldino – GEM

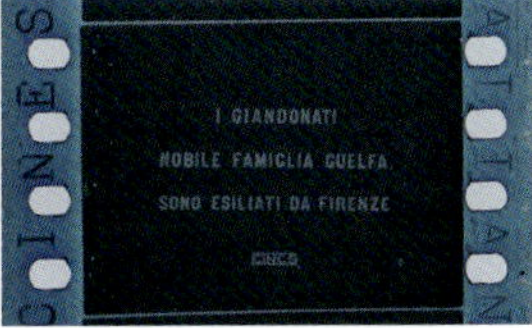

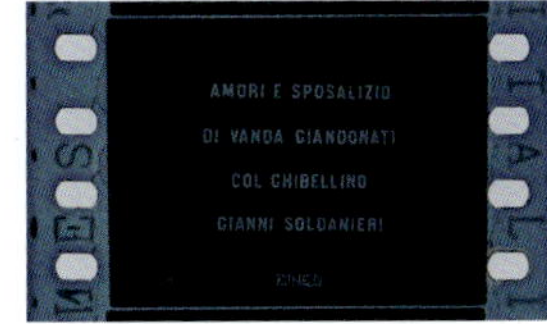

*Intertitles (Italian). *Guelfi e ghibellini* – CdB

1910

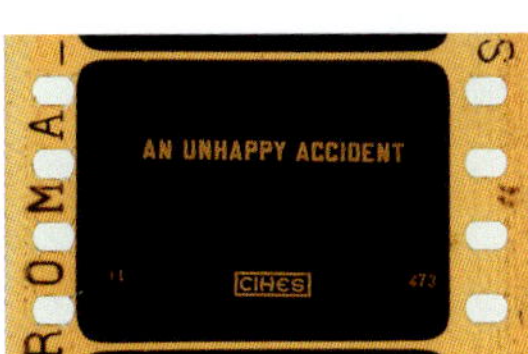

*Intertitle (English).
Amore sentimentale – CdB

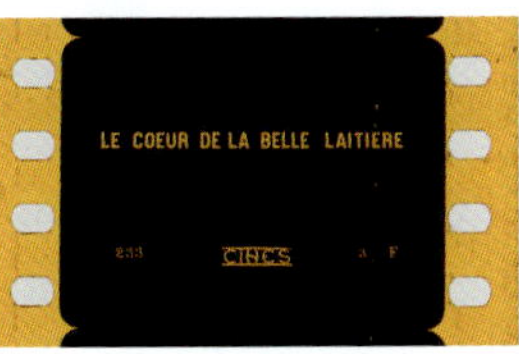

*Intertitle (French).
La bella lattaia – CdB

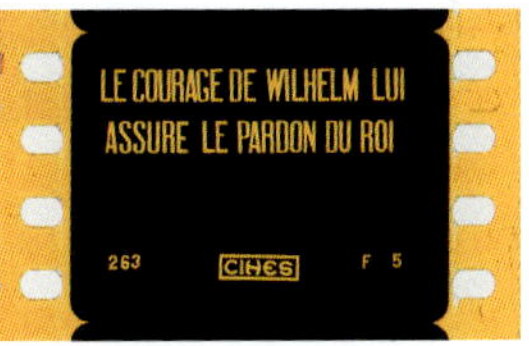

*Intertitle (French).
La spada di legno – CdB

1910-1911

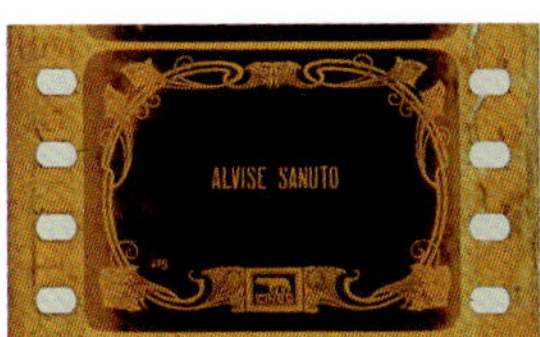

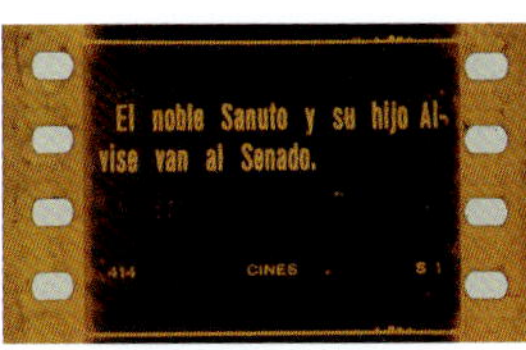

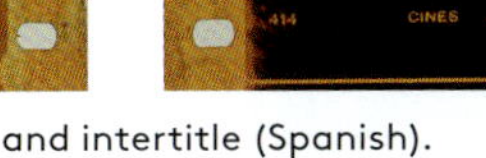

Main title and intertitle (Spanish).
Alvise Sanuto – BFI

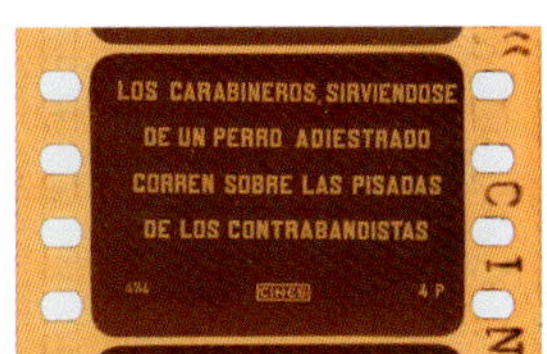

Main title and intertitle (Spanish).
Dramma alla frontiera – BFI

1911

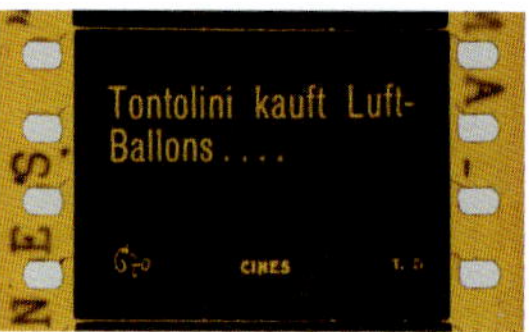

Intertitle (German).
Tontolini e l'asino – BFI

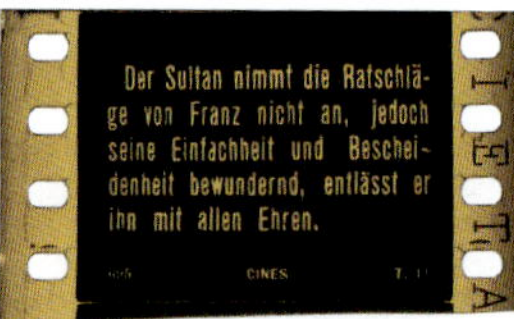

*Intertitle (German).
Il poverello di Assisi – GEM

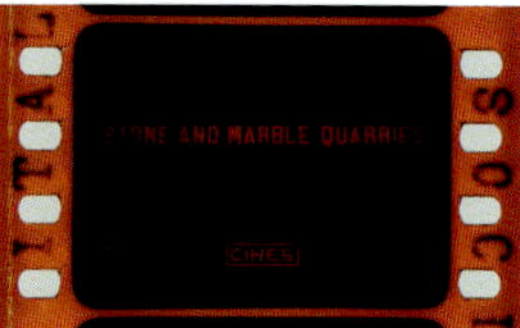

*Intertitle (English).
Massafra – CdB

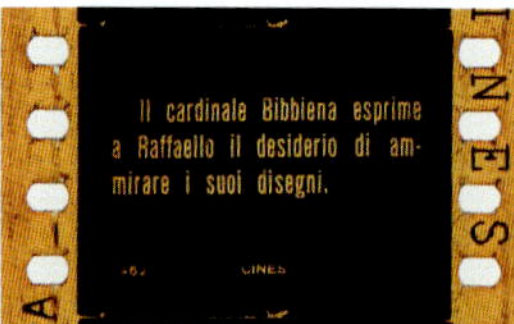

*Intertitle (Italian).
Raffaello e la Fornarina – CdB

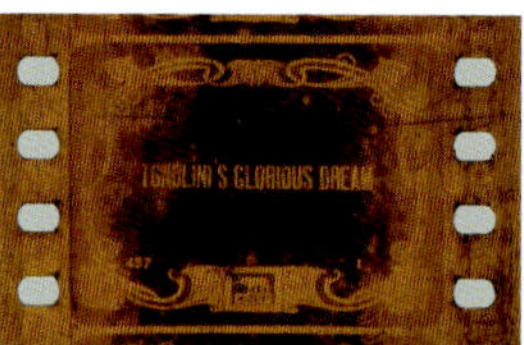

*Main title and intertitle (English).
Un sogno di gloria di Tontolini – CdB

1912

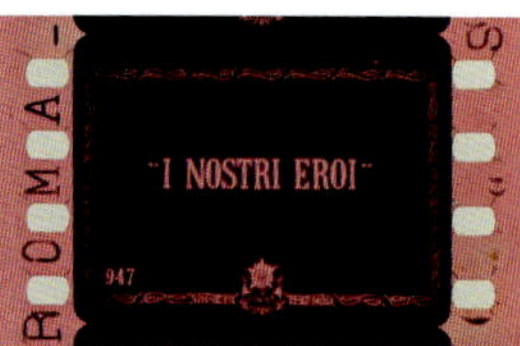

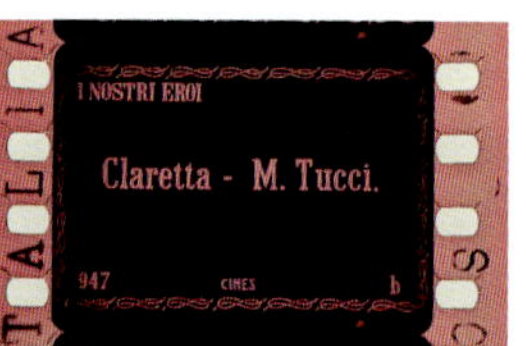

*Main title and intertitle (Italian).
Le medaglie di Bidoni (I nostri eroi) - CdB

*Intertitle (Italian).
Salvata – CdB

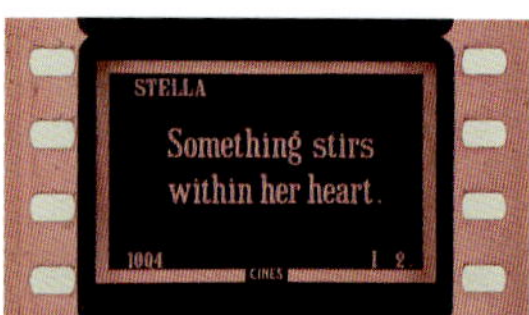

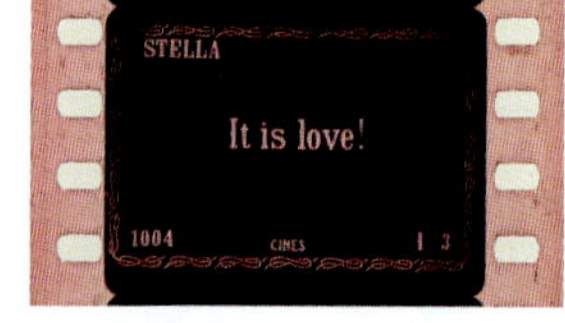

*Intertitles (English).
Stella marina / Stella – CdB

1913

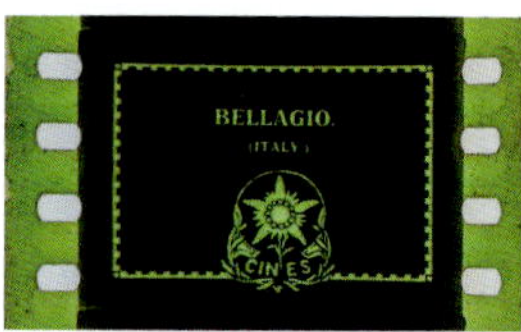

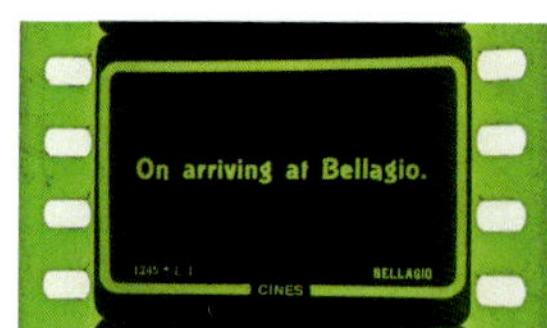

Main title and intertitle (English).
Bellagio – BFI

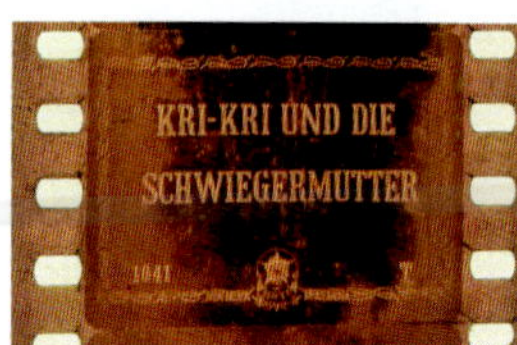

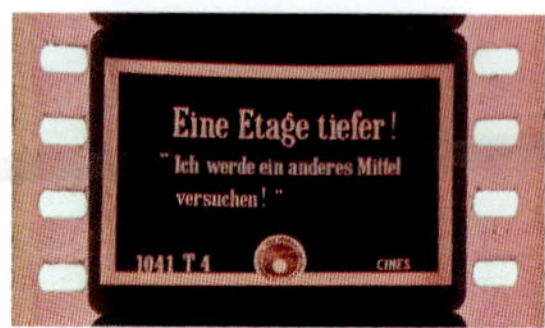

Main title and intertitle (German).
Kri Kri e la suocera – BFI

Vitagraph

1909

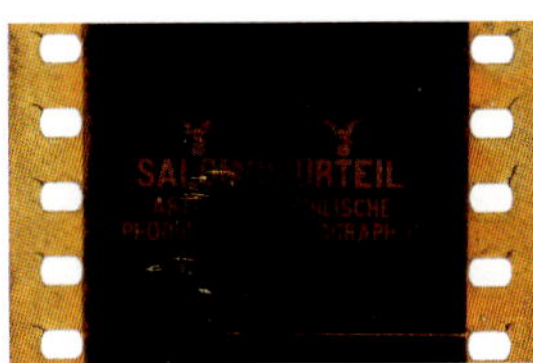

*Main title (German). *Solomon's Judgement* – HB

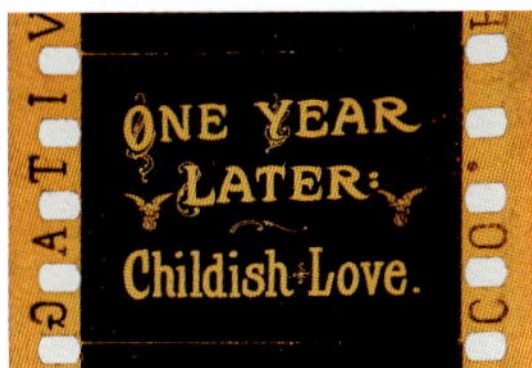

Intertitle (English). *An Alpine Echo* – BFI

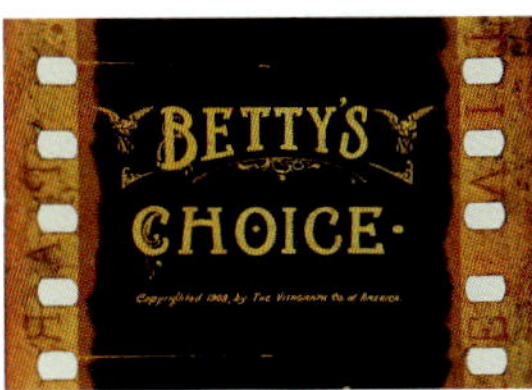

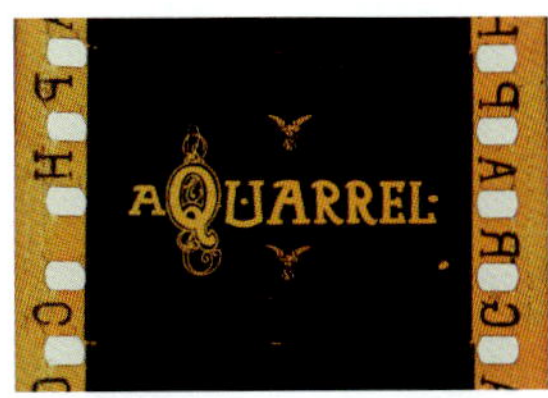

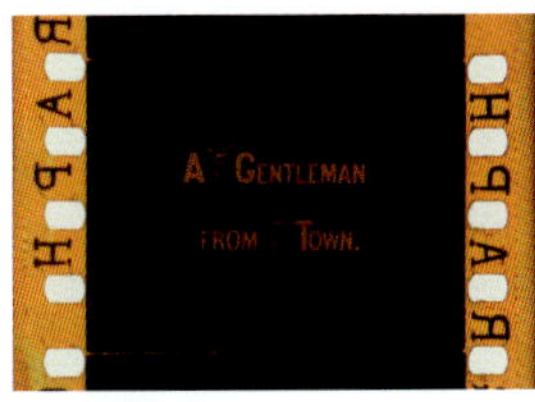

Main title and intertitles (English). *Betty's Choice* – BFI

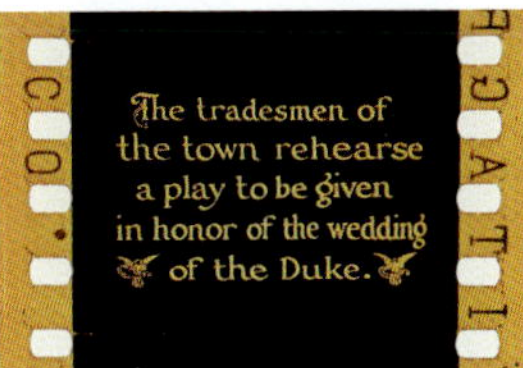

Intertitle (English). *A Midsummer Night's Dream* – BFI

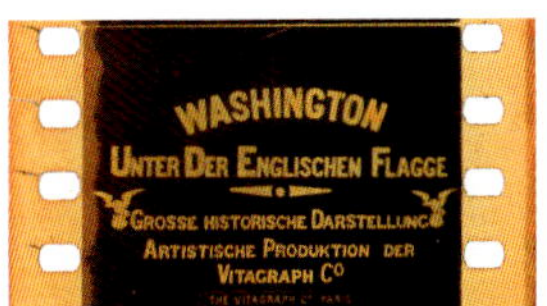

*Main title (German). *Washington Under the American Flag* – GEM

1910

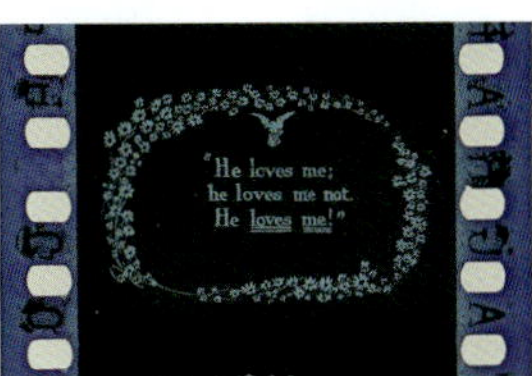

Intertitle (English). *Daisies* – BFI

1911

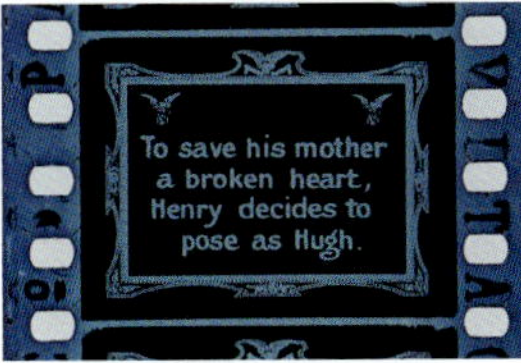

Intertitle (English). *A Dead Man's Honor* – BFI

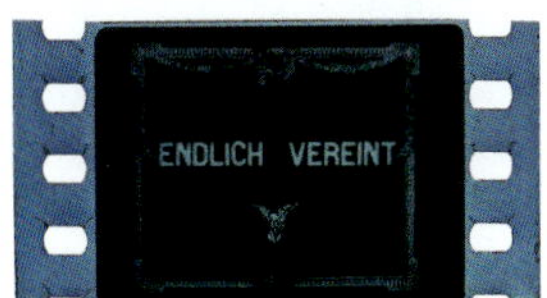

*Intertitle (German). *The Freshet* – GEM

1912

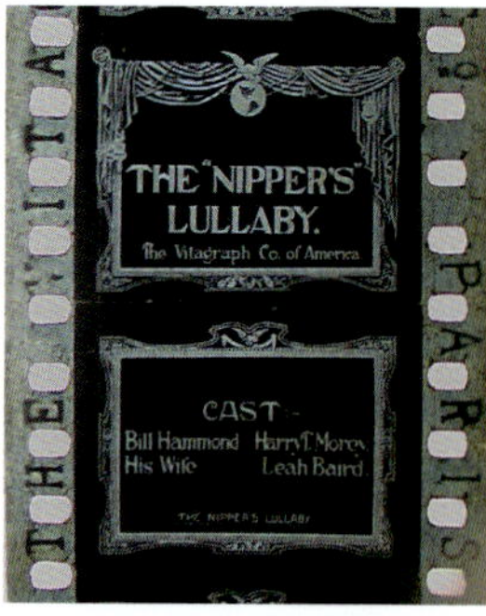

*Main titles (English).
The Nipper's Lullaby – BP

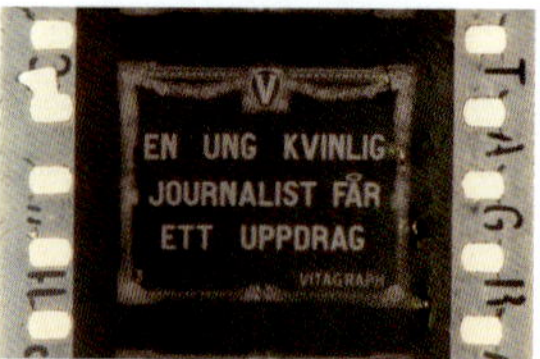

*Intertitle (Swedish).
A Leap-Year Proposal – SFI

1913

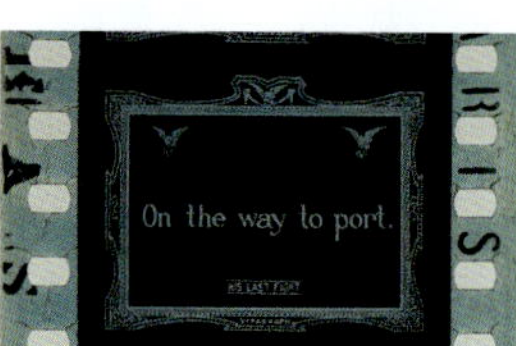

Intertitle (English).
His Last Fight – BFI

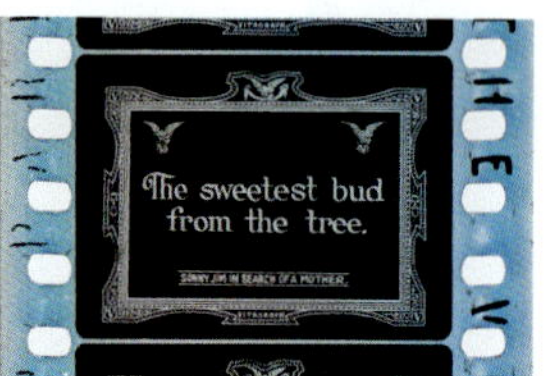

*Intertitle (English).
Sonny Jim in Search of a Mother – AFA

1916

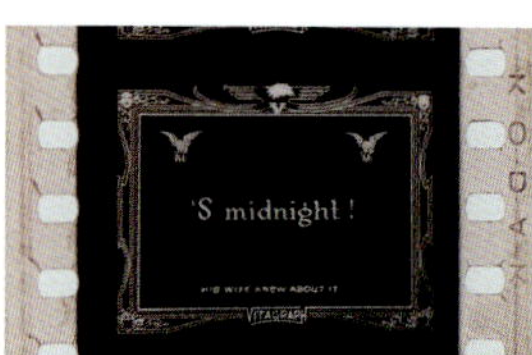

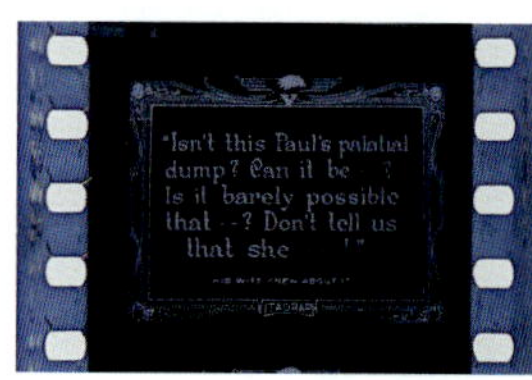

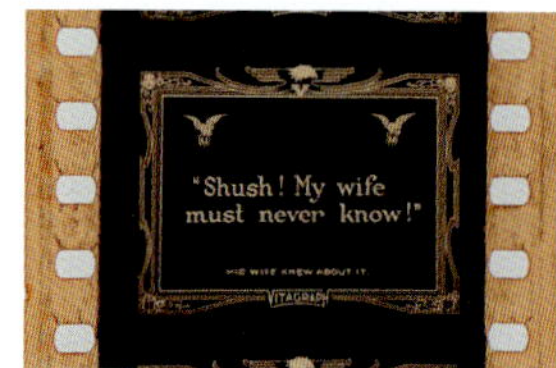

Intertitles (English). *His Wife Knew About It* – BFI.
The titles are tinted the same colour as their related scenes.

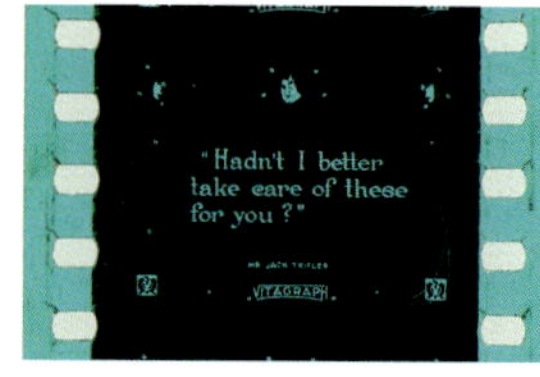

Intertitle (English).
Mr. Jack Trifles – BFI

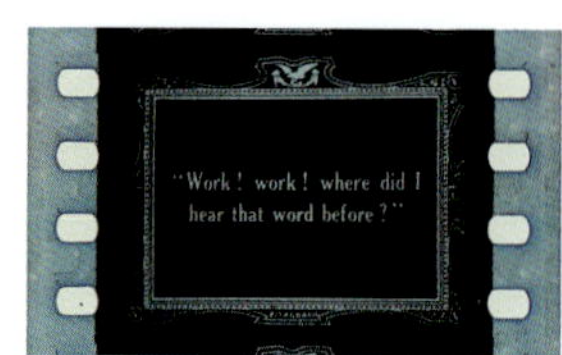

Main titles and intertitle (English). *The New Porter* – BFI

1917

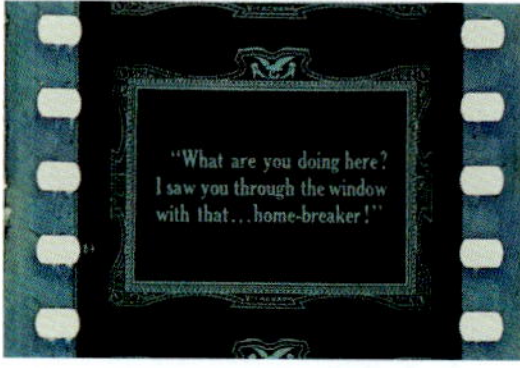

Intertitle (English).
His Lesson – BFI

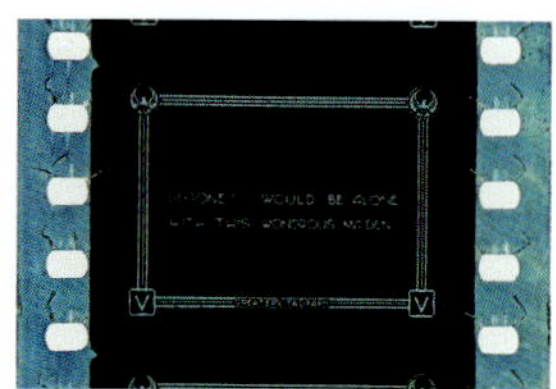

Intertitle (English).
Turks and Troubles – BFI

1918

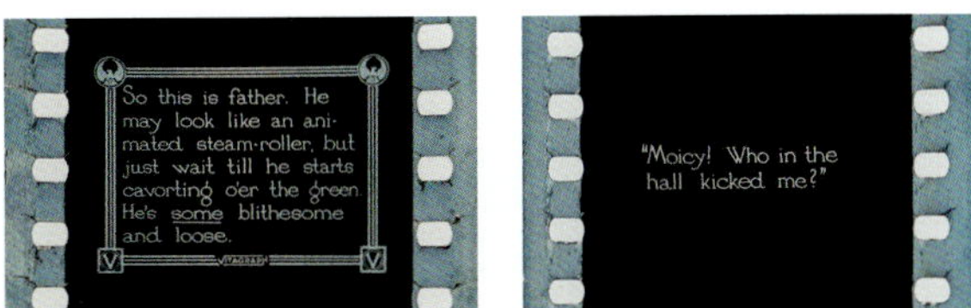

Intertitles (English). *Skids and Skalawags* – BFI

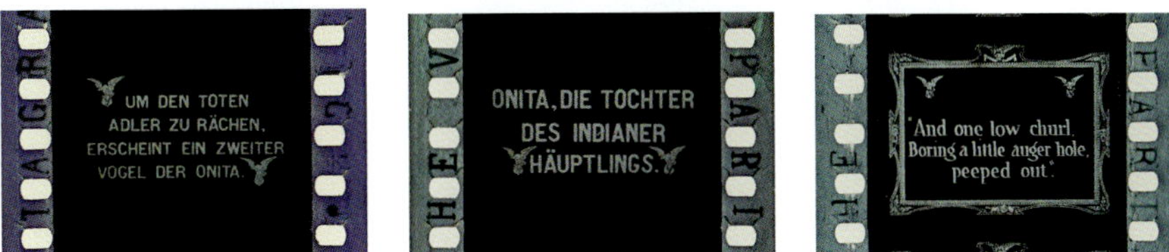

Intertitles (English). *Red Eagle* – BFI

Thanhouser

1910

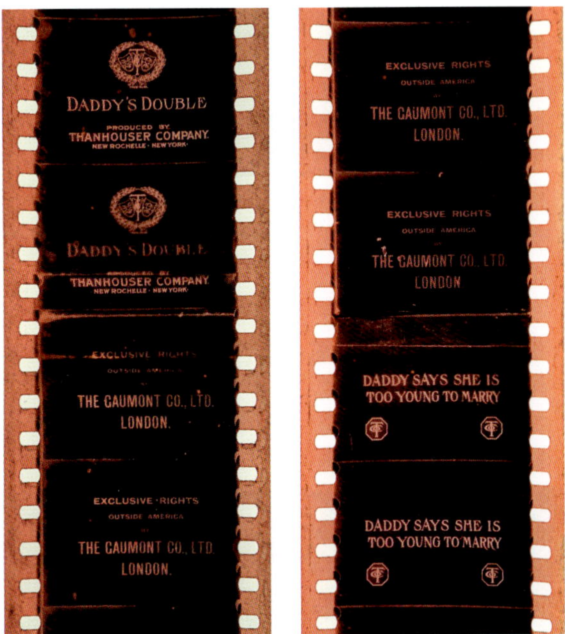

Main titles and intertitle (English).
Daddy's Double – BFI

1910-1911

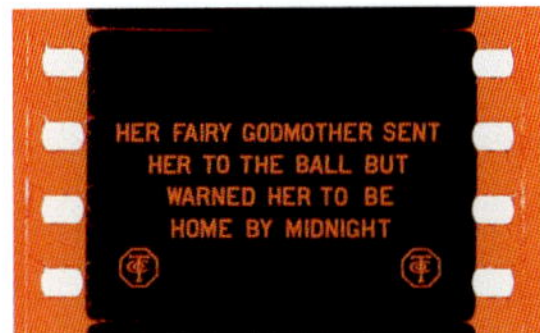

Intertitle (English).
Cinderella – BFI

1912

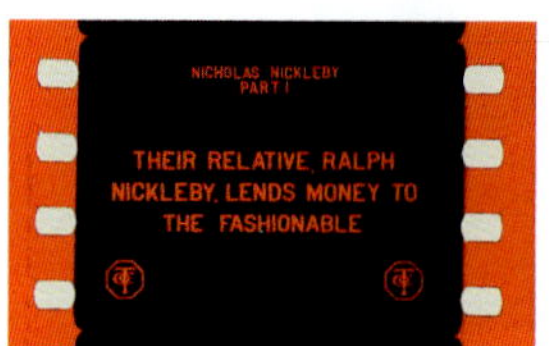

Intertitle (English).
Nicholas Nickleby – BFI

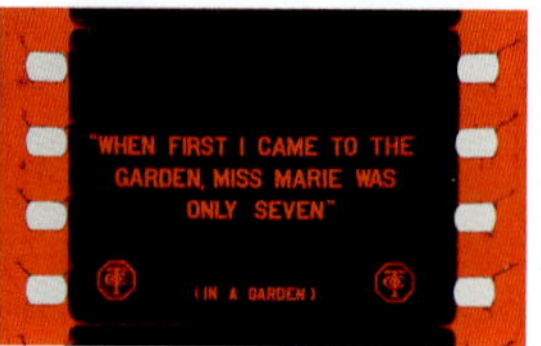

Intertitle (English).
In a Garden – BFI

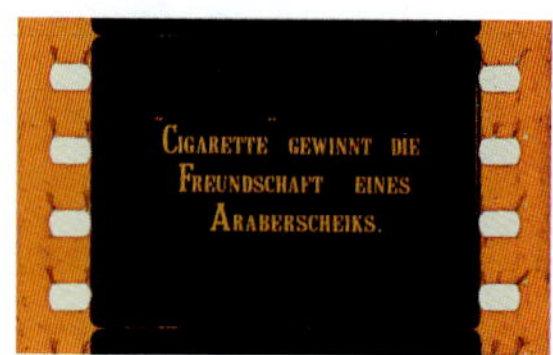

Intertitle (German).
Under Two Flags – BFI

1912-1913

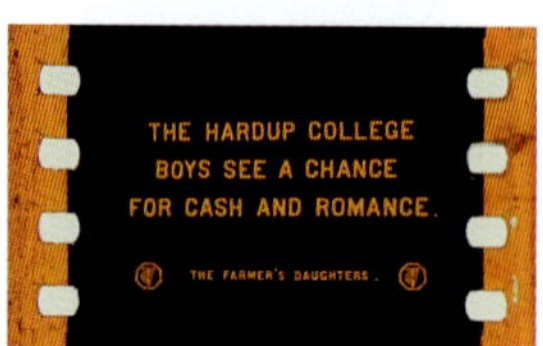

Intertitle (English).
The Farmer's Daughters – BFI

1914

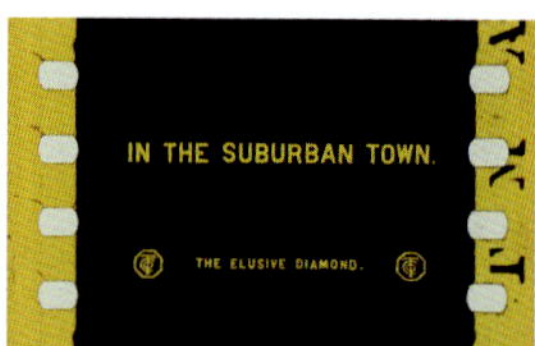

Intertitle (English).
An Elusive Diamond – BFI

Selig

1911

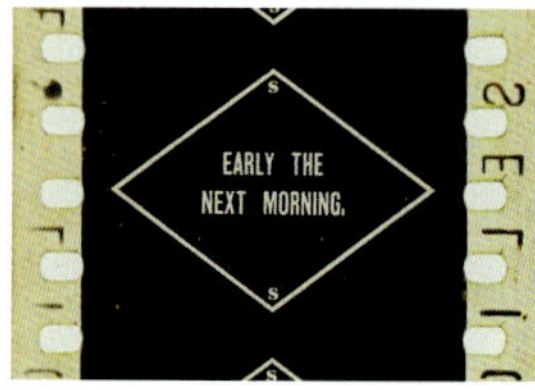

Intertitle (English).
One Hundred Years After – BFI

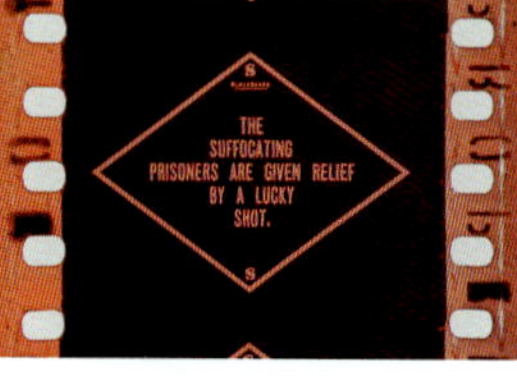

Intertitle (English).
Blackbeard – BFI

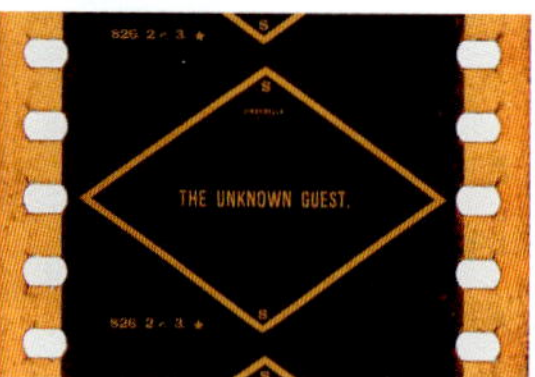

Intertitle (English).
Cinderella – BFI

1912

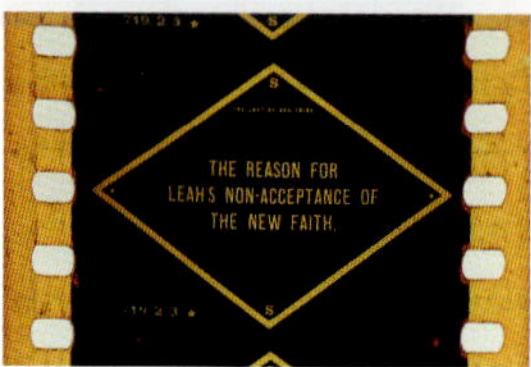

Intertitle (English).
The Last of Her Tribe – BFI

1913

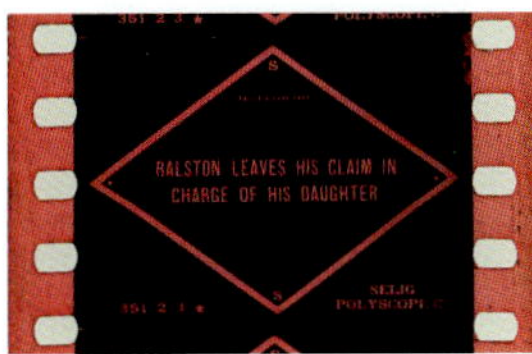

Intertitle (English).
Sallie's Sure Shot – BFI

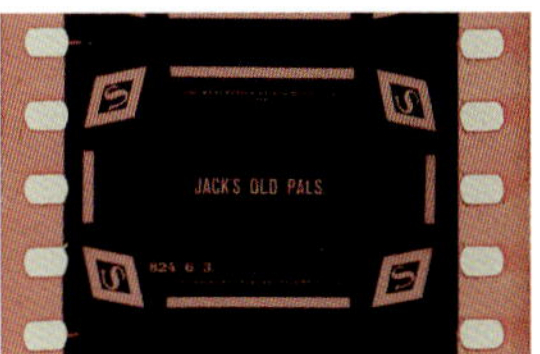

Intertitle (English).
The Redemption of Railroad Jack – BFI

Éclair

1909

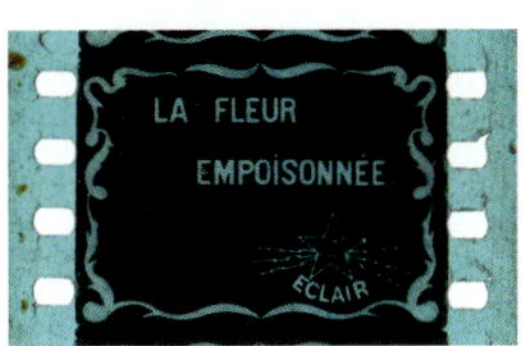

*Main title (French).
La Fleur empoisonnée – CNC

1910

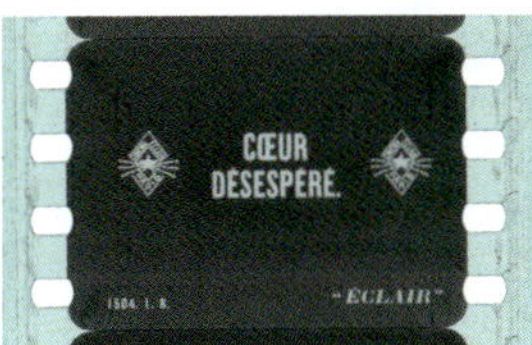

*Intertitle (French).
Eugénie Grandet – CNC

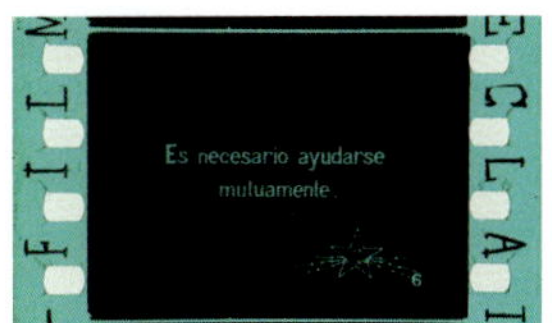

*Intertitle (Spanish).
Une chasse fructueuse – BFI

1911

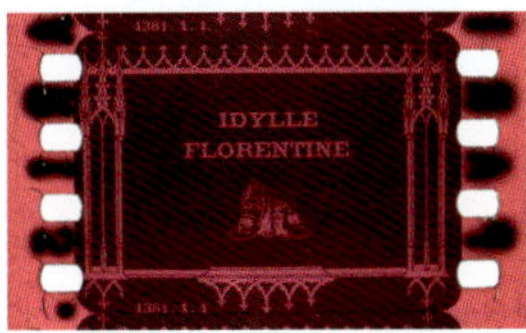

*Main title (French).
Idylle florentine – CNC.
This film was produced by Film d'Art
but distributed by Éclair.

1911-1912

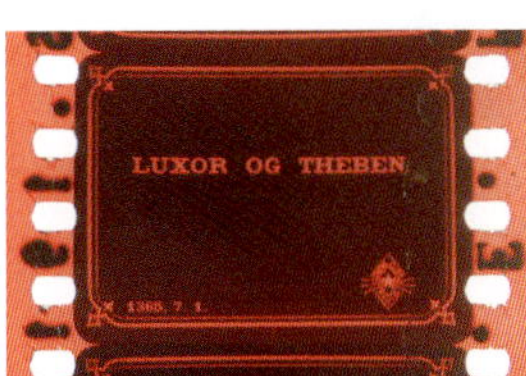

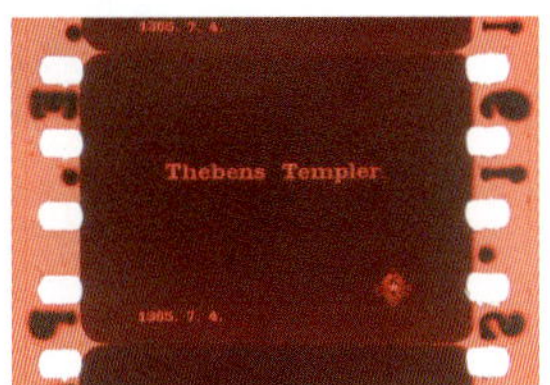

*Main title and intertitle (Danish).
Louqsor et Thèbes – SFI

1912

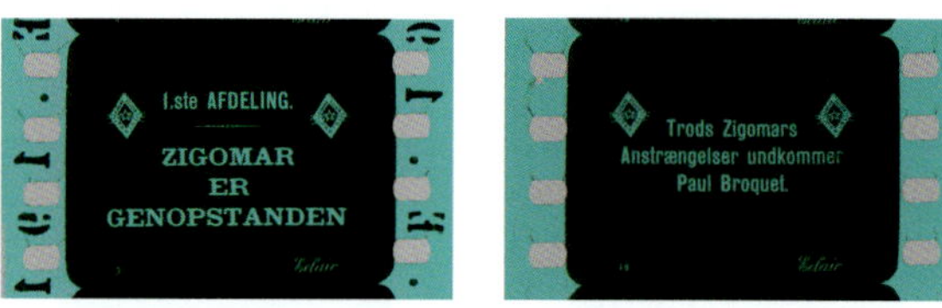

*Main title (episode) and intertitle (Danish).
Zigomar contre Nick Carter – SFI

1913

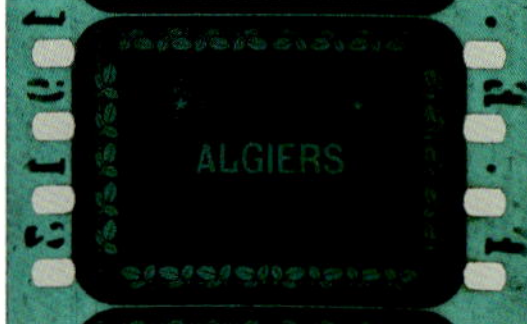

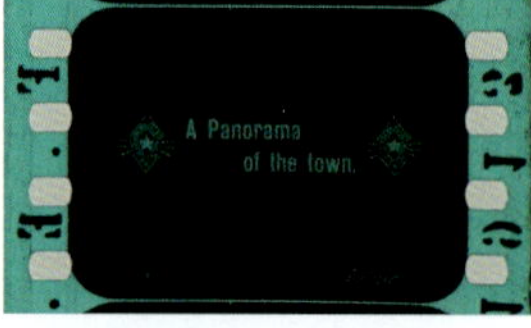

Main title and intertitle (English).
Alger la blanche – BFI

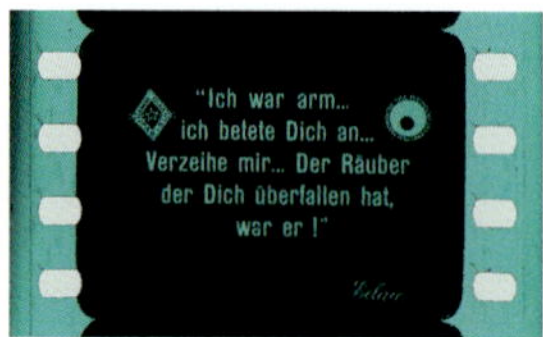

Intertitle (German).
Gontran et son complice – BFI

1914

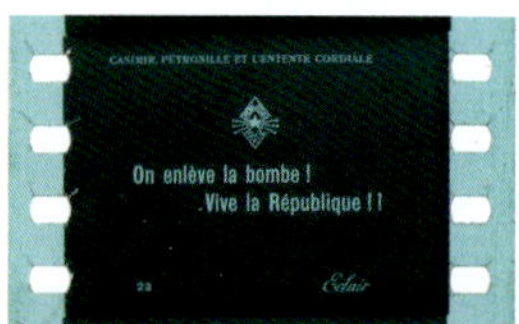

*Intertitle (French).
Casimir, Pétronille et l'Entente Cordiale – CNC

1915

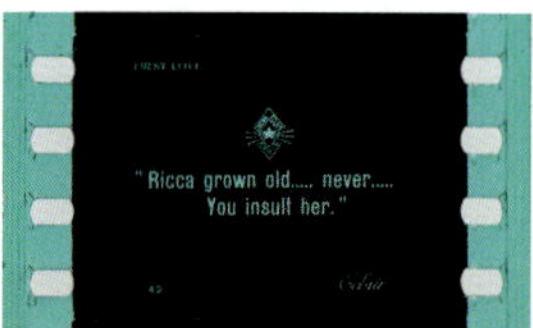

Intertitle (English).
Premier Amour – BFI

*American Standard (Éclair)

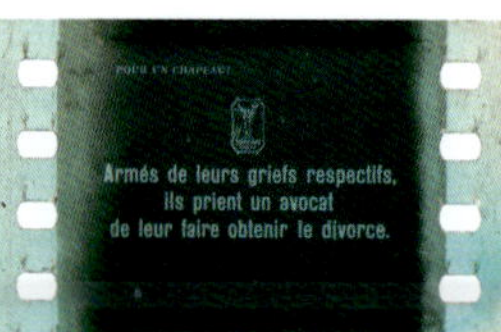

Intertitle (French).
Pour un chapeau (ca. 1913) – CNC

*Scientia (Éclair)

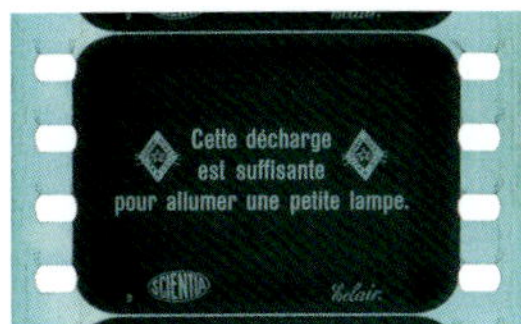

Intertitle (French).
La Torpille (1913) – CNC

Pasquali

1911

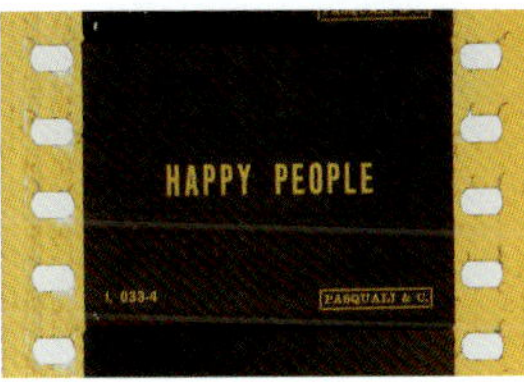

Intertitle (English).
L'isola di Helgoland – BFI

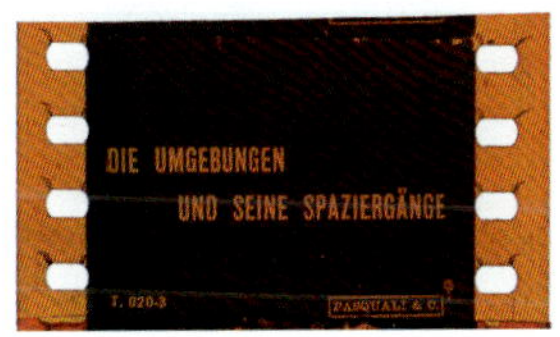

*Intertitle (German).
Primavera a Sanremo – GEM

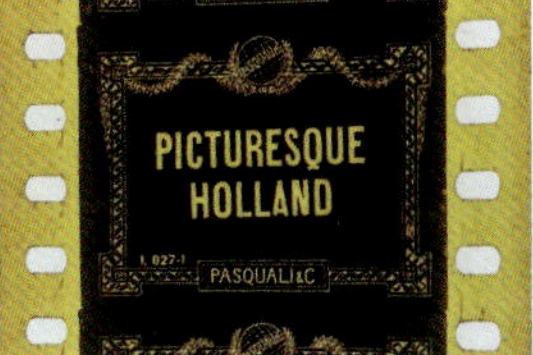

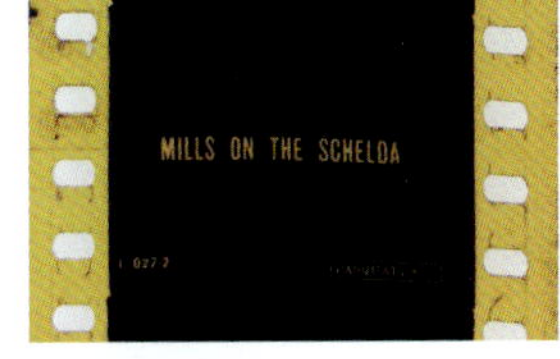

Main title and intertitle (English).
L'Olanda pittoresca – BFI

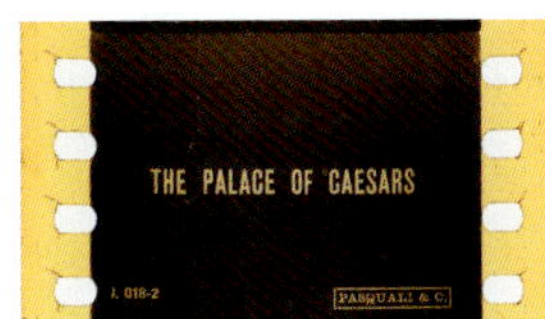

Intertitle (English).
La città eterna – BFI

1912

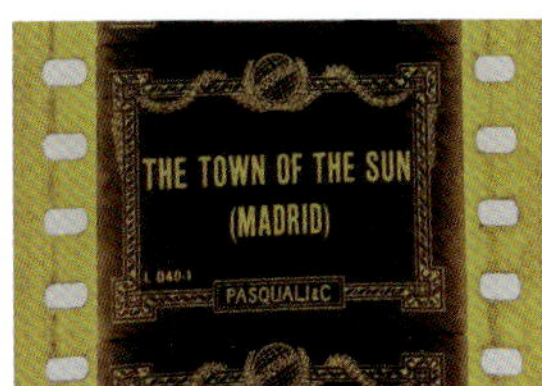

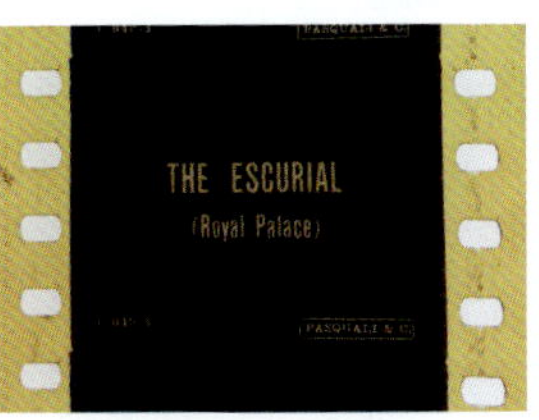

Main title and intertitle (English).
Madrid, la città del sole – BFI

Intertitle (English).
Polidor al club della morte – BFI

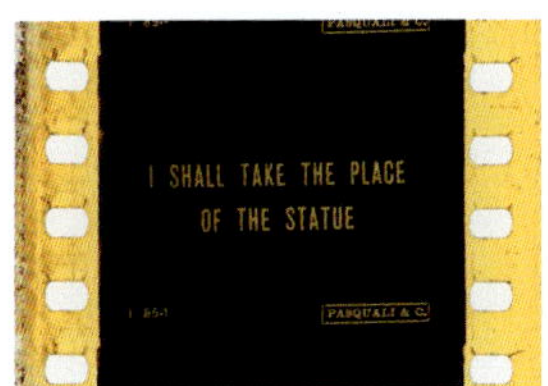

Intertitle (English).
Polidor statua – BFI

*Ambrosio

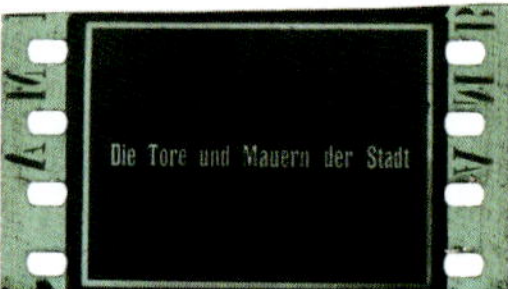

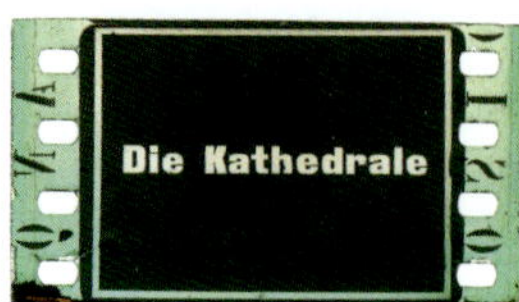

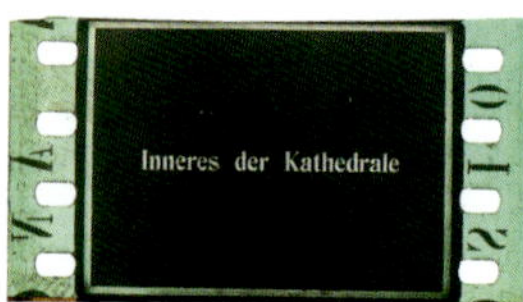

Intertitles (German).
Ani. La città dalle mille chiese – GEM

*Eclipse

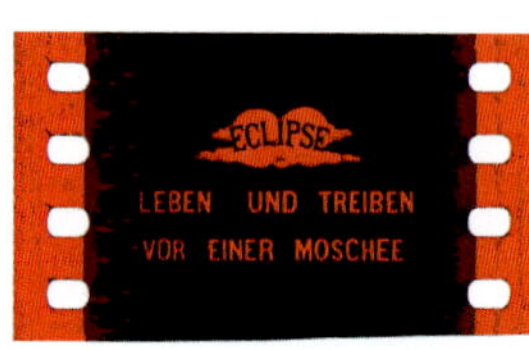

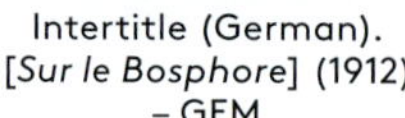
Intertitle (German).
[*Sur le Bosphore*] (1912)
– GEM

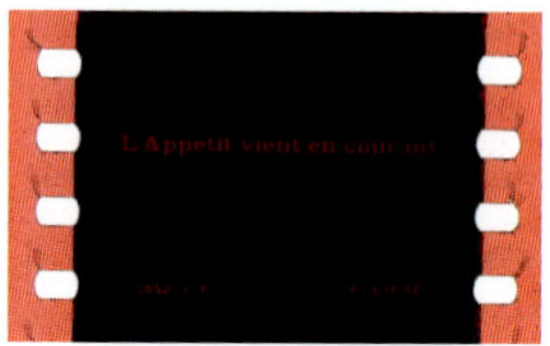

Intertitle (French).
Une idylle montmartroise
(1912) – CNC

*Edison

1907

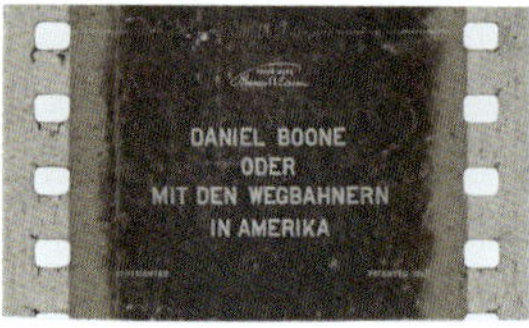

Main title (German).
Daniel Boone (or Pioneer Days in America) – GEM

1909

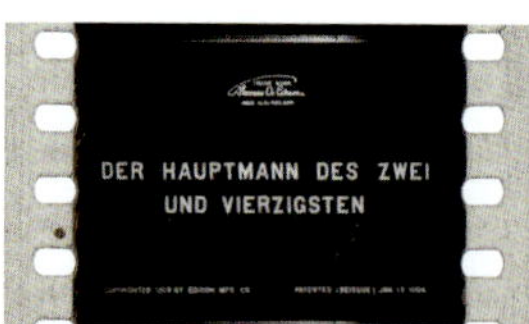

Main title (German).
Fenton of the 42nd – GEM

1910

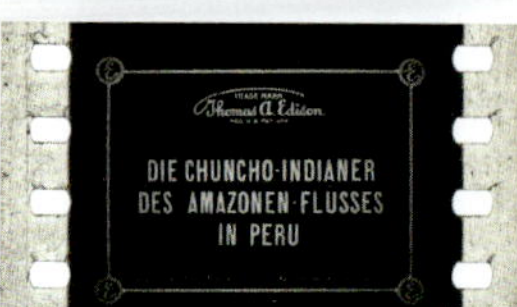

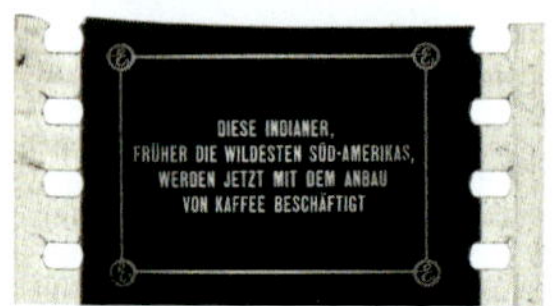

Main title and intertitle (German).
The Chuncho Indians of the Amazon River, Peru (re-issue) – GEM

1911

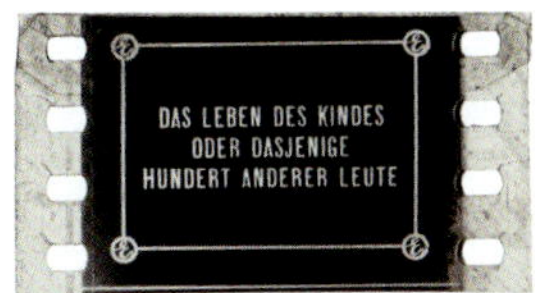

Intertitle (German).
The Switchman's Tower – GEM

1912

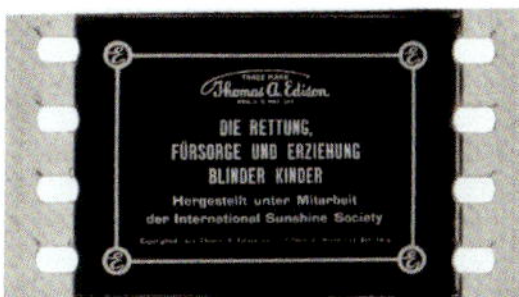

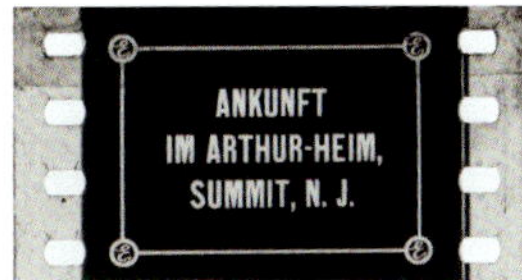

Main title and intertitle (German).
The Rescue, Care and Education of Blind Babies (re-issue) – GEM

*Le Lion

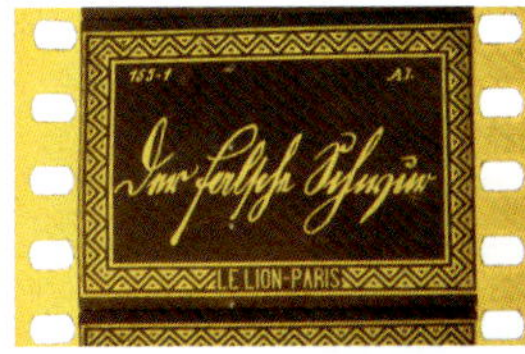

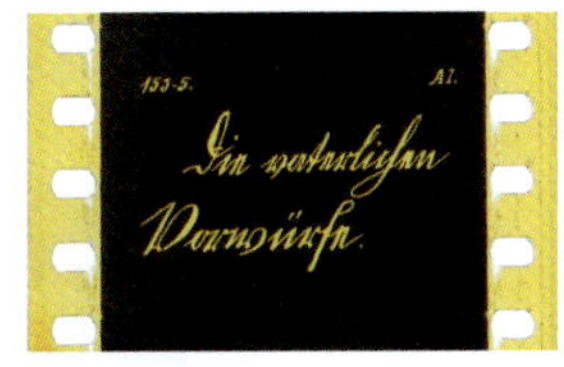

Main title and intertitle (German).
[*Le Faux Serment*] (1908) – SFI

*Lubin

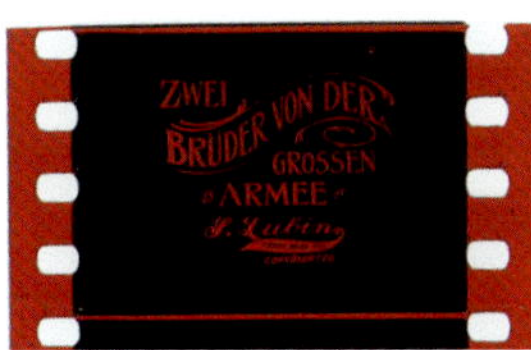

Main title (German).
Two Brothers of the G.A.R.
(1908) – GEM

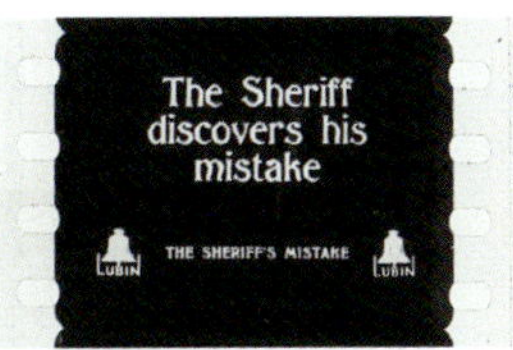

Intertitle (English).
The Sheriff's Mistake
(1912) – HB

*Lux

1907

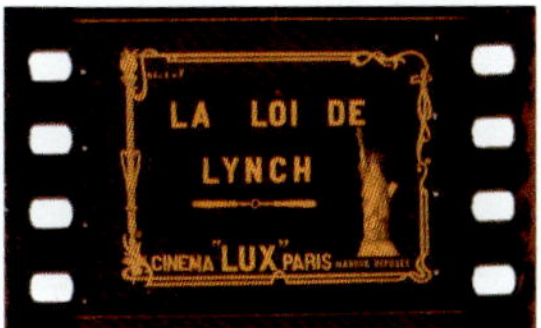

Main title (French).
La Loi de Lynch – CNC

1909

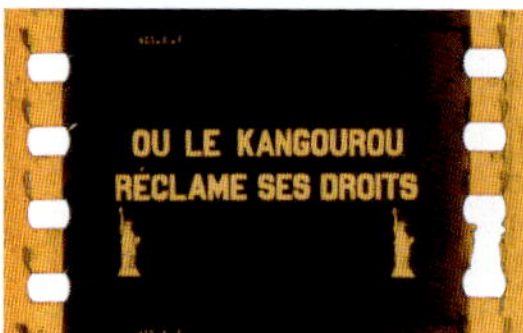

Intertitle (French).
Un monsieur qui a mangé du kangourou – CNC

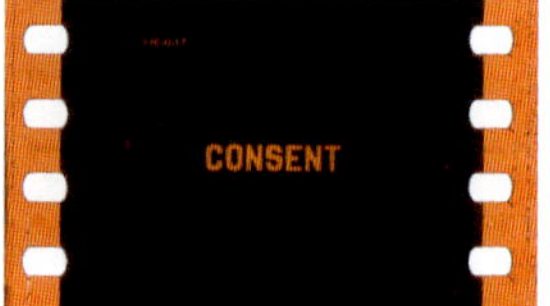

Intertitle (English).
L'Enlèvement – CNC

1913

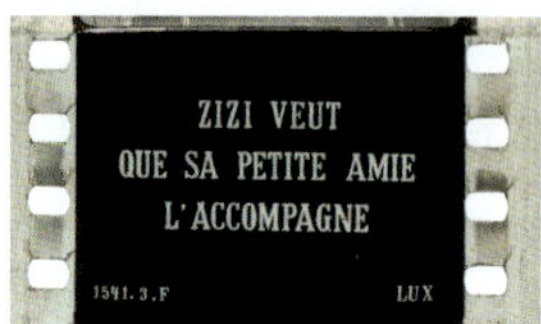

Intertitle (French).
Zizi fait des courses – CNC

*Messter

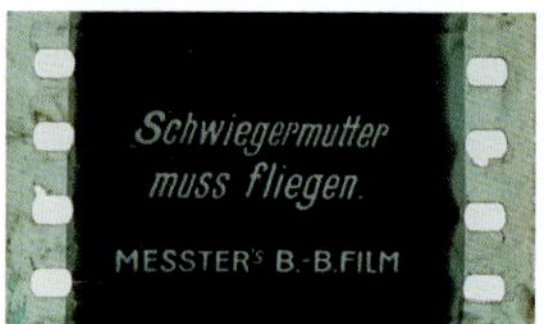

Main title (German).
Schwiegermutter muss fliegen (1909) – BFI

*Mirror Films

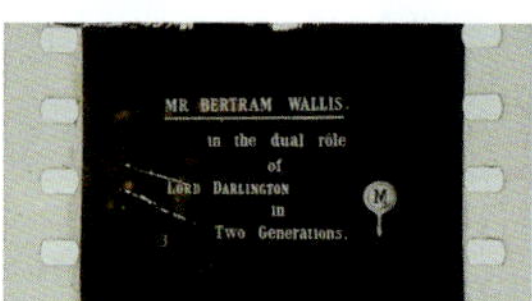

Credit titles (English).
The Cost of a Kiss (1917) – BFI

*Nordisk Films Kompagni

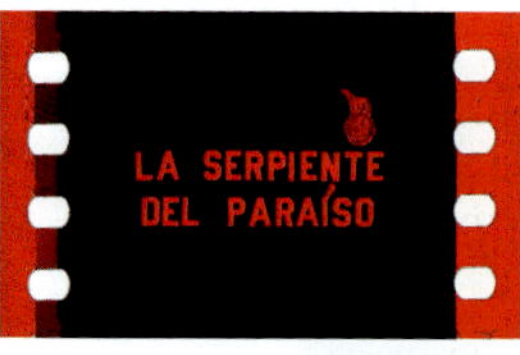

Intertitle (Spanish).
De to Guldgravere (1909) – FE

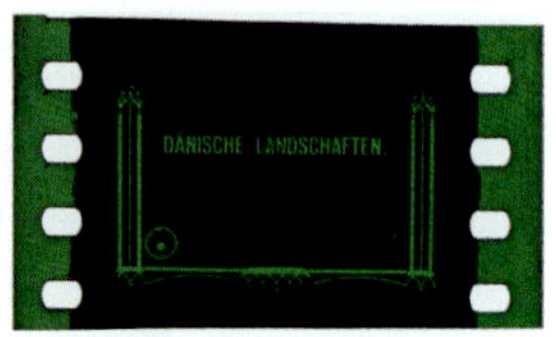

Main Title (German).
Dänische Landschaften (1912) – GEM

*Pathé Frères

1902

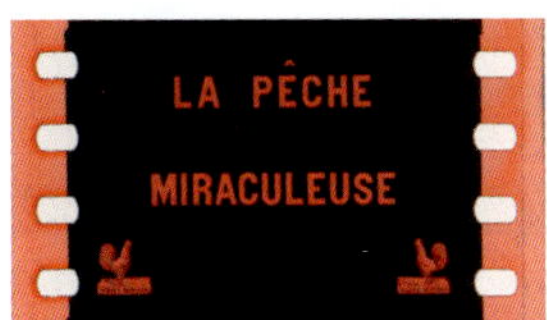

Main title (French).
La Pêche miraculeuse – FE

1903

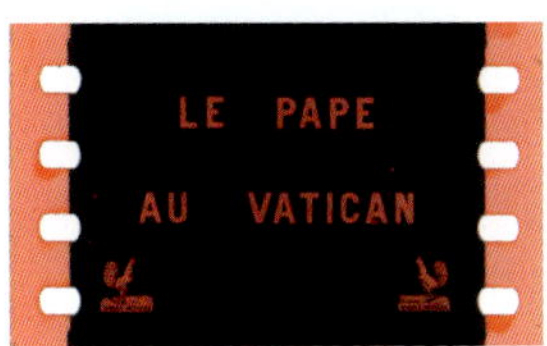

Main title (French).
Le Pape au Vatican – FE

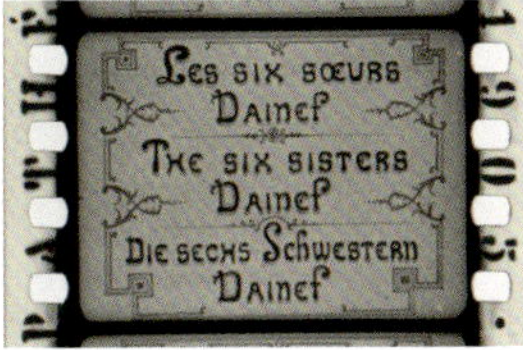

Main title (French, English, German). *Les six soeurs Dainef* – FE. Print dated 1905.

1904

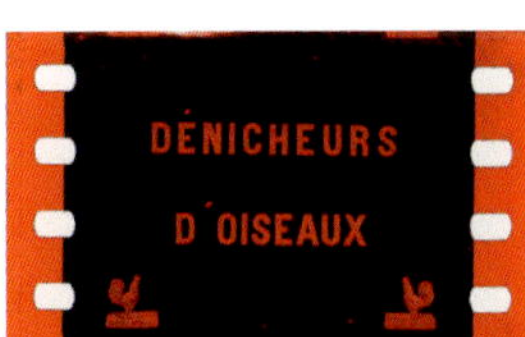

Main title (French).
Les Dénicheurs d'oiseaux – FE

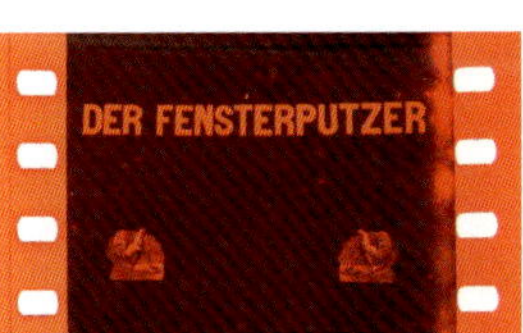

Main title (German).
Le Laveur de devantures – SFI

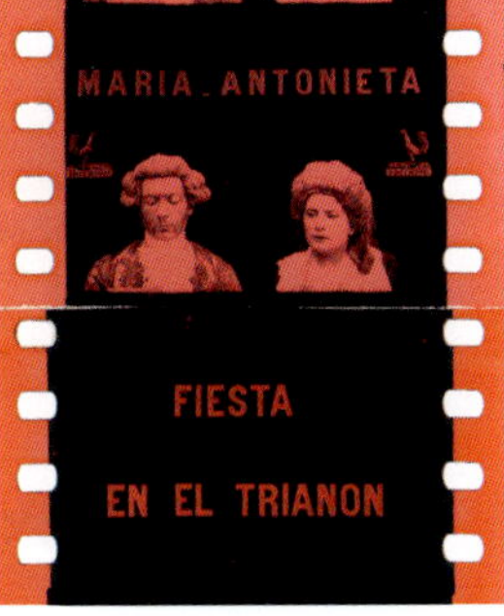

Main title and intertitle (Spanish). *Marie-Antoinette* – FE

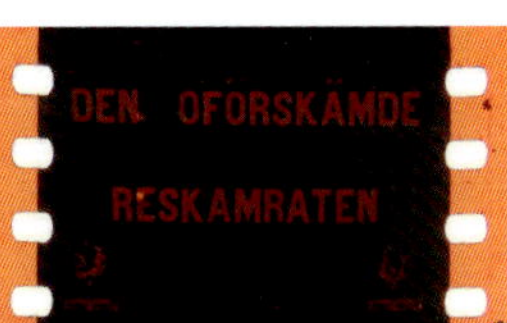

Main title (Swedish).
Voyageur peu gêné – SFI.
Print dated 1905.

1905

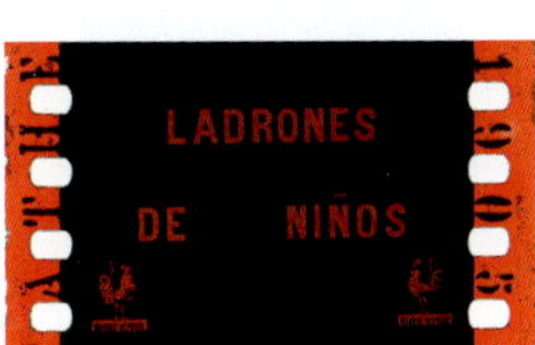

Main title (Spanish).
Les Voleurs d'enfants – FE

1907

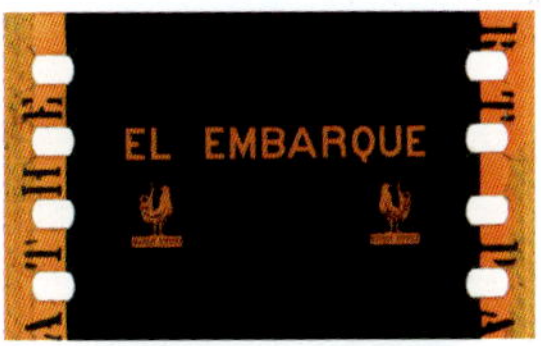

Intertitle (Spanish).
La Veuve du marin – FE

1908

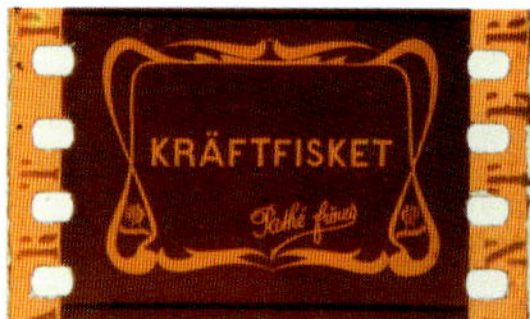

Main title (Swedish).
La Pêcheuse de crevettes
– SFI

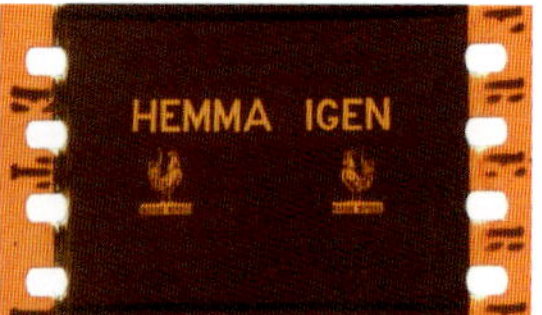

Intertitle (Swedish).
Moine sans vocation – SFI

1909

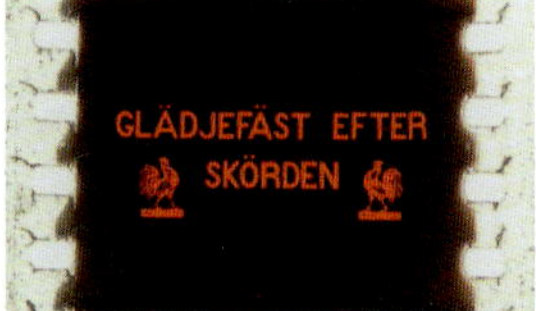

Intertitle (Swedish).
Ruth et Boaz – SFI.
Print dated 1909-1911.

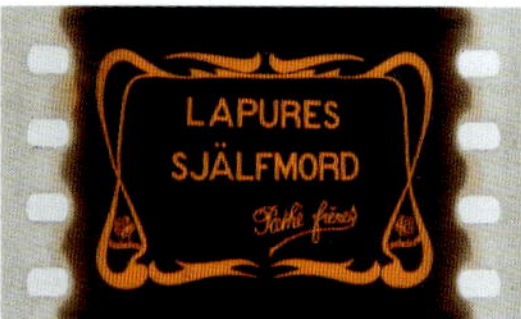

Main title and intertitle (Swedish).
Les Suicides de Lapurée – SFI. Print dated 1909-1911.

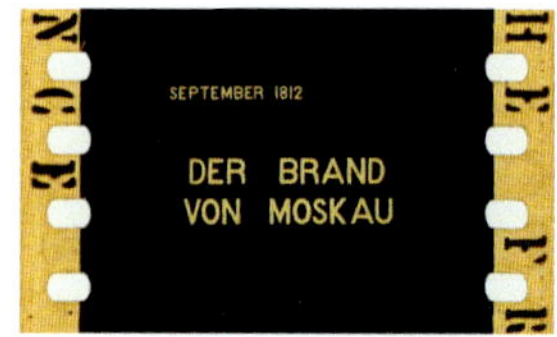

Intertitle (German).
Napoléon – BFI

*Robert W. Paul

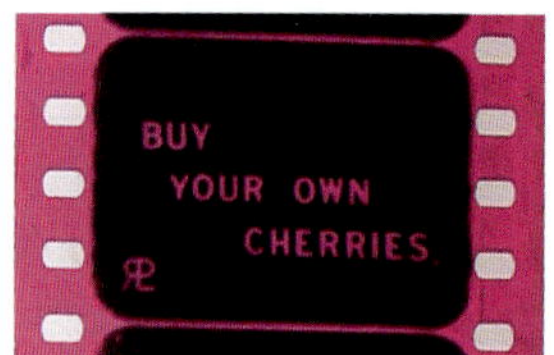

Main title (English).
Buy Your Own Cherries (1904). DC/BP

*Warwick Trading Co.

1906

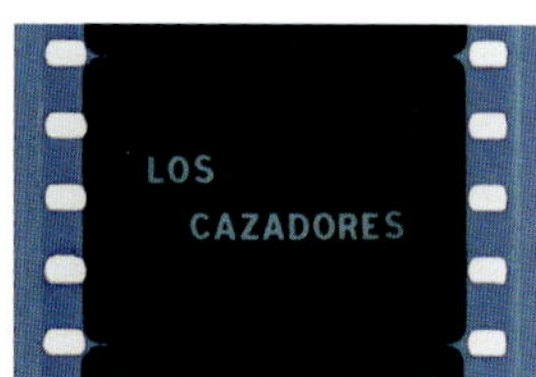

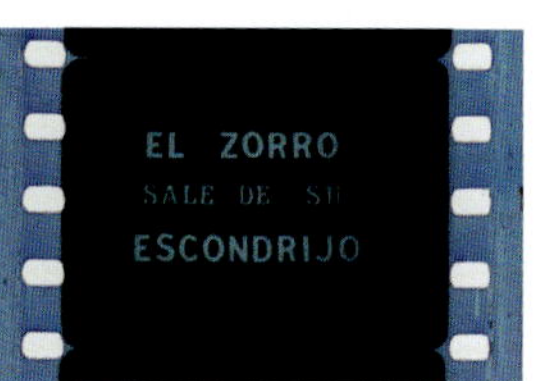

Intertitle (Spanish). *Fox Hunting* – FE

Main title (German).
Tor di Quinto – FE

*Welt-Kinematograph

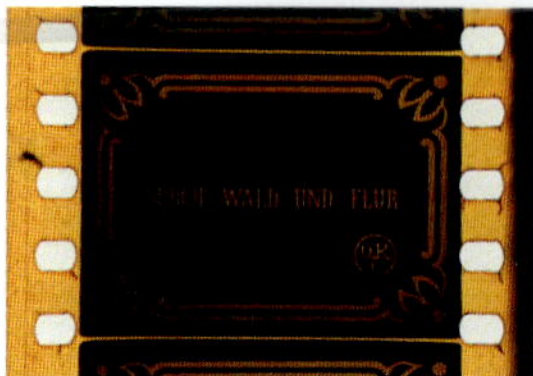

Main title (German).
Durch Wald und Flur (1910) – BFI

PART II

1
Lumière

Text written by Eric Loné,
based on research conducted during the Lumière Project at the CNC in 1995,
with the participation of Jean-Marc Lamotte and Camille Blot-Wellens

History (1892-1905)

1892	Establishment in Lyon of the Société Anonyme des plaques et papiers photographiques Antoine Lumière et ses fils.
1895	13 February: Filing of a patent for a device to obtain and view chronophotographic images. 22 March: Screening at the premises of the Société d'Encouragement pour l'Industrie Nationale in Paris. 10-12 June: Screenings at the Congrès des Sociétés Françaises de Photographie in Lyon. 28 December: First screening to a paying public audience, at the Salon Indien of the Grand Café in Paris.
1896	The Lumière brothers send their operators all over the world and set up a network of distributors for their films.
1897-98	Most of the firm's production is filmed. Appearance of the first lists of films. Gradual abandonment of the concession system. They start selling cameras to independents.
1902	Last theatrical show of the Cinématographe Lumière, in Lyon.
1905	Production ceases.

General Characteristics

Format: 35mm.

Length: approximately 17 metres, which represents between 800 and 900 frames. Towards the end of the Lumière catalogue, some films are about 24 metres long ("Cake-Walk" series), or even longer (the *Vues Fantasmagoriques*: edit of several reels).

Coloration: For negatives, the coloration varies from green to dark yellow. In the case of dark yellow, the negative is more rigid, as it is thicker. For positives, the colouring is most often yellow. In some cases, it is dark red. The reel is usually brittle. It isn't really appropriate to use the term "tinting", which refers to an intentional treatment of the reel to achieve a specific effect. The term "coloration" refers to the natural impact on the medium's original colour of the way it was manufactured.

Perforations

Negatives:

Perforations in the negatives are always round, and of variable diameter, only one on each side of the image and located in the lower third of the frame. This is the main characteristic of Lumière films.

Round perforations indicate that the film was shot using a Lumière camera, which doesn't necessarily mean that it is part of the Lumière productions *per se*, as listed in the catalogue.

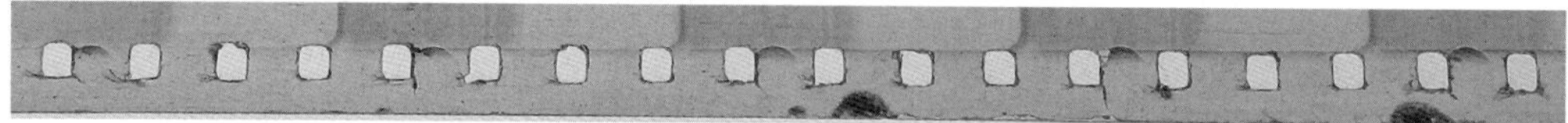

Print of the film *Défilé de cuirassiers* (1896-1897) – SFI

Besides, they sold cameras to independent operators, and other production companies had negatives with Lumière-type perforations.

Positives:

Most positives also have round perforations. But the Lumière brothers also provided prints with "Edison" perforations (4 rectangles on each side of the frame, smaller than standard perforations).

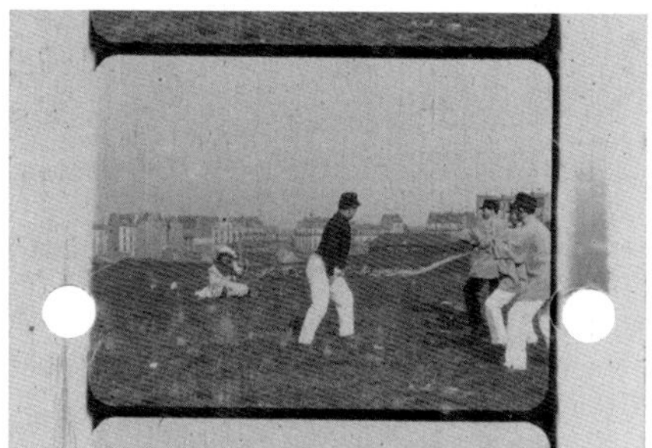

Print of the Lumière film
Les Boxeurs et le spectateur trop curieux puni (1897),
with Lumière perforations – FE

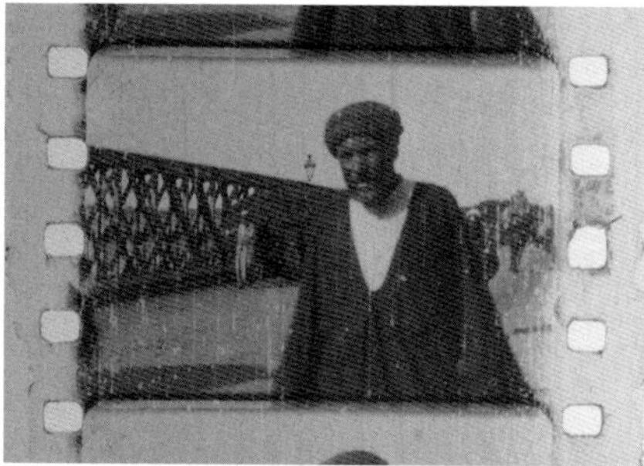

Print of the film
Sortie du Pont de Kasr-el-Nil (1897),
with Edison perforations – SFI

Similarly, there may be positives with round perforations that don't originate from Lumière negatives. The Pathé company produced films with either "Lumière" or "Edison" perforations as early as in its October/December 1901 catalogue. Georges Méliès also distributed copies with round perforations.

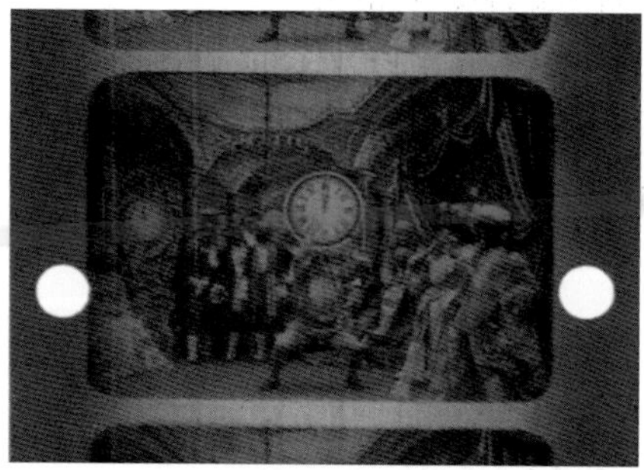

Print of the Méliès film
Cendrillon (1899),
with similar perforations – CNC

It is also common to observe incoming light through the perforations on the edges of prints of Lumière films.

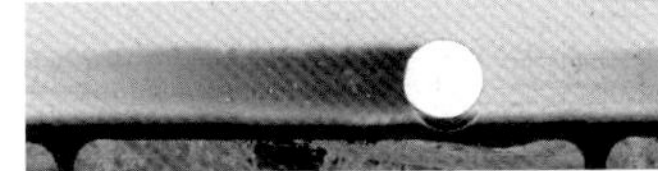

Print of the film
Le Calvaire (1897-1898) – FE

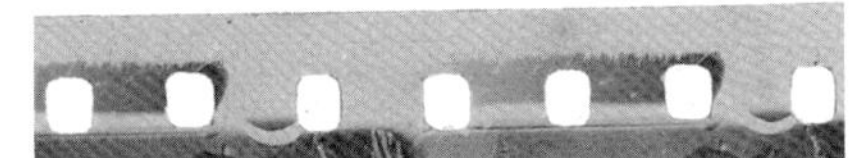

Print of the film
Démolition d'un mur (1897) – FE

The beginning of this Lumière reel shows that the incoming light through the perforations may result from the design of the drive mechanism. As this characteristic is only found on prints, it can be assumed that it appeared in the printing stage rather than during filming. Does this mean that prints bearing this characteristic were made using Lumière equipment?

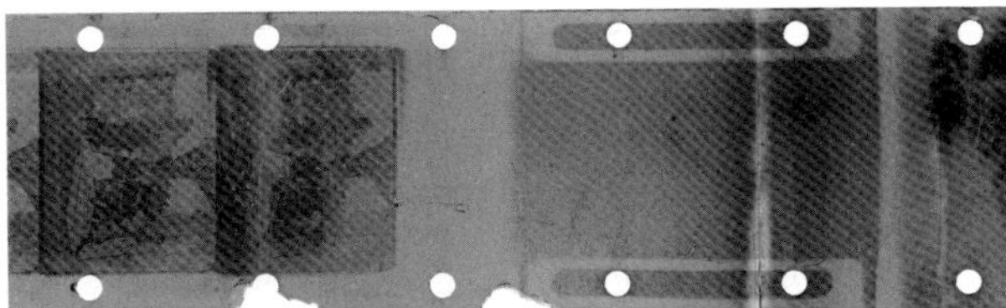

Beginning of a print of the film
Néron essayant des poisons sur des esclaves (1897) – FE

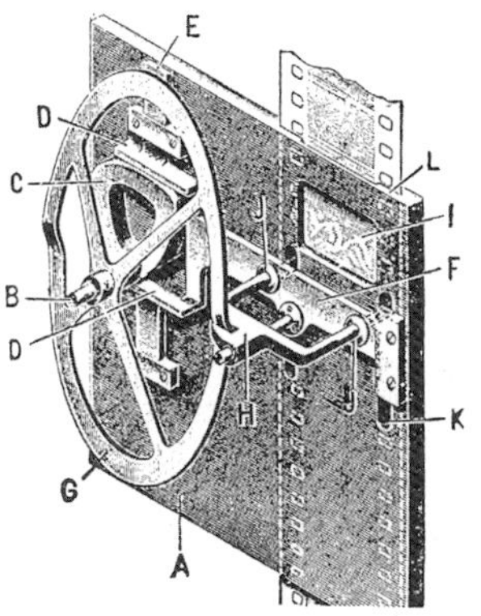

Greiferkonstruktion Lumières

Guido Seeber and Dr. Georg Victor Mendel, *Der praktische Kameramann. Theorie und Praxis der kinematographischen Aufnahmetechnik. Bücher des Praxis* – Band [Vol.] II Berlin, 1927 (reprinted by Deutsches Filmmuseum, Frankfurt am Main, 1980), p.98.

While this characteristic often appears, it is important to note, however, that some prints of Lumière productions don't have any evidence of incoming light.

Inscriptions on the Film Reel

At the beginning of some negatives, the corresponding catalogue number can be found, written in ink.
A special case is the "Fêtes de Nice et de Toulon" series (No. 1312 to 1328), for which the series title and the number corresponding to the film's position in the series are written on the film leader.

On some positives, the catalogue number is written in red ink on the film leader. This is certainly an inscription added after production.

Beginning in 1896, some positives were marked "LUMIERE LYON DEPOSE" on the edges at the beginning and end of the film.

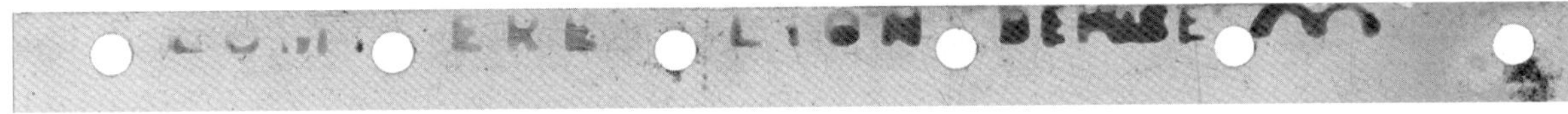

Inscription on a print of the film *Les Boxeurs et le spectateur trop curieux puni* (1897) – FE

Frame Characteristics

Frames, due to the camera gate, have round angles.[1] Since most of the Lumière catalogue was produced over a relatively short period of time, there are no significant variations in the films' technical characteristics.

The frame-line is sometimes irregularly thick, or even non-existent.

Logotype

The "Lumière" trademark cards, inserted in film sets, appear later in time, in films with phantasmagoric sequences. Especially worthy of mention is the case of the film *Colleur d'affiches* (No. 677 in the catalogue), in which the inscription "Cinématographe Lumière Lyon" is visible. It isn't strictly speaking a card, as the inscription is integrated into the fiction.

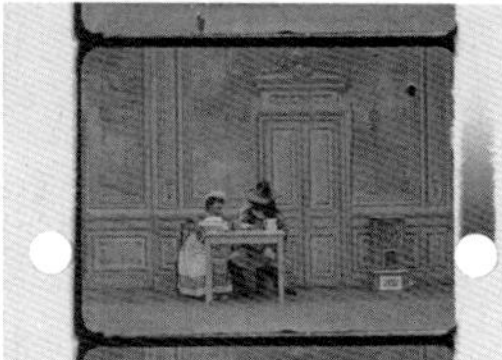

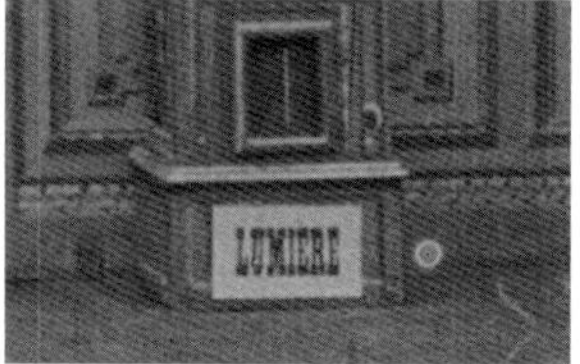

Frame from *Repas fantastique* (1903),
with the Lumière logo in the lower-right corner of the set
(and enlarged, right image) – CNC

Original Cans

The inscription "Cinématographe Auguste & Louis Lumière" is engraved around the lid of the film can. Some cans have a label with either the catalogue title or an approximate title handwritten on it.

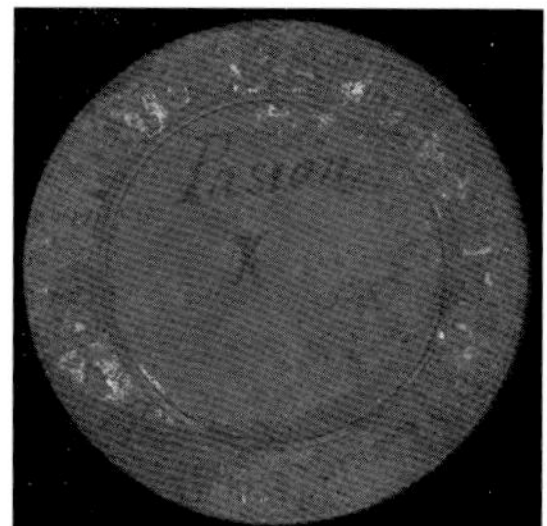

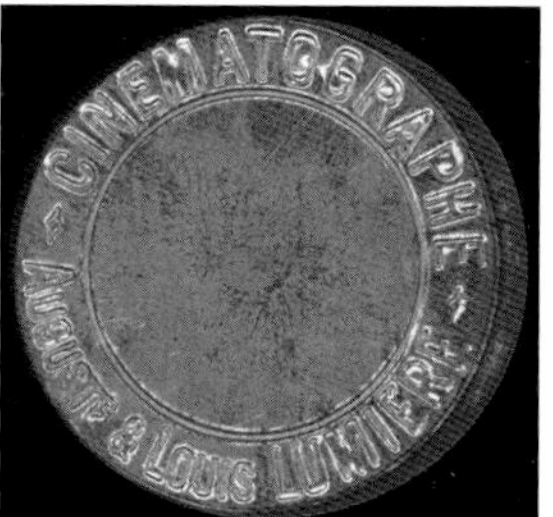

Original film can (exterior and interior) – FE

Bibliography

Bernard Chardère, *Lumières sur Lumière*, Lyon: Institut Lumière / Presses Universitaires de Lyon (1987).

Jacques Rittaud-Hutinet, *Auguste et Louis Lumière. Les 1000 premiers films*, Paris: Philippe Sers Editeur (1990).

Michelle Aubert and Jean-Claude Seguin (eds.), *La Production cinématographique des Frères Lumière*, Paris: Bibliothèque du Film (BiFi) – Editions Mémoires de cinéma (1996).

1 See illustrations above.

2
Identifying a Georges Méliès Film

Jacques Malthête

While it is generally fairly easy to recognize a Méliès film at the mere sight of a promotional photograph or a few frames, the same cannot be said of his 93 non-fiction, "fly-on-the-wall"-style views,[1] which have no observable markings and none of which feature Méliès himself.

Fiction Films

In his transformation views, fantasies, and *féeries* (fairy tales), and even in some of his 23 reconstructed newsreels, Méliès the actor is ubiquitous and easily recognizable, even when he wears make-up, because of his totally unique body language.[2] In addition, many sets and accessories reappear in successive films, a practice which is also helpful in attributing the authorship of views to him, as is the case with the buildings on his property in Montreuil-sous-Bois (Studio A, built in 1897 and extended in 1899-1900; a residential pavilion built in 1901; Studio B, fitted out in winter 1908), visible in some surviving films. Such evidence is of course only really useful if there is no indication in the set that can be observed onscreen, both in the short fantasies and in some longer scenes. Here is where the presence of trademarks comes into play as a useful identification tool.

Trademarks

In Méliès's films, there are about 15 different trademarks placed in sets throughout the production period 1896-1913. These precious clues make it possible both to assess whether films are truly his works, and to determine the date of filming.[3]

The first of these trademarks, used in 1896 and 1897, was a simple 5-pointed star containing the letters M and R, for "Méliès" and "Reulos", the latter having registered the mark on 20 November 1896:

— (top to bottom, centred): MARQUE DÉPOSÉE / [black 5-pointed star] / TRADE MARK:

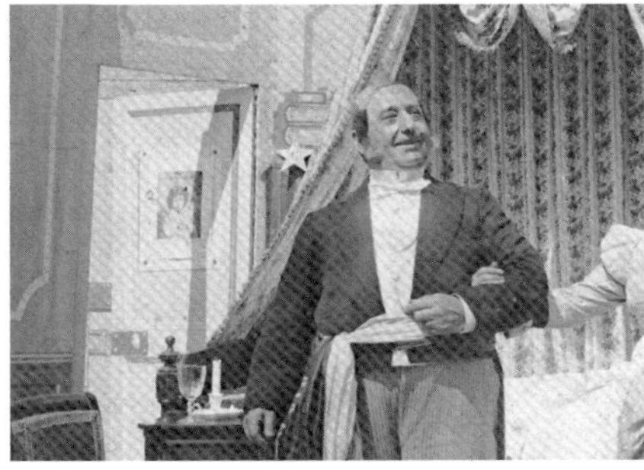

L'École des gendres (No. 102, 1897) – CF

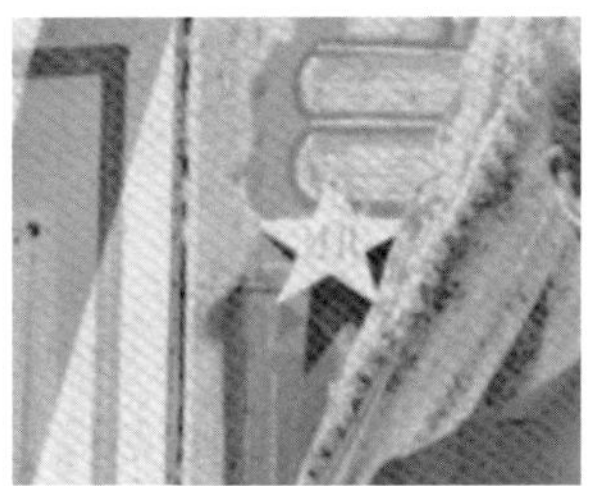

Detail

1 Jacques Malthête, "Georges Méliès, de la non-fiction à la fiction", in Thierry Lefebvre (ed.), *1895*, No. 18 (Summer 1995), "Images du réel. La non-fiction en France (1890-1930)", pp.70-83.

2 It has been estimated that Méliès appears in at least 300 of his films. See Jacques Malthête, "Georges Méliès", in Jacques Richard (ed.), *Dictionnaire des acteurs du cinéma muet en France*, Paris: Éditions de Fallois (2011), pp.603-608.

3 Jacques Malthête, "Pour une véritable archéologie des premières bandes cinématographiques", *1895*, No. 24 (June 1998), pp.9-21, plus Erratum, *1895*, No. 26 (December 1998), p.235. See also Paolo Cherchi Usai, *Silent Cinema. An Introduction*, London: British Film Institute (2000), pp.187-188.

By 1898 Méliès had apparently abandoned this trademark. Puzzlingly, on 7 January of that year, he registered with the notary office of Maître Alfred Chatelain (37, rue Poissonnière, Paris 2) the articles of incorporation of the Public Limited Company "L'Étoile, Société générale de cinématographie",[4] which was apparently very short-lived.[5]

On the other hand, in some films from 1899 and 1900, such as *Le Diable au couvent* (1899), *Les Miracles du brahmine* (1899), and *Jeanne d'Arc* (1900), the following can be read in white letters on a black background, on the side of an element of the set:

— (top to bottom) G. Méliès / PARIS:

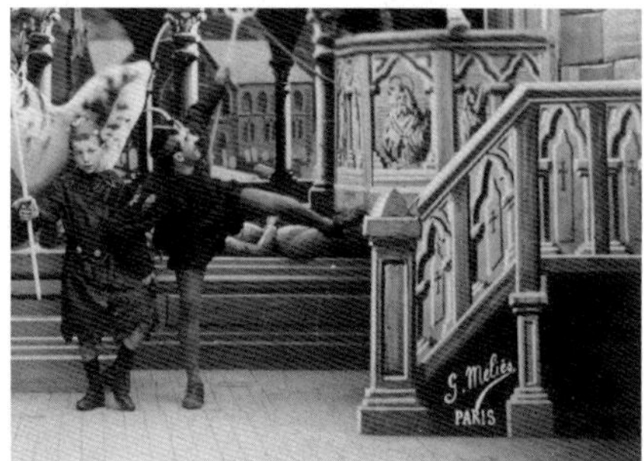

Le Diable au couvent (No. 185-187, 1899) – CF

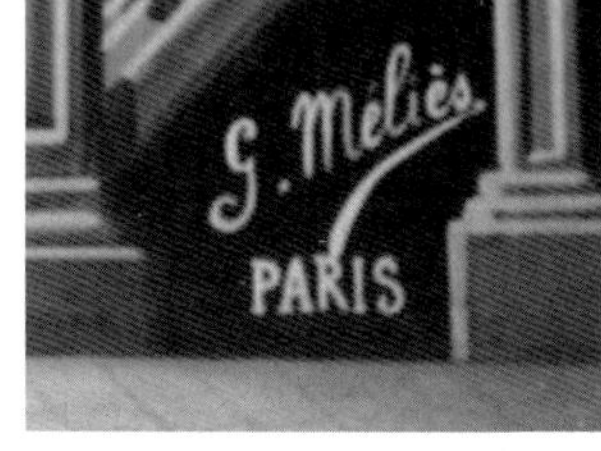

Detail

Some films from the period 1896-1900 contain no marks as such, but instead, relatively self-explanatory signs:

— In *Défense d'afficher* (1896), on the wall in front of which the action takes place hangs a poster for the Robert-Houdin Theatre, which includes an advertisement for *Les Rayons Rœntgen*, a magician sketch which premiered on the stage there in March 1896; it also bears the names of two famous members of the Robert-Houdin Theatre, Édouard Raynaly and Henri Duperrey, who were performing there in this period. The poster also bears the title of another magician sketch, *L'Augure*, which was featured alongside *Les Rayons Rœntgen* in an advertisement published in April 1896 in the daily paper *L'Orchestre*. By then Méliès had been directing the Robert-Houdin Theatre for nine years and had begun showing his own films there.

— In *Entre Calais et Douvres* (1897), a sign attached to the ship's poop deck bears the words: (centred) ROBERT-HOUDIN / [5-pointed star] / STAR LINE (here, the star still contains the letters M and R).

— At the beginning of *Le Portrait mystérieux* (1899), a poster announcing the revue *Passez Muscades* can be seen. This revue was staged at the Robert-Houdin Theatre during 1899.[6]

— In *Le Livre magique* (1900), Méliès's signature appears at the bottom of the drawings that come to life.[7]

In 1901, two new trademarks appear on a rectangular sign placed in the set:[8]

— (1) [5-pointed star] and (top to bottom) G. MÉLIÈS / PARIS (visible in *L'Omnibus des toqués*).

— (2) (top to bottom, centred) G. MÉLIÈS / [5-pointed star] / PARIS (visible in two set shots[9] in *Barbe-Bleue*).

In 1902, a sign placed in the set bears the trademark:

4 Notarial deed dated 22 December 1897.

5 See Jacques Malthête, "Correspondance de Georges Méliès (1904-1937)", in André Gaudreault and Laurent Le Forestier (eds.), *Méliès, carrefour des attractions*, followed by *Correspondance de Georges Méliès (1904-1937)*, Colloque de Cerisy/Presses Universitaires de Rennes (2014), Letter 35, p.349.

6 Jacques Malthête, Frédéric Tabet and Stéphane Tralongo, *"Passez Muscades" (1899), une revue du théâtre Robert-Houdin*, International Symposium *"L'illusion en jeu: techniques, outils, histoire"*, Toulouse, 26-30 March 2018 (publication forthcoming).

7 Another example is the royal galley from Scene 12 (*"Embarquement sur la galère royale"*) in *Le Royaume des fées* (1903), which bears the letters "STAR" after the 5-pointed star.

8 Here, as in all the filmed trademarks that followed, letters, numbers, and stars are white on a black background.

9 The "set shot" is the filmographic unit corresponding to the filming of an action in front of a single uniformly framed set.

— (top to bottom, centred) STAR FILM / [5-pointed star] / PARIS:

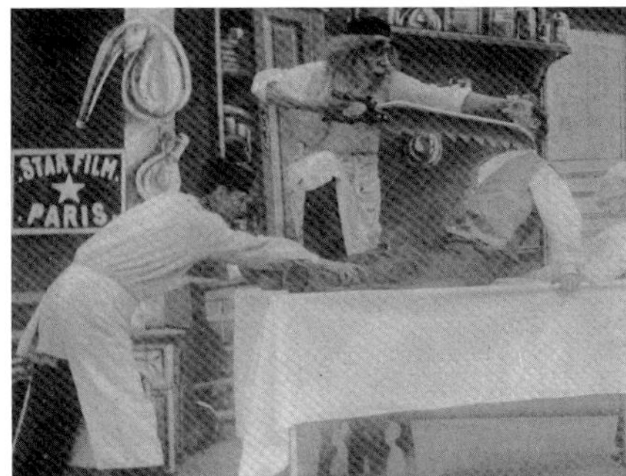

Une indigestion (No. 422-425, 1902) – CF

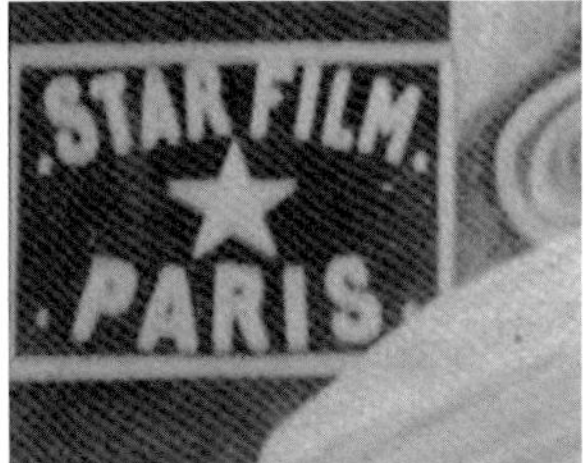

Detail

On 27 December 1902 in Paris, Méliès registered the trademark (top to bottom, centred) MÉLIÈS / [black 5-pointed star] / STAR FILM,[10] which can be seen in several films from December 1902, such as *Les Aventures de Robinson Crusoé*:

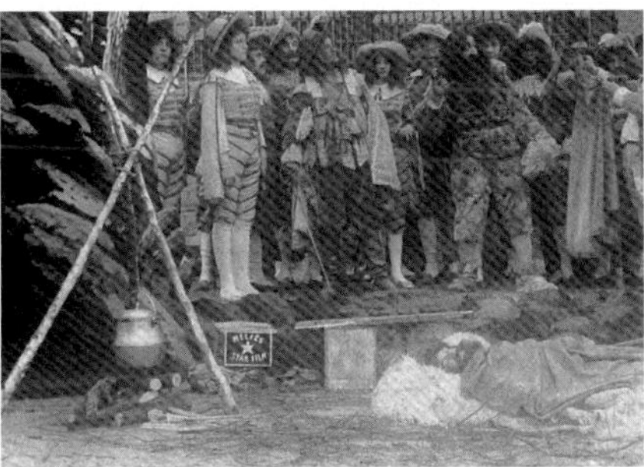
Les Aventures de Robinson Crusoé
(No. 430-443, 1902) – CF

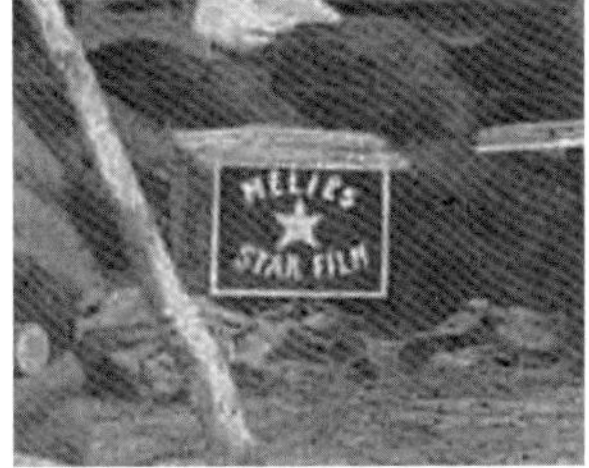

Detail

Georges Méliès sent his brother Gaston to New York to open an overseas branch in March 1903, and the trademark with the star was registered in the U.S.A. Between February and December 1903 (according to the filing date at the Library of Congress), a sign placed in the set bore the following mention:

— (top to bottom, centred) TRADE MARK / [5-pointed star] / STAR / REGISTERED:

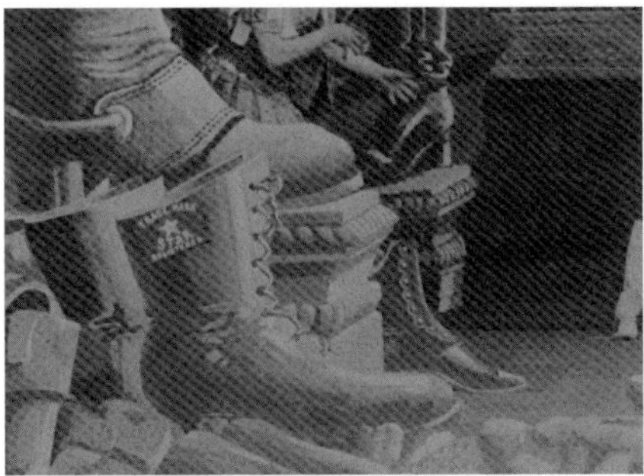
Les Mousquetaires de la reine
(No. 460-461, 1903) – CF

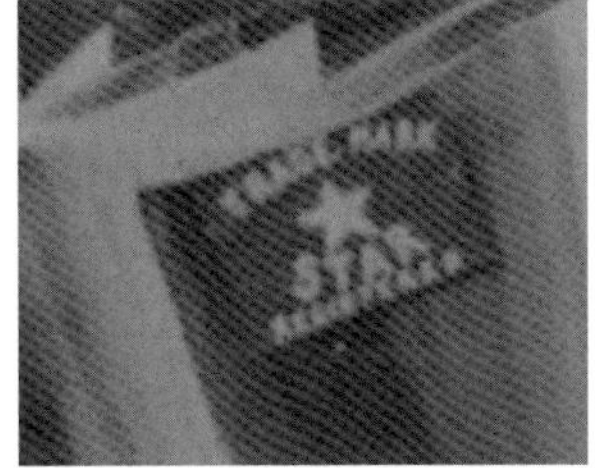

Detail

The first known example of this is *Les Mousquetaires de la reine*; the last known is *Faust aux enfers*.

10 This trademark is often mistakenly assimilated with Georges Méliès's "Manufacture de films pour cinématographes", but Méliès never set up a commercial company, with the exception of the very short-lived "L'Étoile, Société générale de cinématographie" in 1898.

On 25 June 1903, Gaston Méliès deposited the first Star film at the Library of Congress in Washington, D.C.[11] This was *Le Puits fantastique*; the entire reel was transferred to a roll of positive paper ("paper print"). The last Star film to be deposited in this way was *La Sirène* (1904). At the beginning and end of each of these two films, a short strip was pasted with a still copyright image:

- (1) (top to bottom, centred) COPYRIGHT 1903 / BY / Geo MÉLIÈS / PARIS-NEW-YORK
- (2) (top to bottom) COPYRIGHT 1904 / BY Geo MÉLIÈS / PARIS-NEW-YORK (the year shown is the year in which the film was made).

The films were subsequently deposited in the form of promotional pictures that only showed the sets, with a trademark that can be seen in the corresponding films in its definitive form:

- (top to bottom, centred) COPYRIGHTED / BY GEO. MÉLIÈS 1906 / PARIS NEW-YORK / Trade Mark [5-pointed star] Star:

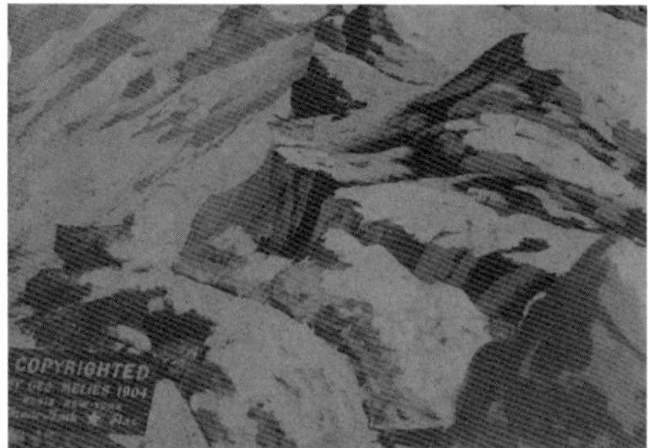

Voyage à travers l'impossible
(No. 641-659, 1904) – CF

Detail

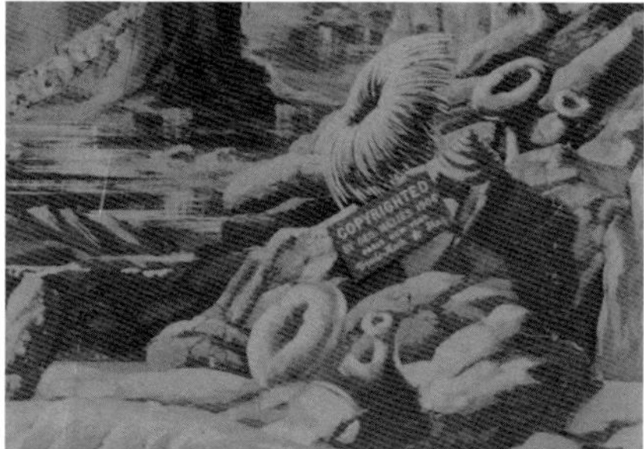

Deux Cents Milles sous les mers
(No. 912-924, 1907) – CF

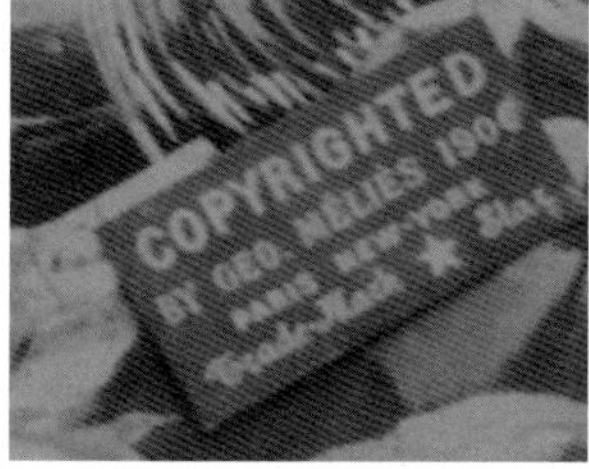

Detail

The last film deposited in Washington, D.C. was *La Poupée vivante* (1908).[12]

While the vast majority of Méliès's films do not seem to have been marketed with a title at the start of the reel, from at least 1903 onwards each reel already had a leader[13] which featured horizontally, parallel to the edges of the print, the film's catalogue number,[14] followed by Méliès's signature in relief:

Document found in the Harold Brown Collection at the BFI, probably copied from the 1905 American Méliès catalogue.

11 For the deposit dates, see Jacques Malthête and Laurent Mannoni, *L'œuvre de Georges Méliès*, Paris: Éditions de La Martinière/Cinémathèque française (2008), pp.345-355.

12 94 promotional photographs I consulted in 1981 are preserved in the Library of Congress, Washington, D.C. (Lot 3047). See Jacques Malthête and Laurent Mannoni, *L'œuvre de Georges Méliès*, in which the other two other major collections of Méliès promotional photographs, held by the Cinémathèque française and the Centre national du cinéma et de l'image animée, are inventoried.

13 A reproduction of this leader can be found in the 1905 American Méliès catalogue: *COMPLETE CATALOGUE OF GENUINE AND ORIGINAL "STAR" FILMS (MOVING PICTURES) MANUFACTURED BY GEO. MÉLIÈS OF PARIS, NO. 204 EAST 38TH STREET, NEW YORK, N. Y., U.S.A. GASTON MÉLIÈS, GENERAL MANAGER*, Washington, D.C.: Library of Congress (1905), p.3.

14 Under this single- or double-digit number are one or two other numbers that are probably billing numbers. Initially handwritten, these numbers were later stamped on the leader, which also bears the trademark "Méliès Star Film" and Méliès's signature in relief.

Before the catalogue number, the first frame of the film is embossed, and in a rectangle that takes up the entire frame, there appears:

— (top to bottom, centred) MÉLIÈS / [5-pointed star] / STAR FILM

A black star takes up almost the entire second frame.[15] However, in the majority of cases, the preserved films unfortunately no longer have these valuable clues.

Between January 1904 (the date of deposit at the Library of Congress) to the end of 1908, however, the copyright sign in the set underwent slight variations:

— (top to bottom) COPYRIGHTED / BY GEO. MÉLIÈS 1904 / TRADE MARK [5-pointed star] STAR:

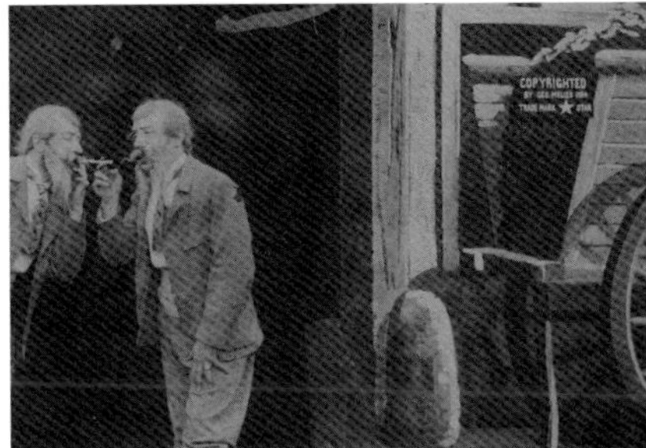

Un peu de feu, S.V.P. (No. 545, 1904) – CF

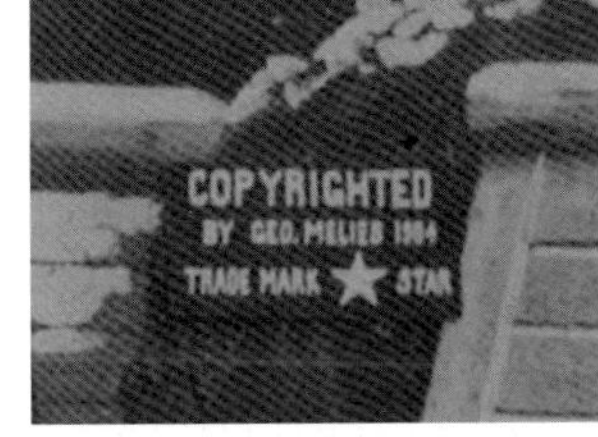

Detail

— (top to bottom, centred) COPYRIGHTED / BY GEO. MÉLIÈS 1904 / PARIS NEW-YORK / TRADE MARK [5-pointed star] STAR[16]

By 1909, Méliès no longer seemed to place trademarks in the sets of his films.

In 1910, no film was released.

In 1910-1911, the promotional pictures for the six films commissioned by Pathé all bear the trademark:[17]

— (top to bottom, centred) STAR FILM / [5-pointed star] / Geo MÉLIÈS / PARIS

Non-Fiction Films

The case of non-fiction films is, as we have seen, much more complicated to address. Consequently, it is best to rely on purely material clues, such as the appearance of the frame-line and edge of the frames, the shape of the perforations, and the craftsmanship of the splices, i.e., everything that can be observed on the Méliès films that have been authenticated with certainty. Admittedly, these clues taken separately are not necessarily characteristic of Méliès's work, but, for specific periods, it should be possible to build up bodies of evidence that allow us to attribute to Méliès certain non-fiction films that so far remain unattributed. In addition, the descriptive texts in the Star-Films catalogue can provide indications that may help us to assign a title to a film possibly made by Méliès. However, again, caution should prevail, as there are sometimes wide discrepancies, at least where fiction films are concerned, between the descriptions in the Méliès catalogue and what can be seen in the existing films.[18]

To our knowledge, only one non-fiction film by Méliès has been saved: *Panorama pris d'un train en marche* (1898).[19] However, among the films on the Paris World's Fair/Exposition Universelle (1900) held by the Direction du Patrimoine — Centre national du cinéma et de l'image animée is a film that could well be one of the 17 that Méliès shot

15 This star was already present, alone, in the early years. Harold Brown notes that it could be found on the first, second, or third frame, and even, sometimes, just before the first frame. See also Brown, p.48 in this volume.
16 See the promotional photograph for *Damnation du docteur Faust* (1904), in Maurice Bessy and Lo Duca, *Georges Méliès, mage*, Paris: J.-J. Pauvert (1961), p.188.
17 The films themselves – all of which survive – are devoid of any trademark.
18 It goes without saying that these are obviously the surviving parts of films described in the catalogue.
19 Held at the British Film Institute.

during the 1900 Exposition.[20] Indeed, in a view of the "moving sidewalk", a man in a suit with a top hat can be seen standing on it who looks a lot like Méliès. If so, it could be the reel entitled *Détail du trottoir roulant* (Méliès No. 249). The Méliès catalogue states: "In the background, the Usines du Creusot exhibition. In the foreground, the moving sidewalk is very clearly visible with its three parts. Part fixed, part running at low speed, part running at high speed (on the sidewalk, many visitors)." There is clearly a need for further study to confirm or refute this hypothesis.

This is where observing some material clues could allow useful comparisons.

Frame-lines and Frame Characteristics

In Méliès's films, frame-lines are white. The appearance of the edge of the frames closely depends on the characteristics of both the camera gate[21] and the printer gate when the negative image is not reproduced in its entirety. For example, in the 3 Méliès films of 1900 found in Switzerland in 1995,[22] the same small faults are apparent in the left vertical part of the edge of the image. On the other hand, the roundness of the upper-left corner of the image has the same recognizable shape in all 3 films, with a white frame-line that is clearly marked, as Brown observed.[23] A film believed to be possibly by Méliès could thus be compared with another from the same year that has been definitely proven to be one of his.[24]

By 1902, it can be noted that the frames' shape has changed. The corners have become more angular, which may be due to a change of camera and/or printer.[25] Unfortunately, little is known about the various cameras and printers used by Méliès, except that he filmed his first views with a Robert W. Paul No. 2 Mark 1 Theatrograph projector transformed into a camera.[26]

Perforation Shapes

The perforations of the different nitrate prints of Méliès's films (35mm format with 4 perforations on each of the two edges of each frame) are hardly homogeneous. The perforations that seem the most authentic can be observed on prints from the Sagarmínaga Collection at the Filmoteca Española: a rectangular perforation, nearly a square, with rounded corners:[27]

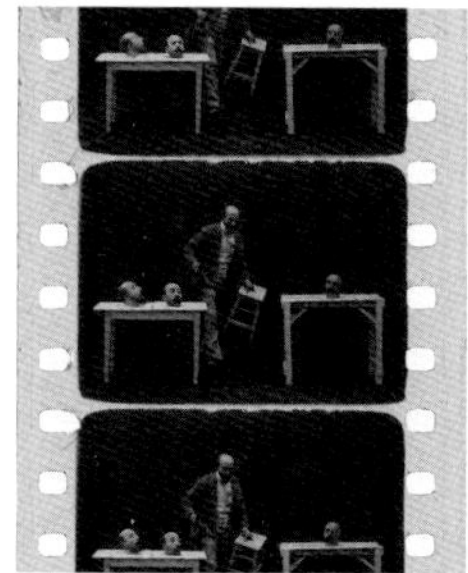

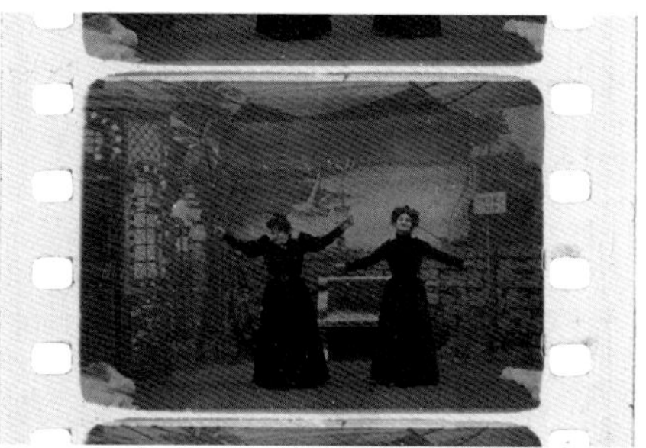

Un homme de têtes (No. 167, 1898) – FE & *Luttes extravagantes* (No. 180, 1899) – FE

20 Jacques Malthête, "*Les Vues spéciales de l'Exposition de 1900* tournées par Georges Méliès", *1895*, No. 36 (February 2002), pp.99-115.

21 And of each of the two cameras from 1902. Indeed, beginning with *Une indigestion* (No. 422-425), Méliès simultaneously used two negatives, corroborated by the *Complete Catalogue of Genuine and Original "Star" Films* (1905), which contains the following statement (pp.20-80): "Class III. The Original negatives for the following subjects are in New York, so that orders for any of the films will be promptly filled." One negative was sent to the New York branch for the American market, the other remained in Paris to print the positives for European customers.

22 They are *Spiritisme abracadabrant* (No. 293), *La Vengeance du gâte-sauce* (No. 243), and *Le Repas fantastique* (No. 311), held by the Swiss Camera Museum in Vevey. See Roland Cosandey, *Cinéma 1900. Trente films dans une boîte à chaussures*, Lausanne: Payot (1996), pp.98-106.

23 See Brown, p.76 in this volume.

24 A similar approach has already been attempted with the Lumière films (see Jacques Malthête, "Poussières d'histoire. À propos du livre de Jacques Rittaud-Hutinet: Auguste et Louis Lumière — Les 1,000 premiers films", in *1895*, No. 19 (December 1995), pp.75-78. This upper-left corner can be seen, with the same distinctive characteristics, in an example provided by Brown (*Visite sous-marine du "Maine"*, 1898, p.76). Another example can be seen in an 1899 hand-coloured film, *Automaboulisme et Autorité*. (See Laurent Mannoni, ed., *Georges Méliès, La màgia del cinema*, Barcelona: Obra Social "la Caixa", 2013, Fig. 3, p.105, and Fig. 75, p.152. In the second Figure, the Kodak brand, visible on the right-hand edge, is the trace of a later restoration of the perforations.) According to Brown (p.76), these characteristics can be observed on Méliès's films from mid-1897 to 1901.

25 See, for example, two fragments of films from 1902: *Voyage dans la Lune* (black & white nitrate) in Thierry Lefebvre, "*Le Voyage dans la Lune*, un film composite", in Jacques Malthête and Laurent Mannoni (eds.), *Méliès, magie et cinéma*, Paris: Paris-Musées (2002), p.185, and *Les Aventures de Robinson Crusoé* (Laurent Mannoni, ed., *Georges Méliès, La màgia del cinema*, pp.144, 147). According to Brown (p.76), this characteristic can be observed on Star films from 1902 to 1904.

26 On Méliès's cameras, see two articles in *Méliès, magie et cinéma* (2002): Laurent Mannoni, "1896, les premiers appareils cinématographiques de Georges Méliès", pp.116-133, and Jacques Malthête, "Les deux studios de Georges Méliès", pp.134-169.

27 Also in *Automaboulisme et autorité* (already cited). A very similar perforation is shown by Brown; see p.76 in this volume.

We are obviously dealing with an early-generation (most likely first-generation[28]) nitrate print, and this type of perforation can therefore be that of the prints originating from Méliès's laboratory,[29] but the situation is made more complex by later-generation nitrates,[30] whose perforations are increasingly diverse: square, rectangular with more or less rounded corners, Bell & Howell... It is also possible to come across first-generation copies whose printer and perforations may be different from those of the first edition of the film. For example, films released in 1899, like *Le Diable au couvent*, or in 1902, like *Voyage dans la Lune*, were still in the Méliès catalogue in 1908. The shape of the perforation is therefore not a determining criterion to identify a Méliès film.

Splices

The style of splices found in Méliès's films is very consistent. Performed on negatives, mainly to adjust camera stops caused by the creation of trick effects such as appearances, disappearances, or substitutions, they take the form of a horizontal white border in the upper part of the image (18 ± 2% of the frame's height):[31]

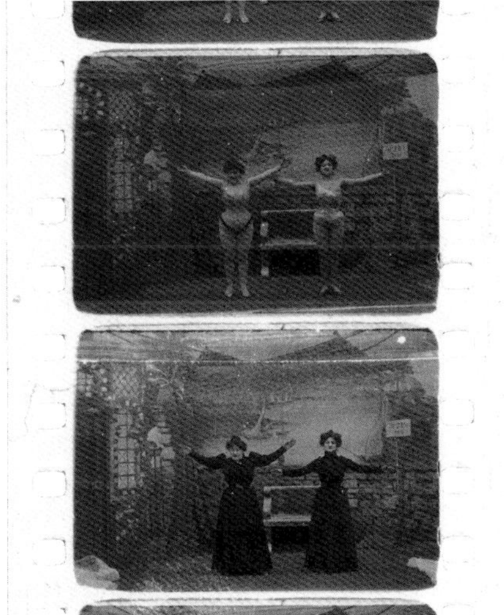

Splice photographed from negative visible on a print of *Luttes extravagantes* (No. 180, 1899) – FE

While similar *collages* can be found in other films of the era, the presence of this type of splice can nevertheless be a precious clue, among others, pointing in Méliès's direction. Of course, in non-fiction films, splices on negatives are rare, except in the case of resuming filming after an on-set incident, a tear in the negative, or an edit of different takes.

Attributing a Film to Méliès Is One Thing, Finding a Title and a Date Is Another

Title

In the luckiest case, occasionally a catalogue number handwritten by Méliès can be found on the back of some of the promotional pictures for fiction films, which makes it easy to find the title of the corresponding film, in the absence of other clues. However, promotional pictures of non-fiction films are unfortunately very rare, and have no catalogue number.

28 No trace of previous generations is detectable.

29 The nitrates kept along with the original invoices at the Swiss Camera Museum in Vevey, and which are certainly first-generation, seem to have the same perforation, although the photographs in Roland Cosandey's 1996 book *Cinéma 1900. Trente films dans une boîte à chaussures* show only a small selection of these.

30 Méliès called these illegal prints *surcopies* ("overprints"). See Jacques Malthête, "Pour une véritable archéologie des premières bandes cinématographiques", *1895*, No. 24 (June 1998), pp.61-62.

31 Jacques Malthête, "Le collage magique chez Edison et Méliès avant 1901", in Réjane Hamus-Vallée (ed.), *CinémAction*, "Du trucage aux effets spéciaux", No. 102 (1st Quarter [January-March], 2002), pp.96-109. Another example showing the splicing of a trick film can be found in Paolo Cherchi Usai, *Silent Cinema. A Guide to Study, Research and Curatorship*, London: Bloomsbury/British Film Institute (2019), Fig. 206a, p.295.

Shooting Date[32]

Dating the making of Méliès's films is not as uncertain an exercise as one has sometimes been led to believe, despite the very unfortunate loss of the "Star" films brand accounting archives.

First of all, an important piece of information. It appears that the numbering system — specific to the Méliès catalogue[33] — follows more or less the chronology of filming. This allows us to determine fairly precisely the period in which a take was shot in relation to well-established milestones.

Scenes filmed in Paris — and included in the Méliès catalogue — have already been assigned precise dates. Between 1897 and 1904, Méliès would stick to current events in a different way, by filming scenes that reconstructed events the press were reporting: the Greek-Turkish war and the Indian revolts against the British in 1897, the Spanish-American War in 1898, the Dreyfus trial in Rennes in 1899, the coronation of Edward VII in 1902. However, seriousness wasn't always the order of the day, and reconstructions sometimes took a comical turn, as with what Méliès shows in *Le Congrès des nations en Chine* (1900), about the great powers' intervention in China, and in *Le Joyeux Prophète russe* (1904), which alludes to the ongoing Russian-Japanese conflict.

Beginning in 1903 and for 6 consecutive years, Méliès's films were copyrighted, with a precise registration date, at the Library of Congress in Washington, D.C. Considering it took over a week for a film to be shipped to New York by boat, it is reasonable to assume that the majority of the material had been completed at least 10 to 15 days before it was deposited in the Library of Congress. As it was possible to submit several films on the same day, it doesn't seem unreasonable to conclude that an average of 3 weeks to one month had elapsed between completion of the editing process of these grouped films and their submission in Washington, D.C. This confirms that filing dates constitute very critical data.

Lastly, announcements of upcoming film releases and newspaper advertisements of the time can obviously be of great assistance. In addition, the consultation of Pathé's accounting ledgers is helpful to establish the chronology of the production of Méliès's last 6 films[34] under Pathé sponsorship.[35] Indeed, comparing the release dates published in the press at the time with those mentioned in the accounting ledgers, one can conclude that it had taken 2 — for the shortest ones — to 4 months to make each of these 6 films, assuming that the tasks of developing, printing, composing the title cards, tinting, toning, and final assembly of the prints, all carried out at Pathé, could take a month. It can thus be established that *Les Hallucinations du baron de Münchausen*[36] was likely filmed in August-September 1911, *Le Vitrail diabolique*[37] in October-November 1911, *À la conquête du Pôle*[38] in December 1911-March 1912, *Cendrillon*[39] in April-July 1912, *Le Chevalier des Neiges*[40] in October-December 1912, and *Le Voyage de la Famille Bourrichon*[41] in January-March 1913.

From this brief introduction to Méliès archaeology, one can conclude by noting, with some satisfaction, that as of 1 January 2019, Méliès remains, after Lumière, the French cinema pioneer best represented in terms of the number of films that survive today, in more or less complete form: 214 titles out of the 520 released, 41% of his total production.

Eight illustrations in this article reproduce details from promotional pictures held by the Cinémathèque française in its Méliès Collection (CF), and three illustrations are from the Sagarmínaga Collection at the Filmoteca Española (FE). One illustration was found in the Harold Brown Collection at the British Film Institute.

32 See Jacques Malthête and Laurent Mannoni, *L'œuvre de Georges Méliès*, pp.22-31.

33 It works according to a simple principle: a title of 'n' digits corresponds to a length between 20'n' and 20'n' + 19 metres of film. For instance, *Un homme de têtes* (released in 1898), which was sold under a unique catalogue number (167), was originally 20 metres long, while *Deux Cent Milles sous les mers* (released in 1907), which was sold under 13 catalogue numbers (912-924), was originally 13x20 metres long, i.e., 260 metres.

34 Méliès is mentioned in relation to the royalties on the sale of his films between 31 December 1911 (accounting ledger Pathé No. 18, p.335) and 30 June 1913 (accounting ledger Pathé No. 22, p.429). The same accounting ledgers also report purchases of Méliès's films between 1899 and 1903, more precisely between 9 November 1899 (accounting ledger Pathé No. 2, p.208) and 14 August 1903 (accounting ledger Pathé No. 4, p.786). These are probably copies acquired to closely study Méliès's techniques (archives of the Jérôme Seydoux-Pathé Foundation).

35 Four letters from Pathé to Méliès (27 November 1911, 22 October 1912, 23 April 1913, and 30 May 1913) refer to the interest on the loan Pathé had granted to Méliès to shoot his last 6 films (archives of the Jérôme Seydoux-Pathé Foundation; we wish to express our thanks for this information to Stéphanie Salmon).

36 Film first circulated in November 1911. For the releases of the 6 films commissioned by Pathé, see Henri Bousquet, *Catalogue Pathé des années 1896 à 1914*, available online on the Jérôme Seydoux-Pathé Foundation website.

37 Film first circulated in January 1912.

38 Film first circulated in May 1912.

39 Film first circulated in January 1913.

40 Film first circulated in February 1913.

41 Film first circulated in May 1913.

3
Parnaland and Éclair

Camille Blot-Wellens and Pierrette Lemoigne

Parnaland (1895-1907)

Camille Blot-Wellens

The name of Ambroise-François Parnaland is almost completely uknown today, apart from his links with the "Doyen case" and the founding of the Éclair company.

Ambroise-François Parnaland, one of the pioneers of animated photography, registered his first patent for a camera in February 1896 and presented his first "filmed views" at the beginning of 1897. From the creation of his first company in 1895, with his brother Louis-Emile, until the premises became the company Éclair, in 1907, Parnaland produced hundreds of films, of all genres: "actualities, historical, scientific, comedies, films that can be run backwards, transformations, trick films, etc.".[1]

Only very few Parnaland films were known until 30 films were identified in the Sagarmínaga Collection at the Filmoteca Española in Madrid.[2] Nevertheless, it is possible that more archives may hold Parnaland prints in their collections, since the firm appeared to be rather active in the first decade of the film industry, and their films were even exported to the United States (by Méliès). The study and comparison of the 30 Sagarmínaga prints in Madrid allowed us (my colleague Encarni Rus and I) to detect characteristics that could be useful for identification.

Some Historical References

1895	Ambroise-François and Louis-Émile Parnaland (two brothers) create "Parnaland frères" in Paris, with the aim of exploiting registered patents.
1896	First cameras patented.
1896	May: First reversal camera, the "Cinépar".
1897	The brand "Parnaland frères" is registered, as well as the acronym "FP". They offer films in both 35mm and 50mm, with different kinds of perforations. They produce their first films.
1898	Collaboration with Eugène-Louis Doyen to film surgeries (together with Clément-Maurice).[3]
1899	The Parnaland brothers open a workshop (98 rue d'Assas) and a store (5 rue de la Santé) in Paris, and intensify their production of films.
1901	Louis-Émile leaves the company, which becomes A.-F. Parnaland. Ambroise-François moves and opens a new workshop (30 rue Le Brun). The (first?) catalogue offers 80 titles, 30 being views from Asia.

1 *Catalogue Parnaland* (1901). For more information on the company and its filmography, see Laurent Mannoni, "Ambroise-François Parnaland. Pioniere del cinema e co-fondatore della società Éclair", *Griffithania*, No. 47 (1993), pp.10-31.

2 For more information on the Sagarmínaga Collection, see Camille Blot-Wellens, *La colección Sagarminaga (1897-1906). Érase una vez el cinematógrafo en Bilbao*, Coll. Cuadernos de la Filmoteca 14 (2011).

3 Parnaland commercialized some of the films made during the operations without the consent of Eugène-Louis Doyen. For more information, see Thierry Lefebvre, *La Chair et le celluloïd. Le cinéma chirurgical du docteur Doyen*, Bar le Duc: Jean Doyen éditeur (2004).

1904 Parnaland joins with Emmanuel Ventujol (a former Lumière operator) to create the company Parnaland-Ventujol.
Parnaland films appear among titles distributed by Méliès in the United States.[4]

1907 The company doesn't thrive. Parnaland and Ventujol look for a new associate. Their new partner is Charles Jourjon. The company becomes Éclair. Parnaland leaves soon after the creation of the new firm.

Frame

The Parnaland frame is very open. There is almost no frame-line, and often the perforations enter into the image. Nevertheless, there are small differences, even for films probably shot in the same period (according to the catalogue numbers).

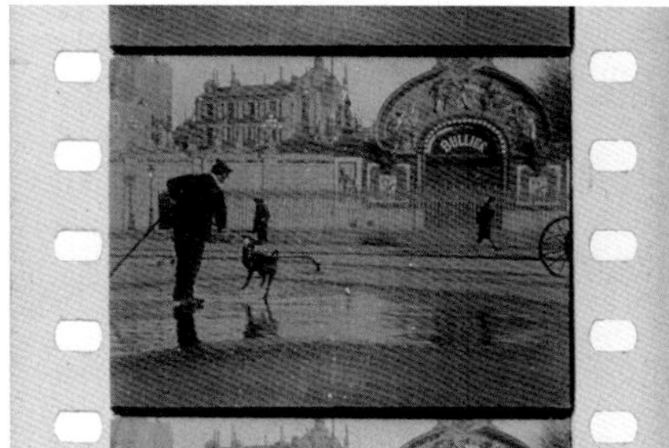

Le Chien et l'arroseur (No. 4, 1897-1899) – FE

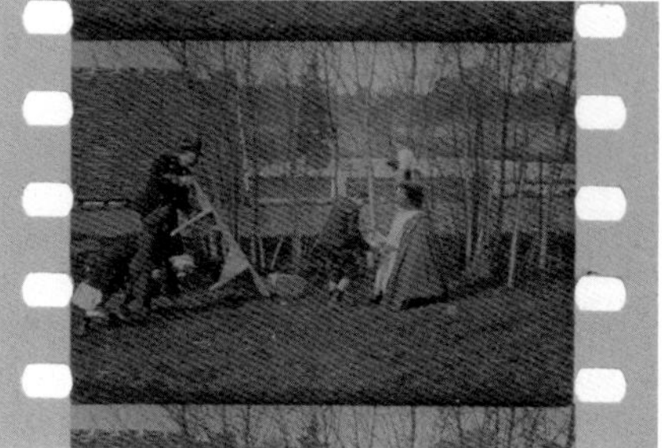

Bonne d'enfants et militaire (No. 9, 1897-1899) – FE

In these two examples, the frame is similar but the perforations are different. Probably as the print of *Le Chien et l'arroseur* is earlier than that of *Bonne d'enfants et militaire*.

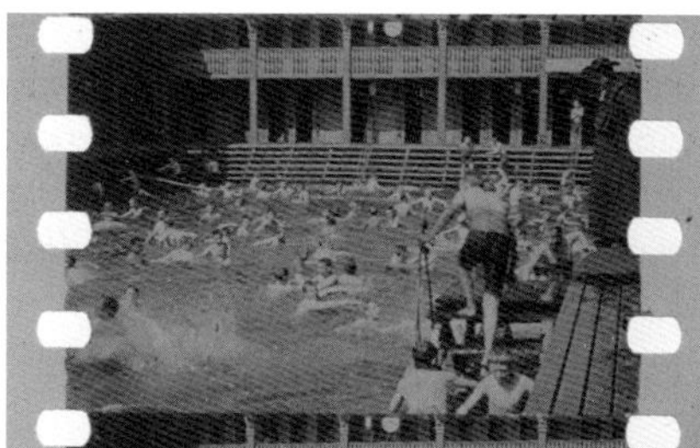

Sauts du tremplin (No. 7, 1897-1899) – FE

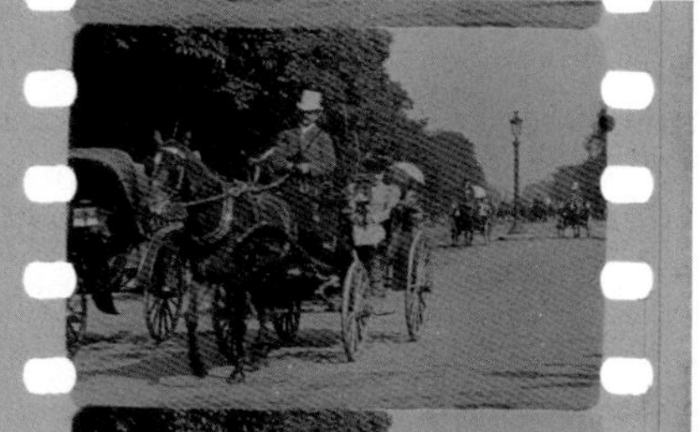

Avenue des Champs-Elysées (No. 20, 1897-1899) – FE

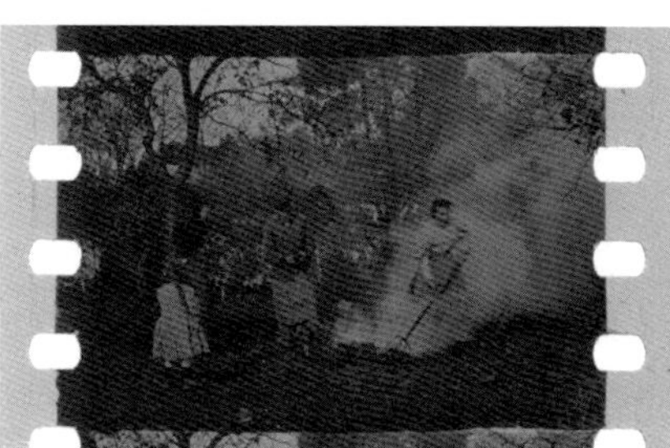

Feu d'herbes (No. 35, ca. 1900) – FE

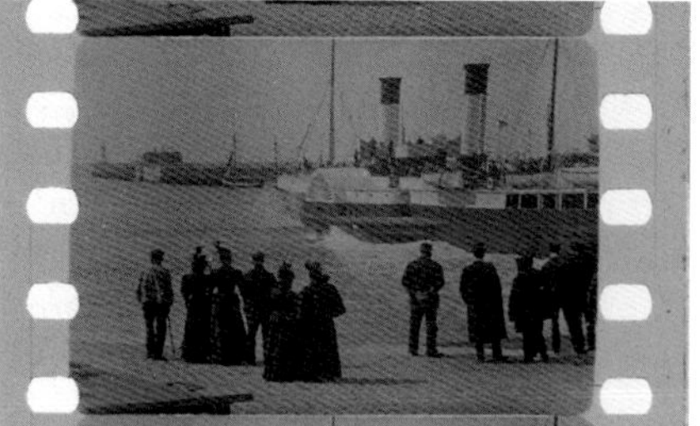

Evolution d'escadre à Toulon (No. 377, 1901) – FE

4 Georges Méliès, *Bulletin No. 2. Choicest "L.* [sic] *Parnaland's" films – "L. Gaumont's" films* (1904).

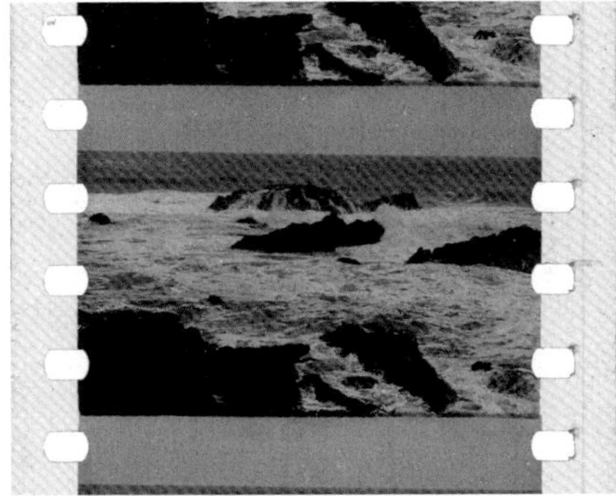

Mer (No. 449, 1902) – FE

The films have very similar characteristics to films produced by Auguste Baron and by Gaumont (but in this case the perforations of the negative are more easily visible):

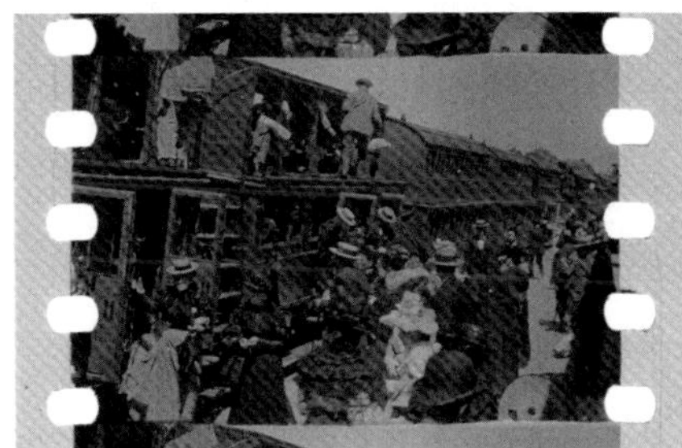

Arrivée d'un train à étage
(Baron, 1897-1901) – FE

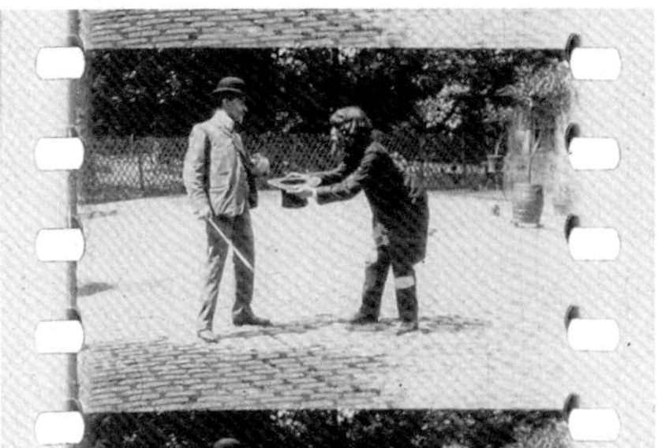

La Charité du prestidigitateur
(Gaumont, 1905) – FE

If the identification of the titles is correct, it seems that several frames (i.e., filmed with different cameras) co-exist. If the frame helps to identify the company, it does not necessarily help date the production.

The films shot in the studio are easier to identify and situate in terms of the production period than the elements of the settings.

Settings

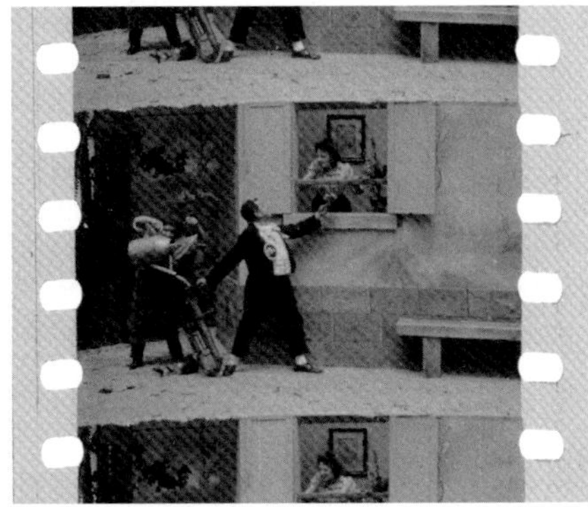

Chanteur des cours (ca. 1900) – FE

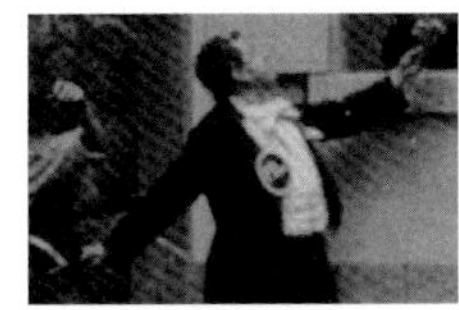

Detail

Chanteur des cours (ca. 1900) is one of the first films preserved that was shot in a studio. The Parnaland brothers were then based at 5 rue de la Santé. It is also possible to see the acronym, FP, on a badge on the man's suit.

From 1901, when Ambroise-François established the workshop at 30 rue Le Brun, the settings become more recognizable, and elements of the settings are repeated in the company's films:

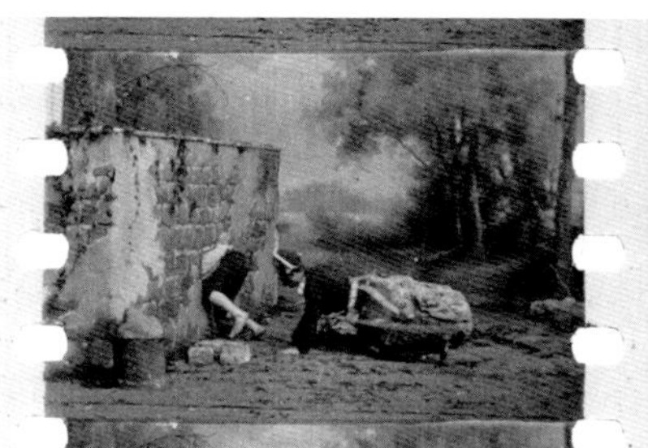

Gendarme et voleur de canard (No. 346, ca. 1901) – FE

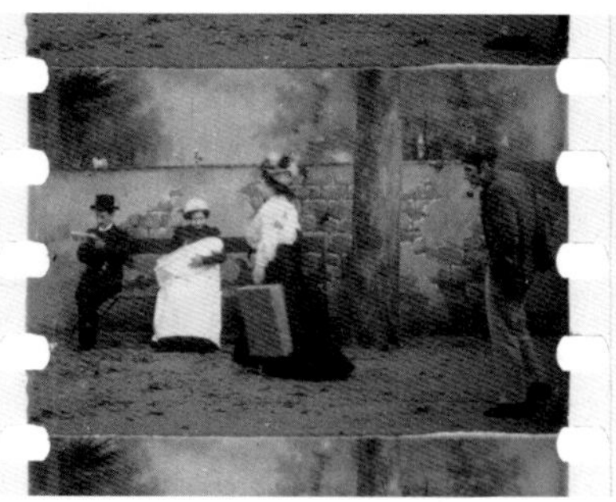

Les Bons payent pour les mauvais (No. 366, 1901) – FE

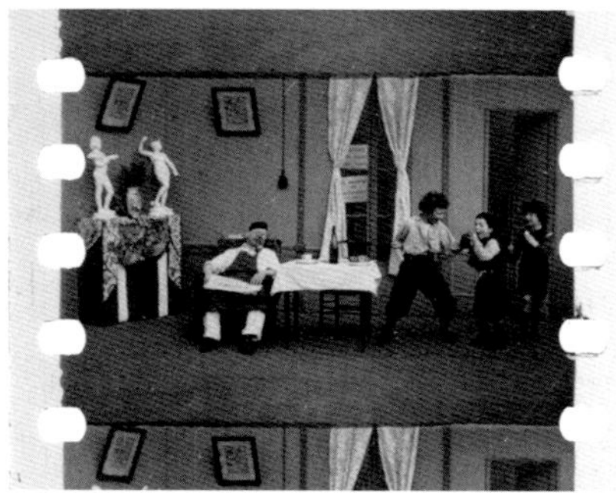

Sieste interrompue (No. 357, ca. 1901) – FE

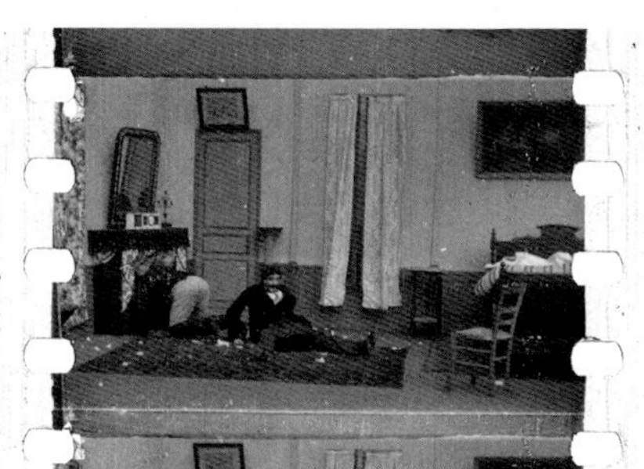

Cambrioleur insaisissable (No. 367, 1901) – FE

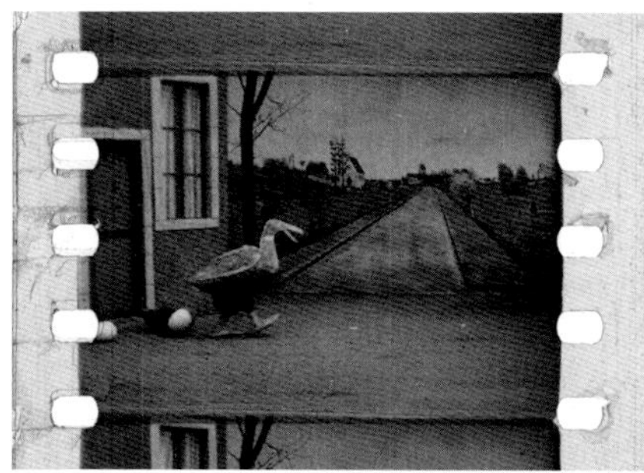

Canards parisiens (No. 462, 1902-1903) – FE

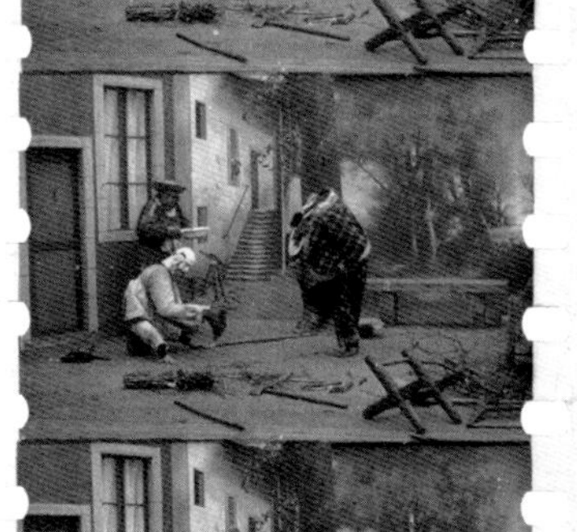

Les Bûcherons (No. 460, 1902-1903) – FE

Titles

Of the 31 prints studied, only 7 had main titles. One title had been produced in 1900, another in 1900-1901, and 5 in or after 1902 (but all the titles were still offered in the 1907 catalogue). On the prints studied, all the main titles were spliced on the negative, and therefore without specific tinting. There is no mention of catalogue number, nor production company.

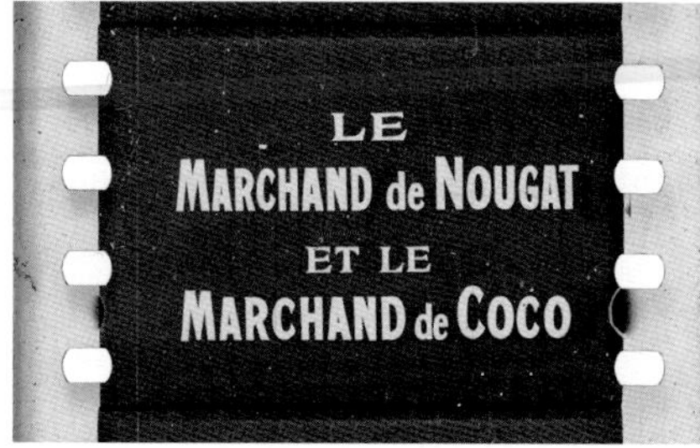

Le Marchand de nougat et le marchand de coco (No. 437, 1902) – FE

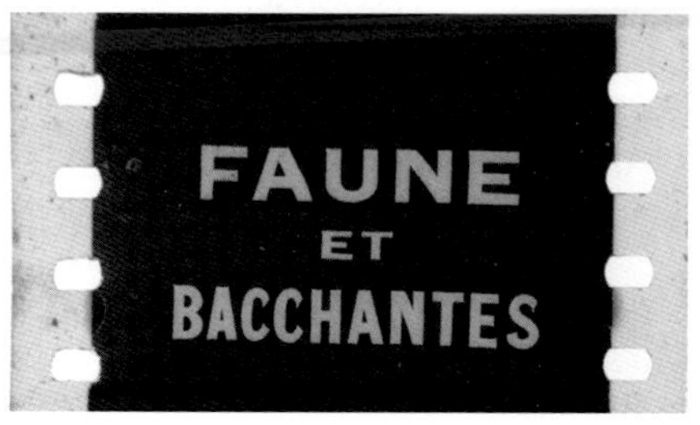

Faune et bacchantes (No. 456, 1902-1903) – FE

Éclair (1907-1931)

Pierrette Lemoigne

History

1907	April: Charles Jourjon and Ambroise-François Parnaland establish the Société Française des Films et Cinématographes "L'Éclair" (S.F.F.C.E.). Its studios are located in Epinay-sur-Seine, at the Château de Lacépède.
1909	November: opening of an office in New York (Éclair Film Company). Distribution of Éclair films abroad begins.
1910	Involvement in the creation of A.C.A.D. (Association Cinématographique des Auteurs Dramatiques), whose works will be Éclair's property. Launch of the science-themed series "Scientia". In the United States, acquisition of a site in Fort Lee, New Jersey, for the building of studios and a laboratory.
1911	New Éclair offices in Moscow and London. Contract with the Buenos Aires Film Society.
1912	New Éclair offices in Barcelona, Milan, and Vienna. Berlin office: Decla Deutsche Éclair. Launch of a weekly filmed newsreel, Charles Jourjon's *Éclair-Journal*, and of "Éclair Coloris" colour films.
1913	Establishment of a new production company, Union Éclair.
1915	March: Éclair produces films shot on the battlefront.
1918	Sale of the Éclair firm's movable property to Serge Sandberg and Louis Aubert. Formation of the Société Industrielle et Cinématographique Éclair (S.I.C.E.). Manufacturing of cameras; mainly laboratory work.
1921	Manufacturing of the "Caméréclair" – cameras designed by Jacques Mathot and Henri Coutant, which remain in use until 1960 – begins.
1931	January: Bankruptcy of S.I.C.E.

Inscriptions on Edges

An article in the 8-14 November 1909 issue of *Ciné-Journal* reports: "In a few weeks' time, films produced by Éclair will bear the following mention on the side: Éclair 1910 Kodak Film; we have been assured that many of our best brands will follow this example."

From that time onwards, Éclair films had inscriptions on their edges[5] (the abbreviation "F.E." stood for "Éclair Films"):

1910:	"PARIS 1910" or FILM "ÉCLAIR" – PARIS 1910 -
1911:	F.E. PARIS 1911
1912:	F.E. 1912
1913:	F.E. 1913

When a film was produced late one year and released early the next year, the edge printing sometimes mentioned both dates. (For example, *Les Deux poltrons*, released in November 1910, bears two mentions: FILMS "ÉCLAIR" PARIS 1910 and F.E. 1911.)

5 See illustrations in "Producers' Edge Marks", pp.63-65.

Edge printings don't appear only on Éclair-produced films. Indeed, in 1910, an agreement was reached between Éclair and the company Le Film d'Art for the processing of films. Consequently some Film d'Art productions bear inscriptions specific to Éclair. To distinguish them, one must refer, if still possible, to the intertitles, which usually have the Film d'Art logo, representing two theatre masks side by side.

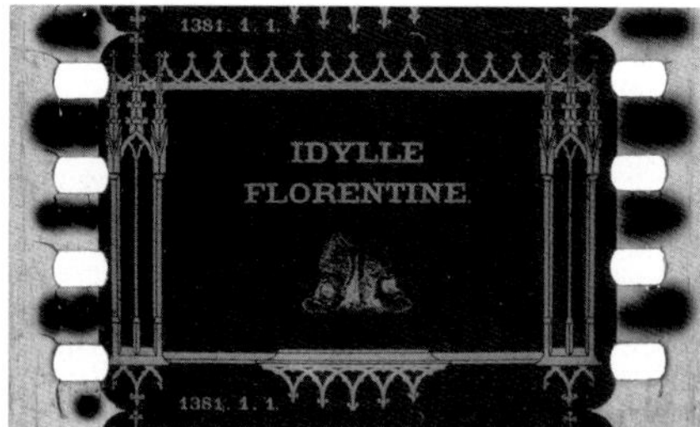

Example of a print made on Éclair-stamped film stock, bearing an Éclair catalogue number, but which turns out to be a 1911 Film d'Art production, *Idylle florentine* – CNC

Intertitles[6]

Éclair films can be identified from several characteristics visible on the intertitles:

Logotype

In 1909-1910, the Éclair logo (one star) appeared on all title cards, with the words "FILMS ÉCLAIR". Between 1910-1911 and 1913, it appeared in the form of a star with a diamond around it, from which lightning bolts spring out, with the words "CINÉMA ÉCLAIR PARIS".

Mention of the Name Éclair

From 1909 to 1912, the name was mentioned on all intertitles. In 1913-1914, the typography was changed (Éclair).

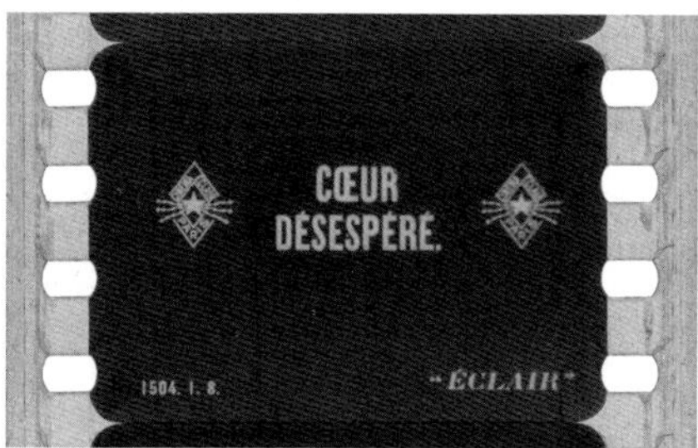

Example of an intertitle on a 1910 Éclair film: *Eugénie Grandet* – CNC

6 See also Éclair intertitles in "Title Styles", pp.103-105.

Catalogue Number

Like other companies, Éclair assigned catalogue numbers to every film. These are sometimes visible at the bottom left on the intertitles. Using the list of catalogue numbers below, one can determine a film's release date, or even find the corresponding title. Other numbers may appear on the intertitles, such as the title card or part numbers, for example.[7]

Catalogue Number	Release Date	Title
1117	22.11.1909	*Le Pays de Foix*
1312	XX.11.1910	*Les Deux poltrons*
1330	22.12.1910	*La Lettre au Père Noël*
1430	20.04.1911	*Le Gilet à pointes*
1460	8.07.1911	*La Machine à attraper les voleurs*
1504	26.05.1910	*Eugénie Grandet*
1550	5.10.1911	*Les Centaures portugais*
1584	16.11.1911	*Copenhague à vol d'oiseau*
1632	26.01.1912	*Un Cri dans la nuit*
1718	31.05.1912	*La Journée d'une Musulmane*
1755	19.01.1912	*Le Bonnet blanc*
1778	16.08.1912	*Une Ville morte: Les Baux de Provence*
1919	31.01.1913	*Gavroche et son fils*
1943	21.02.1913	*La Journée de Lily*
1954	7.03.1913	*Balaoo*
1970	28.03.1913	*Dans la fournaise*
1984	11.04.1913	*Willy diplomate*
1996	25.04.1913	*Gavroche place ses économies*
2011	16.05.1913	*Willy et le paysan pauvre*
2070	18.04.1913	*L'Amblystome*
2085	8.08.1913	*Gavroche et le fils phénomène*
2093	15.08.1913	*Gontran achète un chien de police*
2198	9.01.1914	*Le Bombyx du pin et les chenilles processionnaires*

7 More references on British release dates can be found in "Title Styles", pp.103-105.

Parallel Brand Names

The following names may appear in the intertitles: Scientia, A.C.A.D., A.G.C., American Standard Film, or Savoia. These are all companies affiliated with Éclair.

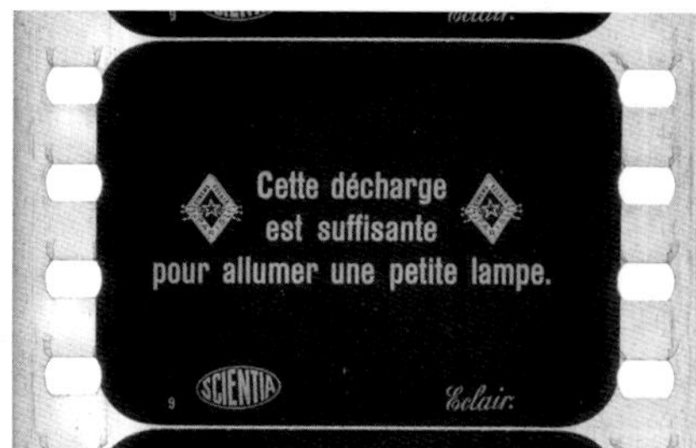

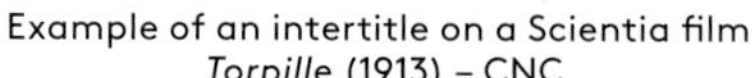
Example of an intertitle on a Scientia film:
Torpille (1913) – CNC

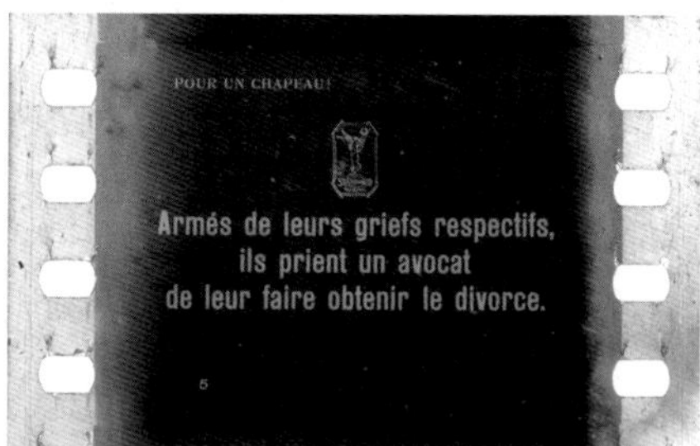

Example of an intertitle on a film produced by American Standard Film:
Pour un chapeau (ca. 1913) – CNC

"Scientia" refers to the series of scientific films produced by Éclair.

A.C.A.D. (Association Cinématographique des Auteurs Dramatiques) was the rival of the Pathé group's S.C.A.G.L., specializing in so-called "artistic" films.

A.G.C. (Agence Générale Cinématographique) was the oldest film rental agency in Paris; it secured an exclusive deal with Éclair, Film d'Art, and Éclipse.

American Standard Films was Éclair's American subsidiary.

Savoia was an Italian company, whose production was partly distributed in France by Éclair.

Film Title

If the first title card is missing, the film's title can sometimes be found in subsequent intertitles.

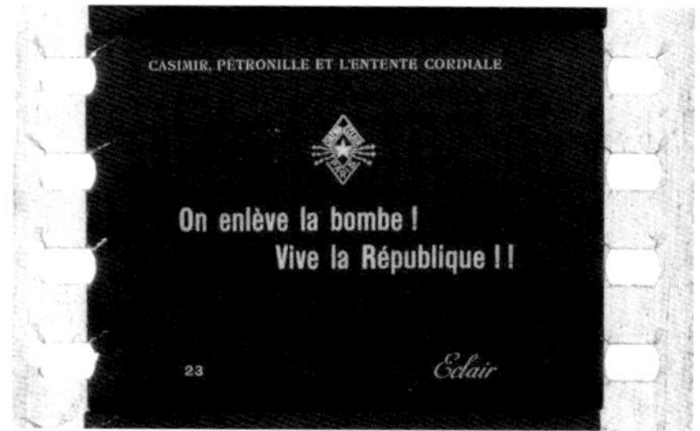

Example of a 1914 intertitle that mentions the film's title:
Casimir, Pétronille et l'Entente Cordiale – CNC

Logotype

To avoid piracy of its films, Éclair incorporated its trademark in their sets. Its logo took the form of a star enclosed in a diamond, from which six lightning bolts spring out. In the diamond are the words "CINÉMA ÉCLAIR PARIS".

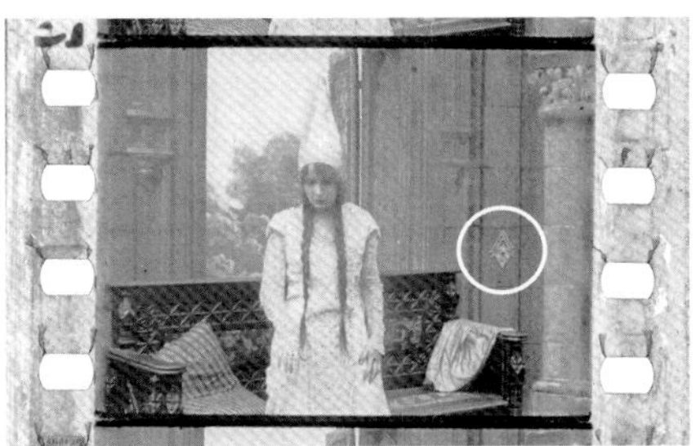

Presence of the Éclair logo in the set, on the right: *Ombre de l'aimée* (1912) – CNC

Series

Bou-Bouf, starring Moret (1916-1917)
Casimir, starring Lucien Bataille (1913-1917)
César (1915-1917)
Charley, starring Villa (1911)
Les Dragonnades sous Louis XIV (1909)
Gavroche, starring Paul Bertho (1912-1914)
Gontran, starring René Gréhan (1910-1916)
La Légende de l'aigle (1912)
Lily (1913-1914)
Meskal le contrebandier (1909)
Morgan le pirate (1909-1910)
Nick Carter, starring Pierre Bressol, and then Charles Krauss (1908-1912)
Pétronille, starring Sarah Duhamel (1912-1917)
Protéa, starring Josette Andriot (1913-1919)
Riffle Bill (1908-1909)
Rouletabille (1913-1914)
Sherlock Holmes (1914-1917)
Teddy, starring Édouard Pinto (1911)
Tommy (1911-1911)
Toto (1911-1914)
Le Vautour de la Siria (1909)
Willy, starring Willy Sanders (1911-1916)
Zigomar, starring Alexandre Arquillière (1911-1913)

Some Bibliographical References

"Société Française des Films et Cinématographes Éclair (1907-1919): A Checklist", *Griffithiana* (May-September 1992), pp.28-89.

"Éclair 1907-1918", *1895,* No. 12 (October 1992).

4
Éclipse

by Eric Loné

History (1906-1923)

1906 August: The firm Éclipse is established as an offshoot of the French subsidiary of the Urban Trading Co.
1908 Éclipse sets up its studios in Boulogne.
Éclipse takes over the firm Radios and produces several films under the label Éclipse-Urban-Radios.
1917 Establishment of Ciné-Location-Éclipse, a film distribution branch.
1923 Éclipse suspends production.

Intertitles

The Éclipse firm's intertitles in the years 1908-1910 were generally tinted red. (See examples in the Colour Section, p.156). The name "Eclipse" appeared at the bottom right, and the production number and shot number, plus a group of numbers or letters (the meaning of which is unclear) appear at the bottom left. The typography of the intertitles produced by Éclipse is easily identifiable. The film title appears at the top centre of the intertitle. The Éclipse logo sometimes appears on the intertitles or at the end of the reel.

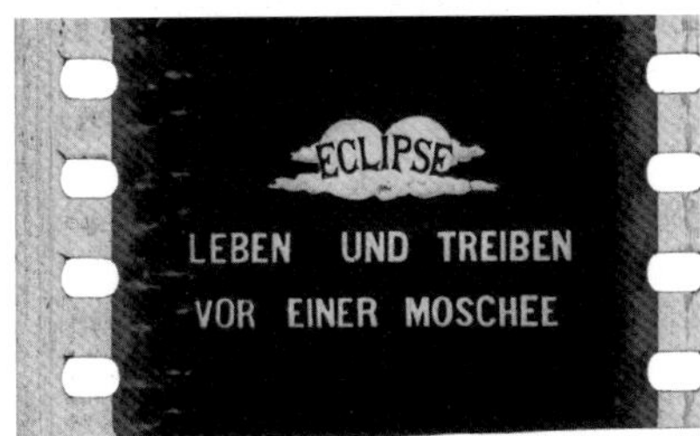

Example of an intertitle on a print of an Éclipse film from ca. 1912: [*Sur le Bosphore*] - GEM

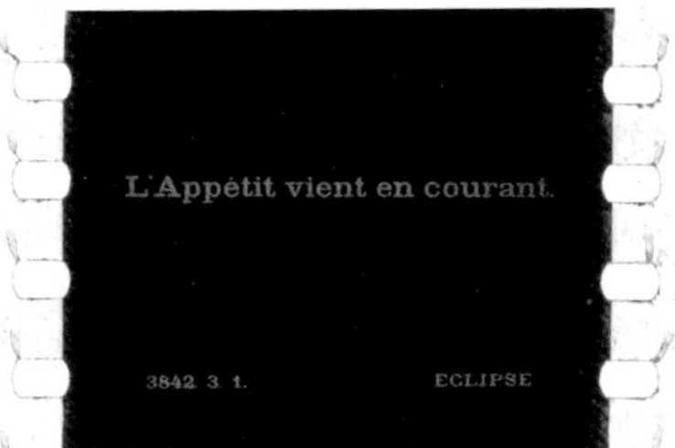

Example of an intertitle on a print of an Éclipse film from 1912: *Une idylle montmartroise* - CNC

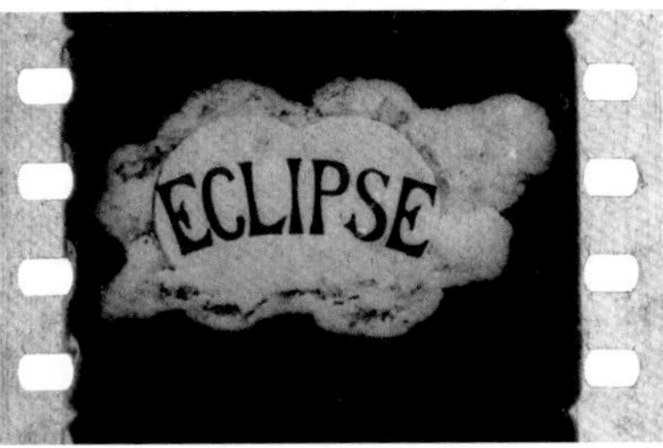

Éclipse title at the end of a print of a film from 1913: *Le Cuirassé Edgar Quinet - La Vie à bord* - CNC

Catalogue Numbers

The presence of catalogue numbers on the intertitles can be helpful; they can be compared with the numbers in the chart below to determine the approximate date of a film's release. It would appear that Éclipse assigned several series of numbers depending on a production's origin, as Urban and Radios were part of the same group. Thus, catalogue numbers below 3,000 were reserved for Urban productions. The films produced by Éclipse itself were assigned numbers in the 3,000s, and later 5,000s. Radios productions were numbered 4,000 to 4,999 (although the latter number was never reached). These catalogue numbers seem to have disappeared after 1915.

Selected catalogue numbers with their corresponding film titles and their release dates:

Catalogue Number	Release Date	Title
3679	1911	*L'Auberge sanglante*
3842	1912 (May)	*Une idylle montmartroise*
3874	1912	*Arthème échappe encore*
3963	1913	*Le Cuirassé Edgar Quinet – La Vie à bord*
3973	1913 (June)	*Statue d'épouvante*
5047	1913	*Papillon prend la mouche*
5056	1914 (February)	*Maud en culottes*
5082	1914 (April)	*Polycarpe fait la morale au centimètre*
5137	1914 (August)	*La Chemise de Polycarpe*
5141	1914 (July)	*Maud clubman*
5194	1915 (December)	*Polycarpe portraitiste*

Series

Arizona Bill, starring Joë Hamman (1911-1913)
Arthème, starring Ernest Servaes (1911-1916)
Barnet Parker, starring Henry Houry (1914)
Cri-Cri, starring Paul Bertho (1910-1911)
Fred, starring René Hervil (1914-1916)
Godasse, starring Raimu (1911-1912)
Maud, starring Miss Aimée Campton (1912-1915)
Nat Pinkerton, starring Pierre Bressol (1910-1914)
Papillon, starring Cauroy (1913-1914)
Polycarpe, starring Édouard Pinto and then Charles Servaës (1912-1917)
Séraphin, starring Charles Servaës (1914-1916)
Teddy, starring Édouard Pinto (1910-1916)

Bibliography

Youen Bernard, *L'Éclipse. L'Histoire d'une maison de production et de distribution cinématographique en France, de 1906 à 1923*. Maîtrise d'Études Cinématographiques et Audiovisuelles (Master's Thesis, Université Paris 8, Vincennes-Saint-Denis. Supervisor: Christian Delage), 1992-1993. (Includes a filmography.)

5
Lux

Eric Loné with the contribution of Camille Blot-Wellens

History (1906-1913)

1906 November: Lux is founded by Henri Joly. Opening of a branch in Berlin and offices in Brussels, Barcelona, Budapest, Buenos Aires, Genoa, London, Moscow, Warsaw, Vienna, and New York. The plants, located in Gentilly, are managed by Léopold Löbel.
1907 Beginning of the shooting of films in the Paris studios.
1911 Lux takes up the European distribution of films by the U.S. company Nestor.
1912 Lux distributes films by Aquila, Flying A, Psyché-Albano, Reliance, and Western Import Co.
1913 December: The liquidation of Lux's assets is announced.

Inscriptions on the Edges

Looking at several Lux films dated 1907 to 1913, there are three main cases, each corresponding to one of three different periods in Lux's productions: the earliest have transparent edges (*Mésaventures d'un réserviste*, 1907), others have black edges (it seems these correspond to the period of late 1907 to early 1908), the later ones bear the inscription "Lux Paris", which appeared regularly on both edges of the film (*Drame sur une locomotive*, 1910). As was the practice of most other production companies of the era, the inscription was printed on Kodak film stock using a special device made by Debrie.

According to the then-director of the Lux plant, Léopold Löbel, "Manufacturers are used to printing their business name on all the positives they edit. The letters of the inscription are placed between the perforations (...). The machines used for this purpose consist of a central drum (...) Inside this drum is an incandescent lamp. The drum's rim is lined with pins to move the film, and between the pins is a foil strip in which the text to be printed is hollowed out. (...) The drum is rotated, and then the lamp's light rays impress the film through the foil's indentations."[1]

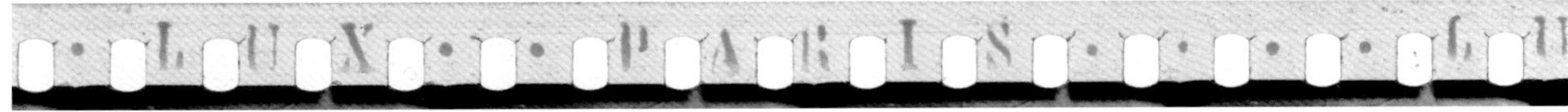

Edgeprint lettering, *Patouillard Crieur de journaux* (1912) – FE

Frame-lines and Characteristics

Examination of the positives' frame-lines in Lux films shows that one of the edges (the thickest) encroaches upon the perforations, while the one opposite borders the edge of the perforations. The image's corners are curved. In 1913, despite some very slight changes, that formula remained characteristic.

1 Léopold Löbel, *La Technique cinématographique. Projection – Fabrication des films*, Paris: H. Dunod & E. Pinat (1912), p.205.

Another characteristic that can be observed in several Lux productions is a notch in the camera aperture.

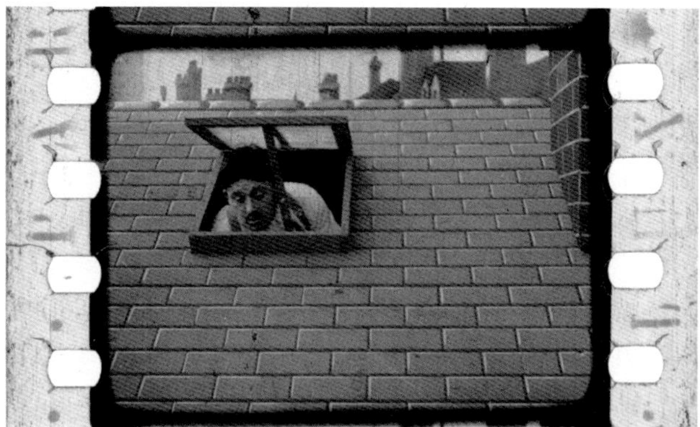

Patouillard Crieur de journaux (1912) – FE

In the print of *Patouillard Crieur de journaux* (1912), the notch is located near the second perforation in the left part of the image.

This notch varies from film to film and does not appear on all reels (it is missing from a 1908 production). This could be due to the use of a Debrie Parvo camera.[2]

Logo

The Lux company's logo depicts the Statue of Liberty. It is sometimes visible in the background of a set, especially it seems for films dated 1907-1910. It should be noted that the same logo also appears, more conspicuously, on the main title.

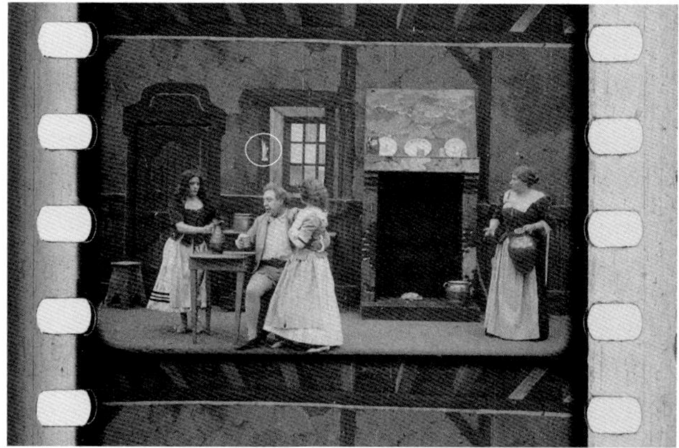

On this set of a film from ca. 1908(?), the Lux logo appears in a window frame. – SFI

2 According to archivist and researcher Encarni Rus, the same type of notch can be seen in a Spanish production from the 1920s filmed with a Debrie Parvo JK.

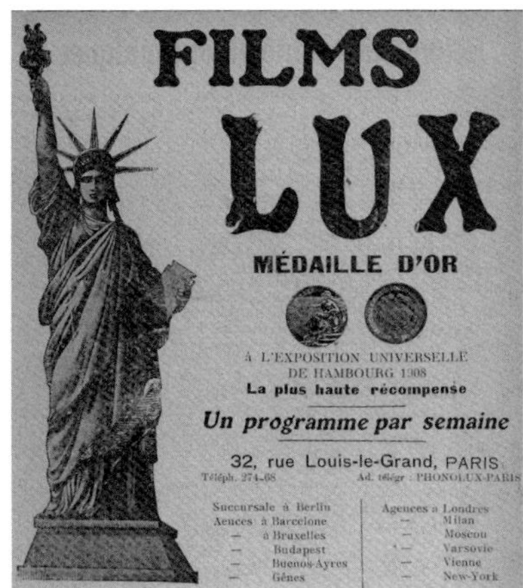

Advertisement for Lux in *Ciné-Journal* (1909) – MoMA (archive.org)

Intertitles

Depending on the films examined, the following characteristics were noted:

Between 1907 and 1910, the main title and intertitles were tinted orange.[3] The production number, shot number, and letter denoting the language version appeared at the top left of the title (separated by dots or stars). On the main title, the Lux logo appears at the bottom right followed by the words CINEMA "LUX" PARIS MARQUE DÉPOSÉE.

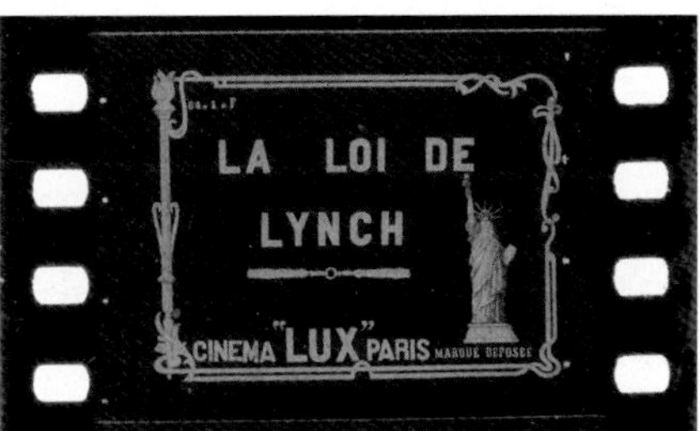

Main title on a print of
La Loi de Lynch (1907) – CNC

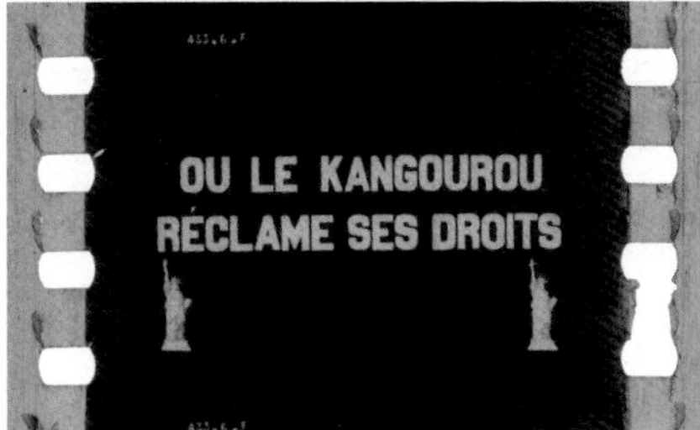

Intertitle on a print of
Le Monsieur qui a mangé du kangourou (1909) – CNC

No example from 1911 could be examined.

In 1912 the intertitles were tinted green. The word "LUX" appeared on the titles at the bottom right. At the bottom left was the production number, the intertitle number, and the version letter.

In 1913 the inserts and intertitles were tinted green. The production number, intertitle number, and version letter were listed at the bottom left. The "LUX" logo appears at the bottom right.

3 See examples of Lux tinted titles in the Colour Section, pp.157-158.

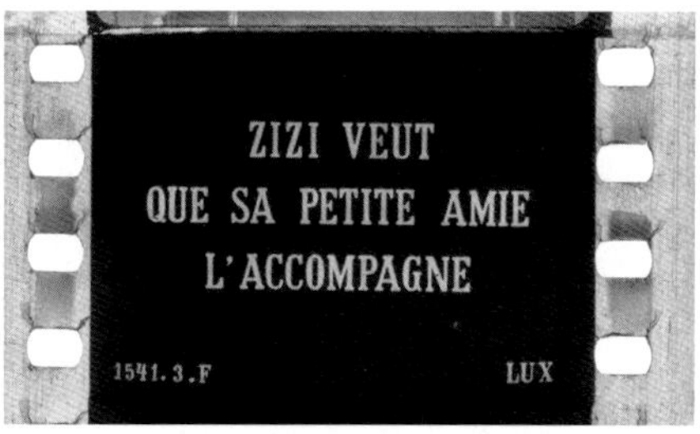

Intertitle on a print of
Zizi fait des courses (1913) – CNC

The sample of films examined remains insufficient to generalize these observations.

Catalogue Numbers

It is common for the production number to appear on the main title card and intertitles (at the bottom, or on the upper left). They are sometimes followed by the shot number, then a letter probably indicating the language version (e.g., F for the French version[4] or AG for the German version).

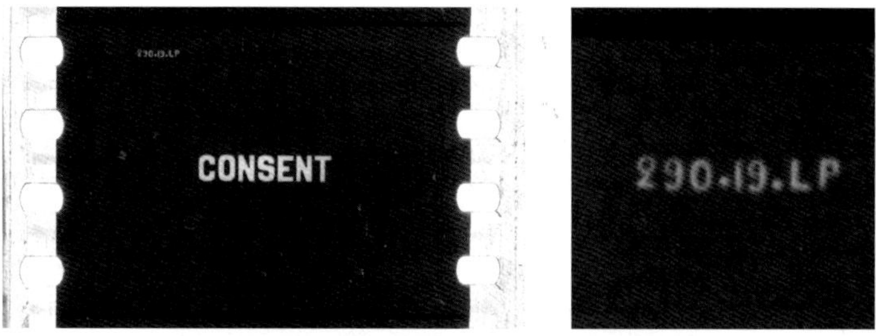

Intertitle on an English print of
L'Enlèvement. Production No. 290 (1909) – CNC

These numbers are more or less assigned chronologically, which allows us, by comparing them with other numbers already known (see the list below), to place a film's date of production in the period 1907-1913. For example, if a Lux film bears the number 992, it is likely to have been shown just a few weeks after film number 976 in the list below, which is known to have been premiered in June 1912.

A few catalogue numbers with the corresponding films and release dates:

Catalogue Number	Release Date	Title
25	1907	*Les Suites d'une nuit d'ivresse*
45	1907	*Mésaventures d'un réserviste*
64	1907	*La Loi de Lynch*
84	1907	*L'Ange de Noël*
104	1908	*La Revanche du chat*
131	01.06. 1908	*Les Tribulations de Pandore*
206	06.10.1908	*Le Costume blanc*
212	06.10.1908	*L'Engrais merveilleux*
228	19.11.1908	*L'Assoiffé*
233	26.11.1908	*La Légende des étoiles*
260	24.12.1908	*Le Jour de l'An d'un pauvre petit*

4 See French intertitles on pp.191-192.

Catalogue Number	Release Date	Title
290	19.03.1909	*L'Enlèvement*
295	11.03.1909	*L'Apôtre des Gaules*
346	27.06.1909	*Sous le drapeau*
382	01.04.1910	*L'Image de l'aimée*
407	06.05.1910	*Le Rêve de la dentellière*
433	03.01.1910	*Le Monsieur qui a mangé du kangourou*
446	10.01.1910	*Le Pneu Machin boit l'obstacle*
461	18.03.1910	*Rico le bouffon*
480	29.04.1910	*Fatima*
556	05.09.1910	*Le Bâton du policeman*
578	24.10.1910	*Le Gros monsieur et la petite baignoire*
592	19.12.1910	*Un drame sur une locomotive*
820	09.02.1912	*Le Chien du chemineau*
861	12.07.1912	*Le Chien voleur*
923	28.06.1912	*Patouillard a une femme qui veut suivre la mode*
944	14.06.1912	*La Maison blanche*
976	28.06.1912	*La Poudrière*
1039	30.08.1912	*Les Femmes députés*
1107	01.11.1912	*Cunégonde ramoneur*
1541	08.08.1913	*Zizi fait des courses*
1544	15.08.1913	*Zizi et le corset-réclame*

Series

Bob (1912)
Cunégonde (1911-1913)
Jacobus (1911)
Moumoune (1913)
Princesse Cartouche (1911-1912)
Patouillard, starring Paul Bertho (1910-1912)
Teddy, starring Édouard Pinto (1912)
Toto (1910-1913)
X le Mystérieux, starring Henri Jullien (1912)
Zizi, starring Zinel (1913)

Bibliography

Youen Bernard, *Les petites maisons de production cinématographiques françaises de 1906 à 1914*. DEA (Certificate of further education) in Film and Audio-Visual Studies (Dissertation, Université Sorbonne Nouvelle – Paris 3, Censier. Supervisor: Michel Marie), 1993-1994.

Eric Loné, "La Production Lux", *1895*, No. 16 (June 1994), pp.59-76 (article and filmography).

6
Pathé

Camille Blot-Wellens

Pathé films represent an interesting subject for several reasons:

- The importance of the production of Pathé Frères and the unprecedented network of the company, which meant that films were distributed almost everywhere on the planet, and now represent a significant part of the collections of archives worldwide;
- The importance of the company, which invested energy and resources in technological developments that are reflected in the film materials;
- The information introduced by the manufacturer which makes the films of the firm easily "recognizable" by archivists;
- The company's being both a producer and film manufacturer, which results in two aspects of identification of film stocks, treated in this publication.[1]

Pathé as Production Company

The concern of Pathé – as of so many film companies during the first years of the film industry – to fight against illegal copies led the French firm to develop several systems to make the materials they produced easily recognizable by their customers.

Logos in the Settings

As pointed out by Harold Brown,[2] Pathé inserted several kinds of trademarks in their settings. In the early productions, it is possible to see "PF" (for Pathé Frères), but the practice, observed mainly in films prior to 1900, soon disappeared.

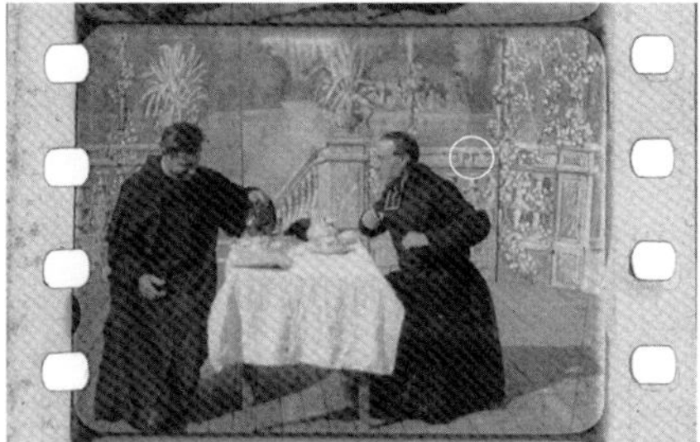

Une bonne histoire (No. 589, 1897-1899) – FE

1 See also Camille Blot-Wellens, "Quelques aspects de la datation des éléments filmiques", Jacques Malthête and Stéphanie Samon (eds.), *Recherches et innovations dans l'industrie du cinéma. Les cahiers des ingénieurs Pathé (1906-1927)*, Paris: Fondation Jérôme Seydoux-Pathé (2017), pp.178-193.
2 See p.110 in this volume.

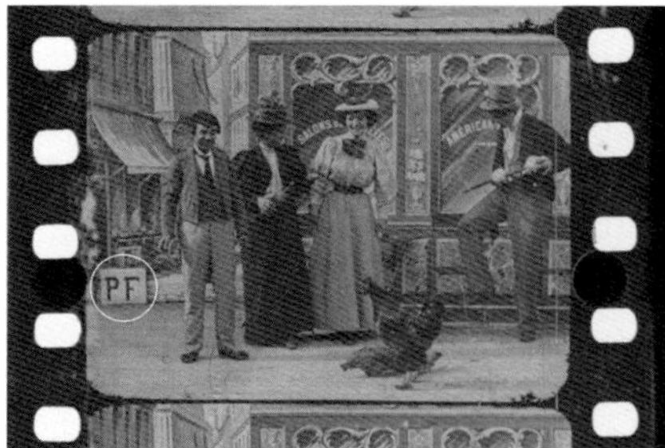
Paysans à Paris (No. 601, 1897-1899) – FE

Later (apparently from 1904[3]), Pathé would introduce the image of a rooster (the famous Pathé coq). This practice continued until at least 1910.

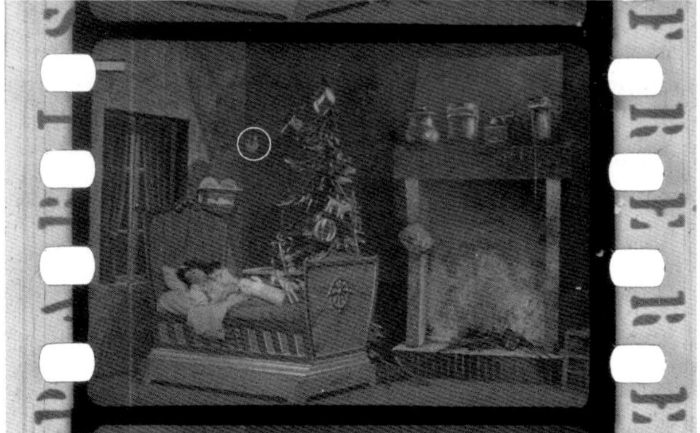
Miracle de Noël (No. 1315, 1905) – SFI

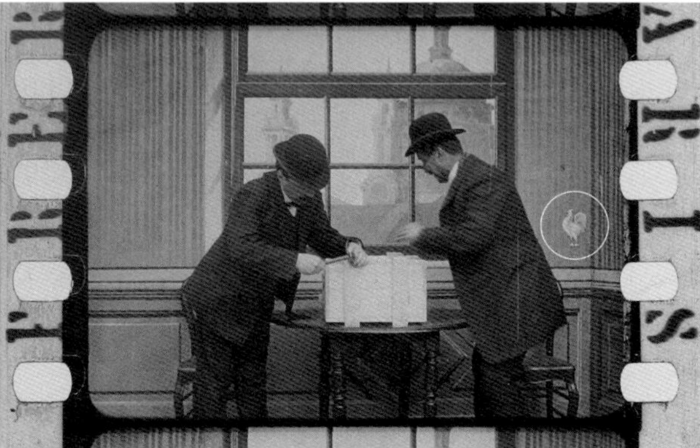
Voleurs de bijoux mystifiés (No. 1366, 1906) – SFI

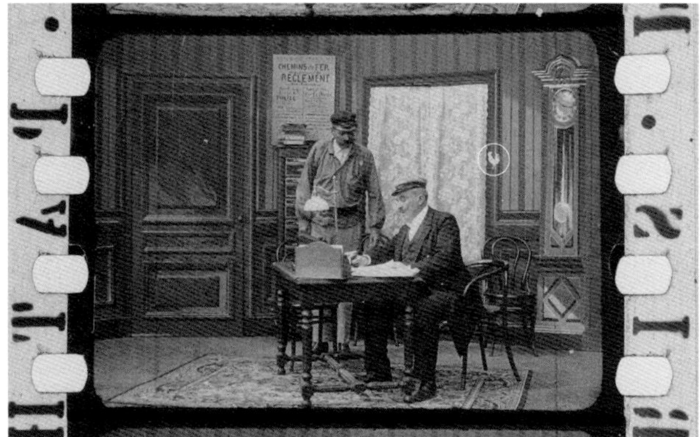
Un attentat sur la voie ferrée (No. 1570, 1906) – SFI

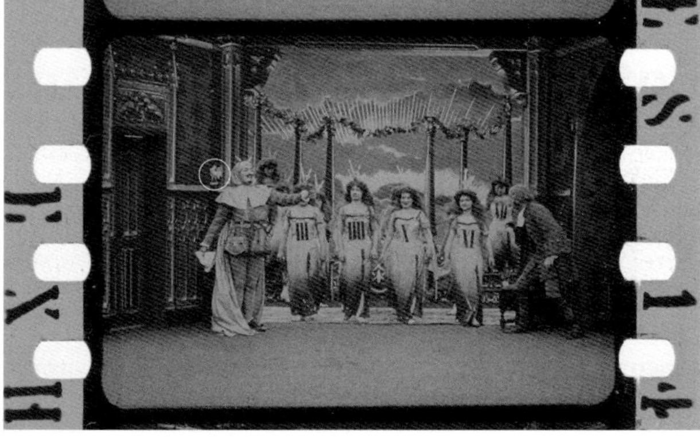
Le Secret de l'horloger (No. 1940, 1907) – FINA

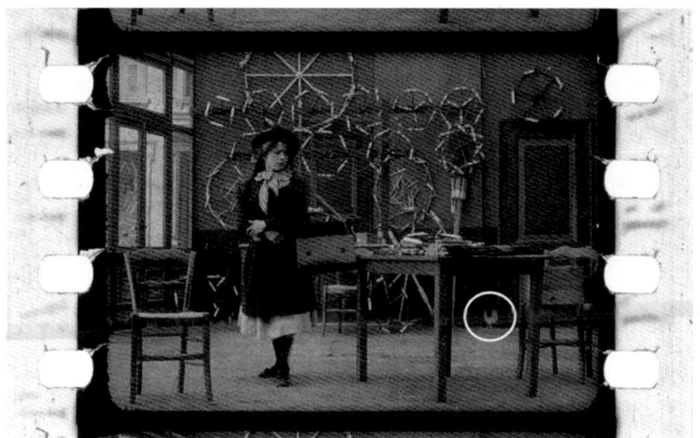
Les Pétards de Léontine (No. 2647, 1910) – FE

3 See Jan Olsson, "Rooster Play: Pathé Frères and the Beginnings of In-Frame Trademarks", *The Moving Image*, Vol. 18, No. 2 (Fall 2018), p.12

However, it is important to note that these trademarks were not systematic, and that they mainly appeared in scenes shot in interior settings and not in exteriors. Note as well that the company's logo appeared only in some scenes of a film; therefore they may have disappeared if the element studied is incomplete. More systematic would be the insertion of titles.

Titles / Intertitles

As a production company, Pathé understood early on the benefits of inserting a main title (at the beginning, this consisted of only a few frames). Like many other producers, Pathé inserted titles (which were optional) as early as 1901. But perhaps more actively than other companies, Pathé used the main title as a trademark (creating a representative image for the company) and as a tool to fight against illegal copies as soon as 1903.[4]

The company even published a "VERY IMPORTANT NOTICE", which appeared in the catalogue of its London branch in May 1903, and in the English supplement dated September—October 1903, with more details[5]: "With all our films, we supply a length of about 1m50 which is attached, and bearing the title of the subject and our trade-mark; this protects us against imitators, and affords our clients the advantage of economising their expenditure as the use of a second lantern for fixed titles is not required. We supply these titles in French, English, German, Spanish or Italian[6], as our clients wish them. Consequently, all those who procure our films through dealers, agents or others, should insist on having the title in red with our trade-mark, the Coq, to the right and left, which is the only guarantee that they are of our manufacture and not merely copies."

The red tinting of the title (a colour that can be found on prints from other companies) made illegal duplication more difficult. Since the emulsion was orthochromatic, it was not sensitive to red, and when duplicated the title disappeared (making the positive image totally black).

This practice is of course very useful to identify the production company and the title of the film — when the title is still on the element. Unfortunately, being at the beginning of the print, titles easily suffered damage, and even loss.

The title's design can also give us information about the date of its manufacture. As Suzanne Richard observed in her researches in the 1980s,[7] the graphic elements vary throughout the years[8]:

Title 1903 (French) - FE

Title 1904 (German) - SFI

Title 1905 (Spanish) - FE

Intertitle 1908 (Swedish) - SFI

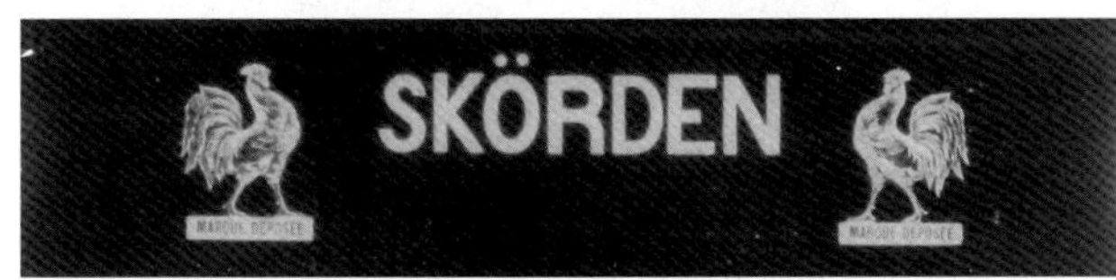

Intertitle 1909 (Swedish) - SFI

4 See Claire Dupré la Tour, "Des titres d'excellente qualité: enjeux et développements chez Pathé, 1903-1908", in Jacques Malthête and Stéphanie Salmon (eds.), *Recherches et innovations dans l'industrie du cinéma. Les cahiers des ingénieurs Pathé (1906-1927)*, Paris: Fondation Jérôme Seydoux-Pathé (2017), p.121.
5 Claire Dupré la Tour, p.122.
6 The languages provided by the company give us interesting information on the evolution of the market.
7 Suzanne Richard, "Pathé, marque de fabrique. Vers une nouvelle méthode pour la datation des copies anciennes", *1895*, No. 10 (October 1991), pp.13-27.
8 For more examples of Pathé titles, refer to the Colour Section, pp.159-160.

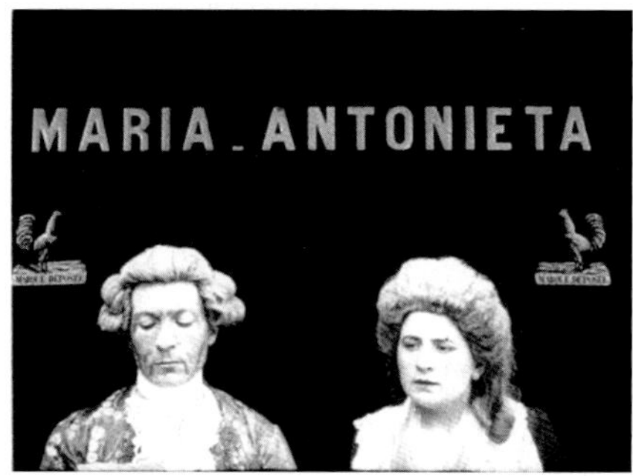

Marie Antoinette (No. 1023, 1903) – FE

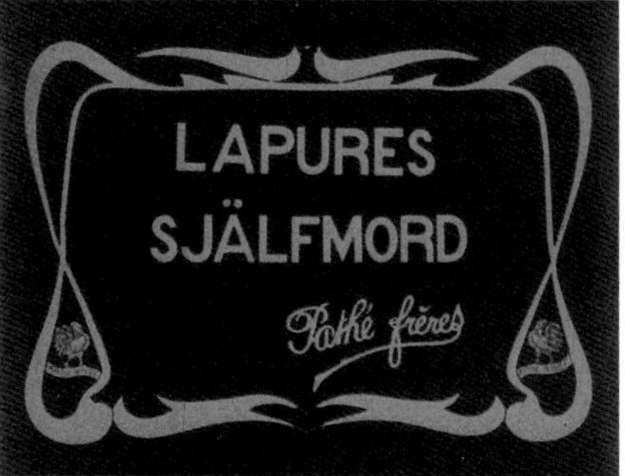

Les Suicides de Lapurée (No. 3224, 1909) – SFI

Nevertheless, it's important to note that similar titles can be found in prints of different years, and that they might be used for several years (as, for instance, in the various versions of *The Life of Jesus*), and that they don't always correspond to the year of a film's production or the making of a print. Therefore, they should not be the unique reference for dating a print or the production of a title.

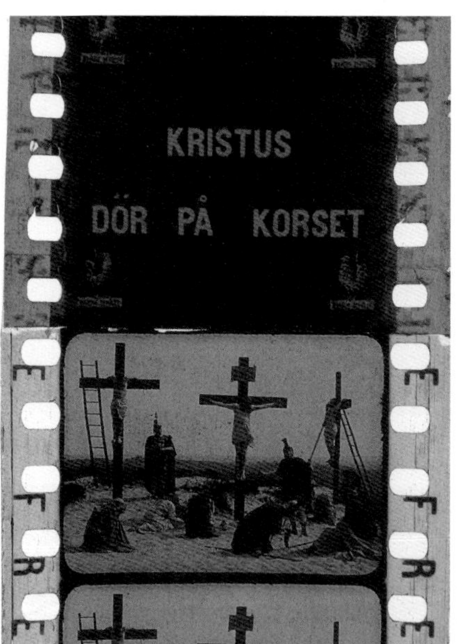

La Vie de Notre Seigneur Jésus-Christ. Passion et mort de Jésus (No. 1607, 1914) – SFI
Diacetate print made in all likelihood around 1914, with titles made on nitrate film stock of 1905.

Edge Printings

From 15 April 1905,[9] Pathé would reinforce its strategy by inserting its name on the edges of the film (edge printings). This time the information was not visible on the screen by the audience. Edge printings were introduced for buyers to certify that the prints had been made at the laboratory of the firm.

The evolution of edge printings is very useful to date prints, as described by Harold Brown in 1967.[10] Gerhard Lamprecht, founder of the Deutsche Kinemathek, was also able to determine the dates of prints according to changes in edge printings by studying nitrate materials in their collections, and could refine the dating of Pathé prints even more than Brown:

1908
PATHÉ FRERES 14 RUE FAVART

EXHIBITION INTERDITE EN FRANCE ET EN SUISSE

1909-1911
PATHÉ FRÈRES 14 RUE FAVART PARIS

EXHIBITION INTERDITE EN FRANCE EN SUISSE EN BELGIQUE (or ET EN BELGIQUE)

Late 1911-1912
PATHÉ FRÈRES 14 RUE FAVART PARIS

EXHIBITION INTERDITE EN FRANCE EN SUISSE EN BELGIQUE ET EN ITALIE

Late 1912-1913
Same edge printings with small, sloping letters.

1913
Same edge printings with small, straight letters.

Late 1913-1914
Same edge printings with narrow, straight, high letters.

1914
PATHE FRERES PARIS (narrow, straight, high letters), or no edge printings at all.

If the indication of the date appeared only in 1905, it is plausible that the changes in the edge printings, notably from 1907, follow the evolution of the politics of Pathé concerning distribution after the establishment of the rental system, since the countries included in the edge printings change.

But because this kind of information was intended primarily for the exhibitors, it is mainly visible on the prints. If this is very useful to date the period of the making of prints, it does not always allow us to date the production of a film. This aspect can lead to a wrong identification of a production title, especially at a period when different versions were still very much present in company catalogues.

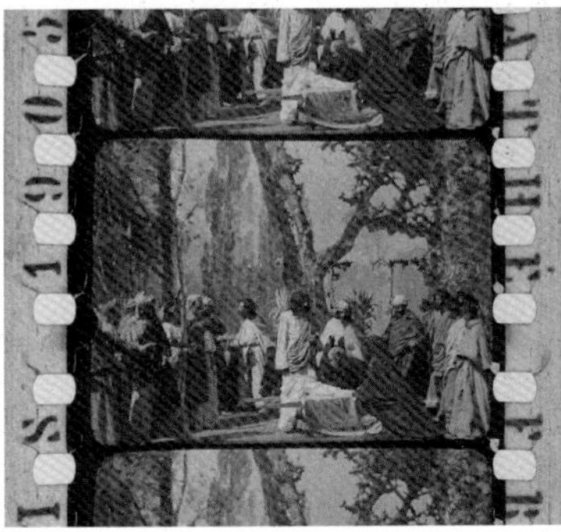

Vie et Passion du Christ. Résurrection de Lazare (No. 854, 1897) – SFI. Print made around 1905.

9 Pathé Frères, *Films. Supplément Avril-Mai 1906*, Paris, 1906, p.5.
10 Harold Brown, *Notes on Film Identification by Examination of Copies* (1967 paper, no pagination). See also Brown, "Producers' Edge Marks" in this volume, pp.50-53.

It is therefore necessary to look for other characteristics in order to date the production of a film, information independent of the making of the print, i.e., derived from examining the negative. Because negatives inform us about the equipment and apparatus used by the company, they allow a much more accurate dating of the period of a film's production.

The study of Pathé films is rather easy, thanks to the important volume of Pathé materials (and recognized as such) in the archives. The comparison of film materials is the most efficient methodology to understand the evolution of these characteristics and to be able to date them.

Pathé as Apparatus Manufacturer

The French firm had an important research department, and the manufacture of equipment represented a significant part of its activities, since its apparatus for shooting, printing, or application of colour were continuously improved.

Application of Colour

For instance, numbering can be observed on every 25th frame, either on negatives or on stencil-coloured scenes on prints (then photographed from negatives).[11]

It is credible that this system was introduced towards 1906 and was very likely linked to the application of colours with stencil.[12] The presence of these numberings helps us not only to situate the production of a film from 1906 onwards, but also to know which scenes were planned to be stencil-coloured from the beginning, even if the element studied is in black & white, or tinted/toned without stencil.

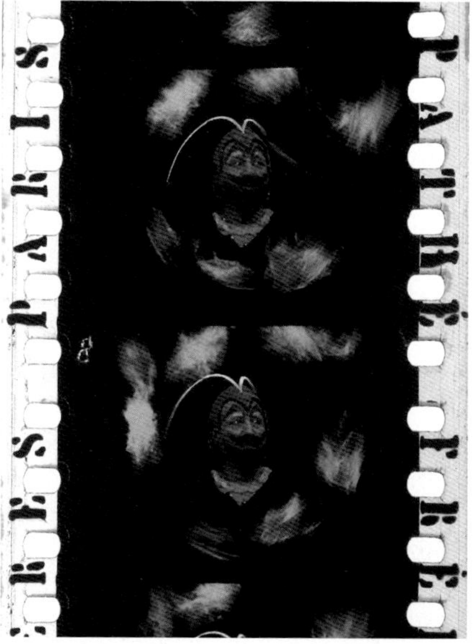

Les Flammes mystérieuses (No. 2174, 1908) – SFI

11 Surprisingly, it was possible to find in the Harold Brown archives some notes on a numbering system introduced between the frames on the firm's negatives, an aspect that he does not treat in his 1990 book.

12 The oldest film on which the author could observe this characteristic was *Le Sorcier arabe* (No. 1446, 1906).

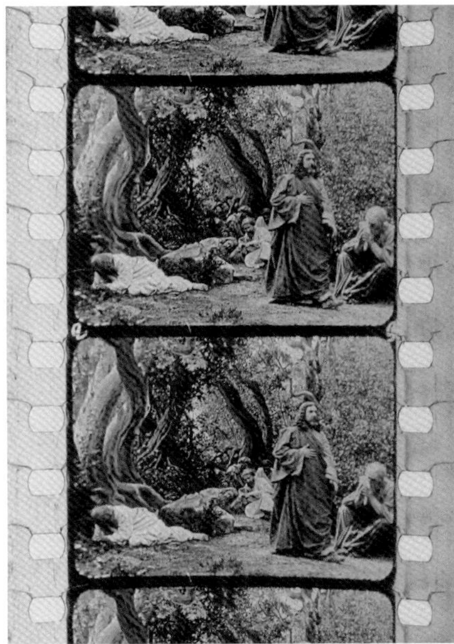

Scenes from *La Vie de Notre Seigneur Jésus-Christ. Passion et Mort de Jésus* (No. 1607, 1914), inserted in *The Photo-Drama of Creation* (Pastor Charles Taze Russell, prod. IBSA, 1914) – SFI
In this case, the number photographed in the frame-line from the negative indicates that this scene, tinted in this print, was meant to be stencil-coloured.

The characteristics of the equipment used for a film's shooting can also be observed on prints, helping a much more accurate dating of a film's production (especially useful for non-fiction films, where other elements – such as logos in the settings – are missing), independently of the date of a print. Harold Brown mentions this aspect in both of his 1967 and 1990 publications.[13]

Frames

The prints of Pathé films from the Sagarmínaga Collection at the Filmoteca Española allowed us to observe significant changes in frames, permitting a more accurate dating of both the making of prints and the production of titles between 1896 and 1904[14]:

1896-1899

The frame is very thin and rounded in the inside of the corners, while the outside is rather square. The position of the frame-line can vary. The frame can be moved towards one of the edges. The edges are either black or transparent.

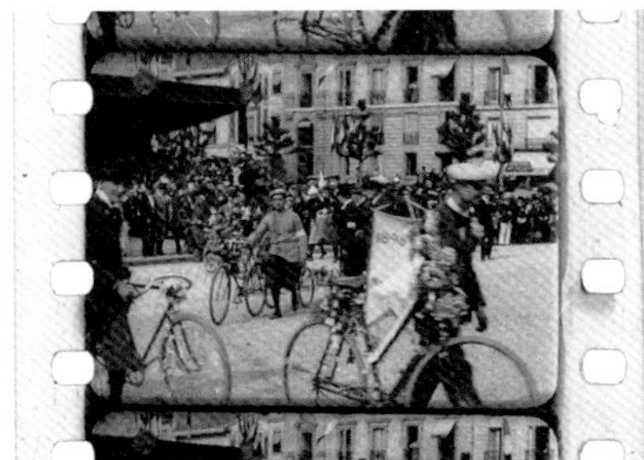

Corso cycliste (No. 318, 1897) – FE

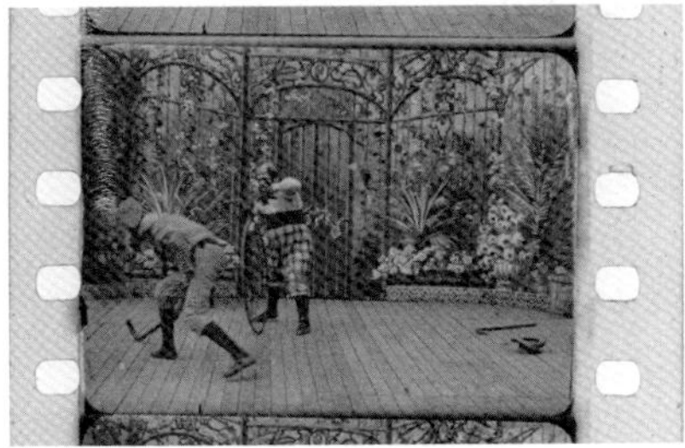

Fantaisie cycliste (No. 346, 1897-1899) – FE

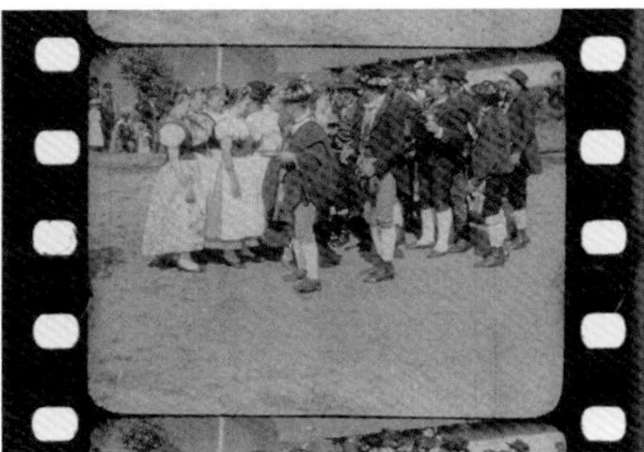

Danse bohémienne (No. 743, 1897) – FE

13 Regarding Pathé films, Harold Brown might not have had access to enough film materials to come to a conclusion in terms of dating, even if there are also notes on negative perforations in his personal archives held at the British Film Institute.

14 See Encarni Rus Aguilar and Camille Blot-Wellens, "Estudio e identificación de películas de los primeros años del cinematógrafo. La colección Sagarminaga", *Journal of Film Preservation*, No. 65, (2002), p.48.

1900-1901

The frame becomes thicker and more rounded outside the corners. The position of the frame-line can vary. The edges are transparent.

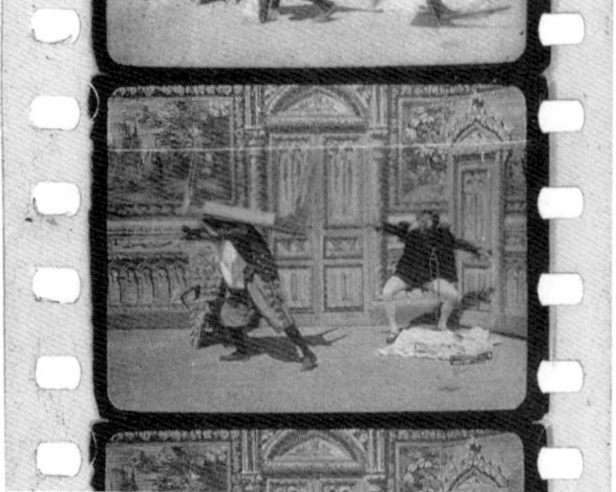

Repas infernal (No. 681, 1901) – FE

From 1902 Onwards

The frame is more regular, rounded on the inside of the corners and more squared on the outside. The position of the frame-line can vary. The edges are transparent.

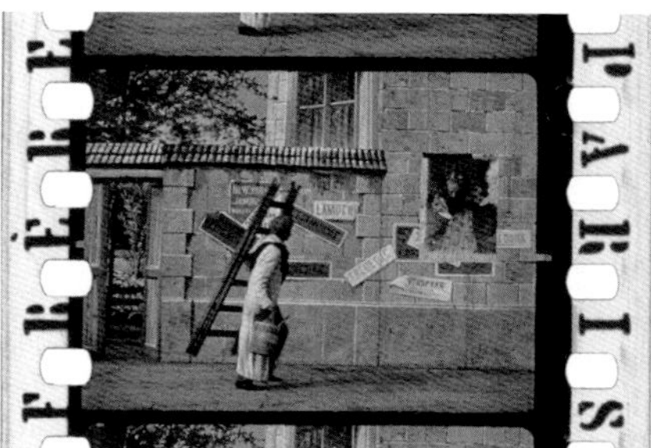

Défense d'afficher (No. 642, 1902) – FE

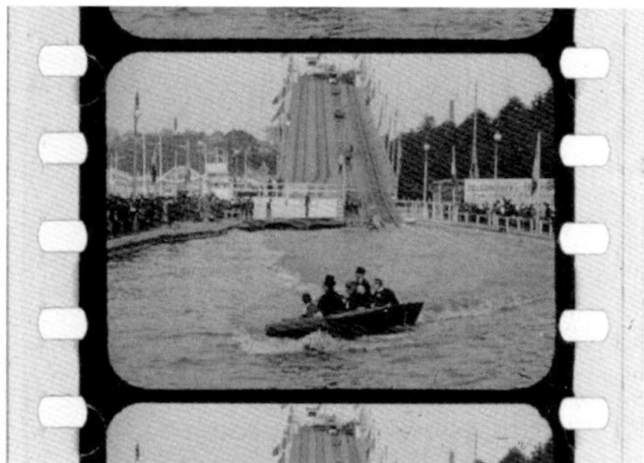

Montagnes russes nautiques (No. 883, 1902) – FE

From 1905 onwards, Pathé prints have edge printings that facilitate the dating of prints.

Perforations

Perforation is crucial at each step of film manufacturing: shooting, printing, and screening. The quality of perforation affects stability, quality of stencilled colours, and resistance during projection, for instance. Pathé engineers worked constantly to improve perforations.

Positive perforations changed a great deal during the company's first years:

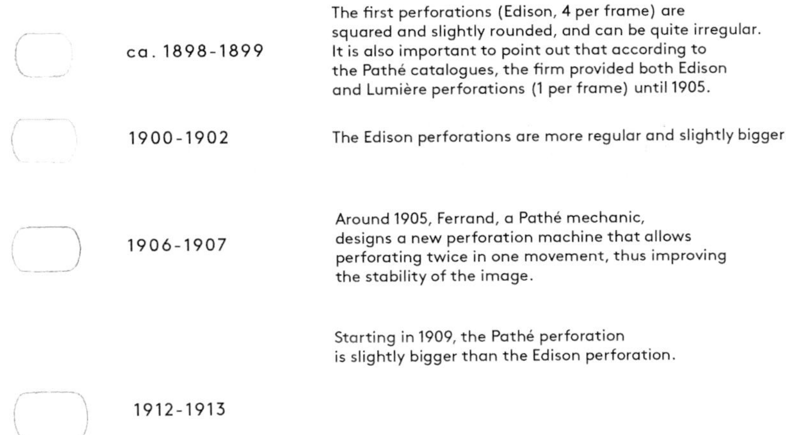

The most relevant piece of data to date the production of a film is probably the perforation of the negative, since it reflects the camera used.[15] As we said earlier, the firm worked ceaselessly to improve its equipment, and the changes can be observed on the materials.

15 Recently, Anne Gourdet-Marès (responsible for the apparatus collections at the Fondation Jérôme Seydoux-Pathé) and I tried to identify the cameras that were used for the shooting of films produced between 1896 and 1909 in order to better understand the intensive use of a unique perforation by the firm, notably between 1906 and 1909. This study was very fruitful, and it was indeed very interesting to connect the film elements with the apparatus used for their production. We believe that this study should be continued in the future to allow a better understanding of the history of film techniques. See also Anne Gourdet-Marès and Camille Blot-Wellens, "Hypothèses sur l'utilisation d'une perforation unique dans les ateliers Pathé (1906-1909)", in Réjane Hamus-Vallée, Jacques Malthête, and Stéphanie Salmon (eds.), *Les Mille et un visages de Segundo de Chomón. Truqueur, coloriste, cinématographiste... et pionnier du cinématographe*, Paris-Villeneuve d'Ascq: Fondation Jérôme Seydoux-Pathé / Presses universitaires du Septentrion (2019), pp.133-151.

Negative perforations observed on Pathé films:

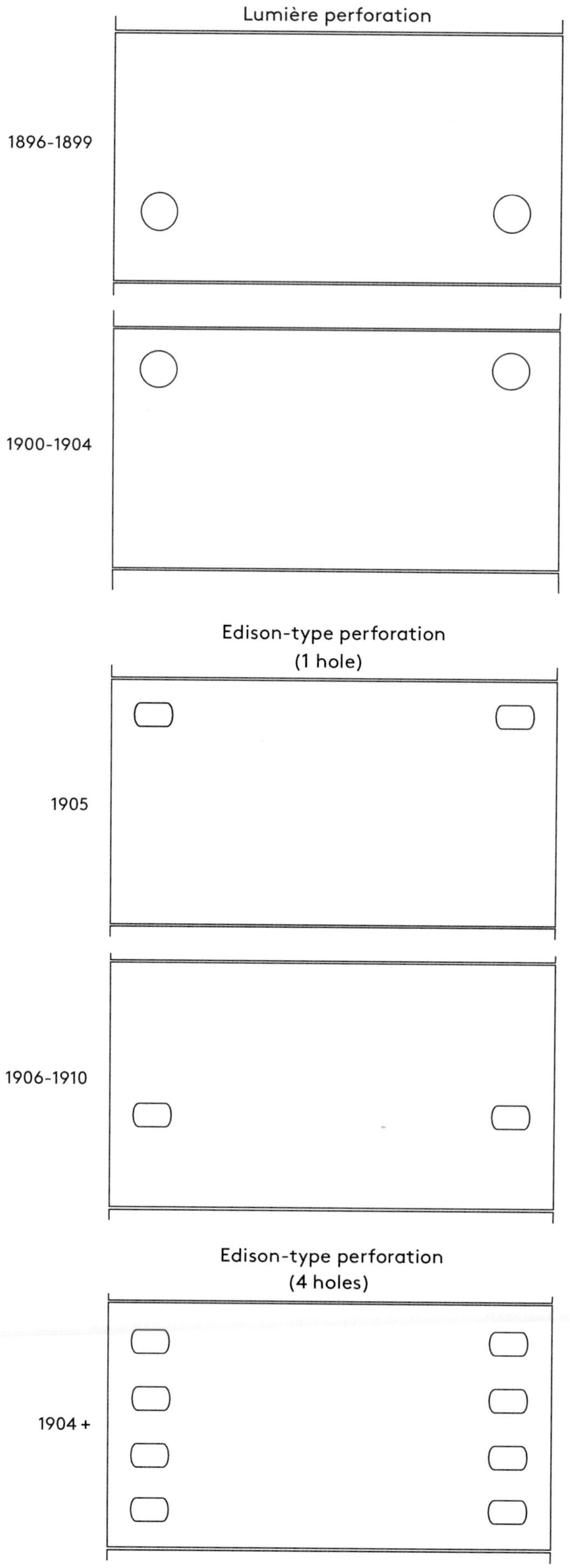

Lumière perforation

1896-1899

1900-1904

Edison-type perforation
(1 hole)

1905

1906-1910

Edison-type perforation
(4 holes)

1904 +

For instance, the Lumière perforation photographed from the negative allows us to identify the episode of the Resurrection of Lazarus as being a production of 1897, even if the print was made in 1905:

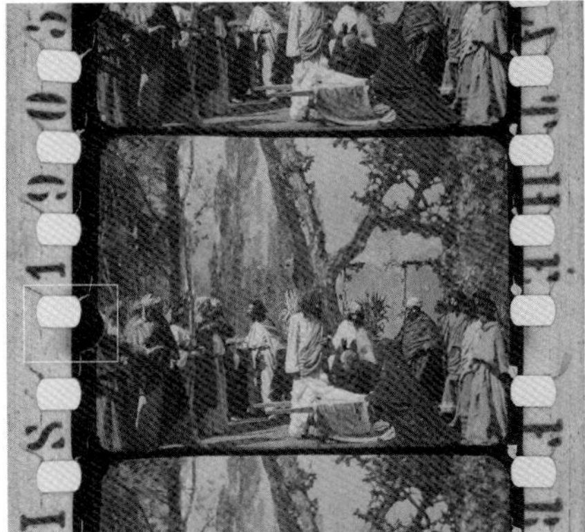

Vie et Passion du Christ. Résurrection de Lazare (No. 854, 1897) – SFI
We can see the Lumière perforation photographed from the negative on the left-hand side of the frame.

A specific Pathé perforation started being used on prints around 1909, for the first time on diacetate prints only. Because of its rounded corners, the Pathé perforation was considered more suitable for prints, made to be screened numerous times.

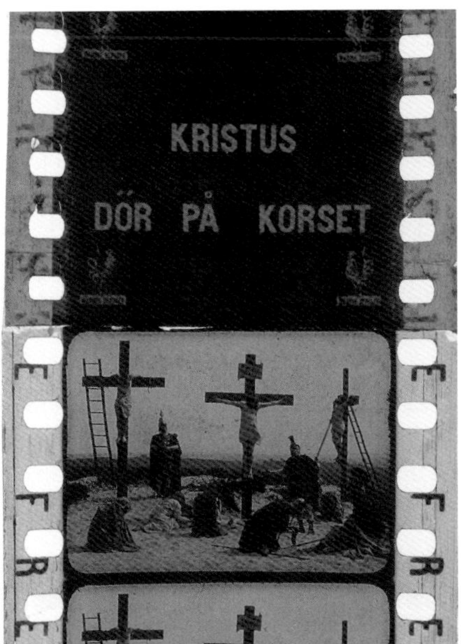

La Vie de Notre Seigneur Jésus-Christ. Passion et mort de Jésus (No. 1607, 1914) – SFI
Here it is possible to appreciate the difference between the 1905 Edison perforations (intertitle) and the Pathé perforations (image).

Therefore, the Pathé perforation would be widely used in the 1920s for positives, and would even be retained after the adoption of the Kodak Standard (KS) perforation[16] by the industry.

16 See Brown 3.2, p.43 in this volume.

Pathé as Film Manufacturer

The procurement of film stock was essential for film companies that had an in-house laboratory and could prepare the prints themselves, such as Pathé. The relations of the French firm with Eastman seem to have been complex, if not strained, and rather soon (around 1906) Pathé considered manufacturing its own film stock.[17] But it would be necessary to wait until 1910-1911 for the company to start production, both of nitrate and diacetate.

Diacetate prints from the mid-1910s are recognizable thanks to the Pathé perforation, but also numbering introduced on the edges of the film:

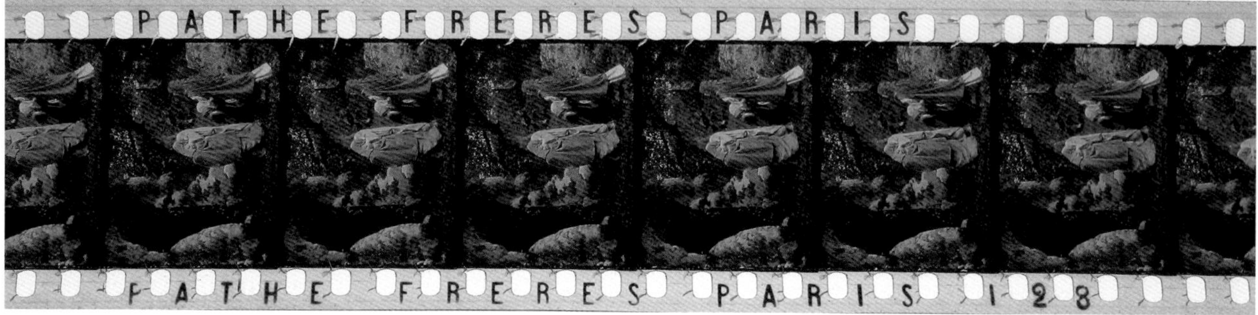

La Vie de Notre Seigneur Jésus-Christ. Passion et mort de Jésus (No. 1607, 1914) – SFI

Unfortunately, this numbering doesn't allow the dating of diacetate prints with certainty (so far), but it can be used for materials from the 1920s, as mentioned by Harold Brown in 1967.[18]

The numbering of the film stock would become systematic on stock manufactured by the firm. In order to better understand what these numbers refer to and how they can help to date the manufacturing of the film stock, it was essential to study the notebooks of the Pathé engineers held by CECIL,[19] made available by the Fondation Jérôme Seydoux-Pathé.

The explanation can be found in a 1922 document by the engineer Sylvain Pouly, from the department in charge of inspection and packing of Positive Film Stock (Vérification de la Fabrication C).[20] Unsurprisingly, the numbering allows tracking the production of film stock: the two numbers on the edges of Pathé film stock refer to the emulsion number and the roll (before slitting).

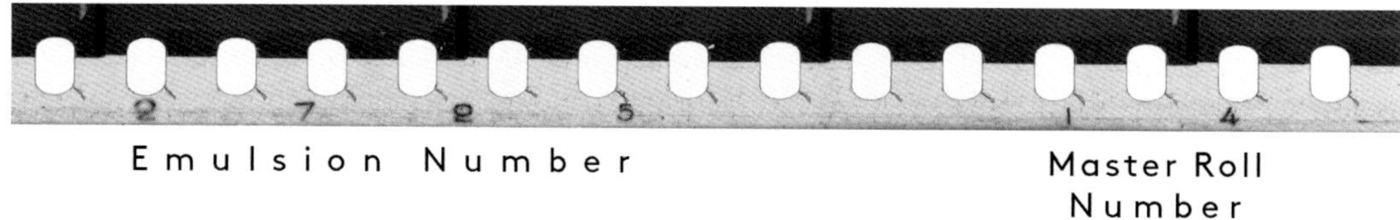

1926-1927

The numbers (reproduced at the end of this chapter) refer only to the manufacturing of the emulsion at the time — there is no code to differentiate the emulsions (nitrate or diacetate, positive or negative, panchromatic or orthochromatic, reversal, studio, etc). There could even be different uses for the same emulsion depending on the roll (since one emulsion was coated on several rolls before slitting).

17 Stéphanie Salmon, *Pathé. A la conquête du cinéma (1896-1929)*, Paris: Tallandier (2014), p.161.
18 Brown, *Notes on Film Identification by Examination of Copies* (1967, no pagination). See also p.73 in this volume.
19 CECIL (Cercle des Conservateurs de l'Image Latente) is an association created in 2009, after the cessation of Kodak's industrial activity in Chalon-sur-Saône, by former engineers from Kodak-Pathé to keep the memory of the factory alive. They were able to gather thousands of documents from Kodak and Pathé, dating back to 1904.
20 "Organisation des services de vérification, métrage, collure & emballage de la bande vierge positive" (22 July 1922), Notebook No. 33716 / HIST-P-735 (CECIL / FJSP).

The emulsions are numbered according to the date of manufacture, and they never repeat (unlike Kodak edge codes, for instance). There is no notable change in the edge printings until autumn 1927, a few months after Pathé merged with Kodak. These changes were noted by Georges Moreau, Head of the Technical Department (Service technique).

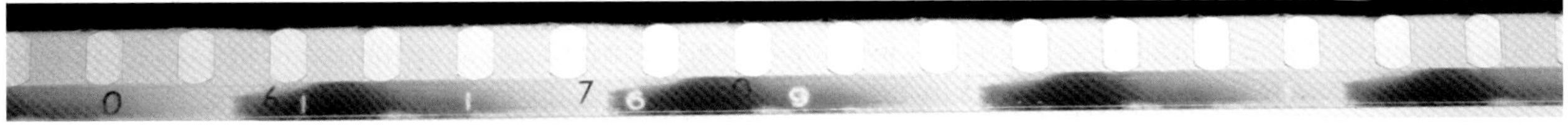

Photographed negative emulsion – CF
Negative emulsion 1169 was coated in October 1924.

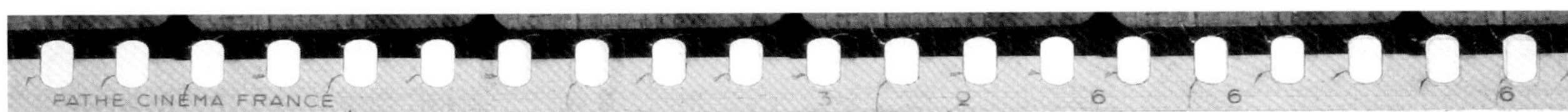

Positive emulsion – FINA
Positive emulsion 3266 was checked in September 1926.

In October 1927, "Pathé-Cinéma" was first replaced by "Pathé" (from emulsion 3823).[21]

And the following month, in November 1927, "Pathé Vincennes France" starts appearing (from emulsion 3914).[22]

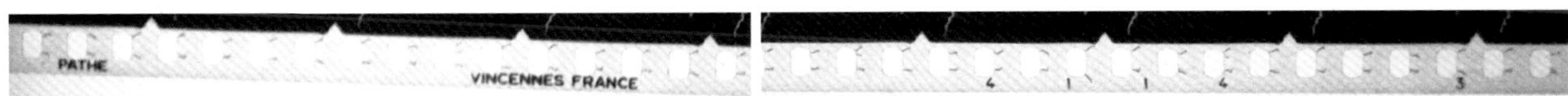

Positive emulsion – CS
Positive emulsion 4114 was coated in April 1928.

Finally, in August 1928, Pathé announces that they started indicating the month of manufacture by a letter: "A represents January 1928. This instruction started with the letter E".[23]

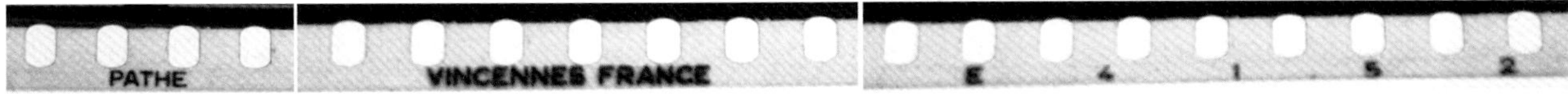

Emulsion number on nitrate print – CS
Positive emulsion E 4152 was coated in May 1928.

The idea that Pathé had totally stopped manufacturing film stock when it merged with Kodak appears to be incorrect. Pathé film stock was apparently still manufactured even after Kodak took control of the plant (in 1931[24]). But it appears clear in the last notebooks I was able to consult (1930-1932) that the manufacture of Pathé stock drastically decreased in favour of Kodak stock. In the last months of 1932, only one positive emulsion was manufactured per month, while in the 1920s production could reach more than 40 emulsions per month. Regarding negative emulsions, the manufacture of Pathé stock seems to have stopped at the end of 1929.

21 "Rapport du mois d'octobre 1927", Notebook No. 33660 / HIST-P-723 (CECIL / FJSP).
22 "Rapport du mois de novembre 1927", Notebook No. 33660 / HIST-P-723 (CECIL / FJSP).
23 "Rapport du mois d'août 1928", Notebook No. 33661 / HIST-P-724 (CECIL / FJSP).
24 Jean-Pierre Martel, "La Recherche et le développement chez Kodak-Pathé (1927-2006)", in Jacques Malthête and Stéphanie Salmon (eds.), *Recherches et innovations dans l'industrie du cinéma. Les cahiers des ingénieurs Pathé (1906-1927)*, Paris: Fondation Jérôme Seydoux-Pathé (2017), p.19.

Emulsion Numbers

More than 1,700 emulsions were referenced in the notebooks (almost 1,400 mentions of positive emulsions and more than 500 mentions of negative emulsions) between January 1920 and November 1932. The emulsions are organized into two lists, one for positives and one for negatives.

It was decided to indicate only the references actually found in the notebooks (when numbers are separated by hyphens, it means that all the numbers in-between were found in the same report). Sometimes the same emulsions appear several times in different reports, but in order to simplify the list it was decided to indicate only one of the references to the emulsions (often the report issued from the Inspection and Packing Department, or the latest date as being closer to the use of the film stock).

The emulsions are ordered by numbers, to facilitate research. Sometimes the chronological order is not logical, since the dates can proceed from different kinds of reports issued by different departments. In some cases the date indicated refers to the manufacture of the emulsion, while in others it refers to the date of the report.

Emulsion numbers which are not listed can be dated by deduction. For instance, the negative emulsion 1740 may have been manufactured between emulsions 1732 (6 December 1927) and 1744 (March 1928).

All additional information found in the notebooks, such as base (acetate, celluloid), destination (Pathé Exchange, Pathé Cinéma France, or without destination), master roll numbers, or pre-tinted stock, are reproduced in the "Comments" column.

The Notebooks have two classification references to facilitate the consultation of the original documents: (1) a 5-digit number corresponding to the number of the notebook in the collections of CECIL, and (2) the last part of the alphanumeric code of the classification at the Fondation Jérôme Seydoux-Pathé (for instance, 642 in the table below should be read as HIST-P-642).

To conclude, in order to make the information in the following tables more understandable, the functions and departments of the authors of the reports (when known or mentioned in the reports) are presented here under their respective departments:

Factory: Usine Vincennes
Roussel (Director 1906-1937)

Technical Department: Service technique
Moreau (Head of Department)

Research: Laboratoire de Recherche
Zelger[25]

Manufacture C: Service technique – Fabrication C (positive emulsion)
Barbier
Ducher

Manufacture B: Service technique – Fabrication B (negative emulsion)
Barbier
Pouly

Casting: Service technique – Fabrication C – Surveillance Générale du Service Coulage
Ducher

Coating: Service technique – Département de l'Emulsionnage
Paolantini (Head of Department)

25 Pierre Clément: *Kodak-Pathé. Histoire et évolution*, [place of publication unknown]: C.I.P. (1987), p.51.

Testing: Laboratoire d'essais
Barbier
Callame
Clerc
De Robert (chemical tests)
Fenal
Ruche (chemical tests)

Quality Control: Service contrôle – Laboratoire (positive and negative emulsions)
Barbier
Callame
Dupoux
Quesnay

Inspection & Packing B: Service technique – Fabrication B – Vérification – Emballage – Expédition négative
Leneveu

Inspection & Packing C: Service technique – Fabrication C – Organisation des services de vérification, métrage, collure et emballage de la bande vierge positive
Pouly

Sensitized: Produits sensibilisés
Pouly

Positive Emulsions

Emulsion	Year	Month	Day	Engineer (Author of the Report)	Department	Notebook	Page	Comments
1247 1258 1262 1283	1920	Feb	7	Ducher	Casting	33011/642	144	
1292	1920	Jan	30	Ducher	Casting	33011/642	143	
1302	1920	Feb	12	Ducher	Casting	33011/642	145	
1334 1337	1920	Mar	18	Ducher	Casting	33011/642	146	
1345	1920	Mar	31	Ducher	Casting	33011/642	147	
1351-1354	1920	Apr	14	Ducher	Casting	33011/642	150	
1364-1366	1920	Apr	23	Ducher	Casting	33011/642	152	
1418-1419 1422-1424	1920	Jul	1	Ducher	Casting	33011/642	153	
1457	1920	Sep		Callame	Quality Control	33174/659	19	
1464	1920	Sep		Callame	Quality Control	33174/659	20	
1473	1920	Sep	14	Barbier	Testing	33111/651	59	
1474-1475	1920	Sep	16	Barbier	Testing	33111/651	60	
1477	1920	Sep	17	Barbier	Manufacture C	33111/651	62	
1539-1544	1920	Dec	18	Callame	Quality Control	33174/659	23	
1545-1548	1920	Dec	24	Callame	Quality Control	33174/659	24	
1576	1920	Sep	16	Barbier	Testing	33111/651	60	
1594-1597	1921	Mar	1	Ducher	Casting	33011/642	155	
1602/1609	1921	Mar	19	Ducher	Casting	33011/642	156	
1643	1921	May	3	Ducher	Casting	33011/642	175	
1660 1662 1664	1921	May	23	Ducher	Casting	33011/642	159	
1669	1921	Sep	15	Zelger	Research	33979/761	218	
1686	1921	Aug	27	Zelger	Research	33979/761	218	
1719	1921	Aug	31	Ducher	Casting	33011/642	164	
1723	1921	Aug	22	Ducher	Casting	33011/642	162	
1730	1921	Sep	10	Callame	Quality Control	33174/659	28	
1733	1921	Aug	22	Ducher	Casting	33011/642	163	
1734	1921	Sep	3	Callame	Quality Control	33174/659	27	
1738	1921	Aug	22	Ducher	Casting	33011/642	163	
1740	1921	Sep	10	Callame	Quality Control	33174/659	28	
1744-1745	1921	Sep	17	Callame	Quality Control	33174/659	32	
1749	1921	Oct	1	Callame	Quality Control	33174/659	34	
1751	1921	Sep	24	Callame	Quality Control	33174/659	33	
1758	1921	Sep	27	Ducher	Manufacture C	33011/642	166	
1759	1921	Oct	8	Callame	Quality Control	33174/659	35	
1760	1922	Aug	9	Paolantini	Coating	33675/731	57	
1763	1921	Oct	15	Callame	Quality Control	33174/659	36	

Emulsion	Year	Month	Day	Engineer (Author of the Report)	Department	Notebook	Page	Comments
1775-1778	1921	Oct	21	Ducher	Casting	33011/642	167	
1789	1921	Nov	30	Ducher	Casting	33011/642	168	
1809	1921	Dec	2	Ducher	Manufacture C	33011/642	169	
1817	1922	Jan	6	Ducher	Casting	33011/642	170	
1825 1827 1829	1921	Dec	30	Marielle		33015/646	125	
1832 1834 1836 1838	1922	Jan	11	Marielle		33015/646	129	
1851-1853	1922	Feb	3	Ducher	Casting	33011/642	171	Celluloid
1854	1922	Feb	3	Ducher	Casting	33011/642	171	Acetate
1855 1857	1922	Feb	3	Ducher	Casting	33011/642	171	
1859-1860	1922	Feb	3	Ducher	Casting	33011/642	171	Celluloid
1861-1863	1922	Feb	3	Ducher	Casting	33011/642	171	Acetate
1864-1866	1922	Feb	3	Ducher	Casting	33011/642	171	1866: Celluloid
1957	1922	Apr	4	Ducher	Casting	33011/642	172	
1978	1922	Apr	28	Ducher	Casting	33011/642	173	
1980-1986	1922	Apr	29	Ducher	Casting	33011/642	174	1980-1981: Pathé Exchange 1982-1986: No Destination
1993-1997	1922	May	11	Callame	Control	33174/659	47	
2003 2007-2008	1922	May	16	Ducher	Casting	33011/642	176	
2012-2013	1922	Jun	28	Paolantini	Coating	33675/731	49	
2016	1922	May	27	Ducher	Casting	33011/642	177	
2081	1922	Aug	9	Paolantini	Coating	33675/731	57	
2083-2084 2090	1922	Jul	27	Paolantini	Coating	33675/731	55	
2094-2095	1922	Aug	9	Paolantini	Coating	33675/731	57	
2117	1922	Aug	29	Callame	Quality Control	33174/659	57	
2128-2129 2131 2138 2143 2149	1922	Oct	17	Callame	Quality Control	33174/659	59	
2155	1922	Oct	17	Callame	Quality Control	33174/659	59	
2192	1922	Nov	15	Callame	Quality Control	33174/659	61	Master roll 1
2216	1922	Nov	27	Zelger		33980/762	137	
2227	1922	Dec	11	Ruche	Testing	33017/648	425	
2235	1922	Dec	21	Ruche	Testing	33017/648	426	
2247 2262	1923	Jan	3	Ruche	Testing	33017/648	428	

Emulsion	Year	Month	Day	Engineer (Author of the Report)	Department	Notebook	Page	Comments
2267-2308	1923	Feb	1	Paolantini	Coating	33675/731	71	
2310 2315 2320 2325 2330 2335	1923	Mar	5	Paolantini	Coating	33675/731	93	
2337 2340	1923	Mar	5	Ruche	Testing	33017/648	432	
2340 2345	1923	Mar	5	Paolantini	Coating	33675/731	93	
2346-2374	1923	Apr	3	Paolantini	Coating	33675/731	97	
2375-2406	1923	May	3	Paolantini	Coating	33675/731	99	
2407-2437	1923	Jun	5	Paolantini	Coating	33675/731	104	2407: Celluloid & Acetate
2441	1923	Jun	29	Paolantini	Coating	33675/731	114	
2442	1923	Jun	18	Ruche	Testing	33017/648	444	
2444	1923	Jun	18	Moreau	Coating	33659/722		2444: Pathé Cinéma France
2458 2463	1923	Jul	2	Ruche	Testing	33017/648	446	2458: Celluloid
2463	1923	Jul	10	Ruche	Testing	33017/648	447	
2469-2470	1923	Aug	7	Paolantini	Coating	33675/731	118	
2471	1923	Jul	10	Ruche	Testing	33017/648	449	Celluloid
2472-2494	1923	Aug	7	Paolantini	Coating	33675/731	118	
2495-2520	1923	Sep	2	Pouly	Inspection C	33716/735	74	
2521-2544	1923	Oct	5	Pouly	Inspection C	33716/735	78	
2545-2570	1923	Nov	8	Pouly	Inspection C	33716/735	82	
2571-2592	1923	Dec	7	Pouly	Inspection C	33716/735	87	
2593 2594	1923	Dec	10	Zelger	Research	33981/763	121	2593: Celluloid 10, Master roll 22 2594: Acetate 34, Master roll 8
2596	1923	Dec	10	Zelger	Research	33981/763	125	Celluloid 10, Master roll 25 Acetate 34, Master roll 27
2597 2599	1923	Dec	28	De Robert	Testing	33009/640	108	
2617	1923	Dec	31	Zelger	Research	33981/763	134	Master roll 2 Master roll 170 Wheel 16
2620	1923	Oct	22	Zelger	Research	33981/763	85	Celluloid Master roll 2
2639 2642-2643	1924	Feb	11	Richard		33820/748	103	
2828	1924	Oct	6	Vaysette		33942/757	98	

Emulsion	Year	Month	Day	Engineer (Author of the Report)	Department	Notebook	Page	Comments
2903-2922	1925	Jan		Roussel	Manufacture C	33017/648	256	2904: Acetate 2908: Manufacture C 2912 / 2916: Manufacture A
3036	1925	Jul	27	Zelger	Research	33982/764	241	
3131	1925	Dec	17	Bossard	Testing	33135/657	291	
3213	1926	May	21	Dupoux	Quality Control	33011/642	281-282	
3238-3258	1926	Aug	3	Dupoux	Quality Control	33011/642	302-303	
3254-3258	1926	Aug	3	Dupoux	Quality Control	33011/642	303	
3259-3275	1926	Sep	5	Quesnay	Quality Control	33761/738	7	
3276-3282	1926	Oct		Quesnay	Quality Control	33761/738	12	
3283	1926	Sep	21	Pouly		33718/737	6	
3284	1926	Sep	23	Pouly		33718/737	8	
3285	1926	Sep	21	Pouly		33718/737	7	
3286	1926	Sep	24	Pouly		33718/737	10	
3287	1926	Sep	27	Pouly		33718/737	14	Celluloid Tinted Green + Acetate 35mm & Baby
3288	1926	Sep	25	Pouly		33718/737	12	Celluloid Tinted Amber, Green & Light Amber
3289	1926	Sep	27	Pouly		33718/737	13	Celluloid Tinted Amber & Violet + for Colouring
3289-3290 3292-3295	1926	Oct		Quesnay	Quality Control	33761/738	12	
3296-3299	1926	Oct		Quesnay	Quality Control	33761/738	13	
3314-3326	1926	Oct	31	Pouly		33718/737	32	3314 / 3320-3322 / 3324-3326: Acetate 3316-3317/ 3320 / 3322-3323: Celluloid
3376 3385-3386	1926	Dec	7	Pouly	Sensitized	33718/737	65	
3392	1926	Dec	31	Pouly	Sensitized	33718/737	79	
3406	1926	Dec	19	Pouly	Sensitized	33718/737	76	
3408 3420	1926	Dec	31	Pouly	Sensitized	33718/737	79	
3425	1926	Dec	21	Pouly	Inspection C	33716/735	129	
3427-3428	1926	Dec	31	Pouly	Sensitized	33718/737	79	
3433-3470	1927	Feb	1	Pouly	Inspection C	33716/735	137	
3499-3500 3502-3503 3511 3521-3522	1927	Feb	21	Quesnay	Quality Control	33761/738	19	
3576-3580	1927	Mar	15	Pouly	Inspection C	33716/735	161	

Emulsion	Year	Month	Day	Engineer (Author of the Report)	Department	Notebook	Page	Comments
3671	1927	May	18	Quesnay	Quality Control	33761/738	41	
3709-3714	1927	Jul	1	Callame		33174/659	283-284	
3736 3741 3744 3747 3750-3753 3759-3760	1927	Aug	2	Callame		33174/659	301-302	
3761-3763 3765 3776-3777 3779 3787-3789	1927	Sep	2	Callame		33174/659	324	
4066-4097 4104 4107	1928	Mar		Moreau	Coating	33661/724	18	
4108-4138	1928	Apr		Moreau	Coating	33661/724	38	4108 (indicated as 4608)
4139-4169	1928	May		Moreau	Coating	33661/724	56	
4170-4205	1928	Jun		Moreau	Coating	33661/724	71	
4206-4227	1928	Jul	31	Moreau	Coating	33661/724	89	
4228-4248	1928	Aug		Moreau	Coating	33661/724	104	
4249-4272	1928	Sep		Moreau	Coating	33661/724	124	
4273-4301	1928	Oct		Moreau	Coating	33661/724	144	
4302-4315	1928	Nov		Moreau	Coating	33661/724	157	
4316-4322	1928	Dec		Moreau	Coating	33661/724	170	4317: Duplicating
4323 4332	1929	Jan	26	Moreau	Coating	33662/725	13	
4333 4356	1929	Feb	29	Moreau	Coating	33662/725	26	
4357 4373	1929	Mar	29	Moreau	Coating	33662/725	37	
4374 4392	1929	Apr	20	Moreau	Coating	33662/725	51	
4393 4410	1929	May	18	Moreau	Coating	33662/725	65	
4411 4415	1929	Jun	15	Moreau	Coating	33662/725	77	Duplicating
4416 4421	1929	Jul	13	Moreau	Coating	33662/725	93	
4437	1929	Sep	10	Moreau	Coating	33662/725	106	
4438	1929	Sep	13	Moreau	Coating	33662/725	106	
4439	1929	Sep	17	Moreau	Coating	33662/725	106	
4440	1929	Sep	24	Moreau	Coating	33662/725	106	
4441	1929	Sep	28	Moreau	Coating	33662/725	106	
4442	1929	Sep	24	Moreau	Coating	33662/725	106	Duplicating

Emulsion	Year	Month	Day	Engineer (Author of the Report)	Department	Notebook	Page	Comments
4443	1929	Sep	26	Moreau	Coating	33662/725	106	
4444	1929	Oct	1	Moreau	Coating	33662/725	106	
4445	1929	Oct	4	Moreau	Coating	33662/725	106	
4446	1929	Oct	6	Moreau	Coating	33662/725	121	
4447-4448	1929	Oct	11	Moreau	Coating	33662/725	121	
4449	1929	Oct	19	Moreau	Coating	33662/725	121	
4450-4451	1929	Oct	22	Moreau	Coating	33662/725	121	
4452	1929	Oct	19	Moreau	Coating	33662/725	121	
4453	1929	Oct	20	Moreau	Coating	33662/725	121	
4454	1929	Oct	31	Moreau	Coating	33662/725	121	
4455	1929	Nov	5	Moreau	Coating	33662/725	138	
4456-4457	1929	Nov	6	Moreau	Coating	33662/725	138	Duplicating
4458	1929	Nov	7	Moreau	Coating	33662/725	138	
4459	1929	Nov	13	Moreau	Coating	33662/725	138	
4460	1929	Nov	15	Moreau	Coating	33662/725	138	Duplicating
4461	1929	Nov	17	Moreau	Coating	33662/725	138	
4462	1929	Nov	16	Moreau	Coating	33662/725	138	
4463-4465	1929	Nov	21	Moreau	Coating	33662/725	138	
4466	1929	Nov	27	Moreau	Coating	33662/725	138	
4467	1929	Nov	26	Moreau	Coating	33662/725	138	"Diaphane"
4468	1929	Nov	29	Moreau	Coating	33662/725	138	
4469-4471	1929	Dec	3	Moreau	Coating	33662/725	148	
4472-4476	1929	Dec	4	Moreau	Coating	33662/725	148	
4477	1929	Dec	11	Moreau	Coating	33662/725	148	
4478	1929	Dec	12	Moreau	Coating	33662/725	148	
4480-4481	1929	Dec	14	Moreau	Coating	33662/725	148	
4482	1929	Dec	17	Moreau	Coating	33662/725	148	
4483-4485	1929	Dec	18	Moreau	Coating	33662/725	148	
4486	1929	Dec	20	Moreau	Coating	33662/725	148	
4487-4488	1929	Dec	21	Moreau	Coating	33662/725	148	4487: Duplicating
4489-4491	1929	Dec	22	Moreau	Coating	33662/725	149	
4492-4494	1929	Dec	23	Moreau	Coating	33662/725	149	
4495	1929	Dec	28	Moreau	Coating	33662/725	149	
4496-4497	1929	Dec	29	Moreau	Coating	33662/725	149	
4498	1929	Dec	29	Moreau	Coating	33663/726	8	"Diaphane"
4499	1930	Jan	5	Moreau	Coating	33663/726	8	
4500	1930	Jan	7	Moreau	Coating	33663/726	8	
4501-4505	1930	Jan	8	Moreau	Coating	33663/726	8	
4506-4507	1930	Jan	16	Moreau	Coating	33663/726	8	
4508	1930	Jan	18	Moreau	Coating	33663/726	8	
4509	1930	Jan	17	Moreau	Coating	33663/726	8	
4510	1930	Jan	18	Moreau	Coating	33663/726	8	
4511	1930	Jan	20	Moreau	Coating	33663/726	8	

Emulsion	Year	Month	Day	Engineer (Author of the Report)	Department	Notebook	Page	Comments
4512-4516	1930	Jan	21	Moreau	Coating	33663/726	8	
4517-4518	1930	Jan	22	Moreau	Coating	33663/726	8	
4519	1930	Jan	23	Moreau	Coating	33663/726	8	
4520-4521	1930	Jan	25	Moreau	Coating	33663/726	8	
4547	1930	Feb	24	Moreau	Coating	33663/726	23	
4548-4550	1930	Feb	25	Moreau	Coating	33663/726	23	4548: Acetate Standard Master roll 1 to 7
4551-4556	1930	Feb	26	Moreau	Coating	33663/726	23	4552: Acetate Standard Master roll 1-5-7
4557	1930	Mar	4	Moreau	Coating	33663/726	23	
4558	1930	Mar	5	Moreau	Coating	33663/726	23	
4559	1930	Mar	30	Moreau	Coating	33663/726	23	
4560-4561	1930	Mar	6	Moreau	Coating	33663/726	23	
4562	1930	Mar	7	Moreau	Coating	33663/726	23	
4563	1930	Mar	12	Moreau	Coating	33663/726	23	
4564	1930	Mar	13	Moreau	Coating	33663/726	23	
4565-4567	1930	Mar	14	Moreau	Coating	33663/726	23	
4568-4570	1930	Mar	15	Moreau	Coating	33663/726	23	4570: "Diaphane"
4571	1930	Mar	17	Moreau	Coating	33663/726	23	
4572	1930	Mar	18	Moreau	Coating	33663/726	23	
4574	1930	Mar	19	Moreau	Coating	33663/726	23	
4575	1930	Mar	20	Moreau	Coating	33663/726	23	
4576-4578	1930	Mar	21	Moreau	Coating	33663/726	23	4577-4578: Duplicating
4579	1930	Mar	22	Moreau	Coating	33663/726	23	
4580-4581	1930	Mar	27	Moreau	Coating	33663/726	35	
4582	1930	Mar	29	Moreau	Coating	33663/726	35	
4583	1930	Mar	31	Moreau	Coating	33663/726	35	
4584	1930	Apr	1	Moreau	Coating	33663/726	35	
4585	1930	Apr	3	Moreau	Coating	33663/726	35	
4586	1930	Apr	4	Moreau	Coating	33663/726	35	
4587	1930	Apr	1	Moreau	Coating	33663/726	35	
4588	1930	Apr	6	Moreau	Coating	33663/726	35	
4589	1930	Apr	8	Moreau	Coating	33663/726	35	
4590	1930	Apr	8	Moreau	Coating	33663/726	35	
4591	1930	Apr	13	Moreau	Coating	33663/726	35	
4592	1930	Apr	14	Moreau	Coating	33663/726	35	
4593	1930	Apr	16	Moreau	Coating	33663/726	35	
4594	1930	Apr	23	Moreau	Coating	33663/726	45	"Diaphane"
4595-4596	1930	Apr	29	Moreau	Coating	33663/726	45	
4597	1930	May	2	Moreau	Coating	33663/726	45	
4598	1930	May	4	Moreau	Coating	33663/726	45	

Emulsion	Year	Month	Day	Engineer (Author of the Report)	Department	Notebook	Page	Comments
4599	1930	May	6	Moreau	Coating	33663/726	45	
4600	1930	May	11	Moreau	Coating	33663/726	45	
4601	1930	May	9	Moreau	Coating	33663/726	45	"Diaphane"
4602	1930	May	10	Moreau	Coating	33663/726	45	Duplicating
4603	1930	May	16	Moreau	Coating	33663/726	45	
4604	1930	May	7	Moreau	Coating	33663/726	45	
4605	1930	May	17	Moreau	Coating	33663/726	45	
4606	1930	May	20	Moreau	Coating	33663/726	53	
4607-4608	1930	May	21	Moreau	Coating	33663/726	53	4608: Duplicating
4609	1930	May	23	Moreau	Coating	33663/726	53	
4610	1930	May	24	Moreau	Coating	33663/726	53	
4611	1930	May	25	Moreau	Coating	33663/726	53	
4612	1930	May	27	Moreau	Coating	33663/726	53	Duplicating
4613-4615	1930	May	28	Moreau	Coating	33663/726	53	
4616	1930	May	31	Moreau	Coating	33663/726	53	
4617	1930	Jun	2	Moreau	Coating	33663/726	53	
4618	1930	Jun	3	Moreau	Coating	33663/726	53	Duplicating
4619	1930	Jun	5	Moreau	Coating	33663/726	53	
4620	1930	Jun	7	Moreau	Coating	33663/726	53	
4622	1930	Jun	5	Moreau	Coating	33663/726	53	
4623	1930	Jun	11	Moreau	Coating	33663/726	53	
4625	1930	Jun	14	Moreau	Coating	33663/726	53	
4639	1930	Jul	16	Moreau	Coating	33663/726	65	
4640	1930	Jul	19	Moreau	Coating	33663/726	65	
4641	1930	Jul	22	Moreau	Coating	33663/726	65	
4642	1930	Jul	31	Moreau	Coating	33663/726	65	
4643	1930	Aug	4	Moreau	Coating	33663/726	65	
4644-4646	1930	Aug	5	Moreau	Coating	33663/726	65	4644: "Diaphane" 4645-4646: Duplicating
4647	1930	Aug	6	Moreau	Coating	33663/726	65	
4648	1930	Aug	11	Moreau	Coating	33663/726	77	
4649	1930	Aug	12	Moreau	Coating	33663/726	77	
4650-4651	1930	Aug	13	Moreau	Coating	33663/726	77	
4652	1930	Aug	28	Moreau	Coating	33663/726	77	
4653	1930	Aug	29	Moreau	Coating	33663/726	77	
4654-4655	1930	Sep	3	Moreau	Coating	33663/726	77	
4656	1930	Sep	4	Moreau	Coating	33663/726	77	
4662	1930	Oct	10	Moreau	Coating	33663/726	88	
4663	1930	Oct	14	Moreau	Coating	33663/726	88	
4664	1930	Oct	15	Moreau	Coating	33663/726	88	
4665	1930	Oct	21	Moreau	Coating	33663/726	88	
4666	1930	Oct	24	Moreau	Coating	33663/726	88	
4667	1930	Oct	25	Moreau	Coating	33663/726	88	

Emulsion	Year	Month	Day	Engineer (Author of the Report)	Department	Notebook	Page	Comments
4668	1930	Nov	3	Moreau	Coating	33663/726	102	
4669	1930	Nov	5	Moreau	Coating	33663/726	102	"Diaphane"
4670	1930	Nov	6	Moreau	Coating	33663/726	102	
4671	1930	Nov	11	Moreau	Coating	33663/726	102	
4672	1930	Nov	17	Moreau	Coating	33663/726	102	
4673	1930	Nov	18	Moreau	Coating	33663/726	102	"Diaphane"
4674	1930	Nov	19	Moreau	Coating	33663/726	102	
4675	1930	Nov	22	Moreau	Coating	33663/726	102	
4676	1930	Nov	25	Moreau	Coating	33663/726	102	
4677	1930	Nov	30	Moreau	Coating	33663/726	102	
4678	1930	Dec	2	Moreau	Coating	33663/726	116	
4679-4680	1930	Dec	3	Moreau	Coating	33663/726	116	
4682	1930	Dec	5	Moreau	Coating	33663/726	116	
4683	1930	Dec	6	Moreau	Coating	33663/726	116	
4684	1930	Dec	8	Moreau	Coating	33663/726	116	
4685-4686	1930	Dec	10	Moreau	Coating	33663/726	116	
4687	1930	Dec	15	Moreau	Coating	33663/726	116	
4688	1930	Dec	16	Moreau	Coating	33663/726	116	
4689	1930	Dec	19	Moreau	Coating	33663/726	116	
4690	1930	Dec	24	Moreau	Coating	33663/726	116	
4691	1930	Dec	27	Moreau	Coating	33663/726	116	
4692	1930	Dec	29	Moreau	Coating	33664/727	10	
4693-4694	1930	Dec	30	Moreau	Coating	33664/727	10	4694: Duplicating
4695	1931	Jan	7	Moreau	Coating	33664/727	10	
4696	1930	Dec	31	Moreau	Coating	33664/727	10	Duplicating
4697	1931	Jan	14	Moreau	Coating	33664/727	10	
4698	1931	Jan	15	Moreau	Coating	33664/727	10	"Diaphane"
4699	1931	Jan	9	Moreau	Coating	33664/727	10	
4700	1931	Jan	20	Moreau	Coating	33664/727	10	
4701	1931	Jan	26	Moreau	Coating	33664/727	22	
4702	1931	Jan	29	Moreau	Coating	33664/727	22	
4703	1931	Feb	3	Moreau	Coating	33664/727	22	
4704	1931	Feb	4	Moreau	Coating	33664/727	22	
4705	1931	Feb	6	Moreau	Coating	33664/727	22	
4706-4707	1931	Feb	13	Moreau	Coating	33664/727	22	4707: Duplicating
4708	1931	Feb	14	Moreau	Coating	33664/727	22	
4709	1931	Feb	21	Moreau	Coating	33664/727	22	
4710-4711	1931	Feb	27	Moreau	Coating	33664/727	35	
4712	1931	Mar	4	Moreau	Coating	33664/727	35	"Diaphane"
4713	1931	Mar	5	Moreau	Coating	33664/727	35	Duplicating
4714	1931	Mar	6	Moreau	Coating	33664/727	35	
4715	1931	Mar	23	Moreau	Coating	33664/727	51	
4716	1931	Mar	18	Moreau	Coating	33664/727	51	Duplicating

Emulsion	Year	Month	Day	Engineer (Author of the Report)	Department	Notebook	Page	Comments
4717	1931	Apr	8	Moreau	Coating	33664/727	51	
4718	1931	Apr	14	Moreau	Coating	33664/727	51	"Diaphane"
4719	1931	Apr	13	Moreau	Coating	33664/727	51	Duplicating
4720	1931	Apr	18	Moreau	Coating	33664/727	51	
4721	1931	May	21	Moreau	Coating	33664/727	65	
4722	1931	May	22	Moreau	Coating	33664/727	65	
4723	1931	Apr	27	Moreau	Coating	33664/727	65	
4724	1931	May	5	Moreau	Coating	33664/727	65	
4725	1931	May	7	Moreau	Coating	33664/727	65	
4726	1931	May	8	Moreau	Coating	33664/727	65	Duplicating
4727	1931	May	18	Moreau	Coating	33664/727	65	Duplicating
4728	1931	May	29	Moreau	Coating	33664/727	78	
4729	1931	Jun	1	Moreau	Coating	33664/727	78	Duplicating
4730	1931	Jun	5	Moreau	Coating	33664/727	78	
4731	1931	Jun	10	Moreau	Coating	33664/727	78	Duplicating
4732	1931	Jun	11	Moreau	Coating	33664/727	78	Dichromia
4733	1931	Jun	17	Moreau	Coating	33664/727	90	
4734	1931	Jul	6	Moreau	Coating	33664/727	90	
4736	1931	Jul	20	Moreau	Coating	33664/727	104	
4737	1931	Jul	25	Moreau	Coating	33664/727	104	
4738	1931	Jul	28	Moreau	Coating	33664/727	104	
4739	1931	Jul	31	Moreau	Coating	33664/727	104	
4740	1931	Aug	3	Moreau	Coating	33664/727	104	
4741	1931	Aug	14	Moreau	Coating	33664/727	117	
4742	1931	Aug	28	Moreau	Coating	33664/727	117	
4743	1931	Sep	1	Moreau	Coating	33664/727	117	
4744	1931	Sep	11	Moreau	Coating	33664/727	130	
4745	1931	Sep	14	Moreau	Coating	33664/727	130	
4746	1931	Sep	15	Moreau	Coating	33664/727	130	Duplicating
4747	1931	Sep	19	Moreau	Coating	33664/727	130	
4748-4749	1931	Sep	25	Moreau	Coating	33664/727	130	
4750	1931	Sep	29	Moreau	Coating	33664/727	130	
4751	1931	Oct	31	Moreau	Coating	33664/727	130	
4756	1931	Nov	4	Moreau	Coating	33664/727	145	
4757	1931	Nov	7	Moreau	Coating	33664/727	145	
4758	1931	Nov	13	Moreau	Coating	33664/727	145	
4759	1931	Nov	17	Moreau	Coating	33664/727	145	
4760	1931	Nov	21	Moreau	Coating	33664/727	145	
4761	1931	Nov	24	Moreau	Coating	33664/727	145	
4762	1931	Nov	26	Moreau	Coating	33664/727	145	
4763	1931	Nov	28	Moreau	Coating	33664/727	145	
4764	1931	Dec	1	Moreau	Coating	33664/727	158	Duplicating
4765	1931	Dec	3	Moreau	Coating	33664/727	158	

Emulsion	Year	Month	Day	Engineer (Author of the Report)	Department	Notebook	Page	Comments
4766	1931	Dec	4	Moreau	Coating	33664/727	158	
4767	1931	Dec	7	Moreau	Coating	33664/727	158	
4768-4769	1931	Dec	10	Moreau	Coating	33664/727	158	
4770	1931	Dec	15	Moreau	Coating	33664/727	158	
4771	1931	Dec	21	Moreau	Coating	33664/727	158	
4772	1931	Dec	23	Moreau	Coating	33664/727	158	
4773	1931	Dec	28	Moreau	Coating	33664/727	158	"Diaphane"
4774	1932	Jan	5	Moreau	Coating	33665/728	12	
4775	1932	Jan	11	Moreau	Coating	33665/728	12	
4776	1932	Jan	12	Moreau	Coating	33665/728	12	Duplicating
4777	1932	Jan	19	Moreau	Coating	33665/728	12	
4778	1932	Feb	2	Moreau	Coating	33665/728	27	
4779	1932	Feb	3	Moreau	Coating	33665/728	27	
4780	1932	Feb	6	Moreau	Coating	33665/728	27	"Diaphane"
4781	1932	Feb	11	Moreau	Coating	33665/728	27	"Diaphane"
4782	1932	Feb	22	Moreau	Coating	33665/728	41	
4783	1932	Feb	23	Moreau	Coating	33665/728	41	
4785	1932	Mar	3	Moreau	Coating	33665/728	41	"Diaphane"
4786-4787	1932	Mar	8	Moreau	Coating	33665/728	41	4786: "Diaphane"
4788	1932	Mar	21	Moreau	Coating	33665/728	55	
4789	1932	Mar	31	Moreau	Coating	33665/728	55	
4790	1932	Apr	8	Moreau	Coating	33665/728	55	
4791	1932	Apr	16	Moreau	Coating	33665/728	55	
4792	1932	Apr	24	Moreau	Coating	33665/728	73	"Diaphane"
4793	1932	May	3	Moreau	Coating	33665/728	73	
4794	1932	May	10	Moreau	Coating	33665/728	73	
4795	1932	May	18	Moreau	Coating	33665/728	91	
4796	1932	May	21	Moreau	Coating	33665/728	91	
4797	1932	May	28	Moreau	Coating	33665/728	91	
4798	1932	Jun	1	Moreau	Coating	33665/728	91	
4799	1932	Jun	3	Moreau	Coating	33665/728	91	Duplicating
4800	1932	Jun	16	Moreau	Coating	33665/728	108	
4801-4802	1932	Jun	23	Moreau	Coating	33665/728	108	4802: "Diaphane"
4803	1932	Jun	29	Moreau	Coating	33665/728	108	
4804	1932	Jul	8	Moreau	Coating	33665/728	108	
4805-4806	1932	Jul	12	Moreau	Coating	33665/728	123	
4807	1932	Jul	15	Moreau	Coating	33665/728	123	
4808	1932	Jul	20	Moreau	Coating	33665/728	123	
4809	1932	Jul	26	Moreau	Coating	33665/728	123	
4810	1932	Jul	28	Moreau	Coating	33665/728	123	
4811	1932	Jul	29	Moreau	Coating	33665/728	123	
4812	1932	Aug	4	Moreau	Coating	33665/728	123	
4813	1932	Aug	10	Moreau	Coating	33665/728	134	

Emulsion	Year	Month	Day	Engineer (Author of the Report)	Department	Notebook	Page	Comments
4814-4816	1932	Aug	11	Moreau	Coating	33665/728	134	4816: Duplicating type Rochester
4817	1932	Aug	23	Moreau	Coating	33665/728	134	"Diaphane"
4819	1932	Aug	26	Moreau	Coating	33665/728	134	Duplicating type Rochester
4820	1932	Sep	6	Moreau	Coating	33665/728	138	Duplicating type Rochester
4821	1932	Sep	8	Moreau	Coating	33665/728	138	
4822	1932	Sep	19	Moreau	Coating	33665/728	138	
4823	1932	Sep	15	Moreau	Coating	33665/728	138	Duplicating type I
4824	1932	Sep	20	Moreau	Coating	33665/728	138	
4825	1932	Sep	21	Moreau	Coating	33665/728	138	
4826	1932	Sep	23	Moreau	Coating	33665/728	138	
4827	1932	Sep	30	Moreau	Coating	33665/728	138	Duplicating type I
4828	1932	Oct	3	Moreau	Coating	33665/728	162	
4829	1932	Oct	4	Moreau	Coating	33665/728	162	
4830	1932	Oct	7	Moreau	Coating	33665/728	162	
4831	1932	Oct	11	Moreau	Coating	33665/728	162	
4832	1932	Oct	12	Moreau	Coating	33665/728	162	
4833	1932	Oct	14	Moreau	Coating	33665/728	162	
4834	1932	Oct	15	Moreau	Coating	33665/728	162	
4835	1932	Oct	22	Moreau	Coating	33665/728	162	
4836	1932	Oct	27	Moreau	Coating	33665/728	162	
4837	1932	Nov	5	Moreau	Coating	33665/728	178	
4838	1932	Nov	9	Moreau	Coating	33665/728	178	
4839	1932	Nov	17	Moreau	Coating	33665/728	178	

Negative Emulsions

Emulsion	Year	Month	Day	Engineer (Author of the Report)	Department	Notebook	Page	Comments
144 174	1920	Nov	20	Zelger	Research	33979/761	120	144: Acetate 174: Anti-halo
180 185	1920	Sep		Callame	Manufacture B	33174/659	17-18	
200-201	1920	Oct	1	Marielle		33015/646	115	
202-205	1920	Sep	16	Barbier	Testing	33111/651	60	203: Grey. Classed 1a ("1 bis"). 204: Even more grey. Classed 2. 205: Seems to be good.
206-216	1920	Oct	1	Marielle		33015/646	115	
227	1926	May	21	Dupoux	Quality Control	33011/642	281-282	Speed
228 237	1920	Nov	15	Marielle		33015/646	138	
242	1920	Nov	20	Zelger		33979/761	120	Ordinary violet
252	1920	Dec	4	Zelger		33979/761	130	
268	1921	Mar		Zelger		33979/761	189	
354	1921	Jul	11	Marielle		33015/646	138	
357	1921	Jul	21	Marielle		33015/646	138	
358	1921	Jul	22	Marielle		33015/646	138	
360	1921	Jul	30	Zelger		33979/761	190	Current. Two layers
372-375	1921	Sep	10	Callame	Quality Control	33174/659	28	
376-384	1921	Sep	17	Callame	Quality Control	33174/659	32-33	
385-389	1921	Oct	1	Callame	Quality Control	33174/659	34	
390-394	1921	Oct	8	Callame	Quality Control	33174/659	35	391: Master roll 1: Acetate. 394: Master roll 2: Acetate.
395-400	1921	Oct	15	Callame	Quality Control	33174/659	36	
428	1922	Jan	25	Paolantini	Coating	33675/731	11	
431	1921	Dec	9	Marielle		33015/646	119	
434	1921	Dec	14	Marielle		33015/646	121	
436	1921	Dec	17	Marielle		33015/646	123	
437	1921	Dec	14	Marielle		33015/646	122	
438	1922	Jan	25	Paolantini	Coating	33675/731	10	A ("bis")
444	1922	Jan	16	Marielle		33015/646	130	Emulsion "large grains"
446	1922	Jan	25	Paolantini	Coating	33675/731	8	
455	1922	Feb	2	Marielle		33015/646	132	Emulsion "large grains"
457	1922	Feb	8	Marielle		33015/646	132	Emulsion "large grains"
458	1922	Feb	10	Marielle		33015/646	132	Emulsion "large grains"
459	1922	Feb	13	Marielle		33015/646	132	Emulsion "large grains"
466	1922	Mar	1	Marielle		33015/646	133	
467	1922	Mar	1	Marielle		33015/646	133	
473	1922	Mar	16	Marielle		33015/646	134-135	
641-650	1922	Dec	11	Ruche	Testing	33017/648	425	
651	1922	Dec	21	Ruche	Testing	33017/648	426	
652-653	1922	Dec	11	Ruche	Testing	33017/648	425	

Emulsion	Year	Month	Day	Engineer (Author of the Report)	Department	Notebook	Page	Comments
654	1922	Dec	21	Ruche	Testing	33017/648	426	
657 660	1922	Dec	26	Ruche	Testing	33017/648	427	
661 665	1923	Jan	3	Ruche	Testing	33017/648	428	
677 679 681	1923	Jan	22	Ruche	Testing	33017/648	429	
682 684-686	1923	Jan	29	Ruche	Testing	33017/648	430	
704-705	1923	Mar	5	Ruche	Testing	33017/648	432	
706	1923	Mar	13	Ruche	Testing	33017/648	433	
707	1923	Mar	5	Ruche	Testing	33017/648	432	
708-709 711	1923	Mar	13	Ruche	Testing	33017/648	433	
718	1923	Mar	17	Fenal	Testing	33371/677	199	
718	1923	Mar	26	Ruche	Testing	33017/648	435	
718	1923	Apr	7	Fenal	Testing	33372/678	Sep-13	
720 722	1923	Mar	26	Ruche	Testing	33017/648	435	
724-727 729-732	1923	Mar	3	Ruche	Testing	33017/648	431	
733-735	1923	Apr	10	Ruche	Testing	33017/648	436	
738 742 747-748	1923	Apr	23	Ruche	Testing	33017/648	438	
749 751	1923	Apr	30	Ruche	Testing	33017/648	439	
753-755 757 760	1923	May	7	Ruche	Testing	33017/648	440	
761	1923	May	15	Ruche	Testing	33017/648	441	
762	1923	May	7	Ruche	Testing	33017/648	440	
764 766-767 769 771-772	1923	May	15	Ruche	Testing	33017/648	441	
773-781	1923	May	22	Ruche	Testing	33017/648	442	
782	1923	May	29	Ruche	Testing	33017/648	443	
783-787	1923	May	29	Ruche	Testing	33017/648	443	
798 809-810 812-813 815-816	1923	Jun	18	Ruche	Testing	33017/648	444	798: Master roll 3: KOK
818-819 821-822 824-825	1923	Jun	27	Ruche	Testing	33017/648	445	

Emulsion	Year	Month	Day	Engineer (Author of the Report)	Department	Notebook	Page	Comments
827-828 830-831 833-834 836-837	1923	Jul	2	Ruche	Testing	33017/648	446	
838-841 843-844 846	1923	Jul	10	Ruche	Testing	33017/648	447- 449	
854 857-860	1923	Jul	24	De Robert	Testing	33009/640	87	
862-865 867-868	1923	Aug	1	De Robert	Testing	33009/640	88	
869-871 873-874 876 878 880	1923	Aug	6	De Robert	Testing	33009/640	89	
881-882 884-885 887-888	1923	Aug	13	De Robert	Testing	33009/640	90	
889	1923	Aug	27	De Robert	Testing	33009/640	92	
891-892 894-896	1923	Aug	20	De Robert	Testing	33009/640	91	
897-899 901	1923	Aug	27	De Robert	Testing	33009/640	92	
903	1923	Sep	3	De Robert	Testing	33009/640	93	
904-905	1923	Aug	27	De Robert	Testing	33009/640	92	
906-907 911-912 914-915	1923	Sep	3	De Robert	Testing	33009/640	93	
916 918 920-921 923 926-927	1923	Sep	11	De Robert	Testing	33009/640	94	
929-931 933-934	1923	Sep	17	De Robert	Testing	33009/640	95	
935-939	1923	Sep	24	De Robert	Testing	33009/640	96	
941-944 946-949 952	1923	Oct	1	De Robert	Testing	33009/640	97	
953 955 957-958 960	1923	Oct	8	De Robert	Testing	33009/640	98	
962 964-965	1923	Oct	15	De Robert	Testing	33009/640	99	
965	1923	Oct	23	De Robert	Testing	33009/640	100	
966-967	1923	Oct	15	De Robert	Testing	33009/640	99	

Emulsion	Year	Month	Day	Engineer (Author of the Report)	Department	Notebook	Page	Comments
968-969 971-973 975-976	1923	Oct	23	De Robert	Testing	33009/640	100	
977-978 980-982	1923	Oct	30	De Robert	Testing	33009/640	101	
983 985-986	1923	Nov	6	De Robert	Testing	33009/640	102	
988 990	1923	Nov	13	De Robert	Testing	33009/640	103	
991	1923	Nov	28	De Robert	Testing	33009/640	104	
992	1923	Dec	3	Zelger		33981/763	117	
996	1923	Nov	28	De Robert	Testing	33009/640	104	
997 999	1923	Dec	7	De Robert	Testing	33009/640	105	
1002-1004	1923	Dec	11	De Robert	Testing	33009/640	106	
1005-1006	1923	Dec	19	De Robert	Testing	33009/640	107	
1007-1008	1923	Dec	28	De Robert	Testing	33009/640	108	
1017	1924	Feb	11	Richard		33820/748	105	
1069	1924	Jun	2	Vaysette		33942/757	46	
1152-1153 1159-1160 1167-1169 1176 1178 1181-1182	1924	Nov	5	Leneveu	Inspection B	33581/709	10	1176, 1181 & 1182: Studio
1184	1924	Dec	4	Leneveu	Inspection B	33581/709	15	
1186-1187	1924	Nov	5	Leneveu	Inspection B	33581/709	10	Studio
1191 1195 1198 1200	1924	Dec	4	Leneveu	Inspection B	33581/709	15	1195 & 1198: Studio (Bossard, 33135, pp.207-208)
1205 1207-1208	1925	Jan	6	Leneveu	Inspection B	33581/709	20	1205: Studio (Bossard, 33135, pp.207-208)
1210-1211	1925	Jan	6	Leneveu	Inspection B	33581/709	20	
1218 1232	1925	Mar	6	Leneveu	Inspection B	33581/709	30	
1239 1241 1243	1925	Apr	3	Leneveu	Inspection B	33581/709	35	1239: Studio
1244 1248 1253 1261	1925	May	4	Leneveu	Inspection B	33581/709	40	
1264 1267 1269-1271	1925	Jun	5	Leneveu	Inspection B	33581/709	46	

Emulsion	Year	Month	Day	Engineer (Author of the Report)	Department	Notebook	Page	Comments
1273	1925	Jul	2	Leneveu	Inspection B	33581/709	56	
1277 1281	1925	Jun	5	Leneveu	Inspection B	33581/709	46	1281: Studio
1289 1297 1302	1925	Jul	2	Leneveu	Inspection B	33581/709	56	
1303	1927	Aug	2	Callame		33174/659	301-302	
1306	1925	Aug	6	Leneveu	Inspection B	33581/709	62	
1309	1927	Aug	2	Callame		33174/659	301-302	
1312 1314-1315	1925	Aug	6	Leneveu	Inspection B	33581/709	62	1314: Acetate Reversal?
1319	1925	Sep	4	Leneveu	Inspection B	33581/709	68	
1320	1927	Aug	2	Callame		33174/659	301-302	
1324	1927	Sep	2	Callame		33174/659	324	
1325	1925	Sep	4	Leneveu	Inspection B	33581/709	68	
1328	1927	Aug	2	Callame		33174/659	301-302	
1328	1927	Sep	2	Callame		33174/659	324	
1330	1925	Sep	4	Leneveu	Inspection B	33581/709	68	
1331	1925	Nov	6	Leneveu	Inspection B	33581/709	81	
1337	1925	Sep	4	Leneveu	Inspection B	33581/709	68	
1342 1345 1347 1349 1357 1358	1925	Oct	8	Leneveu	Inspection B	33581/709	74	1345 & 1357: 35mm Acetate Reversal
1367 1369 1373-1375 1378 1380	1925	Nov	6	Leneveu	Inspection B	33581/709	81	
1381	1925	Dec	4	Leneveu	Inspection B	33581/709	87	
1383	1925	Nov	6	Leneveu	Inspection B	33581/709	81	
1388 1390 1392 1394 1396	1925	Dec	4	Leneveu	Inspection B	33581/709	87	
1399	1926	Jan	8	Leneveu	Inspection B	33581/709	93	
1401	1925	Dec	4	Leneveu	Inspection B	33581/709	87	
1404 1407 1409	1926	Jan	8	Leneveu	Inspection B	33581/709	93	
1413 1419 1425	1926	Feb	4	Leneveu	Inspection B	33581/709	99	

Emulsion	Year	Month	Day	Engineer (Author of the Report)	Department	Notebook	Page	Comments
1431	1926	Mar	6	Leneveu	Inspection B	33581/709	105	
1433								
1435								
1447	1926	Apr	8	Leneveu	Inspection B	33581/709	111	
1454								
1467								
1473	1926	May	10	Leneveu	Inspection B	33581/709	118	
1478								
1485								
1486	1926	Jun	9	Leneveu	Inspection B	33581/709	124	1486: Acetate Reversal (35mm & 16mm)
1489								
1495	1926	Jun	9	Leneveu	Inspection B	33581/709	124	
1499								
1509	1926	Jul	8	Leneveu	Inspection B	33581/709	132	
1513	1926	Jul	8	Leneveu	Inspection B	33581/709	132	
1517								
1523								
1526								
1532	1926	Aug	6	Leneveu	Inspection B	33581/709	139	
1539-1540								
1543	1926	Aug	3	Dupoux		33011/642	303	
1559-1560	1926	Sep	10	Leneveu	Inspection B	33581/709	145	
1562-1563								
1565								
1572								
1575								
1578								
1580								
1582	1926	Oct	13	Pouly		33718/737	24	Celluloid
1583	1926	Sep	10	Leneveu	Inspection B	33581/709	145	
1585-1586	1926	Oct	13	Pouly		33718/737	24	Celluloid
1591								
1593-1594								
1598								
1601	1926	Nov	9	Pouly	Manufacture B	33718/737	46	Celluloid 1611: Master roll 4: Ordinary 59.5mm + Panchromatic (Pouly, p.67) 1613: Acetate Studio
1603								
1605-1607								
1611-1613								
1616-1618								
1623-1624	1926	Dec	11	Pouly	Manufacture B	33718/737	66	1638: Acetate Studio & 28mm KOK
1628-1629								
1631								
1635-1636								
1638								
1640								
1649	1927	Jan	7	Clerc	Testing	33008/639	29	
1652								
1653-1654	1926	Dec	31	Pouly	Sensitized	33718/737	79	

Emulsion	Year	Month	Day	Engineer (Author of the Report)	Department	Notebook	Page	Comments
1659	1926	Dec	21	Pouly	Inspection B	33716/735	130	
1662-1663 1665 1667	1927	Feb	1	Pouly	Inspection B	33716/735	140, 148	
1678	1927	Apr	4	Pouly	Inspection B	33716/735	163	
1678 1682	1927	May	5	Leneveu	Inspection B	33581/709	160	1678: Acetate Studio 1682: D.N. Standard
1692-1693	1927	Jul	1	Leneveu	Inspection B	33581/709	172	1692: 35mm Standard 1693: 35mm Acetate Studio
1707 1710	1927	Sep	3	Leneveu	Inspection B	33581/709	182	
1716-1717	1927	Oct	7	Leneveu	Inspection B	33581/709	187	
1719 1721-1722 1724	1927	Nov	5	Leneveu	Inspection B	33581/709	192	
1728-1729 1732	1927	Dec	6	Leneveu	Inspection B	33581/709	197	
1744-1749	1928	Mar		Moreau	Technical Department	33661/724	13	
1750-1752	1928	Jul	31	Moreau	Technical Department	33661/724	89	
1753-1756	1928	Aug		Moreau	Technical Department	33661/724	104	1753: Cinema
1757-1759	1928	Nov		Moreau	Technical Department	33661/724	157	
1760	1929	Feb	29	Moreau	Technical Department	33662/725	27	
1761	1929	Mar	29	Moreau	Technical Department	33662/725	38	
1762	1929	Apr	20	Moreau	Technical Department	33662/725	52	
1764	1929	Jul	13	Moreau	Technical Department	33662/725	93	
1768-1769	1929	Nov	3	Moreau	Technical Department	33662/725	121	
1770	1929	Dec	28	Moreau	Technical Department	33662/725	139	

7
Eastman Kodak

James Layton

1. Introduction

This survey attempts to be a thorough guide to the edge markings found on motion picture film made by the Eastman Kodak Company from 1913 to the present. This includes the previously well-documented dating symbols in use for over 80 years, but also other clues that can help film inspectors confidently identify the exact year of film manufacture. These clues include the precise text used, its styling and its placement, and other codes and numbers, or their absence. This text is the first time such comprehensive data has been presented in one place.

This chapter draws upon existing research undertaken by Harold Brown, Brian Pritchard, Robin Williams, and staff of the Filmoteca Española, among others, but it also expands upon this important groundwork with a new firsthand study of over 1,500 film reels in the collections of The Museum of Modern Art and the British Film Institute. As a result of this fresh examination, many trends and variations in edgeprint markings can now be utilized as additional dating tools.

In addition to this new analysis, this author has been assisted immeasurably by past and present Kodak employees from around the world, including Stephen Champagne, Jean-Pierre Martel, Alan Masson, John C. Miller, Brian Pritchard, and Robert Shanebrook — all foremost experts in the manufacture and finishing of Eastman Kodak motion picture film. In particular, I would like to thank Frederick Knauf, who has been open in sharing his own research into the production of motion picture film at Kodak, much of which has been gathered over 30+ years from firsthand observations, consultation with colleagues, and surviving corporate documentation.

This chapter would not have been possible without the support and assistance of Camille Blot-Wellens, Céline Ruivo, and Ulrich Ruedel of the FIAF Technical Commission, Theo Harrison and Courtney Holschuh at The Museum of Modern Art, and Jane Fernandes at the British Film Institute. Their contributions have been meaningful and invaluable.

Between 1916 and 1996, the Eastman Kodak Company used dating symbols to indicate the year of manufacture. For stocks made in Kodak's primary U.S. manufacturing base in Rochester, New York, these codes were repeated every 20 years through 1981, when a new set of codes was introduced. Different dating symbols were used for stocks made outside the United States, and for 8mm film.

Despite periodic waves of edgeprint standardization across Kodak's manufacturing plants, discrepancies and variations in style, wording, and symbols remained until all Kodak edgeprint was made consistent in the early 1990s. To best document these variations, this chapter is divided into sections dedicated to different periods of manufacture, countries of manufacture, and film gauges. To further assist the reader, additional context is provided throughout on Kodak's film manufacturing practices and corporate history.

Despite the scale and breadth of this study, many questions still remain. As such, any gaps in documentation and unconfirmed findings are presented openly. It is hoped that this survey will continue after this book is published, further fleshing out our understanding of this specialized topic.

2. 35mm Film Manufactured in the United States

The edge markings on Eastman Kodak 35mm motion picture film typically follow standardized layouts and content. Most significantly for dating a given film, a symbol or group of symbols notate the year of manufacture (see Chart #1 below). These symbols, first introduced with a circle (●) in 1916, repeated in 20-year cycles. The markings also typically indicate the base of the film (nitrate or safety), and in later years, the country of manufacture.

These markings were exposed onto the film's edge at the time the film was slit into 35mm strips from larger rolls. (These rolls were originally 47 inches [1.194 metres] wide, and later 54.5 inches [1.384 metres] wide.) The edgeprint was applied using a mechanism developed in-house by Kodak called a print drum. This hollow metal roller had regular cut-outs around its circumference where an edgeprint stencil could be mounted. Internally, a light source exposed the markings on the film's edge once every rotation (about every 18 inches) as the film came into contact with the stencil. This print drum was located in the film slitter close to the slitter knives.

Due to the demands of continuous production and the lengthy process of changing the stencils, the dating symbols could not be updated on all slitters simultaneously. Each year, this work was most commonly begun in November and continued through to February of the following year.

Edgeprint text is typically black, although it may appear white on a dark background on reversal films, or if printed-through from an earlier-generation element. Be careful not to confuse print-through information when trying to date a film.

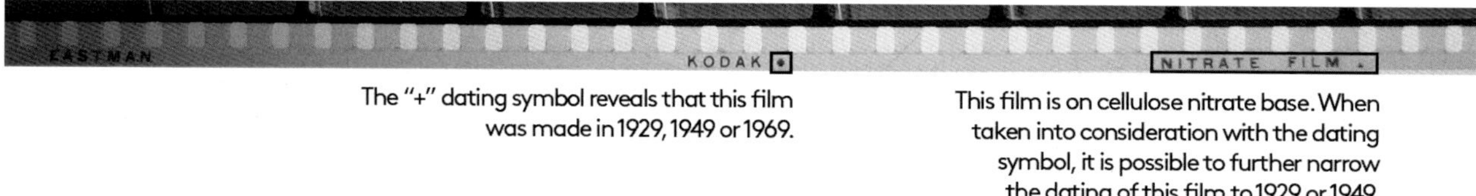

The "+" dating symbol reveals that this film was made in 1929, 1949 or 1969.

This film is on cellulose nitrate base. When taken into consideration with the dating symbol, it is possible to further narrow the dating of this film to 1929 or 1949.

By tracking how these edgeprint markings changed over the years, it is possible to accurately date the exact year of manufacture in nearly all cases. Sections 2.2 to 2.7 chart the variations in edgeprint markings on U.S.-made Kodak stocks from 1913 to the present day. Practically every marking and text used during this period is identified and explained.

Chart #1: Eastman Kodak Dating Symbols, 1916-1996

Year				Symbol	Year	Symbol
1916	1936	1956	1976	●	1982	● ■ ✖
1917	1937	1957	1977	■	1983	✖ ▲ ✖
1918	1938	1958	1978	▲	1984	▲ ■ ▲
1919	1939	1959	1979	● ●	1985	■ ● ▲
1920	1940	1960	1980	■ ■	1986	▲ ● ▲
1921	1941	1961	1981	▲	1987	■ ▲ ▲
1922	1942	1962		● ■	1988	✚ ✚ ▲
1923	1943	1963		● ▲	1989	✖ ✚ ▲
1924	1944	1964		▲ ■	1990	▲ ✚ ▲
1925	1945	1965		■ ●	1991	✖ ✚ ✖
1926	1946	1966		▲ ●	1992	■ ✚ ▲
1927	1947	1967		■ ▲	1993	✚ ▲ ▲
1928	1948			● ● ●	1994	✚ ● ▲
		1968		✚ ✚	1995	✚ ■ ▲

Year			Symbol	Year	Symbol
1929	1949	1969	✚	1996*	✖●▲
1930	1950	1970	▲✚		
1931	1951	1971	●✚		
1932	1952	1972	■✚		
1933	1953	1973	✚▲		
1934	1954	1974	✚●		
1935	1955	1975	✚■		

* Dating symbols were announced by Eastman Kodak through to 2005, but they ultimately ceased being used in 1996, and as such, subsequent year codes are not presented in this document.

2.1 1913 to 1950

Between the 1890s and 1913, Eastman Kodak film had no text or markings on its edge. During this time, motion picture film was made and slit by Kodak in Rochester, NY, but it was typically perforated elsewhere. From 1913 until 1916, the word "EASTMAN" can be found stenciled or exposed on the film's edge, and some rudimentary symbols were used to indicate the year of manufacture. Harold Brown did considerable research on these early years of Eastman Kodak edge markings, and his conclusions (covering 1913-1916) are presented earlier in this publication.[1]

Further standardization was introduced in 1916 with a consistent layout and the formalized use of dating symbols. This coincided with the start of full slitting and perforating of film onsite at the Rochester plant and Kodak's British plant in Harrow. (For more on Kodak Ltd., see Section 3.1.) These dating symbols, as seen in Chart #1, repeated in 20-year cycles. These symbols appear immediately following the word "KODAK".

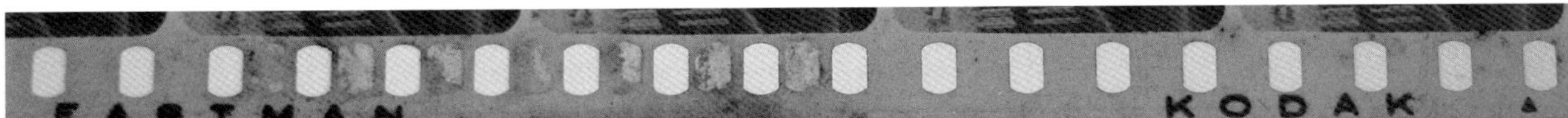

Before 1925, all Kodak prints used Bell & Howell (BH) rounded perforations, typically found on negatives and other pre-print stocks. After 1925, Kodak Standard (KS) or positive perforations were used for prints. At the same time the style of the edge text became more condensed, and the words "NITRATE FILM" or SAFETY FILM" were added to confirm the base. This change coincided with the erection of a new base-manufacturing building at Kodak Park in Rochester and the introduction of new equipment that made film rolls 54.5 inches wide (before the film was slit down into narrower gauges). The change in edgeprint style may be the result of new slitting equipment entering use.

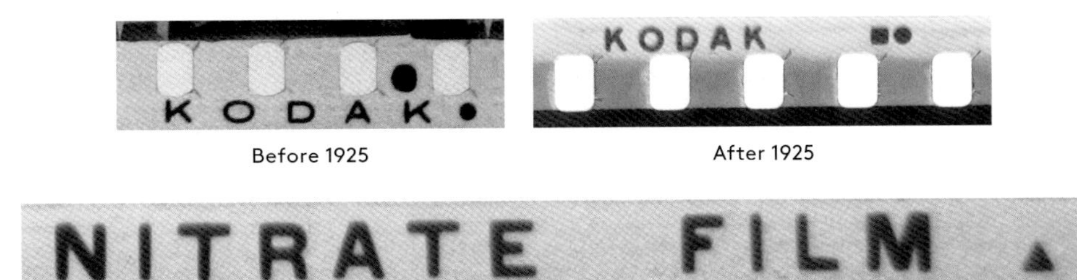

Before 1925

After 1925

"NITRATE FILM" appears on Eastman Kodak stocks following the date code from 1925 to approximately 1951, when nitrate film was eventually phased out by the film industry. The word "FILM" was often followed by a symbol of unknown meaning (which should not be confused as a dating symbol).

1 See "Stock Manufacturers' Edge Marks", p.69.

Beginning in the 1920s, black & white panchromatic negative film was marked "PANCHROMATIC" following the words "NITRATE FILM" or "SAFETY FILM". Orthochromatic negative film was not identified with such markings.

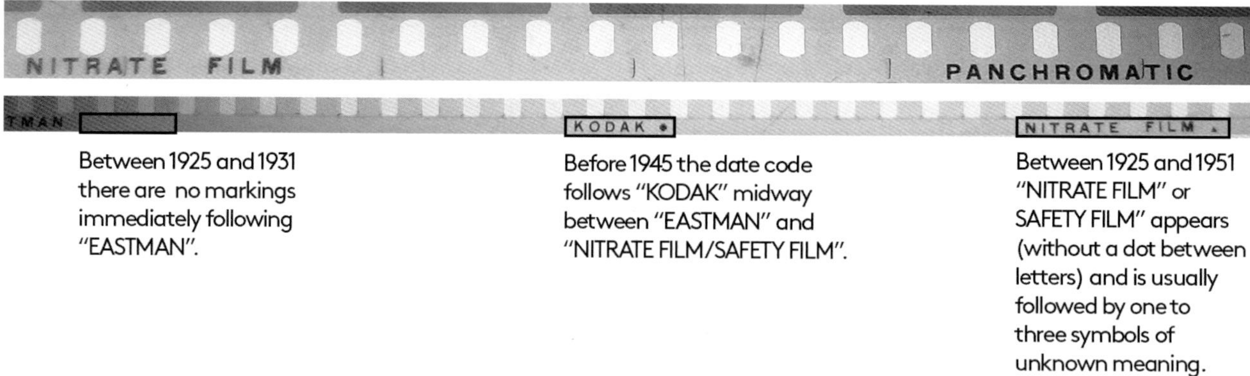

Between 1925 and 1931 there are no markings immediately following "EASTMAN".

Before 1945 the date code follows "KODAK" midway between "EASTMAN" and "NITRATE FILM/SAFETY FILM".

Between 1925 and 1951 "NITRATE FILM" or SAFETY FILM" appears (without a dot between letters) and is usually followed by one to three symbols of unknown meaning.

1945

From 1945, the word "KODAK" no longer appears on the film edge. From this point, the dating symbols occur immediately following a straight vertical line/bar or "pipe" character ("|"). The spacing between this line and the symbols (or between "KODAK" and the symbols, before 1945) indicates whether the film was manufactured within the first or last 6 months of the given year.

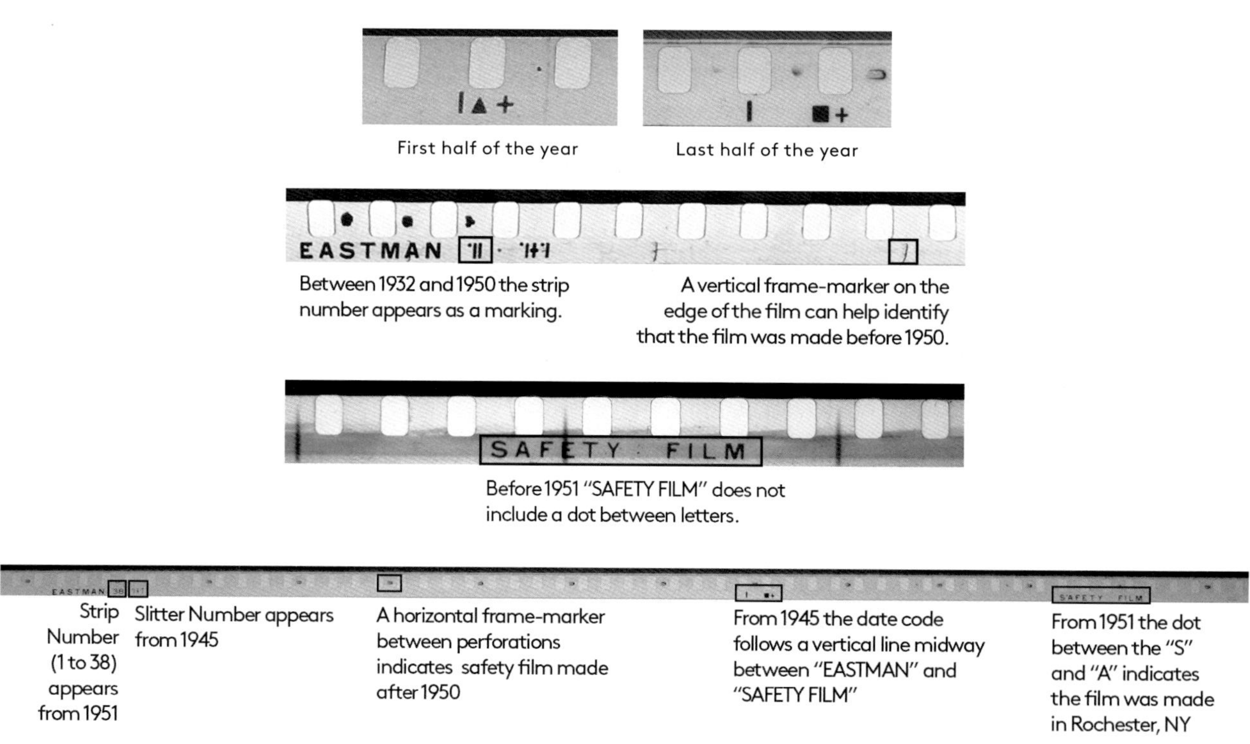

First half of the year

Last half of the year

Between 1932 and 1950 the strip number appears as a marking.

A vertical frame-marker on the edge of the film can help identify that the film was made before 1950.

Before 1951 "SAFETY FILM" does not include a dot between letters.

Strip Number (1 to 38) appears from 1951

Slitter Number appears from 1945

A horizontal frame-marker between perforations indicates safety film made after 1950

From 1945 the date code follows a vertical line midway between "EASTMAN" and "SAFETY FILM"

From 1951 the dot between the "S" and "A" indicates the film was made in Rochester, NY

2.2 Strip Numbers, Frame-markers, and Other Symbols

Eastman Kodak motion picture film has traditionally been manufactured in rolls 54.5 inches wide, then slit down into 35mm-wide strips or different widths. Starting in 1932, Kodak began documenting the strip number on its edge markings, and the slitter identification code was added beginning in 1945. This was largely for quality control purposes, so that a potential technical issue could be traced directly to its source.

For 35mm film, the film slitter cut 38 strips of film from the 54.5-inch-wide roll, so strip numbers range from 1 to 38 (or 1 to 32 for strips cut from 47-inch-wide rolls, which were used in addition to 54.5-inch rolls up to about 1950). For 16mm film, 83 strips were cut (or 84 strips, from the mid-1990s, once slitting control was improved), and for 65mm 20 strips. The slitter identification code or number documents the particular slitter used at the time of manufacturing. It was possible to document this particular detail about strip number and slitter code as the edge markings were recorded onto the film at the time of slitting.

Both the strip number and slitter number were coded to begin with, and these symbols can be decoded using Chart #2, below.

1932-1945

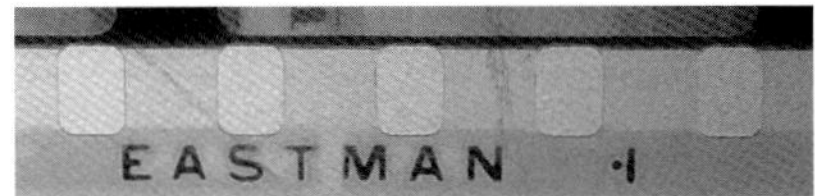

During this period, only the strip number appears, following the word "EASTMAN".

1945-1951

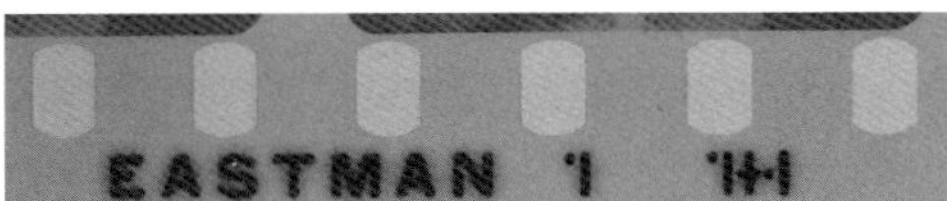

Prior to the end of 1951, "EASTMAN" is followed by two sets of symbols, indicating the strip number (left) and slitter number (right). In the example above, the strip number is 2 and the slitter number is 204.

After 1951

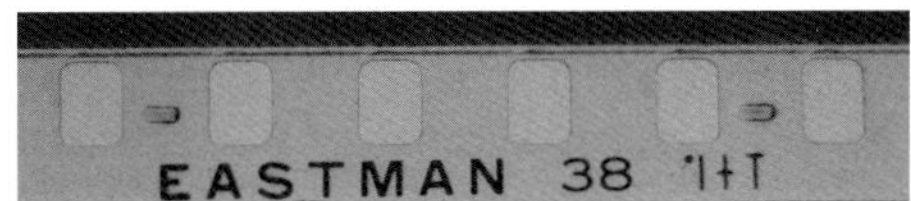

From 1952, numeric characters (from 1 to 38) are used for the strip number, immediately following "EASTMAN." Symbols are still used to identify the slitter number.

Chart #2: Eastman Kodak Strip Number Symbols, 1932-1951

1	I	11	II	21	•II	31	I.I
2	•I	12	I•I	22	•I•I	32	I.•I
3	I.	13	II.	23	•II.	33	I.I.
4	•I	14	I•I	24	•I•I	34	I.•I
5	.I•	15	I.I•	25	•I.I•	35	I..I•
6	I:	16	II:	26	•II:	36	I.I:
7	:I	17	I:I	27	•I:I	37	I.:I
8	T	18	IT	28	•IT	38	I.T
9	⊥	19	I⊥	29	•I⊥		
10	I+	20	•I+	30	I.+		

The frame-marker or frame-line marker was a dash that occurred every 4 perforations on black & white stocks. These began appearing around 1931. According to internal Eastman Kodak sources, the marker was useful for film laboratories to obtain optimal image steadiness during printing. These markers were typically applied in black ink when the film was perforated (and thus were not part of the latent-image edgeprint).

The marker appears perpendicular to the film's edge during the nitrate film era, and parallel to the film's edge between the perforations from 1951. It remained consistently in use until the late 1980s, and appears to have been phased out by about 1992.

The orientation of this marker does not necessarily indicate the film base. For example, safety film made before 1951 will have the same marker orientation as nitrate film: perpendicular to the film's edge. According to Kodak literature from 1956, the parallel frame-markers in use from 1951 identify the base as triacetate. By contrast, a perpendicular frame-marker found on film marked "SAFETY FILM" subsequently indicates diacetate base.

Kodak film made outside the United States does not contain these frame-markers.

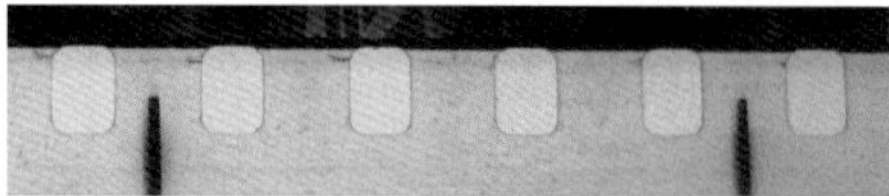

Between 1931 and 1948, the frame-marker extends midway into the perforation area.

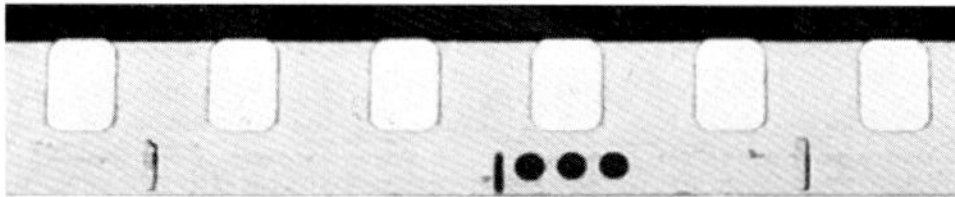

From the middle of 1948, the frame-marker is shortened.

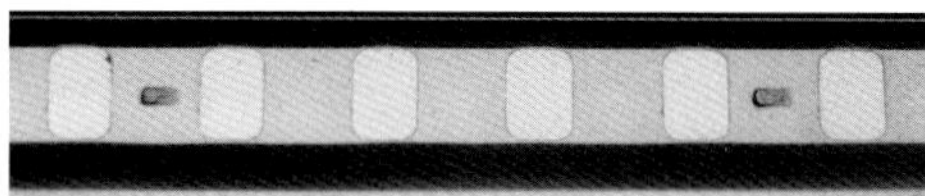

From 1951, the marker appears between the perforation area.

Between approximately 1929 and 1950, "NITRATE FILM" or "SAFETY FILM" on the film edge was frequently followed by a symbol or series of symbols of unknown meaning. These should not be mistaken for dating symbols. It is possible these symbols may identify the building or finishing area where the slitter was in use.

A study of approximately 50 prints from this time period reveals a variation of only 8 symbol combinations: ▲ ■ ●, ■ ■, ● ▲, ▲ ▲, ■, ●, ▲, or F. Sometimes there is no symbol at all.

2.3 Plant of Origin

During the conversion from nitrate to safety film, Eastman Kodak introduced new markings to indicate the plant of origin or country of manufacture. This was coded by the placement of a dot between the letters of "SAFETY FILM" from 1951 to 1996, or between the letters of "KODAK" from 1997 onward (see Chart #3 below). Before this time, different sets of dating symbols were used in different countries of manufacture (see Charts #6, #7, #9, and #10 in Section 3).

Before 1951:

From 1951:

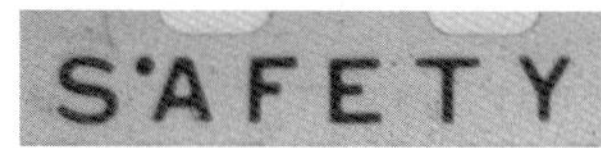

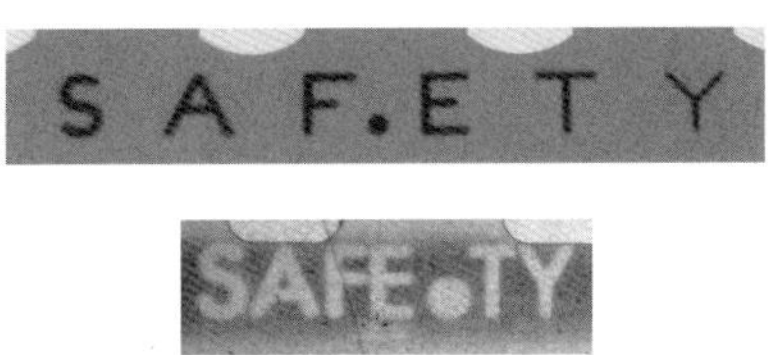

The plant of origin dot does not necessarily indicate where a film was manufactured or coated. Instead, it documents where a film was "finished" (slit, perforated, chopped, spooled, and packaged). In many cases, a film would be coated and finished at the same location, but this is not a given.

Chart #3: Plant of Origin Coding, 1951 to the Present

United States (Rochester, New York)	S·AFETY	K·ODAK
United States (Windsor, Colorado)*	S.AFETY	KOD·AK (35mm) KOD.AK (16mm)
Canada (Toronto)	SA·FETY	KO·DAK
United Kingdom (Harrow)	SAF.ETY	
France (Vincennes or Chalon)	SAFE.TY	KODA·K
Australia (Melbourne)**	SAFET.Y	K.ODAK
Mexico (Guadalajara)	SAFETY.	KO.DAK
Brazil (Manaus)**	SAFETY·	KODA.K

* Eastman Kodak's manufacturing plant in Windsor, Colorado operated from the late 1960s until 2012, when a separate company, Kodak Alaris, took over the operation following Eastman Kodak's bankruptcy reorganization. A large quantity of Super 8 and 16mm finishing moved to Windsor from Rochester in the mid-1970s, with the remainder of 16mm finishing transferring in about 1989, when Keycode edgeprinting was introduced (see Section 2.6). 35mm print film was also finished in Windsor from 1999 to 2006. Before 1989, both 16mm and 35mm finishing were also done at Kodak facilities in Mexico and Toronto.

** Eastman Kodak's operations in Australia and Brazil finished consumer products only, not motion picture film.

2.4 1951-1996

In 1951, Eastman Kodak standardized its dating symbols so that the same codes were used around the world (except for France; see Section 3). Canada and the United Kingdom henceforth used the U.S. dating codes documented in Chart #1, and the country of manufacture was indicated by the placement of a dot between the letters of "SAFETY" (see Section 2.4).

Between 1951 and 1982 the style, markings, and placement of text on Eastman Kodak film remained fairly consistent. As such, it can sometimes be difficult to distinguish some 1950s elements from others made in the 1970s. But a few clues can help determine the exact year in most cases, usually through a process of elimination.

1950-1963

Eastmancolor negative and print film was introduced in 1950. Between 1950 and 1963, these stocks were marked "COLOR", or most frequently "EASTMAN COLOR". Technicolor prints from this era do not use the word "COLOR". Eastmancolor negatives and prints after 1963 use the same standard markings as black & white film.

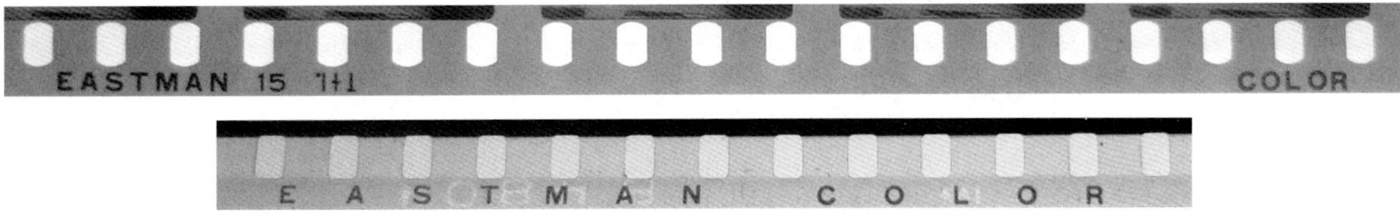

1968-1980

Between approximately 1968 and 1980, many — but not all — black & white stocks featured alternate-style edge text, with condensed markings, and the order of the strip number and slitter number was switched.

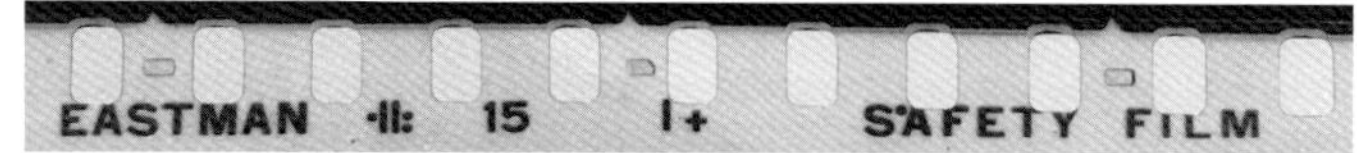

1968-1982

Color reversal intermediates were reversal stocks used to make a duplicate negative directly from the original negative. As such, they have black edges, and the edgeprint is typically orange.

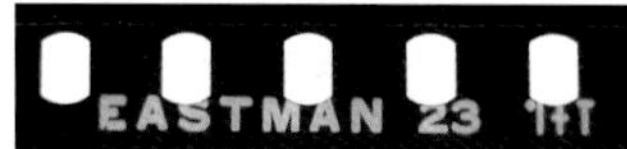

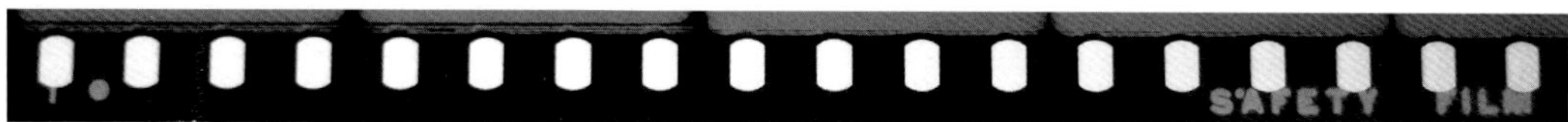

1974-1983

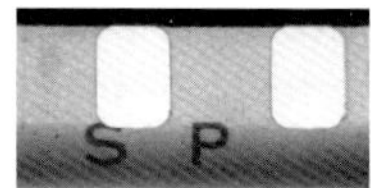

Eastman Kodak introduced Eastmancolor SP color print film in 1974, which offered a significant reduction in processing time compared to previous color print stocks. These prints are marked "SP" (for "Short Process"), usually between "EASTMAN" and the date code. Nearly all color prints marked "SP" have now faded.

1982-1991

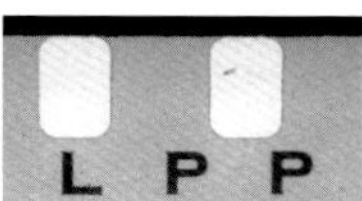

In 1979, Eastman Kodak introduced LF ("Low Fade") and LFSP ("Low Fade Short Process") color print films for 16mm, but it is unclear if these acronyms were used on standard 16mm edgeprint, as no examples have been found during this survey. In 1982, low-fade emulsion became standard for all Eastmancolor print film, and these prints were marked "LPP" ("Low-Fade Positive Print") until 1991. Between 1983 and ca. 1994, it is also possible to see "LC" ("Low Contrast") on 35mm and 16mm prints made for video transfer (television) or from reversal originals.

1982-1996

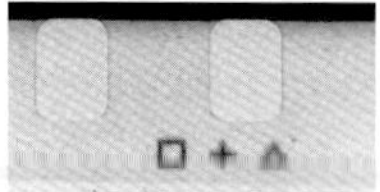

In 1982, Eastman Kodak introduced a new 3-symbol dating system and discontinued the former 2-symbol codes that repeated in 20-year cycles (see Chart #1). The 3-symbol dating system remained in use until 1996.

2.5 Pre-print Letter Codes, 1989 to the Present

1989-2012

In 1989, dating symbols were discontinued from the edge markings on negative and interpositive stocks and were replaced with a 2-character letter code to signify the year of manufacture. Eastman Kodak announced these letter codes through to 2019; however, they were ultimately phased out for 35mm film midway through 2012 and replaced with the full 4-digit numeric year. (The 2-character letter codes continue to be used on 16mm.)

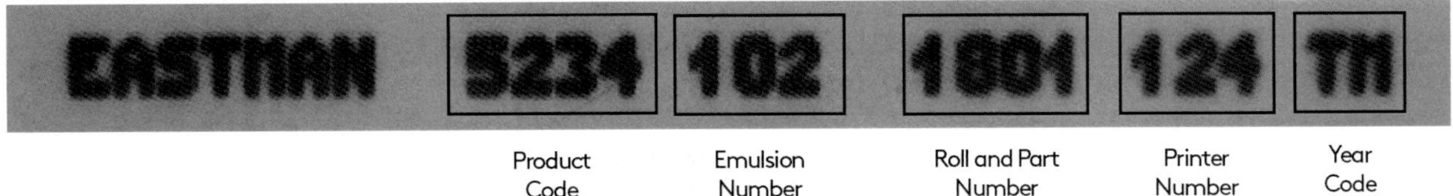

Product Code | Emulsion Number | Roll and Part Number | Printer Number | Year Code

Fine Grain Duplicating Panchromatic Negative Film (5234) manufactured in 1994

Chart #4: Pre-Print Letter Codes, 1989-2019

1989	DE	1997	KD	2005	DS	2013	KL
1990	LE	1998	DF	2006	FT	2014	DE
1991	EA	1999	FL	2007	LM	2015	FA
1992	AS	2000	SD	2008	EN	2016	LS
1993	ST	2001	TF	2009	AK	2017	ET
1994	TM	2002	ML	2010	TK	2018	AM
1995	MN	2003	NE	2011	MD	2019	SN
1996	NK	2004	KA	2012	NF		

Emulsion Type Code

Key Number Count - increments in one per foot

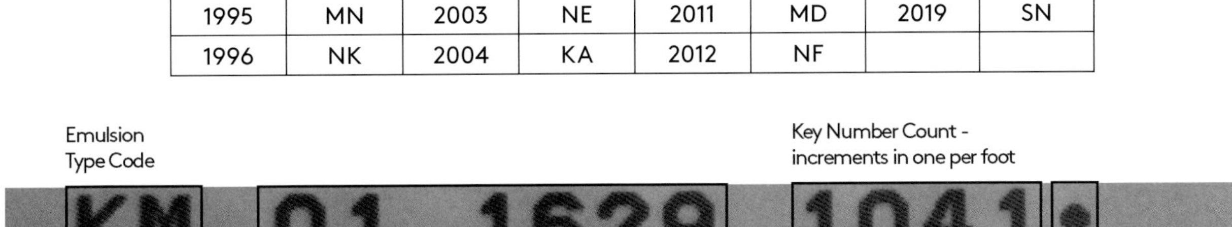

Key Number Prefix - six digits that identify film roll

Zero-Frame Reference Mark - dot identifies frame directly above that Key Number Count references

The Eastman Keycode system has been used on all 35mm negative and intermediate stocks since 1989. In addition to coding stock information in a machine-readable barcode, it also provides the same data in human-readable form. When taken as a whole, these 10-digit formulas effectively provide a unique number for each foot of film manufactured, making it easier to match footage in negatives and workprints during post-production.

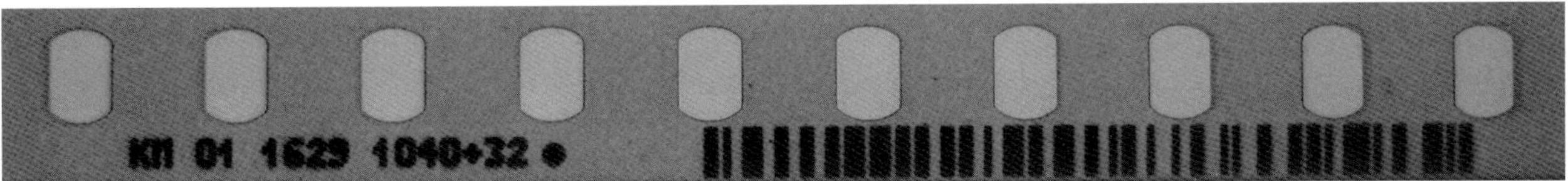

Chart #5: Eastman Keycode Emulsion Type Codes

KA	5243, SO-420, SO-421	KN	7292, SO-175, SO-463	EA	5285	EO	5213
KB	5247, SO-247	KO	5249, SO-455	EB	5229	EQ	5205
KC	5297	KP	5600	EC	7265	ER	5203
KD	5234, 2234, SO-239	KQ	5277	ED	7266	ES	2273
KE	5222	KR	5289, 2374	EE	5263	EU	5260
KF	5295, SO-215	KS	5272, 2272, SO-211	EG	5284	EV	2242, 5242, SO-742, SO-942
KG	5294, SO-185	KT	5298, SO-098, SO-898, SO-262	EH	5218, SO-218	EW	2254, 5254
KH	5231	KU	5279, SO-079, SO-579	EI	5299, SO-080	EX	2366, 2369, 5366, 5369
KI	5246	KV	2244, 5244, SO-443, SO-440, SO-450	EJ	5219, SO-219	EY	2332
KJ	5296, SO-261, SO-290, SO-896	KW	5287, SO-287	EK	5201	EZ	5230
KK	5245, SO-245	KX	5017, 5239, 7239, SO-080, SO-214	EL	5217		
KL	5293, SO-009, SO-893	KY	5620	EM	5212		
KM	5248, SO-848, SO-031, SO-256	KZ	5274, SO-074	EN	5207, SO-207		

Note: SO = Special Order

2012 to the Present

The 4-digit numeric year replaces the letter code on all negative stocks. It remains in use today.

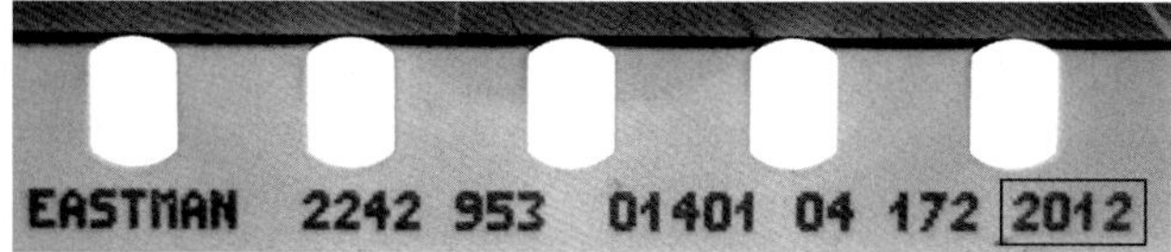

2.6 Color Print Film, 1996 to the Present

The late 1990s saw a mild flurry of edge marking variations, before settling down in the early 2000s. In some instances, layouts appear to have changed every few months. The following applies to color print stocks only — a different stencil was used for text on black & white prints (see Section 2.8).

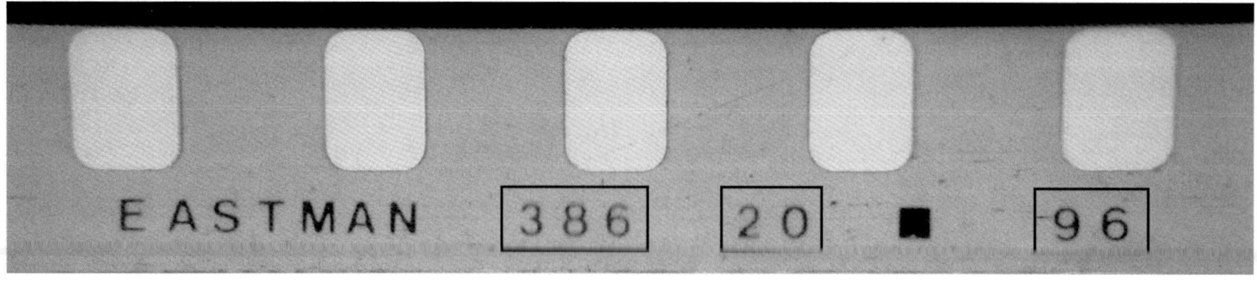

Emulsion Type Code   Strip Number   Year Code

1996

The 3-symbol edge code system was discontinued in 1996 and initially replaced with a 2-digit year reference. "SAFETY FILM" was eliminated entirely.

1997-1999

Early 1997

By 1997, the year reference expanded to 4 digits and the word "EASTMAN" was replaced by "KODAK".

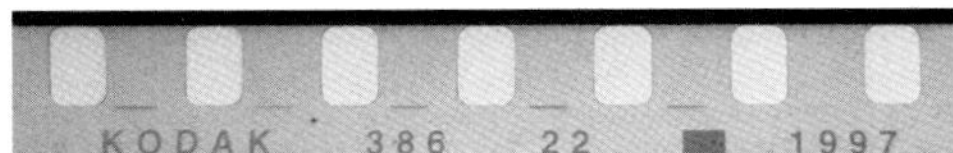

Late 1997 Through 1998

In late 1997, the order was changed: emulsion type code, strip number, KODAK, Roman numerals of unknown meaning, year. But this style was also short-lived.

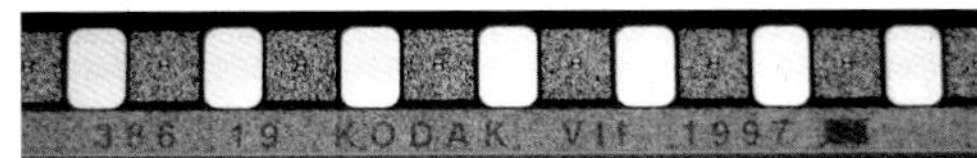

Early 1999

Another shuffling of edge text came in early 1999, followed by a shift to a longer string of text in late 1999 that remains in use today. This change coincided with the phasing out of exposed latent-image edge markings and the beginning of laser-printed and LED-printed text — usually in pink — which was applied at the high-speed perforators rather than at the slitters.

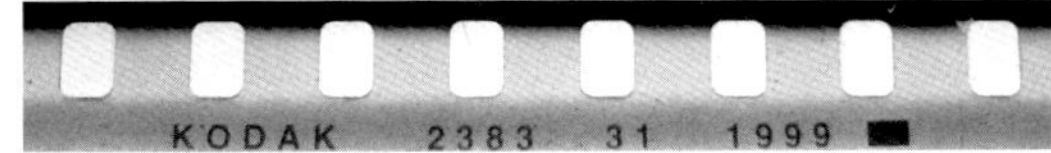

Late 1999 to the Present

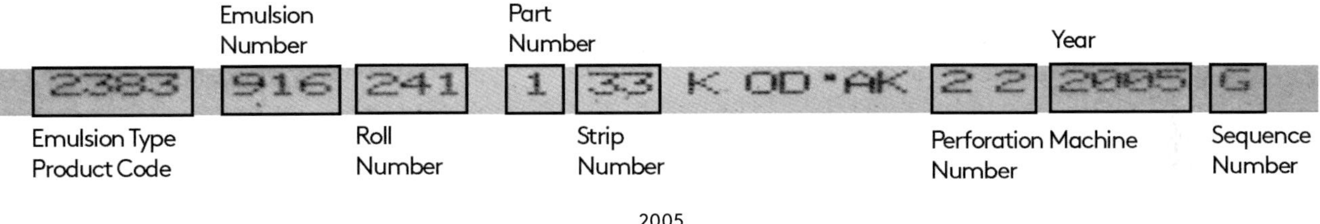

2005

2.7 Black & White Print Film, Late 1990s to the Present

Late 1990s to 2007

From the late 1990s, black & white print film used minimal edge markings: KODAK, strip number, year. This text continued to be exposed as a latent image at the time of slitting, unlike color print film, which used laser-printed characters starting in 1999.

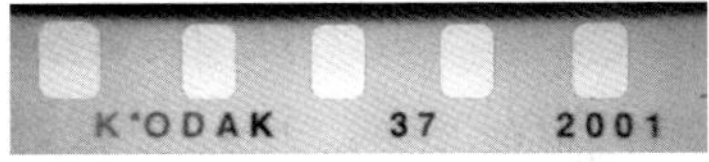

2008-2009

Beginning in 2008, Eastman Kodak launched a multi-year stencil that decreased in length each year. The lowest number on the right indicated the year of manufacture. As each year passed, that number would be eliminated. This allowed the edgeprint stencils to be used for 5 years — rather than 1 year — with the redundant year digits blacked out on the stencil at the end of each year.

Although this stencil format had the potential to run through to 2012 before resetting, it appears to have been discontinued on 35mm products in early 2009, although it remained in use longer on 16mm.

K'ODAK 3 366 12 11 10 9
K'ODAK 3 366 12 11 10 9
K'ODAK 3 366 12 11 10
K'ODAK 3 366 12 11
K'ODAK 3 366 12

2009 to the Present

In mid-to-late 2009, all black & white print film started using the same edge markings that had been used on color print film since 1999. These were applied by laser printing.

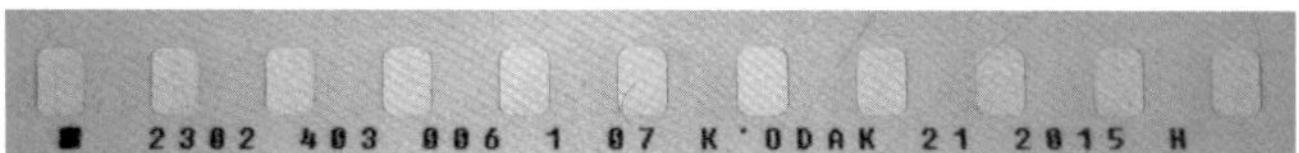

Polyester Black & White Print Film Stock (2302) manufactured in 2015

3. 35mm Film Manufactured Outside the United States

As already noted, Eastman Kodak edgeprints frequently differed between countries of manufacture. Just as with U.S.-made films, it is important to track the subtle variations over the years within each country to be able to better date a film and understand the meaning of its markings.

The study of film elements for this chapter has been undertaken largely in the United States — with limited access to foreign holdings — so it has been especially challenging to document films produced at Eastman Kodak's other manufacturing plants around the world. Fortunately, this author has been supported by colleagues at other international archives, particularly Jane Fernandes at the British Film Institute, who has shared inspection images, and parts of Harold Brown's unpublished research into Eastman Kodak edge markings through to the late 1960s. As such, edge markings on films made by Kodak Ltd. are almost thoroughly documented. The same cannot be said for films manufactured in France, despite considerable digging in the historical archives of both Kodak and Pathé. Some films will remain a challenge to date.

Interestingly, this study has revealed that there can be vast differences in edge markings on films made around the world, some more significant than others. Despite efforts to standardize, many variations remained. These are outlined over the following pages.

3.1 Kodak Ltd. (United Kingdom)

Eastman Kodak's British subsidiary was formed in 1889, and opened its first factory in Harrow, north of London, in 1891.

Motion picture film produced in Harrow started using edge markings and dating symbols in 1917. As "Eastman" was not part of Kodak Ltd.'s business name, film manufactured in the U.K. never includes the word "EASTMAN" on its edge.

Between 1917 and 1950, British Kodak film used its own unique set of dating codes, which were reused in a 19-year cycle. Also particular to Kodak Ltd. film on nitrate base is an "explosion" symbol — often found after the words "NITRATE FILM" — in use from approximately 1920 to 1950.

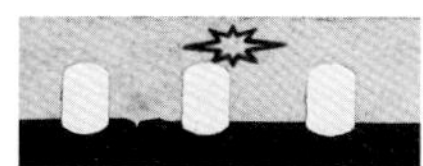

From 1917 "KODAK" appears on all motion picture film manufactured in the U.K.

The U.K. dating symbols follow "KODAK", and between 1917 and 1950, are different from the symbols used in the United States and elsewhere in the world.

From as early as 1920, this explosion symbol appears on nitrate film made in the U.K.

There are no frame index markers on U.K. stocks before 1951.

Beginning in the 1930s or earlier, "NITRATE FILM" precedes the explosion symbol. The slit number follows the symbol.

Chart #6: Kodak Ltd. Dating Symbols, 1917-1950

Year	Symbol
1917 1936	◡
1918 1937	L
1919 1938	▬
1920 1939	◡◡
1921 1940	LL
1922 1941	▬▬
1923 1942	◡ L
1924 1943	▬ L
1925 1944	◡ ▬
1926 1945	▬ ◡
1927 1946	L ▬
1928 1947	L ◡
1929 1948	+
1930 1949	+◡
1931 1950	+L
1932	+▬
1933	◡+
1934	L+
1935	▬+

With the conversion to safety film and the Eastman Kodak Company's international standardization of date codes in 1951, the lettering style and placement of British Kodak film changed. These did not match the standards in place for Kodak film manufactured in the U.S. — the British text is leaner and the kerning (spacing between each character) is wider.

At this time, the strip number appears after "KODAK" and the date symbols occur after "SAFETY FILM." An "S" and the perforator number appear between perforations, one per frame. The orientation of the "S" and perforator number changed around 1955, rotating 90 degrees, but maintaining its position between perforations. In 1965 or 1966, the "S" between the perforations was eliminated.

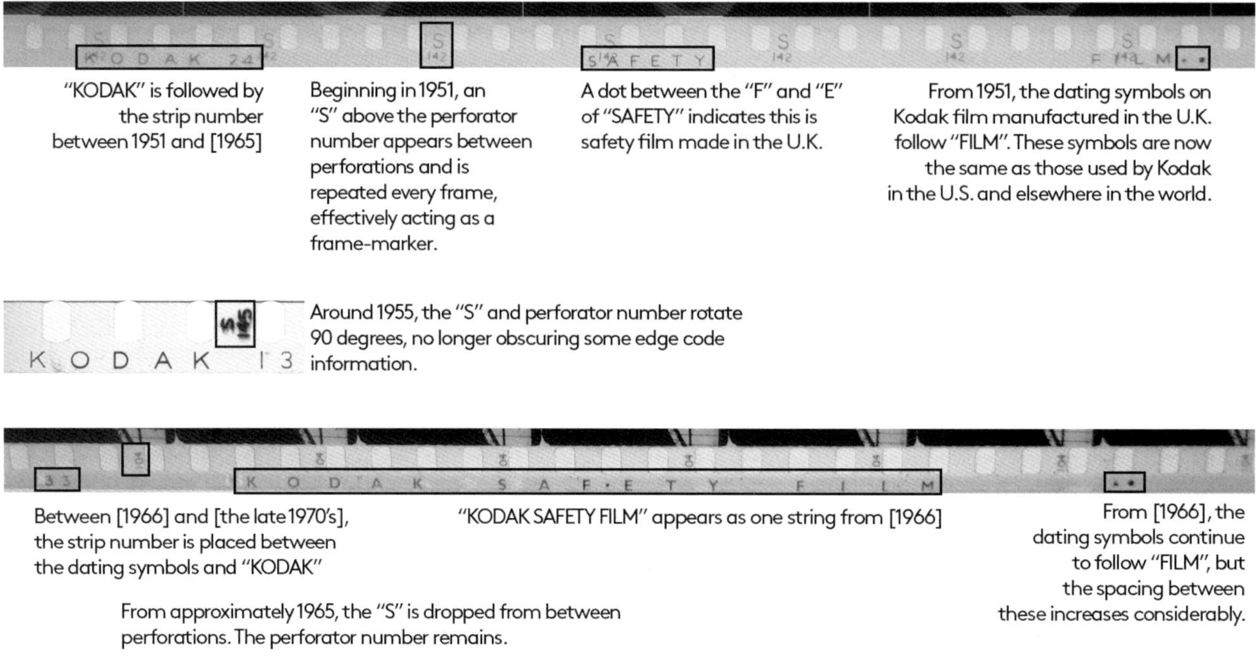

Around 1977, Eastman Kodak introduced a rationalization program to minimize duplication across the company's facilities worldwide. This was known as the Product Interchange Program, or PIP. In Europe, Kodak Ltd. in the United Kingdom manufactured all Color Paper and Graphic Arts products, while Kodak-Pathé in France took all Color Film (Motion Picture and Consumer), Medical Radiography, and Industrial Radiography products. Since the late 1970s or early 1980s, no motion picture film products have been manufactured in the U.K.

3.2 Kodak Canada Ltd. (Canada)

Formed as the Canadian Kodak Company Ltd. in 1899, Kodak Canada Ltd., as it came to be known, operated a manufacturing plant in Kodak Heights, Toronto, from 1915 until its closure in 2005. It is unclear if motion picture film was made at this site before 1925, when Canadian-specific dating symbols were first introduced. These symbols repeated in an 11-year cycle, and remained in use until 1950.

Chart #7: Kodak Canada Ltd. Dating Symbols, 1925-1950

Year	Symbol
1925 1936 1947	● L
1926 1937 1948	● ▬
1927 1938 1949	● ◡
1928 1939 1950	L ●
1929 1940	▬ ●
1930 1941	◡ ●
1931 1942	◆
1932 1943	◆ ●
1933 1944	◆ ▬

Year	Symbol
1934 1945	◆ L
1935 1946	◆ ◡

As part of this survey, samples have been reviewed from the 1920s and 1930s, 1980s, and 2001. It is unclear at present if the significant gap between the 1930s and 1980s indicates a pause in film manufacturing or a lack of samples for review. Kodak Canada Ltd. likely manufactured and/or finished motion picture film during most of this time, but in limited quantities, and only for distribution in Canada. It is known that Kodak Canada finished Eastmancolor negative and print films, and some black & white products, in the late 1980s, but from 1989, only color print films were finished in Canada.

From the samples studied, the content and placement of the edge markings on Canadian stock matches the United States closely, with the exception of the dating symbols. This was most likely due to the fact that the stencils and slitting machines used for films finished at Kodak Canada were made at Kodak Park in Rochester.

It is worth noting that the diamond dating symbol (◆) on Canadian film from before 1951 is easily confused with a plus sign (+), making it possible to misdate a film. For example, this film from 1934 (◆ L) could be misinterpreted as U.K. stock from 1931 or 1950 (+L).

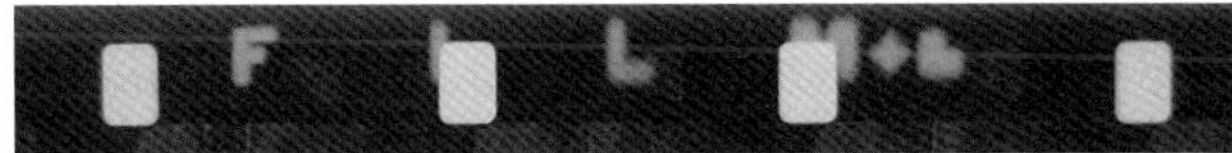

3.3 Société Kodak-Pathé (France)

Although Kodak had operated a subsidiary in France since 1897 — as Eastman Kodak S.A.F. — this company did not enter motion picture production until 1927, when it formed a partnership with its chief competitor Société Française Pathé-Cinéma to become Société Kodak-Pathé S.A.F. This new company took over operations of Pathé's manufacturing plant in Vincennes in the latter half of 1927 and operated the plant until October 1931, when Kodak bought the remaining 49% of the company to make Kodak-Pathé a fully-owned subsidiary of the Eastman Kodak Company.

Prior to Kodak's involvement, Pathé-Cinéma had its own system of edge markings, some of which were retained by Kodak-Pathé after 1927. In this system, "PATHE CINEMA FRANCE" or "PATHE CINEMA PARIS" was followed by a 4-digit number (the emulsion number or batch number) and then by one or two smaller groups of numbers (possibly strip and slitter numbers). The first 2 digits of the 4-digit emulsion number can be used to identify the year of manufacture.[2]

It appears that for a short period in the late 1920s Kodak-Pathé stock used British dating symbols after the word "KODAK." Samples of this have been documented from 1927 and 1928. Perhaps the only way to distinguish French-made stock and British stock during these 2 years is by the absence of an "explosion" symbol on the French-made stocks. Some British stocks at this time do contain groupings of numbers — as on French prints — but not all.

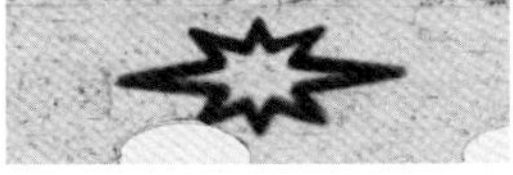

From 1929, the dating symbols on Kodak-Pathé film products were eliminated. However, the groupings of numbers were retained through the end of the nitrate era, until about 1951. It appears that Pathé's emulsion numbering system was expanded from 4 digits to 5 digits in 1927 or 1928.

2 For more details, see Harold Brown, p.73 in this volume, and the chapter on Pathé, pp.206-228.

While researching Pathé edgeprint in the Jérôme Seydoux-Pathé Foundation archives in Paris, Camille Blot-Wellens located manufacture dates for a limited sequence of Kodak-Pathé emulsion numbers. This documentation from 1931 and 1932 reveals that different series of emulsion numbers were used for positive and negative stocks, and for Kodak and Pathé branded products. A sampling of these appear in Chart #8. Continued research in this collection may reveal emulsion numbers used in subsequent years.

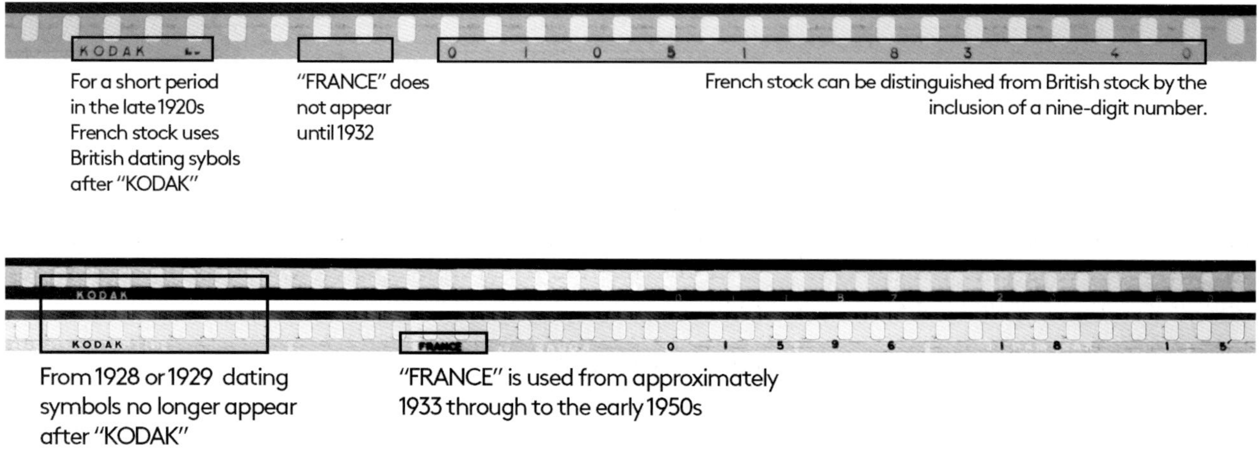

Chart #8: Incomplete Kodak-Pathé Emulsion Number Dates of Manufacture, 1929-1932

Date of Manufacture	Kodak Positive Emulsion Numbers	Kodak Negative Emulsion Numbers
December 1929	01097	
January 1930	01099–01105	
February 1930	01113	01771–01772
March 1930	01114–01115	
April 1930	01116–01122	
May 1930	01123–01127	
June 1930	01128–01129	
July 1930	01139–01146	
August 1930	01147–01151	
October 1930	01157–01166	
November 1930	01167–01180	
December 1930	01181–01200	
January 1931	01201–01216	
February 1931	01217–01223	
March 1931	01224–01227	
April 1931	01228–01232	
May 1931	01233–01237	
June 1931	01238–01243	01781
July 1931	01244–01249	
August 1931	01250–01258	
September 1931	01259–01266	
November 1931	01275–01282	
December 1931	01283–01287	
January 1932	01288	
March 1932	01289–01292	
April 1932	01293–01299	

Date of Manufacture	Kodak Positive Emulsion Numbers	Kodak Negative Emulsion Numbers
May 1932	01300–01308	
June 1932	01309–01319	
July 1932	01320–01327	
August 1932	01328–01334	
September 1932	01335–01345	
October 1932	01346–01357	
November 1932	01358–01365	

From about 1931 — the year Kodak took full control of the Vincennes plant — the word "FRANCE" appears after "KODAK." From this date until the early 1950s, when these markings are present, it is not always possible to date French Kodak stock to an exact year. Some Kodak-Pathé stocks from this period — possibly limited to the 1940s — use British dating symbols after "PATHE".

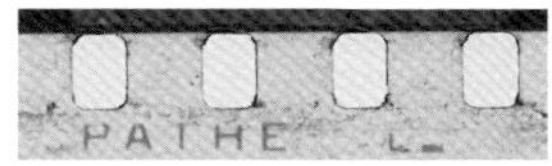

1946

Beginning in the early 1950s, the edge markings on Kodak-Pathé film become more closely aligned with the markings used by Kodak elsewhere around the world. From about 1952, dating symbols appear after "KODAK" and a dot between the letters of "SAFETY" indicates the country of manufacture. Then, from approximately 1959 or 1960, corresponding with a text style change, the dating symbols move after "FILM". But the symbols used by Kodak-Pathé between 1952 and 1982 do not match the standardized symbols used in the United States, United Kingdom, and Canada. The emulsion, strip, and slitter numbers appear to have remained in use until at least the mid-1990s.

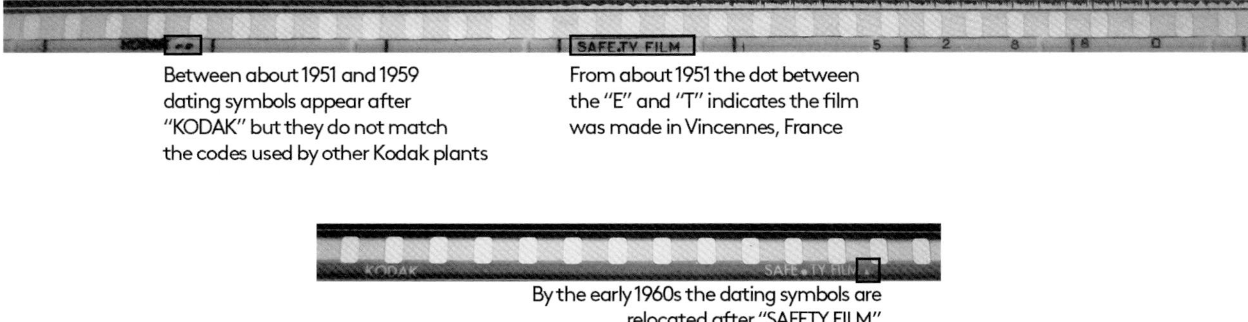

Between about 1951 and 1959 dating symbols appear after "KODAK" but they do not match the codes used by other Kodak plants

From about 1951 the dot between the "E" and "T" indicates the film was made in Vincennes, France

By the early 1960s the dating symbols are relocated after "SAFETY FILM"

Before 1983, Kodak-Pathé's unique dating symbols remain largely a mystery. The archive of L'Association CECIL / Le Musée KODAK (home to the surviving Kodak-Pathé archives) has to date been unable to locate any documentation that relates to dating French stock. Despite this author's study and an earlier survey of edge markings conducted by the Filmoteca Española in the 1990s,[3] it has not been possible to find any correlation of dating symbols to specific years of manufacture. In fact, for some prints surveyed, upwards of three different dating symbols have been found for a single year.

Harold Brown notes that the code for 1982 was ● ■ ▲ on Kodak-Pathé film, and ● ■ ✖ on all other Kodak film around the world, but the source of this information is unknown.

From 1983, Kodak-Pathé film used the same dating symbols as all other Kodak manufacturing plants, but French stocks retained a distinctive text style unlike the stencils used elsewhere. Most noticeable was the use of cyan-colored laser-printed edge markings on color print films, which may have been introduced at the time production started transferring from the Vincennes plant to Chalon-sur-Saône around 1984 or 1985.

3 Jennifer Gallego Christensen, Encarni Rus Aguilar: "La catalogación de las marcas marginales de fábrica como medio para la identificación y conservación de materiales fílmicos", in *Los soportes de la cinematografía 1*, Coll. Cuadernos de la Filmoteca 5 (1999), pp.120-133.

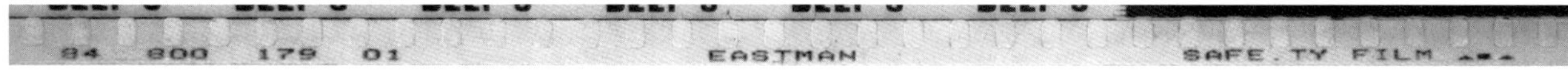

Throughout the 1980s and 1990s, Kodak-Pathé manufactured all color film, both consumer and motion picture, for the territories of Europe, Africa, and the Middle East.

In the late 1990s, the edge markings standardized with all other Kodak plants both in style and content. This continued until 2007, when the Chalon plant ceased production. Chalon stopped coating motion picture print film around 2002, so all subsequent film marked "KODA˙K" would have been manufactured in Rochester, and then slit and perforated in France.

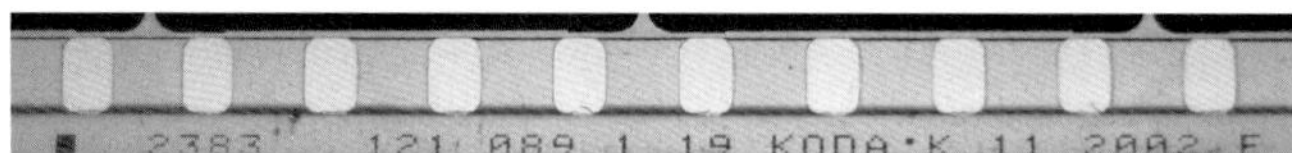

In addition to the above, internal Kodak documentation gathered by Brian Pritchard suggests that unique dating symbols may have been used by Kodak-Pathé on 16mm films for a limited time in the 1930s. No samples containing these dating symbols have been located, however. These symbols presumably appeared after the words "SAFETY FILM".

Chart #9: Kodak-Pathé 16mm Dating Symbols, 1934-1939

1934	⯊ ⯊ ⯊
1935	⯊ ■ ⯊
1936	■ ⯊ ⯊
1937	⯊ ⯊ ■
1938	■ ■ ⯊
1939	⯊ ■ ■

3.4 Kodak A.G. (Germany)

The Eastman Kodak Company's entry into Germany came in 1927 with the purchase of the former Glanzfilm factory in Köpenick, near Berlin. (Glanzfilm stock, manufactured between 1923 and 1927, was typically marked "GLAFI". The German name translates to "glossy film".) Kodak A.G. was formally incorporated in 1931, when Kodak acquired the Nagel Camera Company.

During the 1930s, German Kodak stock does not include dating symbols, but the year of manufacture and other details can be decoded from a 5-digit number string that appears after "KODAK A.G." on the film's edge.

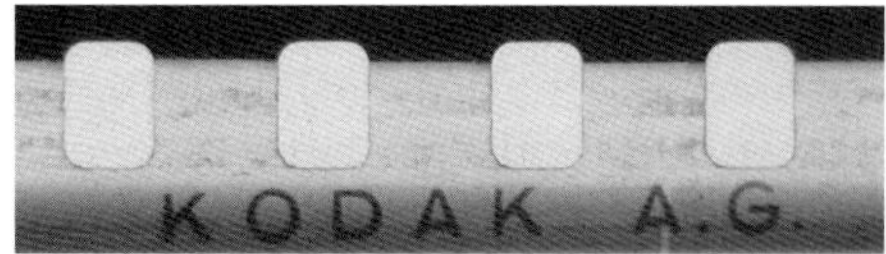

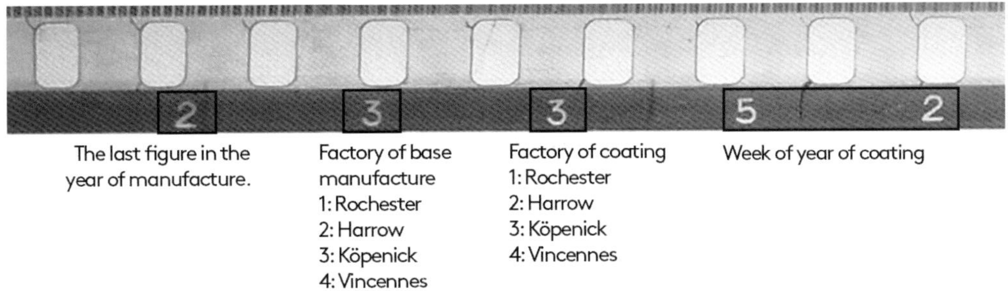

In this case, the code refers to a film stock manufactured in 1932, at the Köpenick plant, and the coating was made in December — very likely 1932 — in Köpenick.

According to documentation uncovered by Brian Pritchard, Kodak A.G. used dating symbols on 16mm film during the 1930s. These do not match symbols used elsewhere in the world by Eastman Kodak. The symbols appeared after "SICHERHEITSFILM" (SAFETY FILM), and the distance between the "M" and the first symbol indicated which half of the year the film was made: 1/16th-inch was January to July, and 3/16th-inch was July to December.

Chart #10: Kodak A.G. 16mm Dating Symbols, 1934-1939

1934	■ ■ ■
1935	● ■ ●
1936	■ ● ●
1937	● ● ■
1938	■ ● ■
1939	■ ■ ●

Despite this surviving documentation about dating symbols, the one sample of Kodak A.G. 16mm film consulted for this survey did not include these symbols, perhaps because it was reversal film rather than print film.

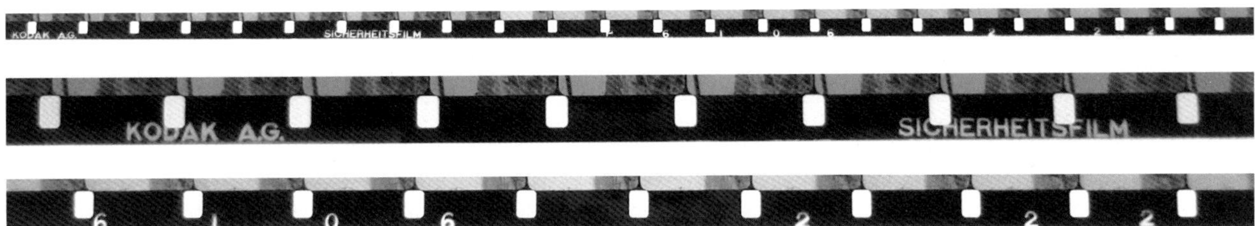

In 1941, the German government declared the Kodak works enemy property and seized the factory. After the war, the factory continued to operate under the Kodak name until 1956, when it transferred to East German management and became VEB Fotochemische Werke Köpenick (FCW). Apparently, Kodak regained ownership of this plant in 1992 following the reunification of Germany, but it closed soon afterwards. It is unclear if the factory produced motion picture film during this whole period or only until the outbreak of the Second World War.

4. 16mm Film Manufactured in the United States

The 16mm format was introduced for amateur photography in 1923. The first product available to the public was camera reversal film. A year later the Kodascope Library began, offering 16mm reduction prints of commercial and educational films for rental.

Unlike 35mm motion picture film, most 16mm color products (as well as all 8mm and Super 8 films) were manufactured by Kodak's Professional Products division rather than the Motion Picture division. Except for Ektachrome Video News Film (launched in 1975), all Ektachrome and Kodachrome films were available as both slide film and cine film.

The style and content of 16mm edge markings on Eastman Kodak film changed more frequently than with 35mm film. Because of this, it is always possible to accurately date 16mm film with certainty, which is not always the case with 35mm film.

Camera films, like negatives and reversal originals, tended to use similar markings, while prints, reversal prints, fine grain intermediates, and interpositives generally used different text, and have been separated accordingly in the following sections.

One interesting peculiarity of 16mm is the variety of edgeprint orientations. It can often be found with the order of text reversed or with upside-down symbols.

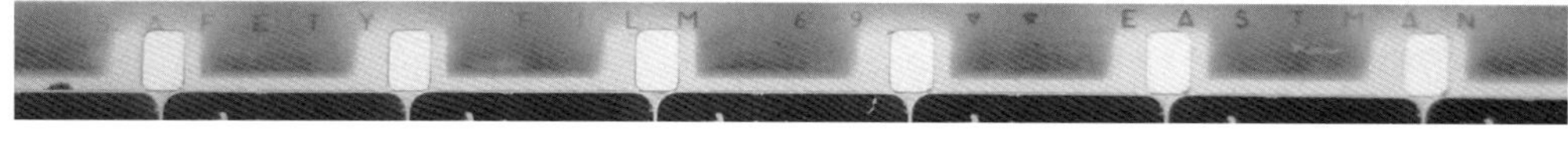

1981

4.1 Camera Originals and Pre-Print

16mm camera reversal film typically used the same edge markings from 1923 through to about 1951, with minor variations. The 2- or 3-symbol dating code appeared after the word "FILM".

1923-1931

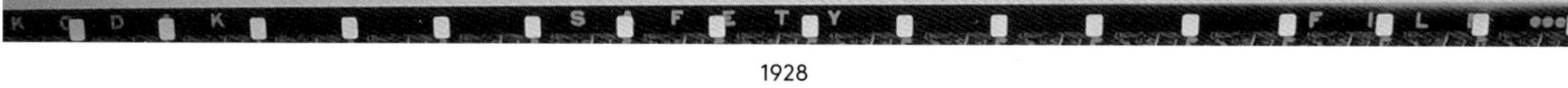

1928

Beginning in 1931, an additional symbol — most frequently a circle (●) — appeared immediately following the dating symbols. This additional symbol should be ignored when trying to date a film. Its meaning is unknown. 16mm film made outside the United States does not include this additional symbol.

Before 1931

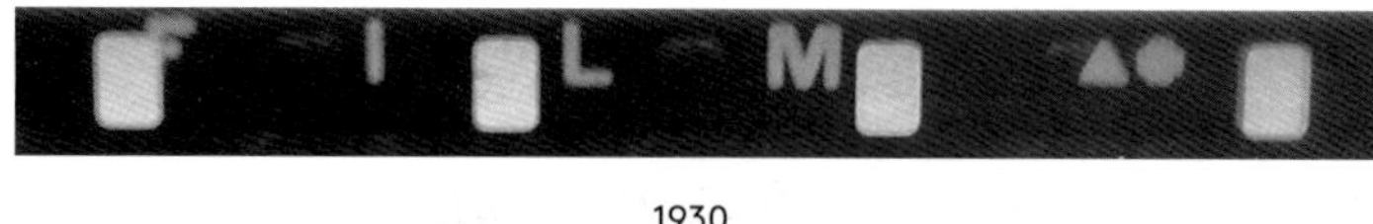

1930

After 1931

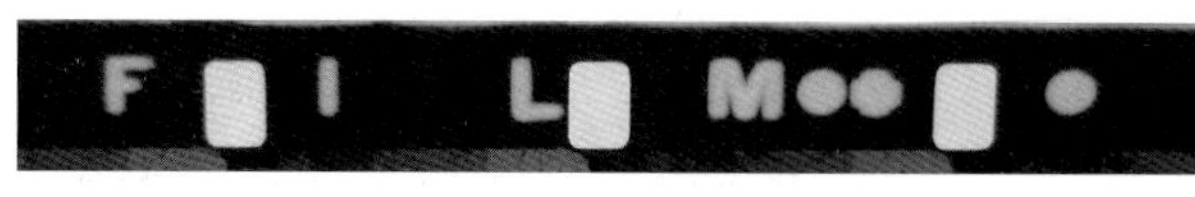

1931

The inclusion of this additional symbol is most confusing when it follows a single-digit dating symbol, like the code for 1936 (●) as seen in the sample below. Film from this year can be easily misdated as 1939 (● ●). (Film manufactured in 1939 would appear as "● ● ●,", which can be confused with 1948.)

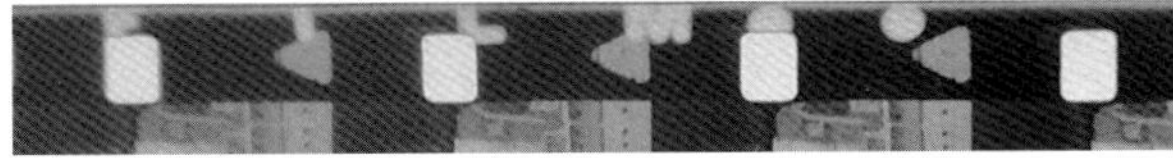

1936

This additional symbol seems to have been retired around 1951.

1928-1935

The only variations to the above come on color films. Kodacolor lenticular film was introduced in 1928 and Kodachrome arrived in 1935. The word "KODAK" was replaced with either "KODACOLOR" or "KODACHROME".

As Kodacolor film stored color information on black & white emulsion, one of the easiest ways to identify this type of film is from the edge markings. A series of vertical lines through the image is also visible under magnification.

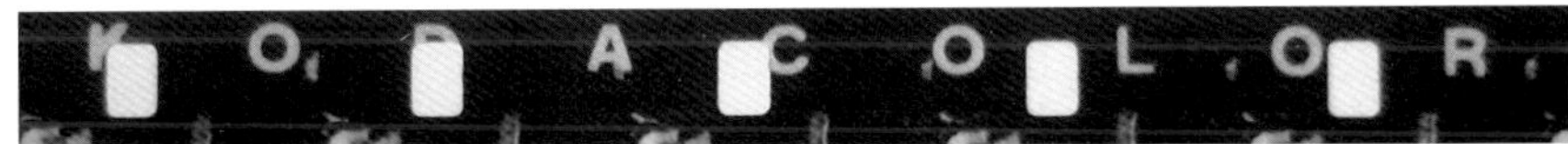

1935-1950

All Kodachrome film before 1951 is marked "KODA CHROME" — with a space in-between. Kodachrome film made between 1935 and 1938 is typically found with its color faded today.

Beginning around 1951, the edgeprint on some camera original films came more in line with the markings found on 35mm, except the placement of the strip number and slitter number was reversed, and "KODAK" was used instead of "EASTMAN". At some point in the mid-to-late 1950s, the dating symbols were relocated before the words "SAFETY FILM" rather than after them.

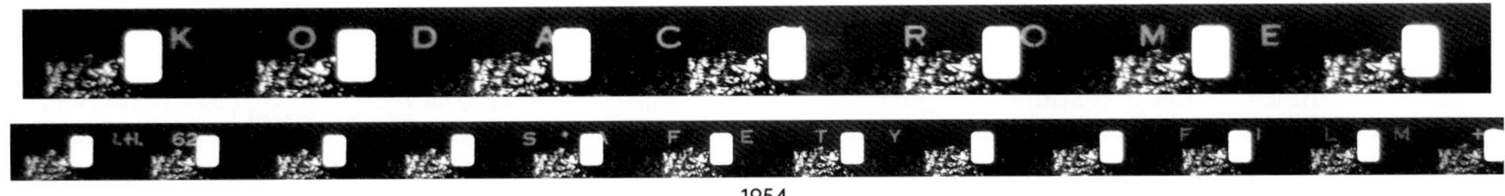

1954

In the late 1950s or early 1960s, a 1-character emulsion code was introduced. This appeared between "KODAK" and the slitter number. To begin with, this appears to have most commonly been "N" for negative, or "R" for reversal, but later — likely around the mid-1960s — 2- and 3-character codes were introduced for more precise identification of the emulsion type.

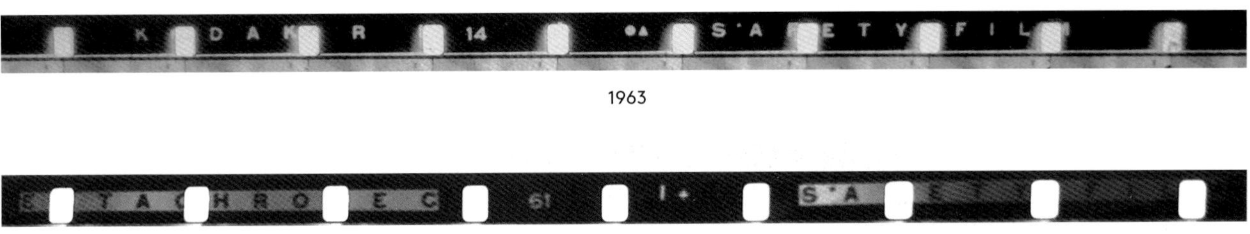

1963

1969

These 2- and 3-character emulsion type codes followed "KODAK" on black & white reversal films, or "KODACHROME" or "EKTACHROME" on color reversal stocks. Some professional Ektachrome products use "EASTMAN" instead of "EKTACHROME". These codes are not found on all negatives.

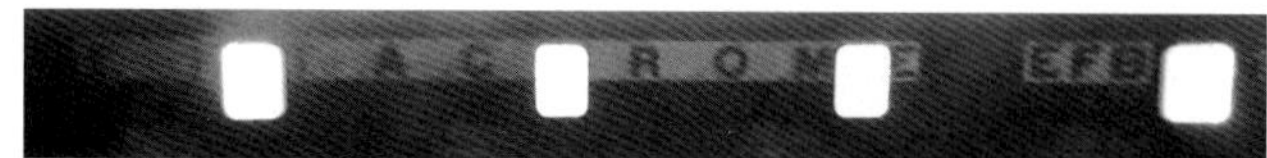

The ability to print a greater variety of product information on the film's edge came in 1959 with the introduction of the edgeprint projection printer, which would come to replace the standard contact drum printer. These projection printers were introduced gradually over many years on a slitter-by-slitter basis, and are still in use today for some 16mm products.

Chart #11: 16mm Edgeprint Markings for Camera Negative and Camera Reversal Films

Product	Edgeprint Designator (in use from mid-1960s to early 1990s)	Product Code	Element Type	Dates Available
Kodacolor	KODACOLOR		Camera Reversal	1928–1935
Kodachrome	KODA CHROME (before 1951); KODACHROME (thereafter)		Camera Reversal	1935–1962
Kodachrome Professional		5267	Camera Reversal	1942–1951
Kodachrome Commercial		5268	Camera Reversal	1946–1958
Plus-X Negative	PXN*	7231	Camera Negative	1949–2010
Plus-X Reversal	KODAK PXR	7276	Camera Reversal	1955–2005
Tri-X Reversal	KODAK TXR	7278	Camera Reversal	1955–2005
Ektachrome Commercial	EKTACHROME C	7255	Camera Reversal	1958–1970
Double-X Negative	DXN*	7222	Camera Negative	1959–Active
Ektachrome ER (News and Space)	ER*	7257	Camera Reversal	1959–1966
Ektachrome ER (News and Space)	ERT*	7258	Camera Reversal	1959–1966
Kodachrome II Movie (Daylight-balanced)	KODACHROME II	7265	Camera Reversal	1961–1974
Kodachrome II Movie (Type-A)	KODACHROME IIA	7266	Camera Reversal	1961–1974
Ektachrome Medium Speed (Daylight-balanced)	EMS*	7256	Camera Reversal	1963–1984
4-X Negative	4XN*	7224	Camera Negative	1964–1990
Ektachrome EF (Daylight-balanced)	EKTACHROME EF	7241	Camera Reversal	1966–1984
Ektachrome EF (Tungsten-balanced)	EKTACHROME EFB	7242	Camera Reversal	1966–1984
4-X Reversal	KODAK 4XR	7277	Camera Reversal	1967–1990
Color Reversal Intermediate (CRI)	KODAK CRI	7249	Intermediate Negative	1968–1991

Product	Edgeprint Designator (in use from mid-1960s to early 1990s)	Product Code	Element Type	Dates Available
Ektachrome Commercial	EASTMAN ECO	7252	Camera Reversal	1970–1984
Ektachrome EG (High Speed 464)	EKTACHROME EG	7248	Super 8 Camera Reversal	ca.1970–1981
Kodachrome 25 Movie (Daylight-balanced)	KODACHROME KM	7267	Camera Reversal	1974–2001
Kodachrome 40 Movie (Type A)	KODACHROME KMA	7270	Camera Reversal	1974–2006
Ektachrome VNF (Tungsten-balanced)	KODAK VNF	7240	Camera Reversal	1975–2005
Ektachrome VNF (Daylight-balanced)	KODAK VND	7239	Camera Reversal	1976–2005
Eastman Color Negative II	KODAK ECN	7247	Camera Negative	1976–1983
Ektachrome VNF High Speed (Tungsten-balanced)	KODAK VNX	7250	Camera Reversal	1977–2005
Ektachrome High Speed (Daylight-balanced)	EASTMAN VXD	7251	Camera Reversal	1981–2005
Eastman Color High Speed Negative	EASTMAN ECH	7293	Camera Negative	1982–1983
Eastman Color Negative	EASTMAN 291	7291	Camera Negative	1983–1989
Eastman Color High Speed Negative	EASTMAN 294	7294	Camera Negative	1983–1990
Eastman Color High Speed Negative	EASTMAN 292	7292	Camera Negative	1986–1992

* No samples of this edgeprint were found during this survey.

During an edgeprint style change in mid-1970, the dating symbols changed from solid shapes to outlines. (This makes it easy to distinguish films manufactured in the 1970s from those with matching dating symbols made in the 1950s.) The text also became smaller and moved closer to the film's edge above the perforations. This meant that the edgeprint text was no longer partially obscured by perforations (although sometimes the text could be printed too close to the edge and was partly cut off).

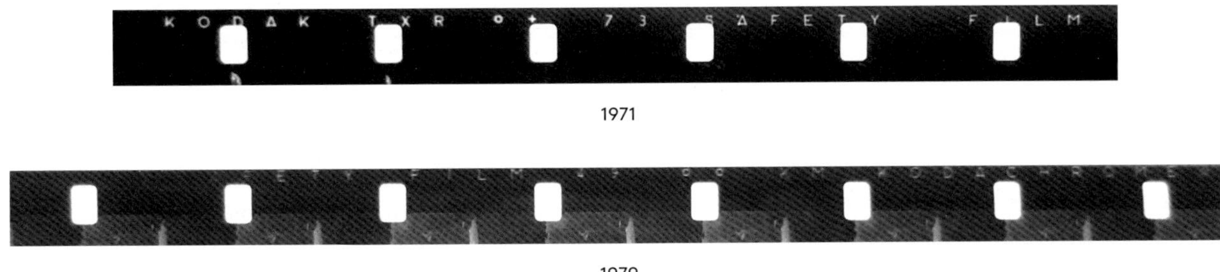

1971

1979

It has been found that camera negative films were typically marked "KODAK", whereas other pre-print elements like track negatives, duplicate negatives, color reversal intermediates, interpositives, etc., were marked "EASTMAN".

Camera negative

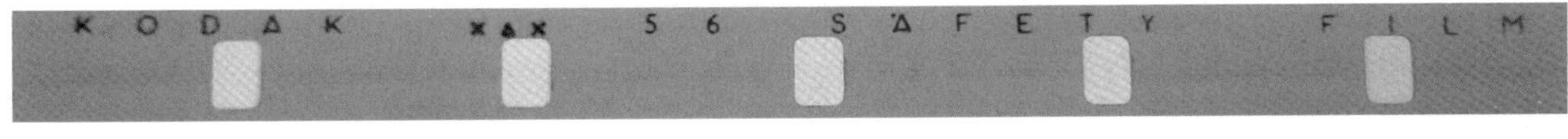

1983

Color reversal intermediate

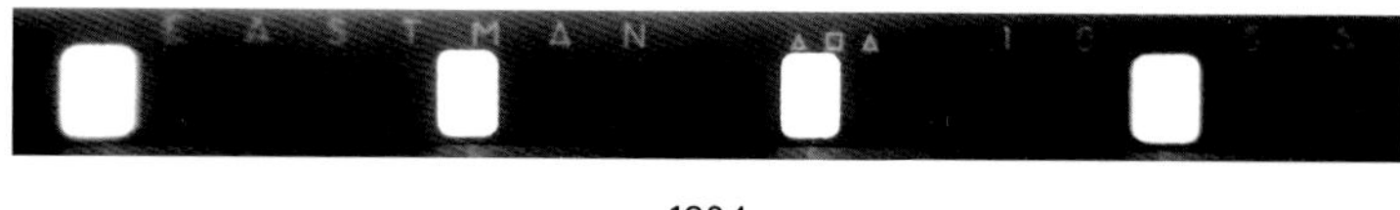

1984

Track negative

1989

From 1989, the edgeprint text on pre-print elements matches the standard found on 35mm film. Chart #4 in Section 2.6 can be used to decode the year of manufacture from the 2-character letter code found at the end of the text string.

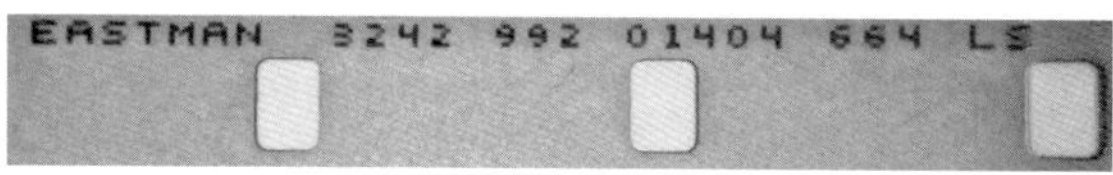

2016

The edgeprint on camera reversal films remained the same until the late 1990s, when it began following the standards introduced at that time for 35mm. In the sample below, "KODAK" is followed by the emulsion product code, the strip number, and the year of manufacture.

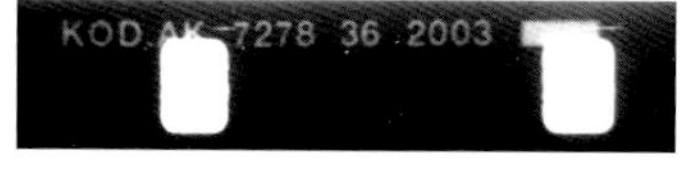

2003

4.2 Prints and Reversal Prints

Edgeprint on 16mm prints during the 1920s followed a standard format throughout the decade, then changed in 1931, with a minor addition a few years later.

1924-1931

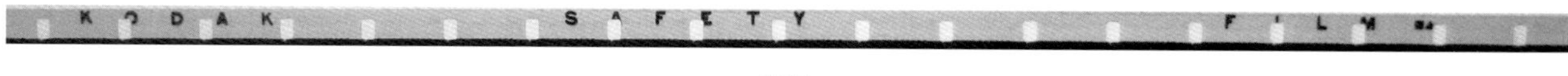

1927

1930 to ca. 1934

From 1930, "KODAK SAFETY POSITIVE" can be found on all prints and reversal prints. The dating symbols appear immediately after "KODAK". "SAFETY POSITIVE" remains on 16mm print film until at least 1957.

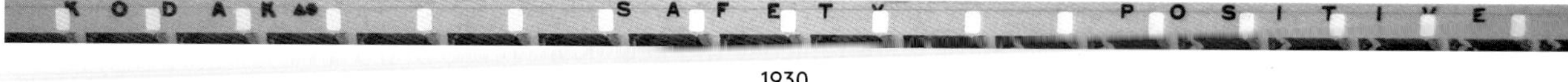

1930

Early 1930s to 1947

Sometime between 1932 and 1934, a symbol of unknown meaning (most frequently a ● or ■) was added between the dating symbols and "SAFETY". This symbol should not be construed as part of the dating symbols, despite its close proximity.

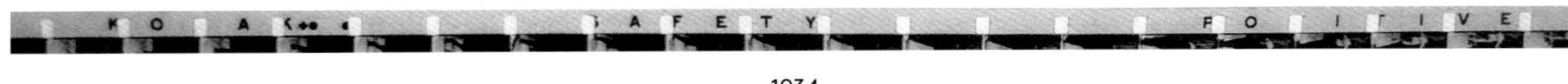

1934

1947-1951

From 1947, the strip number begins to appear between the dating symbols and "SAFETY" (replacing the unidentified symbol).

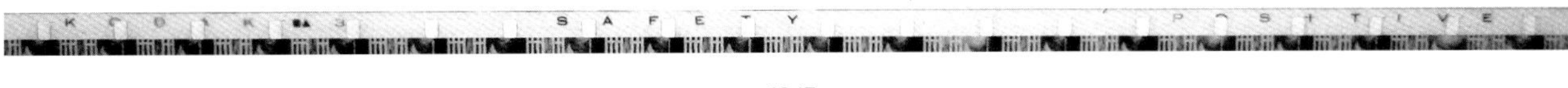

1947

From 1951

Beginning approximately in 1951, the plant of origin dot is added between the letters of "SAFETY". A two-part marking is used for each edgeprint, with the same text used twice and spaced 3 inches apart. In the first half, the strip number appears between "KODAK" and "SAFETY"; in the second half, the dating symbols appear between "KODAK" and "SAFETY".

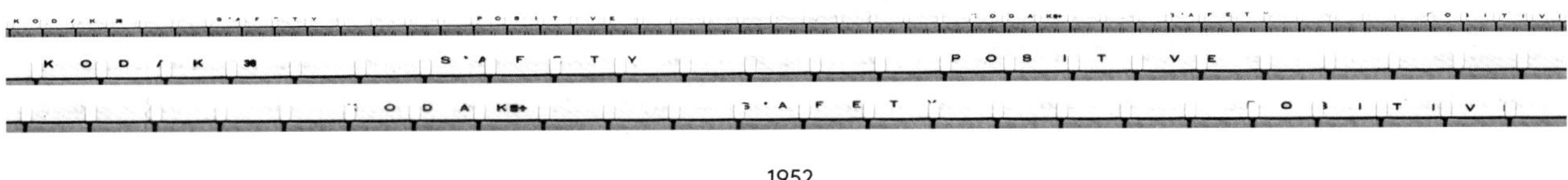

1952

Several parallel edgeprint styles have been noted in use between the mid-1950s and 1963.

ca. 1955 to 1963

Both the unknown symbol and the strip number reappeared between "SAFETY" and "POSITIVE" for a brief period in the mid-to-late 1950s.

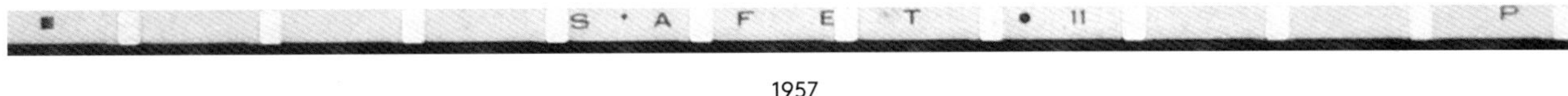

1957

ca. 1958-1963

The slitter number and strip number appear between "KODAK" and "SAFETY". "POSITIVE" is replaced by "FILM" and the dating symbols follow.

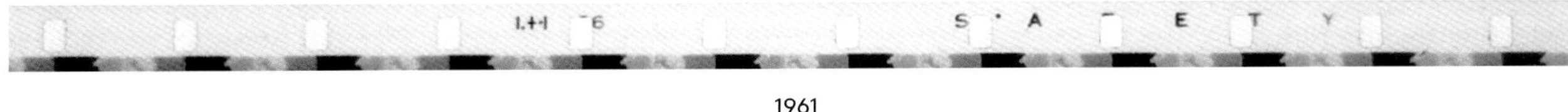

1961

ca. 1959-1963

The dating symbols were sometimes placed before "SAFETY FILM" and sometimes after. This edgeprint layout is close to the 35mm standard in place at the time, except that the order of the strip number and slitter number is reversed.

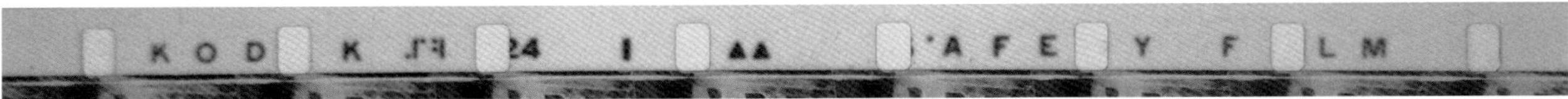

1961

The only known variation to the above was on Eastmancolor prints and Kodachrome reversal prints.

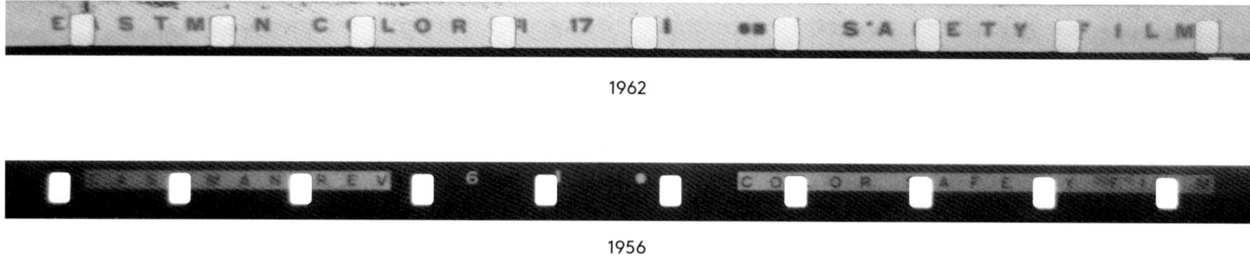

1962

1956

1964-1969

"EASTMAN" replaces "KODAK". All 16mm prints now follow a rigid edgeprint layout. The dating symbols can appear either before or after "SAFETY FILM".

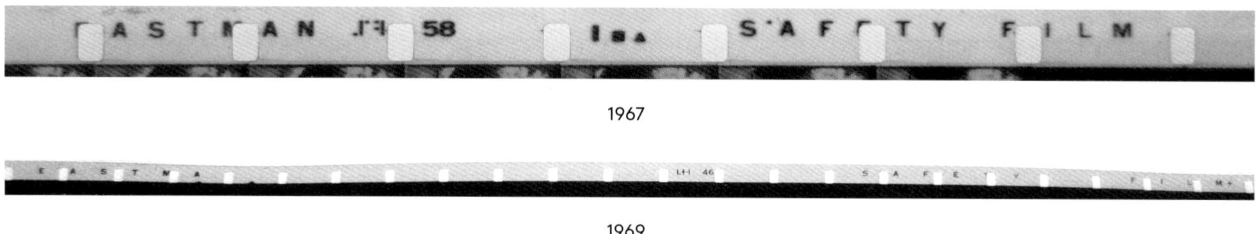

1967

1969

1969 to the Late 1990s

Around 1969, the slitter number is dropped, and the placement of the strip number and the dating symbols seems interchangeable. From 1970 the solid dating symbols become outlines.

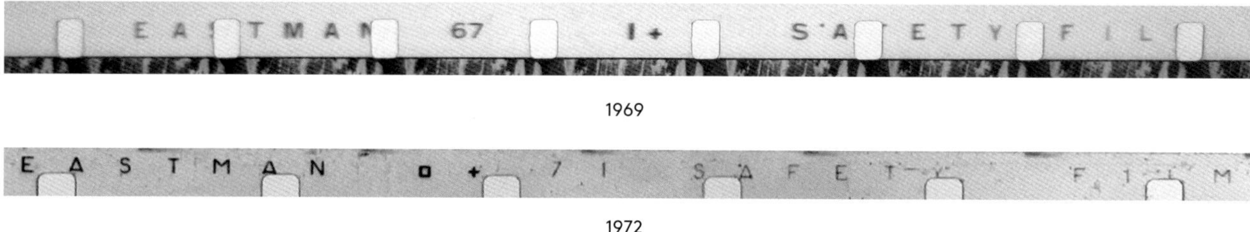

1969

1972

During this time — as was common on most camera films — 2- and 3-character emulsion type codes were sometimes used. These codes appear to have been particular to color stocks, and have not been documented on black & white films.

Chart #12: 16mm Edgeprint Markings for Print and Reversal Print Films

Product	Edgeprint Designator	Product Code	Element Type	Dates Available
Kodachrome Duplicating Film		5262	Reversal Print	1938–1940
Kodachrome Duplicating Film		5265	Reversal Print	1940–1953
Eastman Reversal Color Print Film (Kodachrome)	EASTMAN REV	5269	Reversal Print	1950–1956
Ektachrome EP	EASTMAN EKTACHROME	7386	Reversal Print	1960–1974
Eastman Reversal Color Print Film (Kodachrome)	EASTMAN REV II	7387	Reversal Print	1964–1981
Eastman Ektachrome R Print Film	EASTMAN EKTACHROME R	7388	Reversal Print	1966–1970

Product	Edgeprint Designator	Product Code	Element Type	Dates Available
Ektachrome R Print Film	EASTMAN	7389	Reversal Print	1966–ca.1977
Ektachrome EPF	EASTMAN EP	7390	Reversal Print	1973–1986
Eastman Color SP Print	KODAK SP	7381	Print	1974–1983
Ektachrome VN Print Film	*	7399	Reversal Print	1977–????
Eastman Color Print	EASTMAN LPP	7384	Print	1983–1992
Eastman Color LC Print	EASTMAN LC	7380	Print	1983–ca.1993
Eastman EXR Color Print	EASTMAN 386	7386	Print	1993–2000

* No samples of this edgeprint were found during this survey.

1999 to ca. 2008

16mm edgeprint on prints and reversal prints matches 35mm edgeprint. Sometimes the 4-digit emulsion product code is not present.

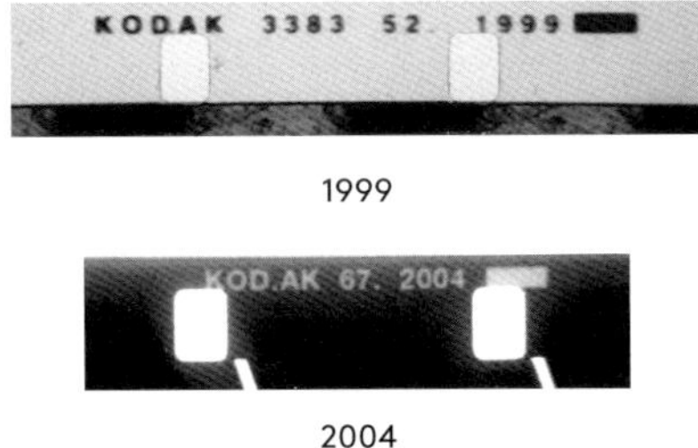

1999

2004

2008 to the Present

The multi-year stencil is introduced. As with 35mm black & white prints, the lowest number on the right of the text indicates the year of manufacture. As each year passed, that number would be dropped from the stencil.

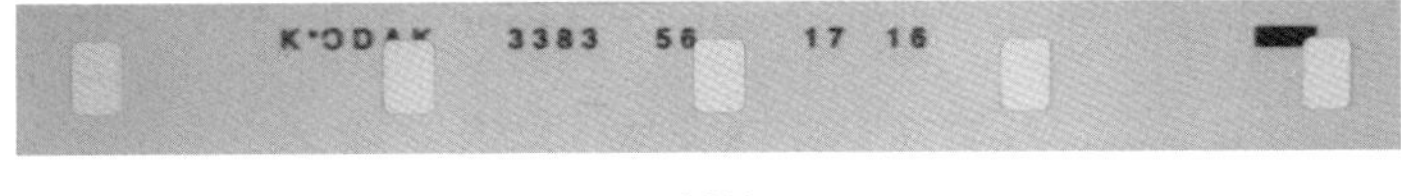

2016

5. 8mm Film

1938 (U.S.)

Standard or Regular 8mm film was introduced to the amateur market in August 1932. From the beginning it was available as a camera reversal film and as prints as part of the Kodascope Libraries rental program. Super 8 film was launched in 1965.

Interestingly, Ektachrome and Kodachrome films made for the 8mm and Super 8 formats were overseen by the Professional Products division rather than the Motion Picture division. As such, the edgeprint standards for 8mm have not always followed the standards in place for 16mm and 35mm film. Very little documentation about 8mm edgeprint has been retained by Eastman Kodak.

The most prominent difference was that a unique set of dating symbols was used for 8mm, changing every 6 months rather than every year. There were different sets of symbols in use in the United States and the United Kingdom. Confusingly, these dating symbols were the same as those used for 35mm and 16mm film, but they changed more frequently, and cannot be used to date the film employing the dating symbols in Chart #1. Kodak documentation for 8mm dating symbols from the decades 1932-1941 and 1954-1964 has been previously disseminated,[4] but the intervening years have remained unconfirmed.

Archivist Robin Williams (formerly of the East Anglian Film Archive) did considerable research into this unknown period, and was able to document many probable symbol combinations. This survey has continued his research and has been able to confirm that the symbols used in the first 10 years did in fact repeat in 10-year cycles.

But contrary to the Kodak documentation previously distributed for the 1954-1964 period, this study suggests that dating codes particular to 8mm were abandoned around 1951, and henceforth the standard dating symbols were used. More research is required to confirm this.

Chart #13: 8mm Dating Symbols, 1932-1951 (confirmed)

	United States: Jan. to June	United States: July to Dec.	U.K.: Jan. to June	U.K.: July to Dec.
1932 / 1942	✚	●	L✚	▬✚
1933 / 1943	■	▲	◡	L
1934 / 1944	●●	■■	▬	◡◡
1935 / 1945	▲▲	●■	LL	▬▬
1936 / 1946	●▲	▲■	◡L	▬L
1937 / 1947	▲●	■●	L◡	L▬
1938 / 1948	■▲	✚✚	▬◡	◡▬
1939 / 1949	✚●	✚■	✚	✚✚
1940 / 1950	✚▲	●✚	✚◡	✚L
1941 / 1951	■✚	▲✚	✚▬	◡✚

Chart #14: 8mm Dating Symbols from Kodak Documentation, 1954-1964 (unconfirmed)

	Jan. to June	July to Dec.
1952	✚	●
1953	■	▲
1954	●●	
1955	▲▲	
1956	●▲	
1957	▲●	
1958	■▲	
1959	✚●	
1960	✚▲	
1961	■✚	
1962	✚	

More work needs to be done on 8mm edgeprint style changes over the years, but they appear to follow closely those used for 16mm (see Sections 4.2 and 4.3).

4 8mm Kodak edge codes were first disseminated online by Brian Pritchard and the East Anglian Film Archive. Some of these findings were later published by Kodak in April 2013.

8
Agfa

Luciano Berriatúa and Camille Blot-Wellens

1920s to 1940s

Luciano Berriatúa

In 2005, I was commissioned by the F.W. Murnau Stiftung to carry out a new restoration of *Nosferatu* (Friedrich Wilhelm Murnau, 1922). In 1925, a court decision had ordered that the negatives and all the original prints be destroyed, because the rights to the novel *Dracula* (Bram Stoker, 1897) on which it was based had not been paid. Hence, we tried to collect all the surviving prints or fragments of prints from various generations preserved in the archives, many of them re-edited materials and duplications with various changes made over several decades. It was essential to determine the origin of these changes and to date all these materials in order to be able to locate the origin of the re-editings detected on the prints of successive generations, and to be able to reconstruct the original version with the best-possible image quality. Many of these materials were on Agfa stock, and its logo appeared on the edge printings, with different graphics that undoubtedly corresponded to different years.

In his 1990 book, Harold Brown only provided the following information about these AGFA logos: "The name 'Agfa' first appears on stock in the early 1920s. Through to 1923 the letters had thick strokes and A's with flat tops. From 1924 the letters are thinner and the A's have pointed tops." [1]

However, many of the other changes that could be noted in the Agfa logos of the different *Nosferatu* materials were also visible in other films by Murnau, or by other filmmakers, and in 2006 I was able to formulate some conclusions.

As to how the AGFA brand was introduced, both in positive and negative elements, I was able to observe that the materials used for the shooting or positives of 1921 invariably had the same graphics, and those of 1922 sometimes had the same (because a stock from the previous year had been used), but I noticed that different graphics appeared that year. The following year, the graphics changed both in terms of design and size, and in 1924 it changed again, increasing much more in size. From 1925 onwards the graphics seemed to stabilize, with the apex of the "A" becoming pointed instead of flat, as Brown said. Clearly Agfa, like Kodak and other manufacturers, introduced visible but encrypted codes into their materials in order to identify their annual stocks. It was necessary to decipher them.

My method of deciphering these codes was very simple. I resorted to comparing the prints and negatives on which I had exact or very approximate information concerning their printing or shooting dates. The parameters to take into account were the graphics or typeface of the edge mark, its size, its position, the distance between two marks, parameters for the numbers of footage, negative or positive material, and perforations — although there is no doubt that other details escaped me, such as changes in the physical or chemical characteristics of bases and emulsions. The comparison of pre-dyed positives could also be used to date materials from the 1920s, since there were catalogues where we could see that the tones of the dyes sometimes changed every year and varied according to the manufacturers.

The deciphering of these codes shows us the year but not the month in which stock was placed on the market.

1 See p.71 in the present volume.

Camille Blot-Wellens recently obtained the following interesting information from Nikolaus Wostry of Filmarchiv Austria, who had received it from Helmut Regel, who had worked at the Bundesarchiv:

AGFA with flat "A" and 8-10mm wide: from August 1920 to August 1924.
Flat "A" and 14mm wide: from November 1924 to January 1926.
With peaked "A": since March 1926.

Regel does not cite his source, but are the dates of August, November, January, and March for the manufacture of the stocks, or for their release to the market? They seem to indicate that this would be information received from the Agfa factory.

Wostry adds other information:

Agfa used B&H perforation on its positive stock until 1927, and Kodak perforation from the end of 1927 onwards.

I have collated this information and it corresponds with my own research, so I have added to my stock lists (which are displayed below) the data on the months I was missing, although questions and hypotheses remain because it is incomplete and not everything seems to match.

1920s

1920-1921 (August 1920 to August 1921?)

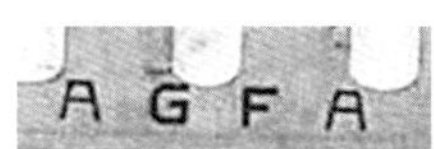

Schloss Vogelöd (F.W. Murnau, 1921).
Negative (shot in February and March 1921).

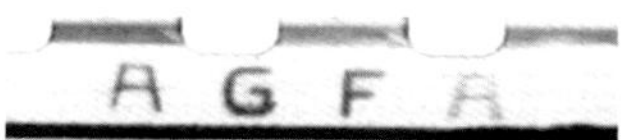

Nosferatu, eine Symphonie des Grauens (F.W. Murnau, 1922).
Short print from Berlin (shooting ended in October 1921).

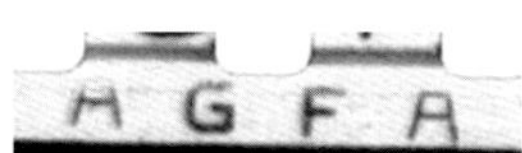

Nosferatu. Short print from Berlin with intertitles and inserts of the diary.

Nosferatu. French nitrate print (made in 1922).

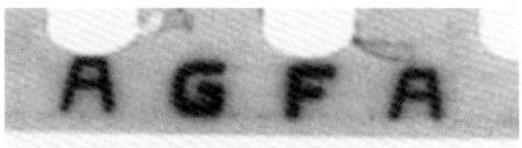

Nosferatu. French nitrate print, with French intertitles with decorations.

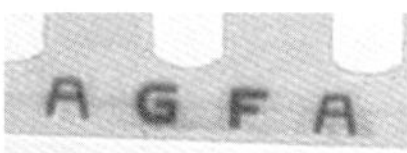

Phantom (F.W. Murnau, 1922).
Negative (shot May to September 1922).

1921-1922 (August 1921 to August 1922?)

The most significant aspects to note are the rounded "G", the short lower bar of the capital "F" in "AGFA", and how the configuration is more expanded than in 1921.

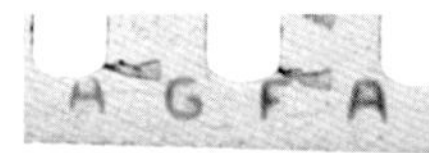

Nosferatu. French nitrate print with French intertitles (released October 1922).

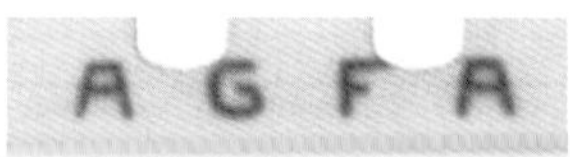

Phantom. Negative (end of shooting: September 1922).

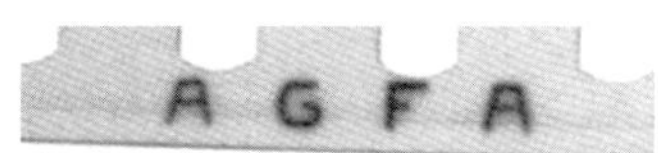

Zur Chronik von Grieshuus (Arthur von Gerlach, 1925).
Negative (shot May 1923 to November 1924).

1922-1923 (August 1922 to August 1923?)

Once again, the ensemble of the "AGFA" lettering expands a bit: the letters stretch horizontally, and the "A" opens up.

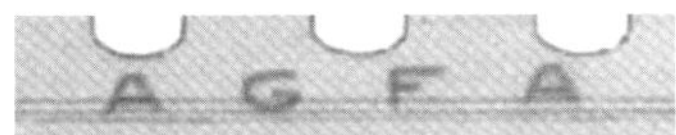

Der letzte Mann (F.W. Murnau, 1924).
Negative (shot March to September 1924).

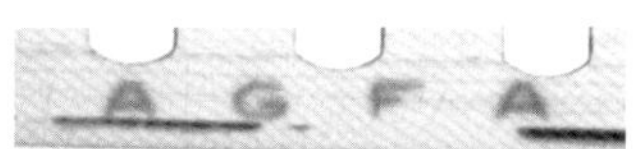

Der Turm des Schweigens (Johannes Guter, 1925).
Negative (shot in 1924).

1923-1924 (August 1923 to August 1924?)

The expansion of the spacing of the letters in "AGFA" is even more pronounced on this new stock.

Zur Chronik von Grieshuus. Negative (shot May 1923 to November 1924).

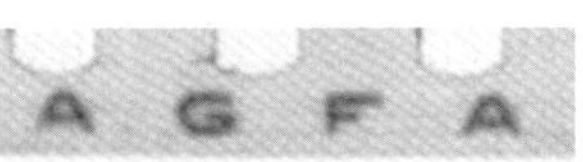

Der letzte Mann. Negative (shot March to September 1924).

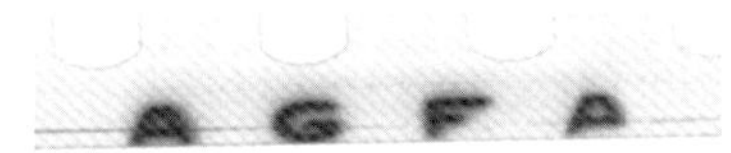

Der Turm des Schweigens. Negative (shot 1924).

1924-1925 (August 1924 to August 1925? According to Regel, November 1924 to January 1926)

The spacing between the letters in "AGFA" expands considerably. The name "AGFA", which previously extended across 3 perforations, started to extend across 4 perforations.

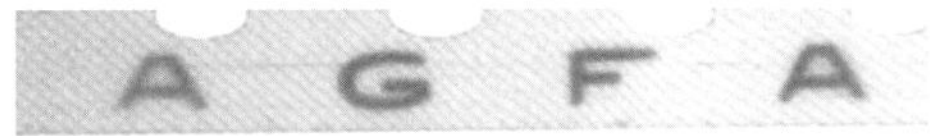

Der Turm des Schweigens. Negative (shot in 1924).

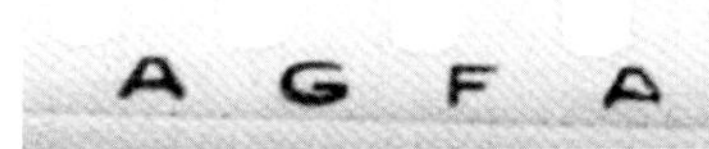

Der letzte Mann. Intertitles negative (released in December 1924).

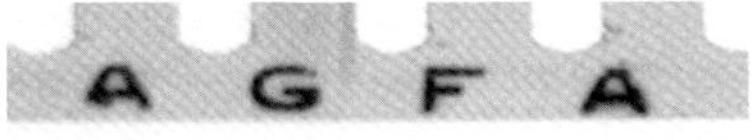

Zur Chronik von Grieshuus. Intertitles negative (released February 1925).

Herr Tartüff (F.W. Murnau, 1925). Swiss print and negative (shot February to May 1925)

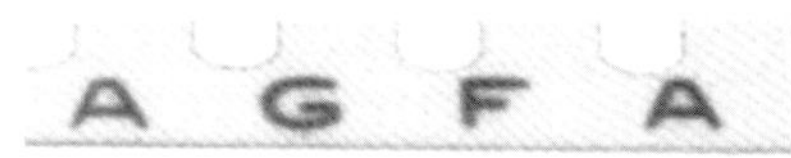

Faust: Eine deutsche Volkssage (F.W. Murnau, 1926). Negative (shot September 1925 to May 1926).
This film stock was used in a few shots only.

1925

Peaked/pointed "A" appears in the word "AGFA".

Regel's dating of this as occurring since March 1926 is not correct —unless it only refers to positive material. In which case it should not be applied to the dating of negative stock.

My conclusions are based on the fact that the pointed "A" is present in "AGFA" in the edge printing in almost all of F.W. Murnau's *Faust* negative, which was started on 10 September 1925 and completed in May 1926.

As we can see in these images of the negative, the shots of the sequence of the sick mother, which correspond to the beginning of the shooting, were filmed with Agfa negative with a pointed/peak "A"; it is well documented that this scene was filmed in September 1925:

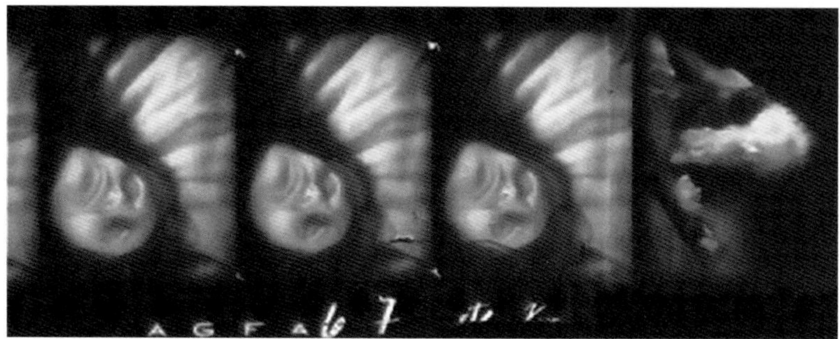

Faust. Negative used for the shooting of the scene in September 1925.

We find Agfa materials with the pointed "A" throughout the *Faust* negative.

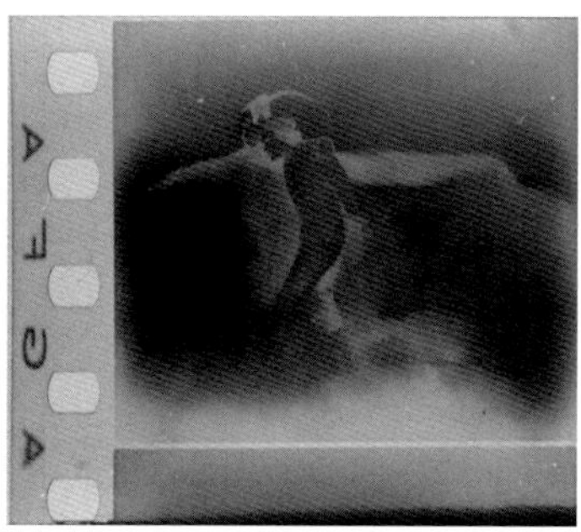

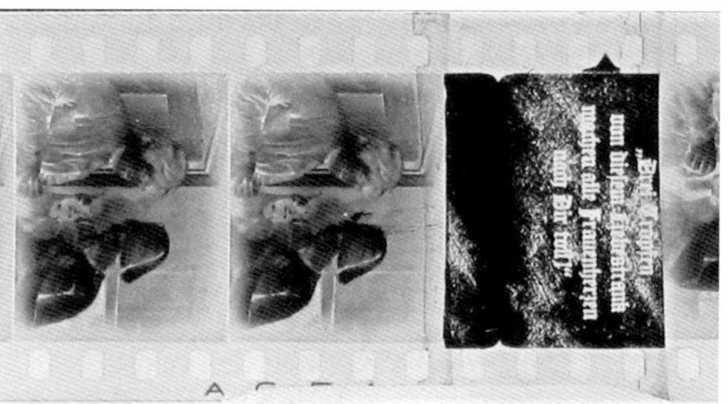

Including the sequences of Lady Martha, played by Yvette Guilbert, who, according to Eisenstein in his *Immoral Memories*, had already finished her participation in the shooting and left Berlin days before he arrived there on 18 March 1926.

According to surviving data from the shooting of *Faust*, Yvette Guilbert arrived in Berlin in January 1926, and shot her scenes between January and February.

It seems evident that the pointed "A" appears on the negative stock in 1925, and at the latest in August of that year.

1925-1926 (since August 1925?)

The "A" in "AGFA" is pointed.

Other signs, such as a plus sign "+" or a circle ●, appear in new low-contrast, fine grain emulsions intended for duplicating or effects.

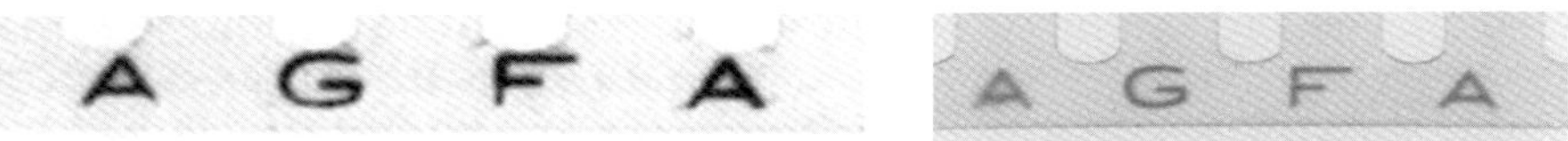

Faust. Negatives (shot September 1925 to May 1926).

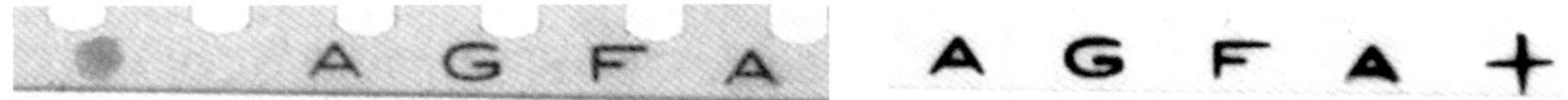

Faust. Fine grain negatives for effects.

1927

The same width and the pointed "A" continue, but the top bar of the capital letter "F" is significantly stretched.

The change of B&H perforation to Kodak appears in positives at the end of 1927.

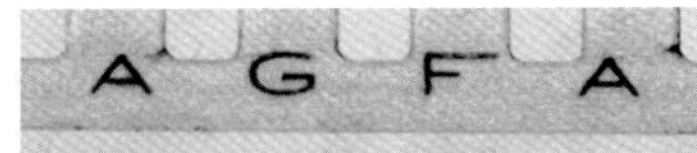

Tartüff. Argentinian print (made in 1928).

1928

The same width and the pointed "A" in the word "AGFA" continue.

A new factor comes into play: The word "AGFA" stops appearing every 15 frames, and instead appears every 8 frames (i.e., every half-foot).

In subsequent stocks, we find "AGFA" every 8 frames + 1 perforation.

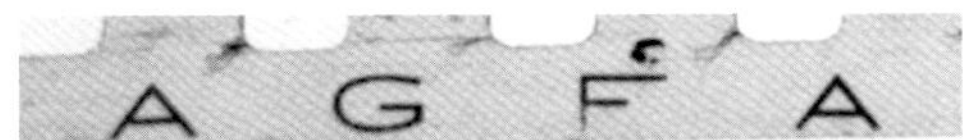

Tartüff. Argentinian print. "AGFA" appears exactly every 8 frames.

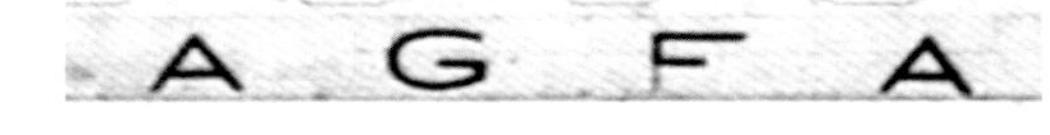

Spione (Fritz Lang, 1928). Print (released in March 1928).

A nitrate print from the Filmoteca de Andalucía preserved in the vaults of the Filmoteca Española is enlightening. This is the feature-length documentary entitled *La Sierra de Aracena*, which was filmed in 1928 (it contains scenes from the pilgrimage procession "Romería de Nuestra Señora de los Ángeles en la peña de Arias Montano", which took place in September 1928), finished in 1929, and premiered at the Pathé Cinema in Seville on 9 June 1929, as part of the Iberoamerican exhibition in Seville (which began on 9 May 1929).

The print was made on Agfa stock, both the image and the intertitles (except for a few main titles printed on "FERRANIA K 6 2 0").

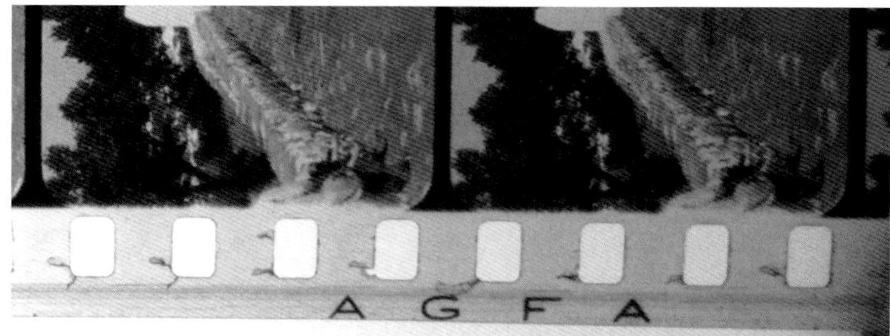

AGFA 1928. Print of *La Sierra de Aracena* (Carlos Emilio Nazari, 1928).
The word "AGFA" appears exactly every 8 frames. Kodak Standard (KS) perforation.

The intertitles were made on two different Agfa stocks, both with B&H perforation. The oldest corresponds to 1927 or 1928, and the newer one can be dated to prior to June 1929.

The intertitles had to be prepared in 1929, before the June Seville premiere.

The 1927 or 1928 element has "AGFA" along one side, exactly every 8 frames, and also in the opposite direction on the other side, but every 15 frames (or to be more precise, every 15 frames minus approximately the width of a perforation).

The 1928 or 1929 element has "AGFA" along one side, exactly every 8 frames.

The most interesting thing is that these intertitles were made on stock from 1927/1928, up to a title which contains 35 frames where it is possible to see that it is the final segment of the reel, as it is "fogged" (i.e., photographic defect caused by extraneous light or inappropriate chemical action). The available stock finished there, and the sequence was completed with 75 frames of the new stock from 1928/1929, and from that point on, all the titles were made on that new stock.

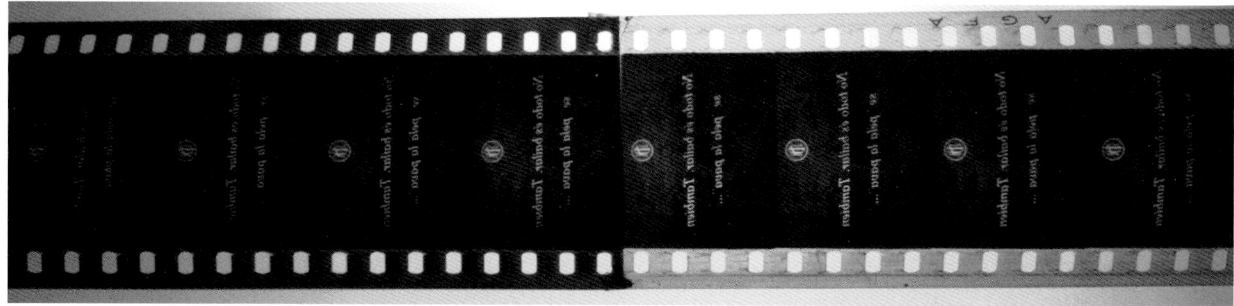

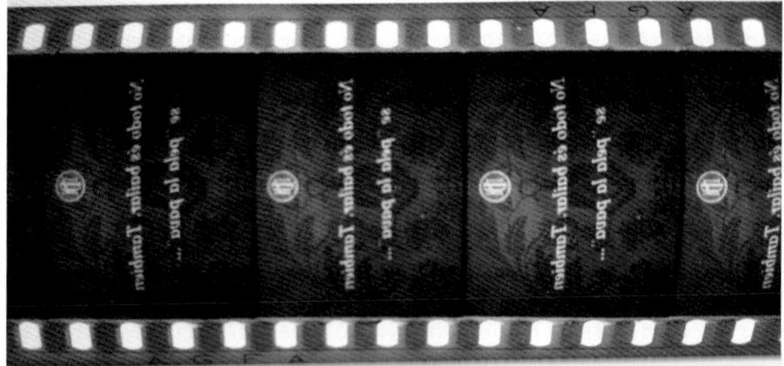

Thanks to this instance, we can clearly see the differences between the two stocks.

Until 1926/1927 the "AGFA" edge mark would appear every 15 frames; in 1927/1928 it would remain, combined with the new distance of every 8 frames; and in 1928/1929, at only the new distance of every 8 frames.

As far as the type of perforation, Kodak was used for prints, while B&H continued to be used for negatives.

It is very curious that in 1927 "AGFA" was printed on both edges and in both directions, especially because in the Argentinian print of *Tartüff*, we find British-produced Kodak stock of that same year, 1927, and the edge mark is also printed on both edges in both directions, with the mark every 19 frames. Does this only happen in 1927? Why? Does it have to do with the arrival of sound, and the combination of image and sound on the print?

1929

The distance between the letters "G" and "F" in the word "AGFA" decreases.

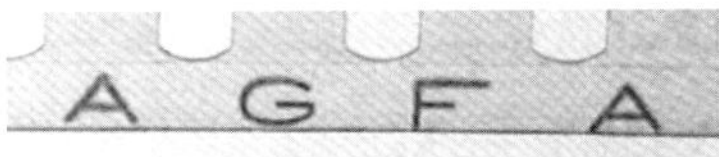

Three Ages (Edward F. [Eddie] Cline & Buster Keaton, 1923). Negative. B&H perforations are still used in the negative; the "AGFA" edge mark appears exactly every 8 frames on the main titles.
The image negative and intertitles negative were shot on KODAK AG 94336 9 (the first "9" corresponds to 1929) and KODAK AG 83348 7 or 83343 1 6 (the initial "8" corresponds to 1928).[2]

1930s

In the 1930s, the distance between the letters "G" and "F" in "AGFA" grows again.

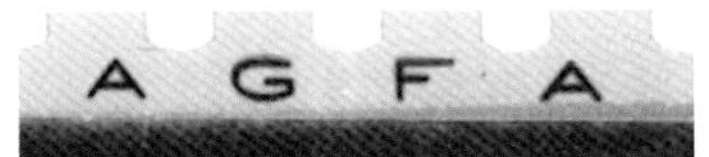

Las Hurdes (Luis Buñuel, 1933). Negative PANKINE

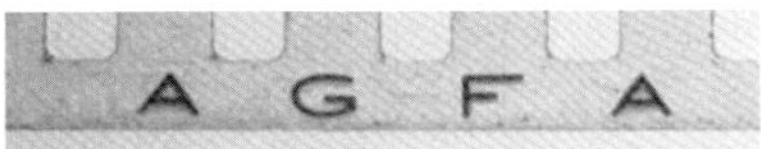

Francisca, la mujer fatal (Ricardo García López 'K-Hito', 1934). Print

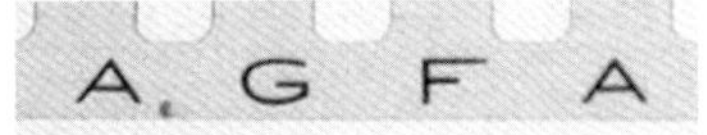

Der letzte Mann. Print (1936)

1940s

Agfa returned to its 3-perforation width, to include new numerical codes which extend to the 4th perforation.

From here onwards, my contribution is limited to showing examples of contractions and expansions of logos.

The subsequent codes of Agfa numbers and letters have been researched and compiled by Camille Blot-Wellens. The rest of this chapter is by her.

2 For Kodak AG film stock, see also the chapter on Eastman Kodak in this volume, pp.246-247.

1940s to 1960s

Camille Blot-Wellens

1940s

Towards the end of the 1930s, and the early 1940s, Agfa began to number their emulsions: each new emulsion produced was identified with a 3-digit number.[3] The same numbers could be re-used after a time, as long as confusion about the emulsions was avoided. This makes it very difficult to use this information to date the year of manufacture of the stock in this period.

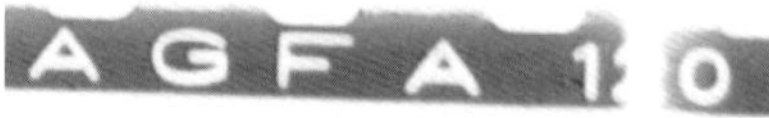

Nosferatu. Negative AGFA 120 of inserts photographed on a Berlin print. – LB

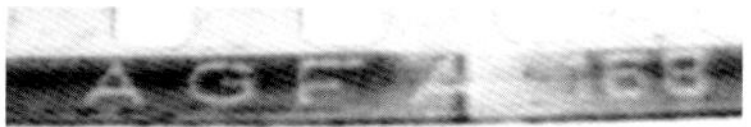

Romancero marroquí (Enrique Dominguez Rodiño & Carlos Velo, 1939). Duplicate Negative AGFA 168, photographed on a print. – LB

Raza (José Luis Sáenz de Heredia, 1942). Duplicate Negative AGFA 302, photographed on a print. – LB

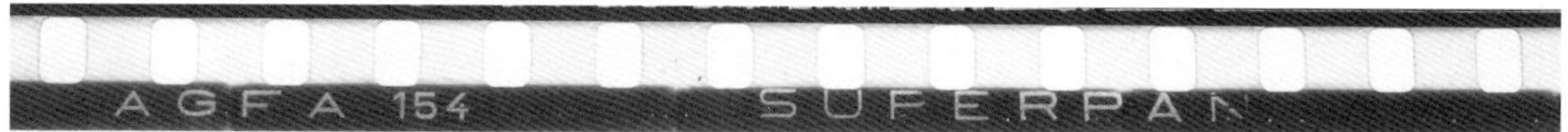

Zdrazieckie serce (Jerzy Zarzycki, 1947). Original Camera Negative AGFA 154 Superpan, photographed on a Duplicate Positive. – FINA

1950s–Mid-1960s

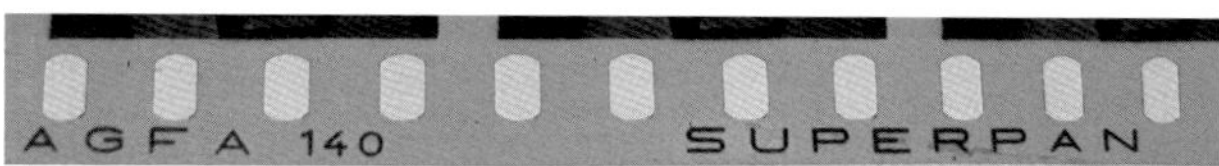

Ciné Journal Suisse 492 (prod. 1951). Original Camera Negative AGFA 140 Superpan. – CS[4]

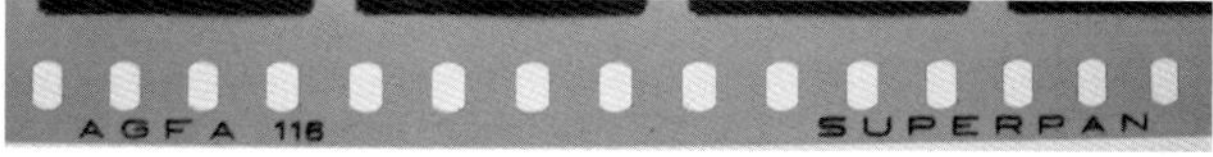

Ciné Journal Suisse 590 (prod. 1953). Original Camera Negative AGFA 116 Superpan. – CS

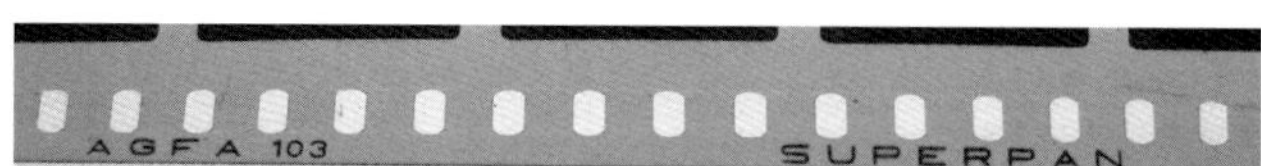

Ciné Journal Suisse 752 (prod. 1957). Original Camera Negative AGFA 103 Superpan. – CS

The edge printings do not have always the same characteristics. For instance, the number of perforations between the different components of the edge printings (brand, emulsion number, and emulsion type) can vary. Unfortunately, the author could not study enough film materials of this period to understand if the information can help to date the manufacture of film stock more precisely.

3 Various Agfa manufacturing documents communicated by Manfred Gill from the Industrie- und Filmmuseum Wolfen indicate this new practice, for example in *Ergänzung zur Entwicklungsvorschrift für Agfa-Color-Umkehrfilm* (June 1939) or *Emulsionerung* (October 1942).

4 The author thanks Noémie Jean for sending her the photographs from the *Ciné Journal Suisse*.

In October 1956, Agfa engineers adopted a system for edge printings for positive motion picture emulsion, in order to be better able to track the manufacture of their film stock:

Abschrift: 6.11.56/Ro

Herrn Ing.Schultze
Entwicklungsstelle
im H a u s e

– – 5221 Aufarb.Kinefilm
He/Ho 29. 1o. 56

Signierung unserer Kine-Positivfilme

Durch Aktennotiz vom 29.9.56 ersucht uns die Prüfstelle, die Kine-Positivfilme mit einer aufsignierten Gußnummer zu versehen, damit eingehende Reklamationen leichter nachgeprüft werden können. Da vorerst diese Arbeit nicht durchführbar ist, wurde in der 11-Uhr-Besprechung vom 26.1o.56 vorgeschlagen, für Positivfilme einen monatlichen Kennbuchstaben zu signieren, wie dies bereits bei der Fußzahlsignierung unserer Negativ- und Tonfilme geschieht. Der leichteren Erkennbarkeit wegen schlagen wir vor, für jeden Monat einen sich in jedem Jahr wiederholenden Buchstaben einzusetzen. Hinter diesen Buchstaben käme jeweils die letzte Ziffer der betreffenden Jahreszahl zu stehen, so daß für 1957 sich folgende Kennzeichnungen für die einzelnen Monate ergeben:

Januar	=	A 7
Februar	=	B 7
März	=	C 7
April	=	D 7
Mai	=	E 7
Juni	=	F 7
Juli	=	G 7
August	=	H 7
September	=	J 7
Oktober	=	K 7
November	=	L 7
Dezember	=	M 7.

Zur Durchführung dieser Arbeit erhalten Sie in doppelter Größe eine Verteilung der verschiedenen Zeichen, welche auf den Steg des Filmes zu signieren sind. Es handelt sich um:

1. Bezeichnung Agfa S
2. Maschinen-Nummer
3. den monatlichen Kennbuchstaben und die Jahreszahl.

Da zwischen den letzten Buchstaben des Wortes Agfa und den ersten Buchstaben der Wiederholung 28 Stege zur Verfügung stehen, schlagen wir vor, die Maschinen-Nummer in den Steg Nr. 7, das "S" -gehörend zu Agfa S- in den Steg Nr. 14, den Monatsbuchstaben in den Steg 25 unddie Jahresziffer in den Steg Nr. 26 einzusetzen. Die Monats- und Jahresziffer müssen auswechselbar sein, und zwar, wenn irgendmöglich, ohne die anderen Signierungen wie Maschinen-Nummer und dergleichen aus der Signiertrommel herauszunehmen.

Wir bitten Sie, uns etliche Muster in der gewünschten Art herzustellen.

Sollten diese Aufgabe leichter zu lösen sein, indem die Signierung des Monatsbuchstabens in die gegenüber liegenden Perforationsreihe

./.

"Edge printing on motion picture positive film" – IFM-W

It was then decided to introduce in positive emulsions the following code system[5]: "Agfa S" ("S" for Safety), perforation machine, and a letter corresponding to the month, together with a number which corresponds to the last digit of the year:

A	January
B	February
C	March
D	April
E	May
F	June
G	July
H	August
J	September
K	October
L	November
M	December

Example:

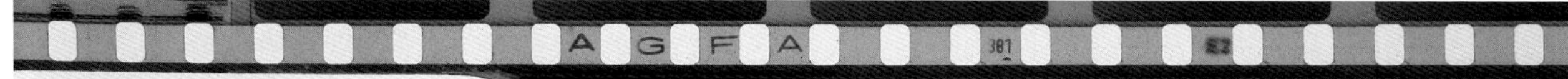

Dupe-Negative 1962 – NFAI

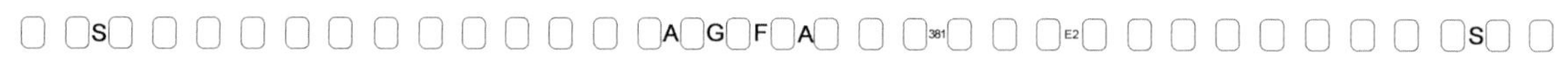

S (Safety) [13 perforations] AGFA [3 perforations] Perforation machine: 381 [3 perforations] Manufacturing Date: E2 (i.e., May 1962) [9 perforations] S (Safety)

The National Film Archive of India (NFAI) holds a duplicate negative of *Sujata* (Bimal Roy, 1959). In this case, the dating is easy: the dupe negative cannot be prior to 1959 or post-1964; therefore, this duplicate negative was made on film stock manufactured in May 1962.

When Agfa merged with Gevaert in 1964, the Gevaert edge printing system continued to be used for Agfa-Gevaert film stock. But the system used by Agfa would be kept by Orwo.[6]

5 According to the document, this edge printing system was already used for image picture negative and sound film stocks.
6 See also the chapters on Gevaert and Agfa-Gevaert, pp.267-276, and Orwo, pp.277-282.

9
Gevaert and Agfa-Gevaert

Camille Blot-Wellens

Gevaert[1]

It is important to note that a significant part of the information presented in this chapter derives from written sources and could not always be confirmed by studying film elements. Given this, it was a hard decision whether to decide to publish the information, not knowing how relevant and useful it would be.

Gevaert was founded in 1894 by Lieven Gevaert, who was working with photographic products. The company was first located in Antwerp on Anselmostraat, and moved to the nearby city of Mortsel in 1905.

Exactly when the company began to manufacture film is unclear. While Carlos Gevaert dates its first research into roll film in 1913, and the bringing to market of roll film, film packs, studio film, X-ray film, graphic film, and motion picture film in 1920[2], several pieces of evidence date the presence of Gevaert film stock on the market much earlier.

The factory manufactured positive film stock and exported it abroad since at least 1910.[3] In 1913 Gevaert provided Safety Positive 35mm film stock.[4] In 1914 Gevaert's agent in the United States, the Raw Film Supply Company, announced a new emulsion, Gevaert anti-halo Negative, and stated: "A majority of the large productions made in Europe during the past six months, including 'Cabiria', have been taken on Gevaert anti-halo negative stock, and the wonderful effects secured in these masterpieces have been heretofore considered impossible. The use of anti-halo negative in combination with the new Gevaert coloured base positive makes it possible to obtain almost any desired lighting or coloured combination as well as countless novel effects."[5] It was natural for an agent to praise the brand highly, but Gevaert was still cited as one of the most important film stock manufacturers, one of "the first craftsmen", alongside Lumière, Pathé, Eastman, and Agfa, by Georges Dureau, the Head and Founder of *Ciné-Journal*.[6] In 1920 the *Moving Picture World* even published an item, "Belgian Film Stock Attracts Attention of Producers Here". Even though this is clearly promotional, it is still interesting to note that the Gevaert factory was then important enough to be described in these terms: "the Gevaert factory in Belgium is so equipped and its organization so flexible that should it suddenly be called upon to increase its output by several millions of feet a week, it could do so without any difficulty,"[7] and in 1922, for the opening of its branch in Hollywood, Gevaert announced it had 18 branches located throughout the United States, Europe, and South America.[8] This success is difficult to confirm through actual film materials, as apparently the company did not then use edge printings, so it is impossible to recognize the emulsion until the early 1920s. Indeed, as far as the author knows, the first Gevaert film stocks able to be identified as such are thus from the early 1920s.

---

1 The author warmly thanks Bruno Mestdagh and David Gruwez from the Cinémathèque royale de Belgique (CRB) for providing almost all the illustrations of Gevaert materials.
2 Carlos Gevaert, "De Fotografische Nijverheid in België", in *Econ. Tijd. Kredietbank* (10.07.1938), p.218.
3 Advertisement in *Ciné-Journal* (13 May 1911), p.13.
4 The Gaumont Collection held at the Cinémathèque française includes a "Film Test" made by a Gaumont engineer on Gevaert Safety Film (Positive 35mm, Emulsion n° 507) dated 24 May 1913 (LG259-B31).
5 *Motography* (1 August 1914), p.160.
6 *Ciné-Journal* (18 July 1914), p.3.
7 *Moving Picture World* (8 May 1920), p.806.
8 *Motion Picture News* (1 April 1922), p.1910.

Harold Brown gives examples of Gevaert 35mm film stock manufactured in 1921, 1922, and 1923.[9] No major changes would be observed from 1923 until the end of the 1920s, the edge printings having the same fonts and being on both sides of the film stock. Furthermore, the space between the letters and their size seem to remain the same, except the shape of the perforations on positives changes from Bell & Howell to Kodak around 1925, and in some cases the word "Belgium" is missing in the edge printings.

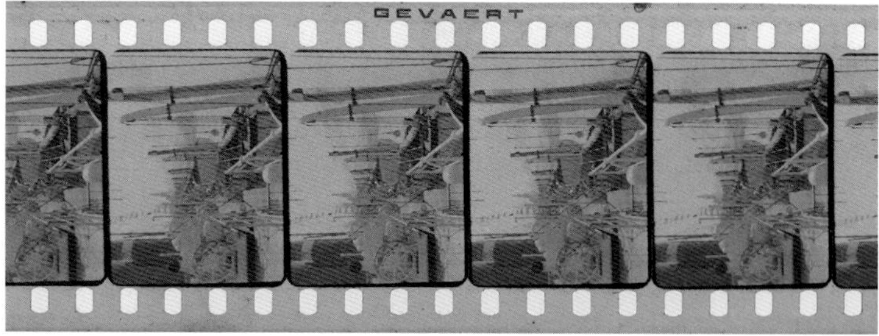

Print ca. 1923 (*Le Baptême du Belgenland par S.E. le Cardinal Mercier*, Willy Druyts, 1923) – CRB

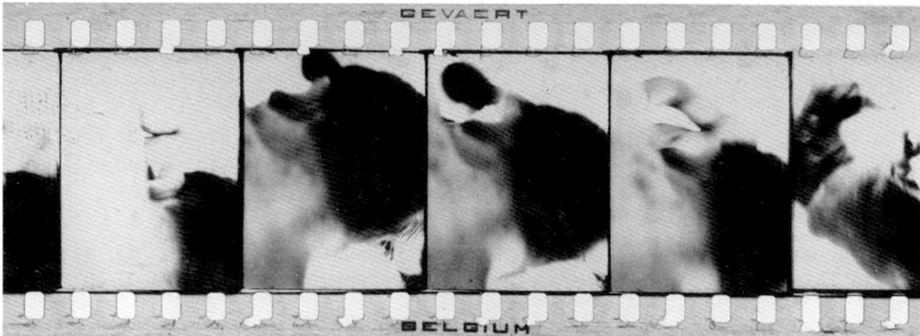

Print ca. 1927 (*Combat de boxe*, Charles Dekeukeleire, 1927) – CRB

Towards the very end of the 1920s,[10] it seems that the words "GEVAERT BELGIUM" started being introduced on only one of the edges of the film and the font became more squared. It would remain similar throughout the 1930s and 1940s.

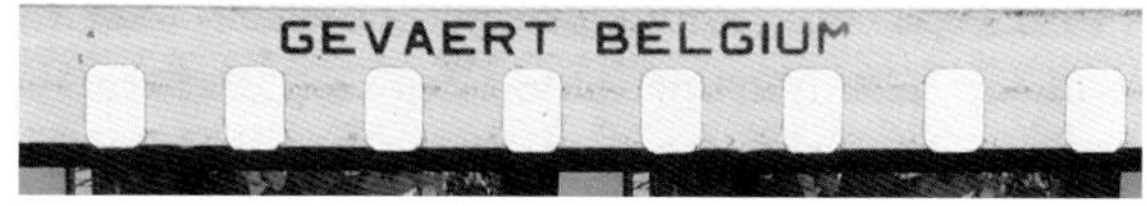

Print ca. May 1931 (*Le Premier mai 1931 à Bruxelles*, Henri Storck, 1931) – CRB

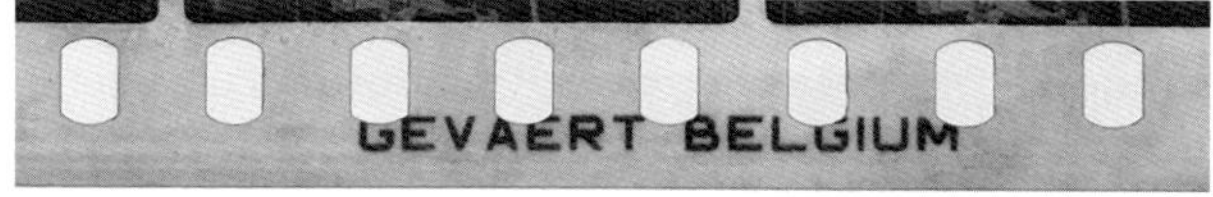

Negative of a film produced in 1932 (*De 100-jarige te Temsche*, Clemens De Landtsheer, 1932) – CRB

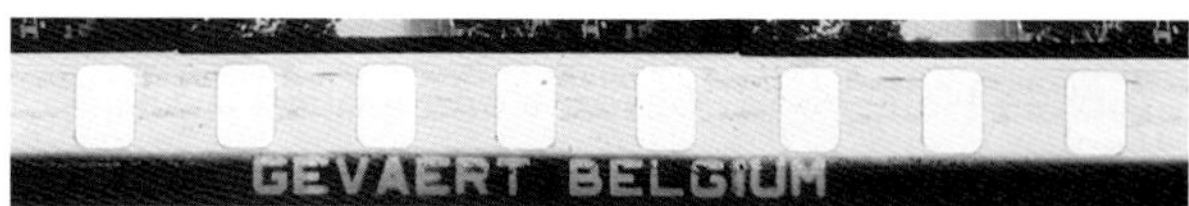

Negative of a film produced in 1939 (*Ceux qui veillent*, Gaston Schoukens, 1939)
[edge printing photographed on the print directly from the negative] – CRB

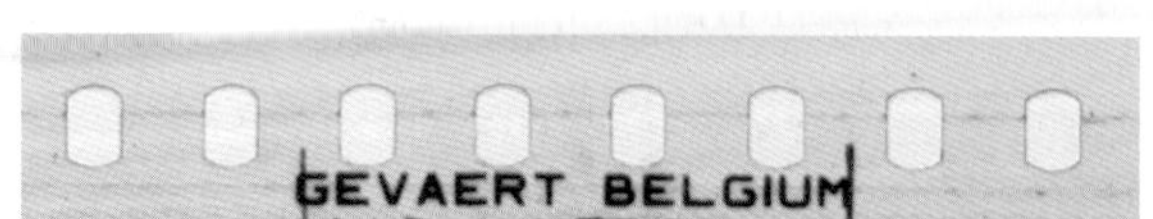

Negative of a film produced in 1943 (*La Naissance d'une cité*, Gaston Schoukens, 1943) – CRB

9 See Brown 6.6., p.72 in this volume.
10 Luciano Berriatúa situates this change around 1929.

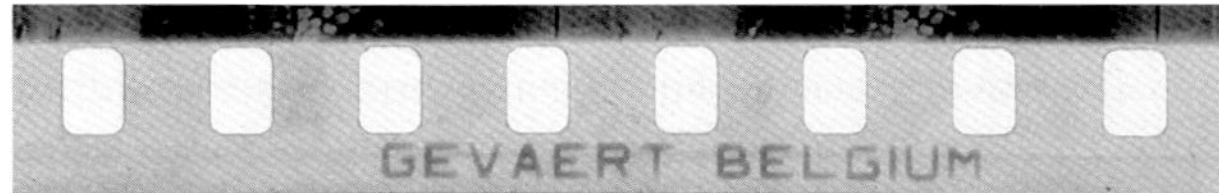

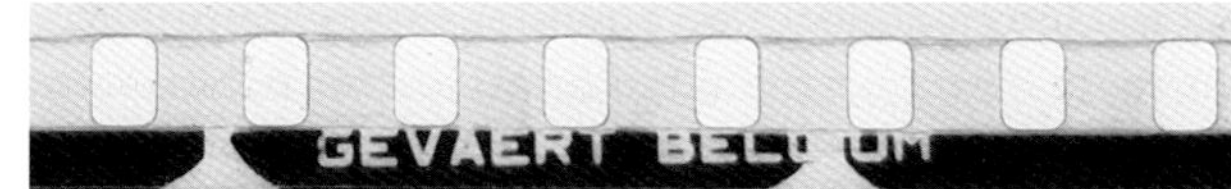

Print and Negative of a film produced in 1945 (*Trois rois*, Edouard Ehling, 1945)
[negative edge printing photographed on the print directly from the negative] – CRB

Negative of a film produced in 1949 (*Au delà des saisons*, Charles Dekeukeleire, 1949)
[edge printing photographed on the print directly from the negative] – CRB

Therefore, it turns out to be very difficult to date the manufacturing of the film stock using the edge printings. One reason for the consistency of the edge printings could be due to another alternative used by the company to date the manufacturing of the stocks.

According to Marc Sutherland,[11] in January 1931 Gevaert introduced a system to identify the emulsion and the date of manufacture of film stock: a 4- or 6-digit numerical code. The first 2 digits refer to the month of manufacture and the next 2 digits to the emulsion. In the information provided, the cycle for the month of manufacture is 8 years; i.e., the numbers repeat after 8 years (from 1 to 96, 1 being January and 96 being December).[12]

Notwithstanding my attempts to gather information, I could not have access to enough elements made on Gevaert film stocks to corroborate the presence and regularity of this data before the 1960s.[13] It is therefore necessary to be cautious when using this information.

In 1934 Gevaert could provide all kinds of emulsions for motion pictures, including sound film (ST1 and ST2) and panchromatic film, "which without showing colours is able to create different relative densities per colour, and as such reduces the need for make-up to an absolute minimum, making it a great success in the film studios".[14]

According to technical documents published by the company in 1950,[15] the firm changed the edge printing of its negative film stock: "GEVAERT BELGIUM PANCHROMATIC" was added during the manufacture of the film stock in latent image (i.e., it was not visible until the film was processed at the laboratory) on 35mm panchromatic negatives, while "GEVAERT A.B. SAFETY" was added on 16mm negatives. Also, the emulsion number and an order number were perforated at the beginning of each reel. If the exact moment of the modification of the edge printings is unknown, elements manufactured in the 1950s (and perhaps even earlier) should have these characteristics.

A similar edge printing and perforation system was adopted for the duplicating emulsions: on 35mm film stock, "GEVAERT BELGIUM" (Duplicating Positive, Duplicating Positive Fine Grain, and Duplicating Negative) or "GEVAERT BELGIUM PANCHROMATIC" (Duplicating Negative Fine Grain Panchromatic).

Duplicate Negative of a film produced in 1942, but probably duplicated later (*Actualités mondiales* n° 582, Ufa, 1942) – CRB

11 The former General Manager of Classics at Agfa, who communicated this information to the author in January 2016.
12 See the table with the codes at the end of this chapter.
13 For the numbering system, see the charts on pp. 273-275.
14 *De Gevaert-fabrieken en haar Produkten* (1934). Translation by Marc Sutherland.
15 *Pellicules cinématographiques négatives* (December 1950): Panchromosa Type 47, Gevapan 27 and Gevapan 33. This document also mentions that the negatives could be perforated both in BH and KS. See also *Pellicules cinématographiques négatives* (April 1951).

All the 16mm duplicating emulsions were identified with the words "Gevaert A.B. Safety".[16] A few months later, Gevaert modified the edge printings for positive emulsions (Positive, Positive Fine Grain, and Positive Super Contrast): the 35mm nitrate film stock was still identified as "GEVAERT BELGIUM", while safety 35mm film stock was identified with the same wording as its 16mm stock, "GEVAERT A.B. SAFETY".[17]

In the early 1950s, the company also changed the edge printings by introducing small hyphens and points around "Gevaert Belgium":

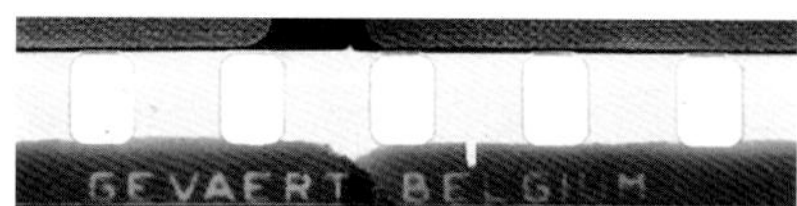

Negative of a film produced in 1952
(*Een eeuw ging voorbij*, Raphael Algoet; Maurice Delattre; Marcel Roothooft; Gaston Vernaillen; Georges Lust, 1952)
[edge printing photographed on the print directly from the negative] – CRB
In this case, we see a double hyphen of different lengths between "Gevaert" and "Belgium".

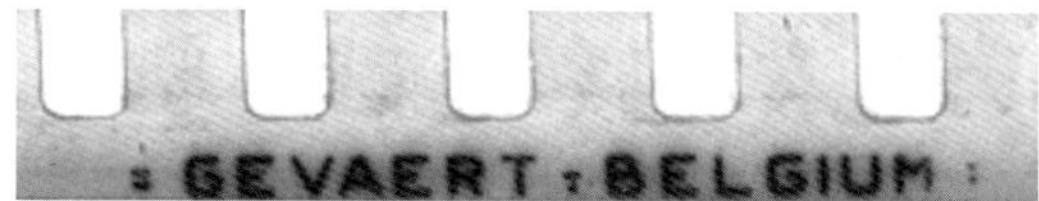

Print dated 1952 – LB
On this print, we can observe two hyphens
(like an equal symbol [=] before "Gevaert", a double hyphen of different lengths
between "Gevaert" and "Belgium", and two dots after "Belgium".

If the meaning of this system still needs to be clarified, it is possible to situate the manufacturing of film materials with these characteristics in the 1950s. This edge printing can be observed both on positive and negative emulsions.

The information that follows concerns only negative and duplication emulsions, and was found in the technical documents of the company, but for lack of examples, it was not possible to verify it on film stocks. Therefore, the information should be regarded with caution.

Towards the end of 1952, according to the technical sheets published by the firm, the identification system of the emulsions changed: negative and duplicating film stock were identified by a letter preceding the footage number. All 35mm emulsions were indicated by "GEVAERT BELGIUM SAFETY", while 16mm was indicated by "GEVAERT SAFETY", and if there was no letter preceding the footage number on these film stocks, the emulsion number was perforated at the beginning of the reel.

In 1957, Gevaert announced that "all 35mm Gevaert cinefilms carry between the perforations the mark S GEVAERT BELGIUM S, followed by a 3-figure number indicating the perforating machine number. The opposite rebate carries the number of the reel (1-31).[18] The majority of Gevaert 35mm cinefilms also have on the rebate of the film the footage in feet (6 figures) preceded by a different letter of identification for each type of film."[19]

Below is a chart with the codes found in the documents, as well as the date of publication of the document in which it was found:[20]

16 *Pellicules cinématographiques pour contretypage* (June 1950). This document also indicates that the duplicating emulsions could be perforated both in BH and KS.
17 *Pellicules cinématographiques positives* (November 1950). This document also indicates that the positive emulsions could be perforated both in BH and KS.
18 As it was important for the manufacturer to be able to know on which part of the jumbo roll the reel was before slitting.
19 *Motion Picture Films for Professional Use* (1957).
20 All the documents consulted proceed from the Madrid Project (a project developed by Alfonso del Amo gathering hundreds of technical documents, available on the FIAF website – member access only).

Code	Emulsion	1952	1957[21]	1964[22]
Ø	Gevacolor Negative – Type 6.53			X
Ø	Negative Cine Film Gevapan 36 – Type 1.91 B&W. Ultra-fast emulsion. 35mm/16mm.		X	
D	Duplicating Positive Fine Grain – Type 3.62 B&W. Yellow dyed film for printing master positives. 35mm/16mm.		X	X
E	Duplicating Negative – Type 4.51 B&W. 35mm/16mm.		X	X
G	Duplicating Negative Fine Grain Panchromatic – Type 4.63 B&W. Works with Duplicating Positive Fine Grain. 35mm/16mm.		X	
G	Duplicating Negative Fine Grain Panchro – Type 4.64			X
H	Negative Cine Film Gevapan 33 – Type 1.83 B&W. Recommended for newsreels. 35mm/16mm.		X	
H	Gevapan 36 Negative – Type 1.92			X
H	Gevapan 36 Reversal – Type 8.80			X
M	Negative Cine film Gevapan 27 – Type 1.52 B&W. Very fine grain emulsion. 35mm/16mm.		X	
M	Telerecording Reversal – Type 8.50			X
N	Negative Cine Film Gevapan 30 – Type 1.64 B&W. Best emulsion for production work. 35mm/16mm.		X	
N	Gevapan Negative – Type 1.65			X
N	Gevapan 30 Reversal – Type 8.63			X
S	S.T. 1 – Type 2.81 For variable density recording (odd number). 35mm/16mm.	X[23]		
S	S.T. 3 – Type 2.83 For variable density recording (odd number). 35mm/16mm.	X[24]	X	X
S	S.T. 4 – Type 2.54 For variable area recording (even numbers). 35mm/16mm.	X[25]	X	X
T	S.T. 0 – Type 2.50 Universal for sound recording. 35mm/16mm.	X[26]		
T	S.T. 6 – Type 2.56 For variable area recording (even numbers). 35mm/16mm.	X[27]	X	X

This system seems to have been used on all negative and duplicating emulsions for motion picture films (at least 35mm and 16mm).

The information related to the emulsion is only on negative and duplicating film stock to be used by cinematographers and laboratories. The edge printings introduced on positive emulsions were used by the manufacturer to better track their production.

According to an article published in *Le Technicien du film*,[28] in 1953 Gevaert was considered as the major film manufacturer in Europe.[29] In 1964 Gevaert merged with Agfa, and from this period the film stock was identified as Agfa-Gevaert.

21 *Motion Picture Films for Professional Use* (1957).
22 *Peliculas profesionales de cine* (1964).
23 *S.T. 1. Pellicule ciné pour l'enregistrement du son – 35mm et 32mm. Type 281* (November 1952).
24 *S.T. 3. Pellicule ciné pour l'enregistrement du son – 35mm, 32mm et 16mm. Type 283* (November 1952).
25 *S.T. 4. Pellicule ciné pour l'enregistrement du son – 35mm, 32mm et 16mm. Type 254* (November 1952).
26 *S.T. 0. Pellicule ciné pour l'enregistrement du son – 35mm, 32mm et 16mm. Type 250* (November 1952).
27 *S.T. 6. Pellicule ciné pour l'enregistrement du son – 35mm, 32mm et 16mm. Type 256* (November 1952).
28 "Gevaert 60 années d'expérience", *Le Technicien du film*, No. 7 (June 1955), pp.18-19.
29 To learn more about the Gevaert factory in Mortsel in the late 1950s, see also the documentary *Germes de lumière* (1958), directed by Charles Dekeukeleire for the company. This film has been made available online by the Cinémathèque royale de Belgique on the European Film Gateway portal.

Agfa-Gevaert

Agfa-Gevaert was established in 1964, when Agfa, Aktien-Gesellschaft für Anilin-Fabrikation (1867) and Gevaert Photo-Producten N.V. (1894) merged. The head offices were located in Leverkusen (Cologne, West Germany) and Mortsel (Antwerp, Belgium). The main factories were in Mortsel, Leverkusen, and Munich, but other factories were also functioning in Germany, Belgium, Argentina, France, India, Portugal, Spain, and the U.S.A.[30]

The numbering system introduced by Gevaert in the early 1930s was kept after the two firms merged. As before, the first 2 digits refer to the month of manufacture, and the second 2 digits refer to the type of emulsion:[31]

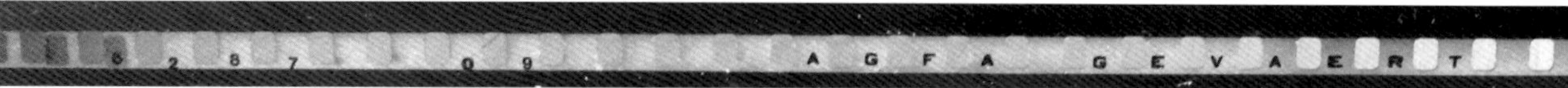

Positive manufactured in 1984 – CS

The example above is a leader on Agfa-Gevaert stock found on a release print for Switzerland of *A Passage to India* (David Lean, 1984) struck on Eastman Kodak US 1984 LPP film stock.

The first 4 digits of the Agfa-Gevaert stock are "6287". When referring to the information provided by Marc Sutherland (reproduced at the end of this chapter), "62" corresponds to February 1984 for the date of manufacture, and "87" to the emulsion CP2 colour print film. Therefore, the edge printing seems to conform with the element to which it belongs.

This number should not be mistaken for a footage number. It's important to check that it repeats exactly the same way throughout the reel in order to differentiate it from a footage number (which is usually situated on the edges and not between the perforations, in order to be readable once photographed to the next generation).

Often, the edge printing on Agfa-Gevaert film stock only consists of the name of the firm and a 3-digit number which refers to the perforation machine:

Positive (no date) – MoMA

From possibly the late 1970s until the early 1990s, Agfa-Gevaert also appears as "AG". In the early 1990s, the name of the manufacturer starts being indicated as "AGS": Agfa Gevaert Safety Film.

During the same period, Agfa-Gevaert adopted a barcode system for colour negatives which included the manufacturer identification code, film type, film roll specification, and a 4-digit footage number (every 64 perforations on 35mm and every 20 perforations on 16mm).

In the example below, the manufacturer identification code is "A" (for Agfa-Gevaert), the film type is "S" for the emulsion XTS 400 (other codes for film types are "N" for XT 100, "M" for XTR 250, and "F" for XT 320), the film roll specification is "01 0038", and "6055" is the 4-digit footage number, followed by a zero-frame reference.

Agfa

30 Agfa-Gevaert: *Películas Cine y TV*, Mortsel: Agfa-Gevaert N.V. (1981).

31 See the table for manufacturing dates and kinds of emulsions and the table of identification codes at the end of this section. According to the firm's published technical information, *Gevaert – Películas profesionales de cine* (1964), Gevacolor negatives 6.53 and 9.53 also have an emulsion number between the perforations, even if so far I have only been able to read it on positive film stock.

This example (from a technical sheet from 1994[32]) allows us to see the information which helps to date the film stock. After the edge code "AGFA XTS 400", there is a 6-digit number, "818415". Referring to the information provided by Marc Sutherland, "81" corresponds to September 1993 and "84" to the emulsion XTS Color Negative. The information seems to conform with that included in the edge printing (AGFA XTS 400) and the date of publication of the technical sheet (1994). The meaning of the last 2 digits (in this case, "15") is still unknown.

The same 6-digit code that allows dating the film stock also seems to appear on 16mm materials.

Manufacturing Month/Year Codes for Gevaert and Agfa-Gevaert[33]
Source: Marc Sutherland

	Jan.	Feb.	Mar.	Apr.	May	June	July	Aug.	Sep.	Oct.	Nov.	Dec.
1931	1	2	3	4	5	6	7	8	9	10	11	12
1932	13	14	15	16	17	18	19	20	21	22	23	24
1933	25	26	27	28	29	30	31	32	33	34	35	36
1934	37	38	39	40	41	42	43	44	45	46	47	48
1935	49	50	51	52	53	54	55	56	57	58	59	60
1936	61	62	63	64	65	66	67	68	69	70	71	72
1937	73	74	75	76	77	78	79	80	81	82	83	84
1938	85	86	87	88	89	90	91	92	93	94	95	96
1939	1	2	3	4	5	6	7	8	9	10	11	12
1940	13	14	15	16	17	18	19	20	21	22	23	24
1941	25	26	27	28	29	30	31	32	33	34	35	36
1942	37	38	39	40	41	42	43	44	45	46	47	48
1943	49	50	51	52	53	54	55	56	57	58	59	60
1944	61	62	63	64	65	66	67	68	69	70	71	72
1945	73	74	75	76	77	78	79	80	81	82	83	84
1946	85	86	87	88	89	90	91	92	93	94	95	96
1947	1	2	3	4	5	6	7	8	9	10	11	12
1948	13	14	15	16	17	18	19	20	21	22	23	24
1949	25	26	27	28	29	30	31	32	33	34	35	36
1950	37	38	39	40	41	42	43	44	45	46	47	48
1951	49	50	51	52	53	54	55	56	57	58	59	60
1952	61	62	63	64	65	66	67	68	69	70	71	72
1953	73	74	75	76	77	78	79	80	81	82	83	84
1954	85	86	87	88	89	90	91	92	93	94	95	96
1955	1	2	3	4	5	6	7	8	9	10	11	12
1956	13	14	15	16	17	18	19	20	21	22	23	24
1957	25	26	27	28	29	30	31	32	33	34	35	36
1958	37	38	39	40	41	42	43	44	45	46	47	48
1959	49	50	51	52	53	54	55	56	57	58	59	60
1960	61	62	63	64	65	66	67	68	69	70	71	72
1961	73	74	75	76	77	78	79	80	81	82	83	84
1962	85	86	87	88	89	90	91	92	93	94	95	96
1963	1	2	3	4	5	6	7	8	9	10	11	12
1964	13	14	15	16	17	18	19	20	21	22	23	24

32 Agfa – Motion Picture Division: *Machine-readable BARCODE and human-readable edge information on 35mm and 16mm colour negative films*, Mortsel: Agfa-Gevaert N.V. (1994).

33 The author reproduces here the entire list of codes supplied by Marc Sutherland (former Manager of Classics at Agfa), covering the period 1931-2018 (independently of the period of manufacturing by the company).

	Jan.	Feb.	Mar.	Apr.	May	June	July	Aug.	Sep.	Oct.	Nov.	Dec.
1965	25	26	27	28	29	30	31	32	33	34	35	36
1966	37	38	39	40	41	42	43	44	45	46	47	48
1967	49	50	51	52	53	54	55	56	57	58	59	60
1968	61	62	63	64	65	66	67	68	69	70	71	72
1969	73	74	75	76	77	78	79	80	81	82	83	84
1970	85	86	87	88	89	90	91	92	93	94	95	96
1971	1	2	3	4	5	6	7	8	9	10	11	12
1972	13	14	15	16	17	18	19	20	21	22	23	24
1973	25	26	27	28	29	30	31	32	33	34	35	36
1974	37	38	39	40	41	42	43	44	45	46	47	48
1975	49	50	51	52	53	54	55	56	57	58	59	60
1976	61	62	63	64	65	66	67	68	69	70	71	72
1977	73	74	75	76	77	78	79	80	81	82	83	84
1978	85	86	87	88	89	90	91	92	93	94	95	96
1979	1	2	3	4	5	6	7	8	9	10	11	12
1980	13	14	15	16	17	18	19	20	21	22	23	24
1981	25	26	27	28	29	30	31	32	33	34	35	36
1982	37	38	39	40	41	42	43	44	45	46	47	48
1983	49	50	51	52	53	54	55	56	57	58	59	60
1984	61	62	63	64	65	66	67	68	69	70	71	72
1985	73	74	75	76	77	78	79	80	81	82	83	84
1986	85	86	87	88	89	90	91	92	93	94	95	96
1987	1	2	3	4	5	6	7	8	9	10	11	12
1988	13	14	15	16	17	18	19	20	21	22	23	24
1989	25	26	27	28	29	30	31	32	33	34	35	36
1990	37	38	39	40	41	42	43	44	45	46	47	48
1991	49	50	51	52	53	54	55	56	57	58	59	60
1992	61	62	63	64	65	66	67	68	69	70	71	72
1993	73	74	75	76	77	78	79	80	81	82	83	84
1994	85	86	87	88	89	90	91	92	93	94	95	96
1995	1	2	3	4	5	6	7	8	9	10	11	12
1996	13	14	15	16	17	18	19	20	21	22	23	24
1997	25	26	27	28	29	30	31	32	33	34	35	36
1998	37	38	39	40	41	42	43	44	45	46	47	48
1999	49	50	51	52	53	54	55	56	57	58	59	60
2000	61	62	63	64	65	66	67	68	69	70	71	72
2001	73	74	75	76	77	78	79	80	81	82	83	84
2002	85	86	87	88	89	90	91	92	93	94	95	96
2003	1	2	3	4	5	6	7	8	9	10	11	12
2004	13	14	15	16	17	18	19	20	21	22	23	24
2005	25	26	27	28	29	30	31	32	33	34	35	36
2006	37	38	39	40	41	42	43	44	45	46	47	48
2007	49	50	51	52	53	54	55	56	57	58	59	60
2008	61	62	63	64	65	66	67	68	69	70	71	72
2009	73	74	75	76	77	78	79	80	81	82	83	84
2010	85	86	87	88	89	90	91	92	93	94	95	96
2011	1	2	3	4	5	6	7	8	9	10	11	12
2012	13	14	15	16	17	18	19	20	21	22	23	24

	Jan.	Feb.	Mar.	Apr.	May	June	July	Aug.	Sep.	Oct.	Nov.	Dec.
2013	25	26	27	28	29	30	31	32	33	34	35	36
2014	37	38	39	40	41	42	43	44	45	46	47	48
2015	49	50	51	52	53	54	55	56	57	58	59	60
2016	61	62	63	64	65	66	67	68	69	70	71	72
2017	73	74	75	76	77	78	79	80	81	82	83	84
2018	85	86	87	88	89	90	91	92	93	94	95	96

Emulsion Codes

Source: Marc Sutherland

Code	Emulsion
01	BW Print T561 on triacetate base
03	BW Print T561 on polyester base
04	BW T553
05	BW T554
06	AGFA Sound ST8 on triacetate
08	ST9 AGFA Sound Negative inclusive SDDS
09	AGFA Sound ST8 on polyester
10	AGFA Sound ST8D on polyester for digital purposes
11	Duplicating Positive T362 on triacetate
12	Duplicating Positive T362 on polyester base
14	Duplicating Negative T464 on polyester base
17	AGFA Pan 250 Negative on triacetate base
20	XT100 Color Negative
21	XT100 Color Negative improved
24	XTR 250 Color Negative
65	XT500 Color Negative
78	BW Print T782
84	XTS400 Color Negative
85	CP1 Improved Color Printfilm
86	CP1BP Color Printfilm for Technicolor
87	CP2 Color Printfilm
88	CP1P Color Printfilm on polyester base
89	CP10P Color Printfilm on polyester base
90	CP1B Color Printfilm
92	Color Intermediate film T643
93	CP1B Color Printfilm for Technicolor
94	CP10B Color Printfilm
95	CP10BP Color Printfilm on polyester base
96	CP10 Cold Color Printfilm for cold processing
97	CP20P Color Printfilm on polyester base
98	CP10P Cold Color Printfilm on polyester base for cold processing
99	CP30 Color Positive Printfilm on polyester base

Chronology of Some Gevaert and Agfa-Gevaert Emulsions[34]

Year	Gevaert	Agfa
1925	Negative Pan 25	
1929	ST 1	
1935	Pan Special	
1936	Panchromosa Micrograin	
1937	Positive Fine Grain	
	Fin Grain Super Contrast	
Post- 1940	Gevapan	
	Gevapan Reversal (Television)	
1947	Gevacolor Positive 9.51	
	Gevacolor Reversal R5	
1948	Gevacolor Negative 6.51	
1954	Gevacolor Negative 6.52	
	Gevacolor Positive 9.52	
1958	Gevacolor Negative 6.53 /	
	Gevacolor Duplicating Reversal 16mm 9.00	
1962	Gevacolor Positive 9.53	
1964	Gevacolor Duplicating Reversal 16mm 9.01 (replaces 9.00)	
1966	Gevachrome Duplicating Reversal 16mm 9.02 (replaces 9.01)	
1966-1967	Gevacolor Positive 9.54	
1968	Gevacolor Negative 6.55 (replaces Agfacolor CN5)	
	Gevachrome Original Speed Film 6.05	
1969	Gevachrome Gevachrome Original High Speed 6.00	
1969-1970	Gevacolor Print Film 9.85	
1972	Gevachrome Original Daylight HS 6.15	
1974	Gevacolor Print Film 9.86 (replaces 9.85)	
1974-1975	Gevacolor Negative 6.80	
	Gevachrome II Studio 7.00	
	Gevachrome II Newsreel 7.10 "artificial light"	
	Gevachrome II Newsreel 7.20 "daylight"	
1975	Gevachrome Print Film 7.80 (Television)	
1978	Gevacolor Negative 6.82	
1979	Gevacolor Print Film 9.82	
	Gevachrome II 7.30	
1982	Gevachrome 7.32	
	Gevachrome 7.02 (replaces 7.00)	
	Gevachrome 7.22 (replaces 7.20)	
1984	Gevachrome II Print Film 7.82	Agfa XT

34 *Source: 60 jaar Cine – 60 ans de cinéma – 60 years in motion pictures – 60 Jahre Kine*, Mortsel: Agfa-Gevaert N.V. (1985).

10
Orwo

Camille Blot-Wellens, with the contribution of Martin Koerber and Florian Wrobel

Its Origins and Creation

After the Second World War, Farbenfabriken Bayer AG in Leverkusen (in West Germany) founded Agfa AG, wishing to inherit the legacy of the Agfa factory.[1] In 1952, a district court of Frankfurt recognized them as the legal successor of Agfa, instead of Wolfen (in East Germany, then under Soviet control).

Nevertheless, the Leverkusen factory was not able to produce enough film stock, while the Wolfen factory had problems purchasing raw materials to manufacture film stock. Therefore, although the two factories were in competition, they reached an agreement in order to co-operate. The first time was in March 1954, to share trademarks and work together in order to assure production. This decision was followed by negotiations which led to a formal 5-year agreement, signed on 16 April 1956 between Aktien-Gesellschaft für Anilin-Fabrikation, Deutsche Innen- und Aussenhandel Chemie Berlin, and VEB Filmfabrik Wolfen, for complete co-operation (sharing the trademark and joint sale of products, apparently in different markets). But the solution was finally not seen as sustainable, and the idea of creating a new trademark was born.

The determining factor would be a political decision towards the independence of the photochemical industry in the GDR from 1958 and the desire to be self-sufficient from 1963. The first stage was the registration of the trademark ORWO (for "Original Wolfen"). The preparation to launch the new brand started in 1962, and in March 1963 the council of ministers of the GDR decided to end the contract with Leverkusen, effective 1 April 1964.

That same day, the Agfa film stock produced in Wolfen officially became ORWO. Agfa itself decided to merge with the Belgian company Gevaert.

Edge Printings

The change of name on the edge printing was made in March-April 1964.[2] The newly created Orwo kept the system that had been used since 1957 by Agfa. The edge printing was composed of:

"S" for Safety / the name of the manufacturer / the perforation machine number (2 or 3 digits) / a code made of a letter and a number, which corresponded to the month and the last digit of the year of manufacture:

A	January
B	February
C	March

1 All the historical information on the creation of Orwo comes from Rainer Karlsch, "50 Jahre ORWO. Die wechselvolle Geschichte eines Warenzeichens". Lecture delivered 12 November 2014, at Städtisches Kulturhaus Wolfen. On the film factory in Wolfen (one of the world's biggest film manufacturing plants), see also Erhard Finger, *100 Jahre Kino und die Filmfabrik Wolfen*, Wolfen: GÖS – Druckhaus Köthen GmbH (1996), Rainer Karlsch, *Von Agfa zu Orwo. Die Folgen der deutschen Teilung für die Filmfabrik Wolfen*, Wolfen: Vorstand der Filmfabrik Wolfen (1992), and Early Cine-Film Stock", *Film History*, Vol. 20, No. 1 (2008), pp.59-76.

2 Letter from Jürgen Dunkel and Horst Kühn to Martin Koerber, 27 March 1995.

D	April
E	May
F	June
G	July
H	August
I or J	September
K	October
L	November
M	December

Unfortunately, unlike Agfa, changes in the fonts or position of the edge printings do not aid the dating of the manufacture of film stock, and most perforation machines were used for decades.[3] However, some changes in the characteristics of some emulsions can be observed by studying documentation and film materials. Nevertheless, this information should be treated carefully, since the volume of materials studied does not allow the establishment of a clear system.

The Mid-1960s

When Orwo started, it seems that the edge printings were mainly situated between the perforations. This characteristic is observable in a document held at Industrie- und Filmmuseum Wolfen, which gathers samples of Orwo film stocks of different emulsions and gauges with their corresponding edge printings: *Signier-Mappe Aufarb.-Kinefilm* (which can be translated as "Portfolio of edge printings. Moving Picture Stocks Collection").[4] Even if this document lacks a date, it seems to have been composed relatively early – since some emulsion numbers[5] correspond to the 1965 version of the document, *Kenn-Nummern-Tabelle* (which translates as "Identification Number Table").

Colour Positive Film Stock (Type PC 5), emulsion number 441, manufactured in March 1964 and perforated on machine 410:

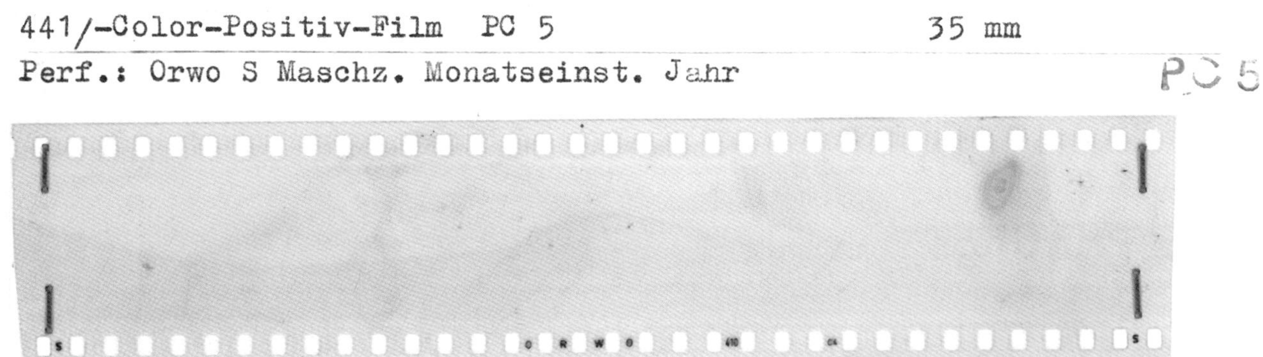

Signier-Mappe Aufarb.-Kinefilm – I-FMW

Signier-Mappe gathers 76 samples of motion picture film stock (8mm, 9.5mm, 16mm, 35mm, and 70mm) and 14 samples of still-photography film stock. 23 samples are 35mm emulsions (negatives, duplicates, positives, special). All the samples have edge printings between the perforations, except for Negativ-Film, Titelkopierfilm, and some special emulsions (Negativ-Zeitlupenfilm, Fernseh-Spezialfilm) that have the footage number, and, on negatives, emulsion types on the edges.

3 E-mail from Frank Böhme (Marketing and Sales Manager at FilmoTec) to the author (21 May 2015). Mr. Böhme could not find extra documentation on the use, implementation, or modification of the perforation machines.

4 The author warmly thanks Uwe Holtz, Director, and Manfred Gill, Archivist, of the Industrie- und Filmmuseum Wolfen for their invaluable help, and providing her with several documents quoted in this chapter.

5 As for Agfa, each manufacturing of emulsion was numbered. If the edge printings give accurate dates, the more recent film stocks of the document were manufactured in January 1967.

Examples found in different collections:

Image Negative 1966

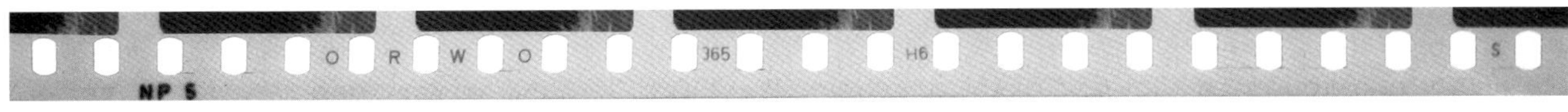

NFAI

Film stocks of original negatives are easier to date, since the release of the film can be used as a reference: *Ek Baon Bara Bhangadi* (Anant Mane, 1968) was released in 1968, and therefore "H6" is necessarily August 1966.[6] In this example, only the type of emulsion is indicated on the edge, NP 5, Image Negative (Picture Negative).

On the contrary, on still-photography film stock, the main part of the edge printing was situated on the edges (except for the codes denoting the date and the perforation machine).

The Late 1960s Until the Mid-1970s

On duplicate negatives dated 1968 and 1974, "ORWO S" is situated on the edge. It is possible that towards the end of the 1960s, the Wolfen engineers decided to differentiate negative and duplication emulsions from positive emulsions. Perhaps this was in order to enable the duplication of part of the edge printing during the printing process and differentiate the emulsions, since Orwo provided negatives and duplicates with both Bell & Howell and Kodak Standard perforations. Caution: These are the only examples the author could find for these years, so this information could not be confirmed.

Examples found in different collections:

Duplicate Image Negative 1968

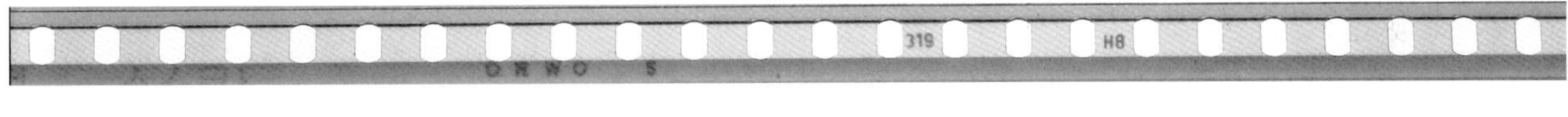

NFAI

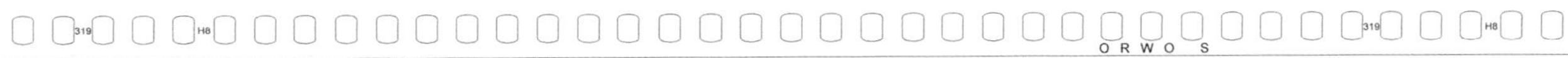

The dupe negative of *Sujata* (Bimal Roy, 1960) was first inspected by the NFAI on 26 September 1977 (per a document in the film can). The stock was necessarily manufactured between the introduction of Orwo in 1964 and the inspection of the film in 1977: "H8" corresponds to August 1968.

Duplicate Image Negative 1974

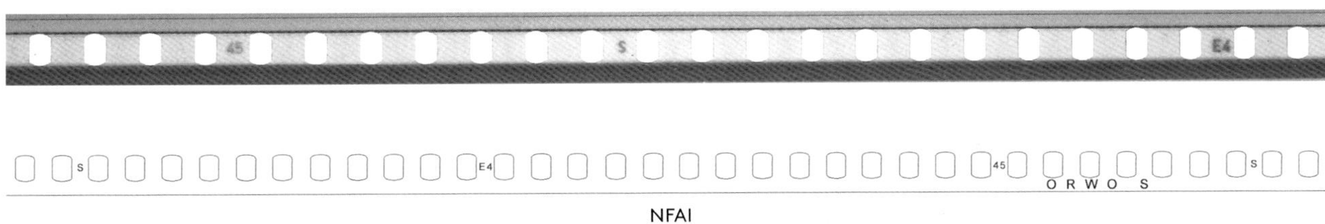

NFAI

The dupe negative of *Nidahanaya* (Lester James Peries, 1972) was first inspected by the NFAI on 11 February 1976 (per a document in the film can): "E4" corresponds to May 1974.

6 According to Rainer Karlsch, India was an important market for the Wolfen factory, even before the creation of Orwo ("50 Jahre ORWO. Die wechselvolle Geschichte eines Warenzeichens").

From 1975 to 1998

Apparently towards the mid-1970s the position of the edge printings started to vary, depending on the type of emulsion. For this chapter, Florian Wrobel studied and compared 13 film elements on Orwo film stock (corresponding to 80 emulsions between 1975 and 1990) from the collections of the Bundesarchiv (mainly DEFA productions). He observed that part of the edge printings (Orwo, footage number, and emulsion type) were situated on the edges on the original image negatives, while they were situated between the perforations on the other elements (sound negatives, duplicates, and prints).

Examples:

Print 1977

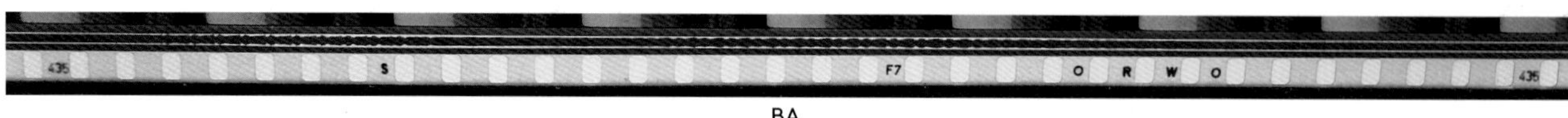

BA

On this image of a print of the documentary *Porzelliner* (Heinz Müller, 1977-1978), struck on film stock manufactured in June 1977 (F7), each element of the edgecode is printed between the perforations.

Duplicate Negative 1977

BA

Duplicate Negative of the documentary ... *und sie bewegt sich doch* (Kurt Tetzlaff, 1978), made on stock manufactured in June 1977 (F7). All of the edgecode is located between the perforations. It is also possible to read the emulsion type of the Image Negative (NP 55) photographed from the Original Negative (edges).

Note the code preceding the footage number (987469 on the edge), 35G, which corresponds to the machine used for the footage number (35) and the month of the insertion of the footage (in this case G, i.e., July).

Image Negative 1981

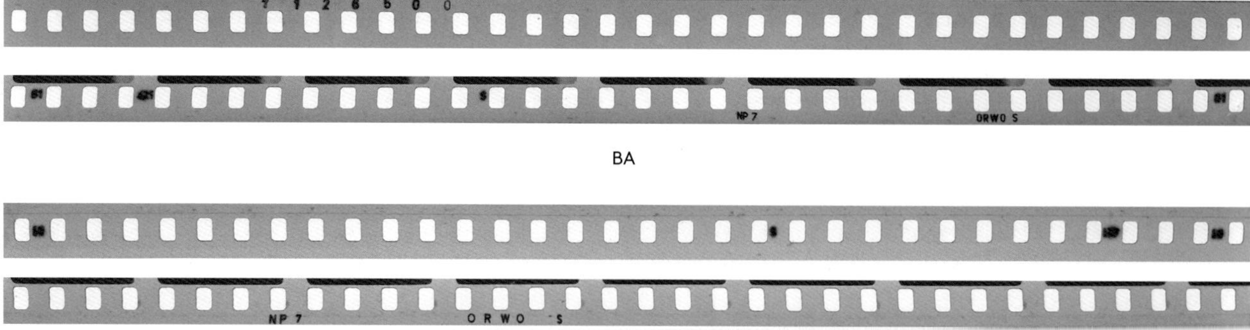

BA

BA

Both image negatives belong to the documentary *Walter Ballhause – einer von Millionen* (Karlheinz Mund, 1983); they were both manufactured in July 1981 (G1), and even perforated on the same machine (421). Despite being the same emulsion, manufactured the same month, and the same emulsion type, the edge printings "NP 7" and "ORWO S" were made with different stencils and were situated differently on the stock (the edge printings are not all on the same edge). According to Frank Böhme, "masks with different types of letters or numbers were used for optical signing. Today there are no scripts or other notations about them, especially regarding in which timeframes they were used."[7]

Therefore, it appears that the shape of the fonts, as well as their position on the film, cannot be used to date film stocks.

Also, perhaps we should note again here the code before the footage number (40 G), which corresponds to the machine used, and the month of the insertion of the footage (which usually coincides with the month of manufacture, as it does here).

Positive 1986

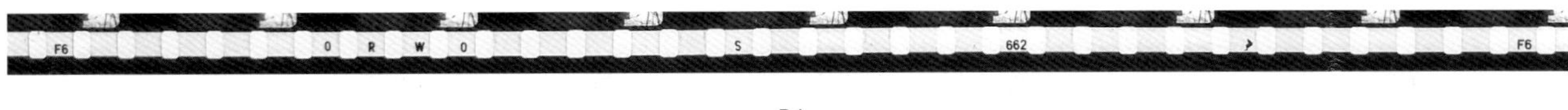

BA

Dailies of *Der Himmel über Berlin* (Wim Wenders, 1987), on film stock manufactured in June 1986 (F6). All of the edge printing is situated between the perforations.

Dupe positive 1989

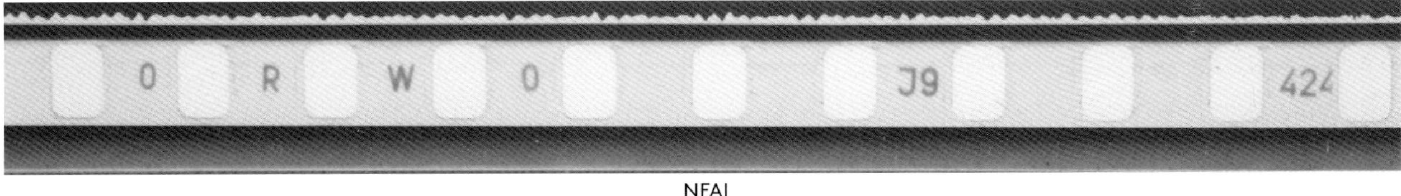

NFAI

424 S O R W O J9 424

This dupe positive of *Meghe Dhaka Tara* (Ritwik Ghatak, 1960) was made by the Bundesarchiv in 1991 (according to a document in the film can): "J9" very probably refers to September 1989. In this case also, all edge printings are inserted between perforations.

Sound Negative 1990

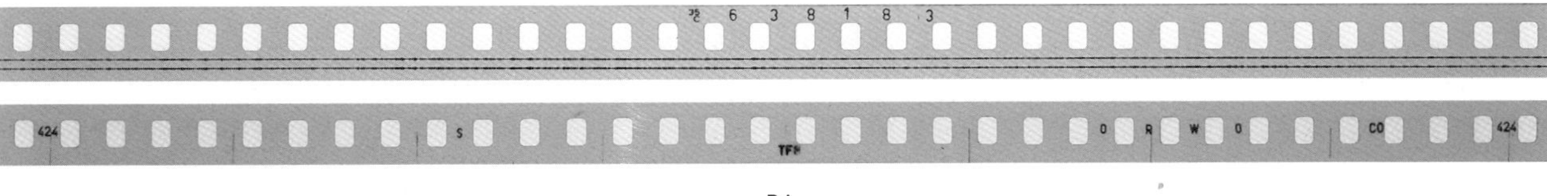

BA

Sound negative of the documentary *Kehraus* (Gerd Kroske, 1990), on film stock of March 1990 (C0). As with the duplicate positive of *Meghe Dhaka Tara*, the edge printings Orwo, perforation machine number, and date of manufacture are between the perforations, but the emulsion type (TF8) and footage numbers are on the edges.

The radical political changes of 1989 that resulted in the reunification of Germany in 1990 offered an opening for a return to a unified Agfa, with the potential market of both former Eastern and Western bloc countries. There were even discussions with Agfa-Gevaert, but this option was not taken up because the Wolfen factory was not considered productive enough.

7 E-mail from Frank Böhme to the author (21 May 2015). Different stencils were also used in the document *Signier-Mappe Aufarb.-Kinefilm* for Negative Picture emulsions.

In 1993 the Wolfen factory was reduced in size, and some of the buildings were demolished.[8] In December 1994 production was halted. But at the end of 1995, the creation of ORWO GmbH Wolfen, and of ORWO AG in December 1996, allowed production to begin again (with important changes, since the films were no longer manufactured in the Wolfen factory), but some of the market had been lost (after the refusal of Agfa Leverkusen to supply part of the market). In March 1998 ORWO AG went bankrupt.

1998 to the Present

In 1998, the former employees of ORWO decided to create a new company from ORWO AG, FilmoTec GmbH Wolfen, to manufacture black & white film (first UN 54, PF 2, then N 74 and all other motion picture film stocks) using the technology of Orwo. FilmoTec uses the same edge printings system as Orwo, and the films have "Orwo" and "FilmoTec" edge printings (except for TF 12 – sound-recording film – which has none).[9]

---

8 Herbert Bode, "Geschichte der Filmfabrik Wolfen 1909 bis 1994", in *Mitteilungen, Gesellschaft Deutscher Chemiker / Fachgruppe Geschichte der Chemie* (Frankfurt/Main), Bd. [Vol.] 13 (1997), p.160. E-mail from Frank Böhme to the author (21 May 2015). Therefore 5, 6, and 7 cannot correspond to 1995, 1996, and 1997.

9 The information about the last years of Orwo, from 1994 to 1998, and the transition to FilmoTec, plus its technology and edge markings, is from an e-mail from Frank Böhme to the author (11 June 2019).

11
Fuji

Hidenori Okada

The Foundation of Fuji Photo Film

Fuji Photo Film was founded in January 1934. While the domestic production of motion picture films was desired, its predecessor, the Dainippon Celluloid Company (founded in 1919; now Daicel) started research on nitrate-base motion picture films in 1928. It also took over research on film emulsion from Toyo Dry Plate, which would be merged into Fuji Photo Film following the failure of the Dainippon Celluloid Company's proposal of technical co-operation with Eastman Kodak in 1924.[1] Given the stimulus of huge industrial subsidies by Japan's Ministry of Commerce and Industry in 1933, the establishment of a domestic manufacturer was strongly encouraged, which indicates the government sought national control over the film industry by taking the opportunity of Fuji's founding.

However, Japan's major film studios, including Nikkatsu and Shochiku, announced in a unified statement a boycott of the products of Fuji Photo Film, even before the new company's first shipment. Fuji then reluctantly sold their film stock to minor feature productions and newsreel companies. The boycott ended in a year, however, and the majors gradually started to use Fuji products. Although the statement officially attributed the boycott to their poor quality, insisting "the domestic products are still unqualified for use", it is possible that the Japanese film community felt repelled by the government's intention to neutralize domestic competition and monopolize the infrastructure of the film industry.

Eventually, the Japanese film industry finally could not help depending on the products of Fuji Photo Film, in almost every case. In 1939 the government's policy of import suppression led to the suspension of imports from the United States, and even the banning of Agfa products from Germany, a friendly nation at that time.

Edge Codes of Fuji Photo Film Products (35mm motion picture films)

In Fuji edge codes, the rule of referring to the date the film was manufactured is quite simple. 64-JM, for example, indicates that the film was manufactured in the first quarter (January to March) of 1964. The first 2 numbers refer to the last 2 digits of the year (as the date of manufacture approached the end of the 20th century, the use of the full 4-digit notation increased), and the last 2 letters refer to one of the quarters of the year, as follows: JM: January to March; AJ: April to June; JS: July to September; OD: October to December.

1 See also Hidenori Okada, "Nitrate Film Production in Japan: A Historical Background of the Early Days", *Journal of Film Preservation*, No. 62 (April 2001), pp.22-24.

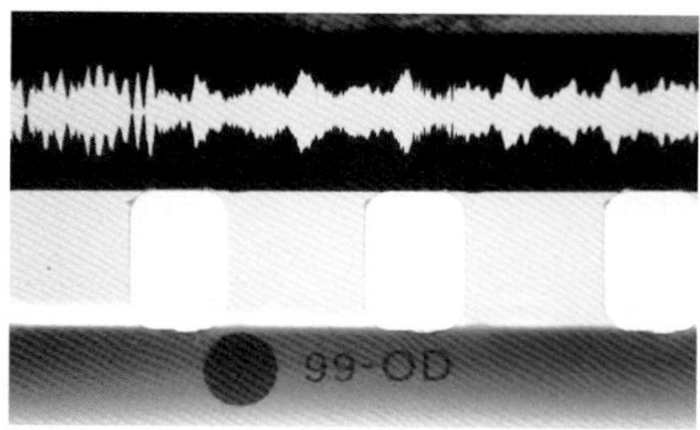

Colour positive film stock manufactured between October and December 1999 – NFAJ

As far as we can identify, studying material from the film collection of the National Film Archive of Japan (NFAJ), however, a new system of code notation has been applied to monochrome film stocks since May 1996 at the latest, and to colour film stocks since 2003.

For colour films, the following information and numbers are noted on the edge: film type, emulsion number (3 digits), roll number (3 digits), slitter number (3 digits), perforator number (3 digits), manufacturer number (6 digits).

For monochrome films: light sensitivity, emulsion number (3 digits), roll number (3 digits), manufacturer number (6 digits).

The date of manufacture is included in the manufacturer number. For colour film, the first digit refers to the year of manufacture (e.g., "4" refers to the year 2004); the second digit refers to the month of manufacture (the numbers 1 to 9 refer to the months from January (1) to September (9), after which the letter A refers to October, B to November, and C to December). For monochrome films, negative films, and intermediate films, the second digit refers to the year of manufacture, and the third digit to the month of manufacture (encoded in the same way as colour films).

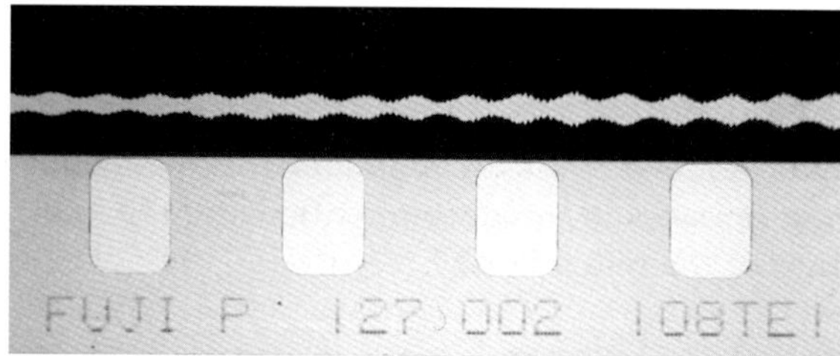

Black & white positive film stock manufactured in August 2000 – NFAJ

According to the applied new rules, we can identify for the first time the month of manufacture as well as the year. (The only exception is ETERNA-RDS, 3-colour separation recording intermediate negative film for long-term archiving, which has revived the rule of referring to the last 2 digits of the year, for example, 4791 14-OD.)

However, there is no date reference on motion picture film stocks manufactured in the early days of Fuji Photo Film after they began production in 1934. Only the company name FUJI appears on the edge. Judging from study of the NFAJ film collection, it seems the notation of the manufacturing date started in 1948. The oldest date reference identifiable in the NFAJ film collection is 48-JM. Fuji Film — the corporate name changed from Fuji Photo Film in October 2006 — discontinued the production of motion picture films for shooting and screening at the end of March 2013, 79 years after the company's founding. The only motion picture film stock they are still manufacturing is the above-mentioned ETERNA-RDS.

12

Preliminary Notes on Soviet Nitrate Film Stock and Other Aids to Identification of Russian and Soviet Films

Peter Bagrov

These notes are indeed preliminary, since I am only in the early stages of my research. This particularly refers to the field of trademarks and title fonts: much information has still to be collected in order to compile a proper guide. Acquiring a comprehensive set of data would take years, maybe decades. But this article is a first step.

It is extremely difficult to make any generalizations about Soviet nitrate prints, because so few of them are available. At Gosfilmofond of Russia, the Russian state film archive, there was a practice of preserving films on safety stock and destroying the nitrates afterwards. This was typical of many film archives at the time, but Gosfilmofond was particularly active in this respect. In the 1960s and 1970s the bulk of its collection was preserved in this manner, and the process slowly continued all the way until the early 2010s. The remaining selection is hardly representative – there are about 50 Russian nitrate prints from the silent era and less than 50 from the sound era, and many of these are fragments, outtakes, etc. But I also had access to materials from the Russian State Archive of Film and Photo Documents (RGAKFD) in Krasnogorsk, and the State Central Film Museum (GTsMK) and the All-Russian State Institute of Cinematography (VGIK) in Moscow, as well as at the British Film Institute, Bundesarchiv, The George Eastman Museum, The Museum of Modern Art, Österreichisches Filmmuseum, Svenska Filminstitutet, and Marina Umanskaia's private collection of nitrate film frames in Kiev. I would like to thank all those who made these materials available to me, in particular Olga Beliakova, Natalia Chertova, Olga Chizhevskaia, Olga Dereviankina, Bryony Dixon, Dirk Förstner, Karl Griep, Natalia Kalantarova, Egbert Koppe, James Layton, Marianna Kushnerova, Michael Loebenstein, Marina Umanskaia, Alisa Nasrtdinova, Larisa Solonitsyna, Jon Wengström, and Alexander Zöller. While working on this study I was privileged to consult with some of the veteran cinematographers who worked on Soviet nitrate film stock at the beginning of their careers, namely Dmitrii Masurenkov, Mark Osepian, Levan Paatashvili, and Anatolii Petritskii. I wish to express my deep gratitude to them.

1. A Brief History of Soviet Film Stock Manufacture in the Nitrate Era

The Russian Empire produced neither cameras nor film stock. Everything was imported before World War I, mostly from France, and after the war from the United States (shipping via Asia turned out to be easier and cheaper). After the wars – both World War I and the Russian Civil War – were over, it again became more convenient to obtain film stock from Europe.

The collection of original Russian nitrate prints from the 1910s at Gosfilmofond is too small to talk about in terms of statistics. But the few existing nitrates meet all the expectations: a print from before 1914 would most likely be on Pathé stock; that of 1915-1920 on Eastman Kodak (though, of course, there could have been remains of old supplies), and in the 1920s one could easily find Pathé, Gevaert, Agfa, and DuPont, occasionally running across Kodak (mostly in imported U.S. films). A good example is *Tsvety zapozdalye* [Belated Flowers]

(1916) – an extremely rare case when two (albeit incomplete) nitrate prints of an early Russian film survive. One of them is on Agfa stock, the other on Eastman Kodak. That alone should help us to determine that the latter is an original print, whereas the former is a Soviet re-release (this is also confirmed both by the orthography and the content of the intertitles).

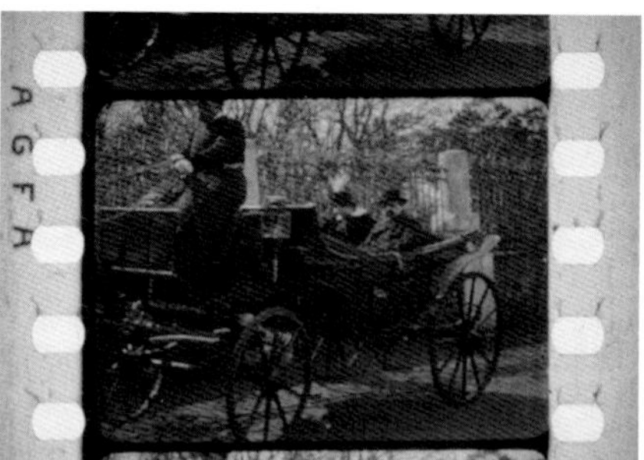

Tsvety zapozdalye [Belated Flowers] (Boris Sushkevich, 1916).
Soviet re-release print – GF

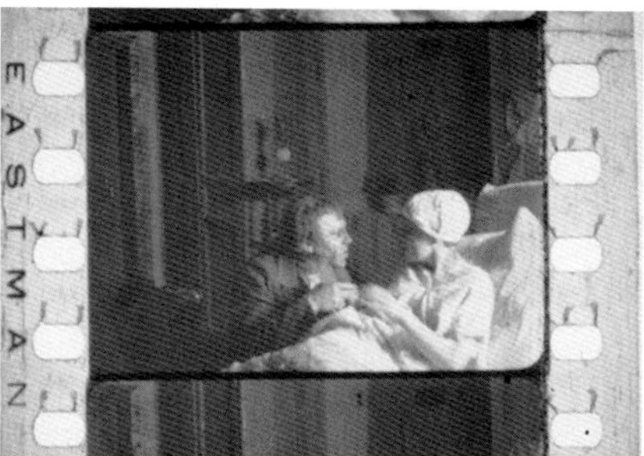

Tsvety zapozdalye [Belated Flowers] (Boris Sushkevich, 1916).
Original release print – GF

The history of Russian stock begins only in the Soviet period.

"The appeal for the USSR's independence from foreign countries in economic construction has to be reflected in the field of cinema as well. The Soviet film industry has to get increasingly emancipated from the foreign market, it has to launch production of film stock, equipment, chemicals,"[1] states the resolution of the First All-Union Party Film Conference held 15-21 March 1928. Production of original Soviet film stock was considered an important part of Industrialization, the main state policy of 1928-1941.

But the first serious step was made back in 1925, when the Commission for the Organization of Photo-Chemical Production at the Factories of the All-State Council of the National Economy was founded.[2] By that time several factories had already been experimenting with the production of negative film stock.[3]

The construction of the first Soviet film stock factory was launched the following year, in 1926, in the little town of Pereslavl-Zalessky. These were still the days of the NEP (New Economic Policy) — one of the most liberal eras of Soviet history, when the private market was re-animated for a short period (alongside the public/ nationalized one) — and this mixed economy led to active foreign investments, albeit short-lived. But since the goal of the Soviets was to avoid dependence on foreign markets, none of the leading manufacturers, such as Eastman Kodak or Agfa, were engaged. The Chemical Warfare Trust (which for some reason was in charge of film stock manufacture) had close connections to France, a country with vast experience in the production of chemical weapons and gas masks. Therefore, the commission for the construction of the film stock factory in Pereslavl-Zalessky was entrusted to the French company S.I.M.P. (Société Industrielle des Matières Plastiques), which was created specifically for that purpose. S.I.M.P. was supposed to reconstruct one of the factories of the state Rubber Trust into a film stock plant, but, according to Soviet sources, it ended up producing plastic

1 *Puti kino: Pervoe vsesoiuznoe partiinoe soveschanie po kinematografii* [Ways of Cinema: First All-Union Party Film Conference] ([Moscow:] Tea-kino-pechat', 1929), p.444.
2 Evsei Goldovsky, *30 let sovetskoi kinotekhniki* [30 Years of Soviet Film Technology] (Moscow: Goskinoizdat, 1950), p.64.
3 "Pervaia russkaia plionka" ["First Russian Stock"], *Kino* (Moscow), No. 11 [91] (2 June 1925), p.2.

buttons instead.[4] In reality, though, it seems that the French had basically completed their reconstruction of the factory, but were frozen out shortly before the work was done.[5] In September 1930 the factory was nationalized, and began operating on 1 July 1931 under the name "Factory No. 5" of the Chemical Warfare Trust.

A second factory ("Factory No. 6") opened shortly afterwards, on 1 October 1931, in Shostka, a small town in Ukraine (even though it took several years to get it up and running, the first batches of film stock were released there as early as October 1929, which makes it a pioneer).[6] Two more film plants went into service in 1935: one in Kazan ("Factory No. 8"), the other in Leningrad ("Factory No. 9").

The construction of the Shostka factory, which began in 1928, was also supervised by the French. This time a contract was signed with the Lumière factory in Lyon. Lumière was supposed not only to build the factory but to provide technical support for 10 years, both in Shostka and at the French facilities, and train Soviet engineers and technicians.[7] But, just as in the case of S.I.M.P., this collaboration lasted for only a few years: in early 1931 the work was taken over by Soviet technicians.[8]

The factories in Kazan and Leningrad were modeled after the first two and were managed without direct foreign input.

Factory No. 9 in Leningrad specialized in X-ray film only (it was a modification of an older X-ray film workshop that was established back in 1931) and was closed on 17 January 1939. But the other three factories functioned actively until the end of the nitrate era (which in the USSR lasted all the way up to the late 1960s–early 1970s) and beyond.

Eventually the Pereslavl-Zalessky factory started focusing on special sorts of film – for radiographic testing, aerial photography, X-rays, etc. – but kept producing traditional 35mm as well. In 1964 Factory No. 5 was reorganized into Pereslavskii khimicheskii zavod ("Pereslavl Chemical Factory"), which produced all sorts of material. The last batch of film stock was manufactured there in 1973. Consequently, factories in Shostka and Kazan catered to the whole Soviet film industry until the last days of the USSR. In December 1943 Factory No. 6 changed its name to Factory No. 3, and later, in October 1959, to Shostkinskii khimzavod ("Shostka Chemical Factory"), and finally, in 1964, to SVEMA (an abbreviation for "Svetochuvstvitel'nye materialy", which translates to "Photosensitive Materials"). Factory No. 8 was renamed Khimicheskii zavod imeni V.V. Kuibysheva ("V.V. Kuibyshev Chemical Factory") in 1958, and in 1974 it became TASMA ("Tatarskie svetochuvstvitel'nye materialy" / "Tatarian Photosensitive Materials").

Induktsionnye pechi [Induction Furnaces] (1932) was most likely the first Soviet motion picture filmed on domestic stock. A feature-length training film, it was directed by Gino De Marchi, an Italian Communist who immigrated to the USSR. As an experiment, at first it was decided to create two negatives (this happened for the first time in Soviet cinema after the production of blockbusters in the late 1920s): one on domestic stock, the other on Agfa.[9] But the last half of this work was done entirely on Soviet stock, due to the fact that "the results turned out to be brilliant for the young Soviet film stock; its sensitivity [was] one and a half times higher than that of the foreign one, it [was] highly orthochromatic, the coating [was] smooth."[10]

In the early 1930s training films and documentaries were considered perhaps even more politically significant than features. De Marchi and some of his colleagues from the non-fiction field urged others to follow their example – namely Dmitrii Surenskii, Leonid Kosmatov, and Vasilii Pronin, three promising young cameramen.

However, fiction filmmakers were much less enthusiastic. On New Year's Eve, 31 December 1933, filmmakers from Leningrad, led by Grigorii Kozintsev, released a *stengazeta* ("wall newspaper" – such newspapers posted on bulletin boards were common in all the Soviet institutions). But this one was a parody of all the clichés of the genre. Among other things, it contained the following announcement:

---

4 Vladimir Fefer and Iu. Konovalov, *Rozhdenie sovetskoi plionki* [The Birth of Soviet Film Stock] (Mosow: Gizlegprom, 1932). pp.7-10.
5 Yevgenii Zhirnov, "Tseluloidnoe iskusstvo" ["Celluloid Art"], *Kommersant Den'gi* [Kommersant Money], No. 37 (2004), p.122.
6 Pavel Kozlov and Boris Tolstoguzov, "Kinoplionochnaia promyshlennost'" ["Film Stock Industry"], *Kino-foto-khim promyshlennost'* [Cinema-Photo-Chemical Industry], No. 11-12 (1939), p.42.
7 Efaim Lemberg, *Kinopromyshlennost' SSSR: Ekonomika sovetskoi kinematografii* [Film Industry in the USSR: Economics of Soviet Cinema] ([Moscow:] Teakinopechat', 1930), pp.156-157.
8 Kozlov and Tolstoguzov, "Kinoplionochnaia promyshlennost'" ["Film Stock Industry"], p.42.
9 "Tol'ko na sovetskoi plionke" ["Only on Soviet Stock"], *Kino* (Moscow), No. 3 [474] (15 January 1932), p.4.
10 [Gino] De Marchi, "Ispytanie vyderzhano" ["The Test Has Been Passed"], *Kino* (Moscow), No. 8 [479] (16 February 1932), p.4.

"For the attention of comrade cameramen.
'Emulsion-123', the best film stock in the Soviet Union, has the following essential qualities:
1) a variety of emulsions in one can!
2) a smaller amount of metres than indicated on the label – which is particularly important for the economy effort.
3) does not require costly and complicated shooting in foul weather (rain, hail, fog, etc.).
Due to limited supplies, hurry to stock up for a year.
For every four sprocket holes, bulk customers get one extra hole punched for free.
A novelty! We have received new samples of film stock: (a) without emulsion, (b) dactyloscopic (with fingerprints), (c) contaminated with sewage."[11]

In October 1932 Aleksei Kaliuzhnyi, the founding father of the Ukrainian school of cinematography, who had recently been invited to work in Moscow, published an article entitled "Soviet Stock and Its Defects" in the leading Soviet film periodical *Proletarskoe kino* [Proletarian Cinema]. Among the standard characteristics of domestic stock, Kaliuzhnyi pointed out mechanical defects (caused by scratches and electrical discharges), multiple splices within a standard 300-metre reel of raw stock (most of these were clumsily made and contained traces of glue), inaccurate perforating (sprockets were often misplaced or skipped), uneven slitting of the film (34.6 or 35.2mm, instead of the standard 34.9), and insufficient strength of the emulsion, which would easily peel off. On top of all this, negative and positive stocks had different contrast ranges, which led to the overall gray look of Soviet prints.[12] New testing standards were to be developped by a commission, which included Kaliuzhnyi himself and Andrei Moskvin, arguably the best Russian cameraman of his time. Most of these experiments were conducted at NIKFI (the Cinema and Photo Research Institute) in Moscow, which was founded in December 1929,and for the first two years was entirely focused on the issues of film stock (later it started dealing with a whole range of topics, such as projection, sound recording, optics, etc.).[13]

By 1933 the Soviets stopped releasing new prints on foreign film stock.[14] Or rather, claimed to have done so. Film historian Jamie Miller writes that in April 1935 Boris Shumyatsky, head of the Chief Directorate of the Film and Photo Industry (GUKF), "informed Stalin that film stock was being produced at approximately two times below demand, but also at roughly two times more than was being imported from abroad. This indicated that film stock imports were still very significant".[15] On the other hand, the numbers published in the trade press do look impressive. If in 1930 46,000,000 metres were imported and only 588,000 metres produced, by 1936 the situation had reversed: 1,900,000 metres were imported and 115,029,000 metres were produced.[16]

In 1938 released positive film stock totalled 112,003,000 metres.[17] (If we take 2,200 metres as the average length of a 1938 Soviet film, this would be enough for about 50,000 prints.)

Replacing foreign negative film stock took a long time, but by 1938 the transfer to domestic negative was practically completed, and most of the Soviet pictures from then onwards were shot on domestic stock.[18] Percentage-wise, in 1932 only 4.5% of total Soviet-produced film stock was negative; by 1937 it had reached 18%.[19]

Ostensibly, by 1937-1938 there was no more foreign film stock imported. But, since quite a number of camera negatives from the early 1940s turn out to be filmed on Agfa, the situation might have changed *sub rosa*. The Soviet annexation of Eastern Poland in September-October 1939 provided, among many other things, a large supply of films. Much has been written about prints that were never legally acquired by the USSR, but ended up being screened in Moscow and Leningrad. But there must have been a certain amount of raw stock as well.

---

11 Iakov Butovskii (ed.), "Smeshno o kino. Satiricheskii vzgliad na sovetskoe kino, god 1933-i" ["About Cinema with Humor. A Satirical Perspective on Soviet Cinema, the Year 1933"], *Kinovedcheskie zapiski*, No. 40 (1998), p.207.
12 Aleksei Kaliuzhnyi, "Sovetskaia plionka i eio nedostatki" ["Soviet Stock and Its Defects"], *Proletarskoe kino* [Proletarian Cinema], No. 17-18 (1932), pp.46-51.
13 Goldovsky, *30 let sovetskoi kinotekhniki* [30 Years of Soviet Film Technology], p.92.
14 Goldovsky, *30 let sovetskoi kinotekhniki* [30 Years of Soviet Film Technology], p.92.
15 Jamie Miller, *Soviet Cinema: Politics and Persuasion under Stalin* (London and New York: I.B. Tauris & Co. Ltd., 2010), p.31.
16 Pavel Kozlov and Boris Tolstoguzov "Sovetskaia kinoplionochnaia promyshlennost' k dvadtsatoi godovschine" ["The Soviet Film Stock Industry at Its 20th Anniversary"], *Kino-foto-khim promyshlennost'* [Cinema-Photo-Chemical Industry], No. 10 (1937), p.21.
17 Veniamin Vishnevskii and Vladimir Fefer, *Ezhegodnik sovetskoi kinematografii za 1938 god* [Soviet Cinema Year Book 1938] (Moscow: Goskinoizdat, 1938), p.286.
18 Goldovsky, *30 let sovetskoi kinotekhniki* [30 Years of Soviet Film Technology], p.107.
19 Vishnevskii and Fefer, *Ezhegodnik sovetskoi kinematografii za 1938 god* [Soviet Cinema Year Book 1938], p.285.

By 1934, 20 kinds of Soviet film stock were known to exist (including film for duplicate negatives, sound recording, and panchromatic negatives).[20] By 1941 this number had doubled – due to a whole set of special emulsions (such as fine grain for rear projection).[21] According to Soviet sources, in 1937 the USSR became the world's third-largest manufacturer of film stock.[22]

Colour was still a serious issue. And even though both negative and positive stocks for dye-transfer were produced domestically, most Soviet colour films (and there weren't too many) were still filmed and released on Agfa or DuPont.

On 23 March 1938, when a new governing body for cinema, the Committee for Cinema Affairs of the USSR Council of the People's Commissars, was founded, a special branch of it was dedicated to the manufacturing of film stock: Glavnoe upravlenie kinoplionochnoi promyshlennosti [Main Directorate of the Film Stock Industry].[23]

By the late 1930s (it is hard to tell when exactly) the film stock manufacturers came up with a name for their brand. From then on Soviet positive film was called Soiuz ("Union"). There was also a whole range of negative film stocks: "Ortokhrom", "Izopanhkrom", "Infrakhrom", "SChS-1", "SChS-2", "SChS-4", and "SChS-5", as well as stocks for the two Soviet sound systems: "ZT" for Aleksandr Shorin's "transversal" (variable area) and ZI for Pavel Tager's "intensive" (variable density) soundtracks.[24] Multi-layer colour stock was introduced in 1948.[25]

There were several attempts at creating a narrow-gauge format. One of the first experiments was by Nikolai Kosmatov. In 1931 he proposed to divide a 35mm frame into 4 equal parts, measuring 9mm x 12mm each. Thus, every 35mm reel in fact contained 4 reels of film, which could be demonstrated by projecting the reel in two opposite directions. This format was considered time-saving, because the reel didn't need to be rewound.[26] In 1931 the majority of Soviet films were still silent. So a necessary modification, due to the advent of sound, was introduced in 1939. This time a 35mm frame was divided into two, with 2 soundtracks arranged on opposite sides of the film[27]. Neither of these models were ever released commercially, though.

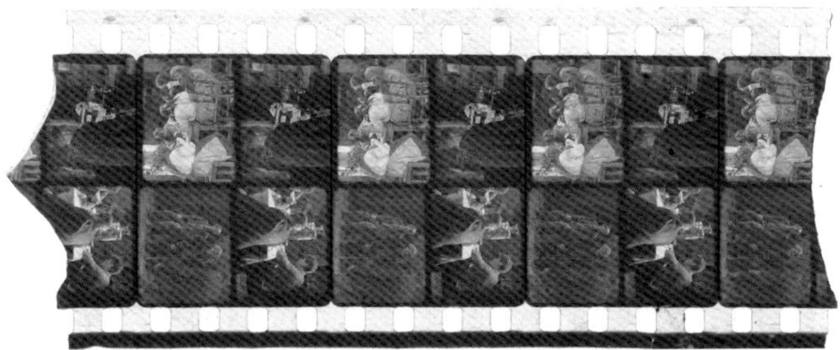

Unidentified experimental film (Nikolai Kosmatov, ca. 1931) – GTsMK

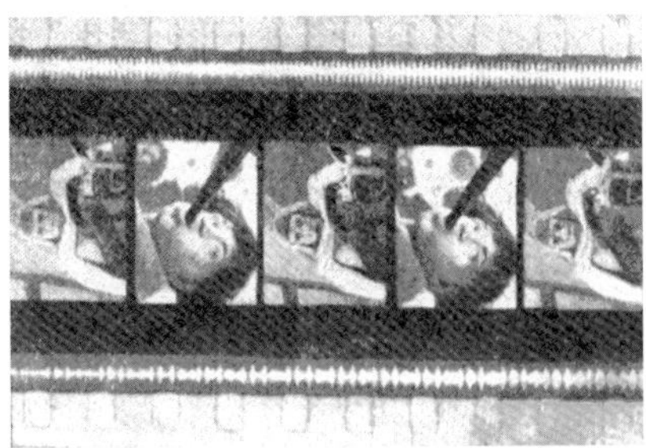

Aleksandr Nevskii [Alexander Nevsky] (Sergei Eisenstein, 1938).
An experimental 35mm format, bearing twice the usual number of images – *Kino-foto-khim promyshlennost* [Cinema-Photo-Chemical Industry], No. 6 (1939), p.20.

---

20 Goldovsky, *30 let sovetskoi kinotekhniki* [30 Years of Soviet Film Technology], p.92.
21 Goldovsky, *30 let sovetskoi kinotekhniki*, p.107.
22 Kozlov and Tolstoguzov, "Kinoplionochnaia promyshlennost'" ["The Film Stock Industry"], p.42; "Kinematografiia za 20 let suschestvovaniia (Tsifry i fakty)" ["Filmmaking in the 20 Years of Its Existence (Numbers and Facts)"], *Kino-foto-khim promyshlennost'* [Cinema-Photo-Chemical Industry], No. 11-12 (1939), p.8.
23 Vishnevskii and Fefer, *Ezhegodnik sovetskoi kinematografii za 1938 god* [Soviet Cinema Year Book 1938], p.285.
24 Nikolai Anoschenko, *Obschaia kinotekhnika* [General Film Technology] (Moscow: Goskinoizdat, 1940), pp.93-94.
25 Goldovsky, *30 let sovetskoi kinotekhniki* [30 Years of Soviet Film Technology], p.138.
26 Evsei Goldovsky, *Uzkaia kinoplionka* [Narrow-Gauge Film Stock] (Moscow: Goskinoizdat, 1944), p.5.
27 A. Gert, "Ob udvoenii chisla kadrov na 35-millimetrovoi fil'mokopii" ["On the Doubling of the Number of Frames on the 35mm Film Print"], *Sovetskaia kino-photo-khim promyshlennost'* [Soviet Cinema-Photo-Chemical Industry], No. 6 (1939), p.20.

The production of non-inflammable acetate film stock was launched almost simultaneously with nitrate. It was developed in 1930-1934 at NIKFI by engineers Pavel Kozlov and N. Kuzminskii. But pure acetate base was considered costly and deficient, so one of the goals of Soviet technicians was to introduce a sort of non-inflammable stock with as much nitrate cellulose as permissible. After years of experimenting, it was decided that film can remain non-inflammable if it contains up to 20% of nitrate cellulose, providing the composite materials are properly processed.[28] Regular triacetate stock was introduced only after the Second World War.[29] But all this was done on 16mm; 35mm remained nitrate until the mid-1950s.

The first batch of Soviet 16mm positive stock was released in late 1933.[30]

16mm projectors appeared the previous year, and by 1934 as many as 4,605 were released.[31] This opened a market for 16mm prints. The mass production of 16mm was launched in 1934, and by 1 January 1935 the state distribution office "Soiuzekran" could offer a choice of 86 titles and 1,069 prints: 26 features, 34 documentaries, 22 children's films, and 4 animation films.[32] 1938 was considered a turning point in the development of 16mm film in the USRR. That year more than 37,000 16mm projectors were manufactured by the Odessa film equipment factory, and a large portion of the blockbusters of 1934-1937 were released as 16mm prints.[33] Let alone that since the mid-1930s the majority of educational films were distributed exclusively on 16mm.[34] All types and genres of film were covered by 16mm. This is an important thing to keep in mind, considering that until recently not a single Soviet 16mm print from the 1930s was known to exist in Russian archives. (Only in 2018 were about half-a-dozen culture films and educational films from the 1930s donated to Gosfilmofond as part of the collection of Rudolf Kots.)

16mm negative stock was introduced in 1934. There is evidence that it was first manufactured on nitrate base – a unique case for this gauge.[35] But this must have proven to be unsustainable, because by 1939 (or maybe even earlier) there was still virtually no 16mm negative stock production.[36] Until the late 1940s even those films that were intended to be released exclusively on 16mm were struck from standard 35mm negatives.[37]

Russia never produced 17.5mm or 9.5mm. 8mm was introduced only in the late 1950s.

As a matter of curiosity, I should mention an amateur camera called "Pioner" which filmed on 17.5mm stock that had to be prepared manually by slitting 35mm in two.[38] The information about this camera and format was published in June 1941, right before the outbreak of war in the Soviet Union, and the invention was soon forgotten.

---

28 Goldovsky, *30 let sovetskoi kinotekhniki* [30 Years of Soviet Film Technology], pp.92-93.
29 Goldovsky, *30 let sovetskoi kinotekhniki* [30 Years of Soviet Film Technology], p.138.
30 Evsei Goldovsky, *Uzkoplionochnaia kinematografiia* [Narow-Gauge Cinematography] (Moscow: Kinofotoizdat, 1935), p.184.
31 Goldovsky, *Uzkoplionochnaia kinematografiia* [Narrow-Gauge Cinematography], p.204.
32 Goldovsky, *Uzkoplionochnaia kinematografiia* [Narrow-Gauge Cinematography], p.204.
33 Evsei Iofis, *Tekhnologiia obrabotki kinoplionki* [Technology of Film Porcessing] (Moscow: Goskinoizdat, 1939), p.220.
34 Evsei Goldovsky, *Izgotovlenie uzkikh kinofilmov* [Production of Narrow-Gauge Films] (Moscow: Goskinoizdat, 1945), p.27.
35 Goldovsky, *Uzkoplionochnaia kinematografiia* [Narrow-Gauge Cinematography], p.188.
36 Iofis, *Tekhnologiia obrabotki kinoplionki* [Technology of Film Processing], p.221.
37 Goldovsky, *Izgotovlenie uzkikh kinofilmov* [Production of Narrow-Gauge Films], p.27.
38 Teodor Bunimovich, "Kinoapparat 'Pioner'" [" 'Pioner' Film Camera"], *Tekhnika molodiozhi* [Engineering for Youth], No. 6 (1941), p.60.

2. Identification of Soviet Nitrate Film Stock

From 1933 to 1941, Factory No. 8 in Kazan released a newspaper, *Sovetskaia kinoplionka* [Soviet Film Stock] on a weekly (sometimes 5-day) basis. This might sound like an excellent source for film-stock historians. But, browsing through the issues (a nearly complete collection exists at the Russian National Library in St. Petersburg), one may find articles on a whole variety of social, disciplinary, and political issues – and very little about film itself. At least, nothing that would help in film identification: there isn't a word about perforation shapes, edge codes, etc.

The vast literature on film stock doesn't provide much help either.

One of the reasons is that throughout the 1930s there were no standards or norms for film stock in the Soviet Union. As late as 1937, technicians were complaining that "the applied norms amount to a patchwork from old German standards that have already been abolished and American ones from 1934 that are either abolished as well or have since been revised."[39] At last, on 27 August 1939 the Committee for Cinema Affairs of the USSR Council of the People's Commissars introduced a national standard for film stock: OST-KINO 1.

Perforations

The document OST-KINO 1 ratified two perforation formats. (1) *Standard* perforation (rectangular with slightly rounded corners, height 1.98mm, width 2.8mm) was, in fact, the Kodak Standard (KS) shape. (2) *Special* perforation (round sides with straight top and bottom edges, diameter 2.8mm, height 1.85mm, width 2.8mm) matched the Bell & Howell (BH) format.

According to OST-KINO 1, from now on "mass production of all sorts of 35mm film [was] to be carried out with Standard perforation. The use of Special perforation [was] temporarily allowed until 1 October 1940 upon special requests, and in each case with the approval of the Committee."[40]

Indeed, all the pre-1940 Soviet camera negatives which I have had the opportunity to examine have BH perforations, and all post-1940 ones have KS.

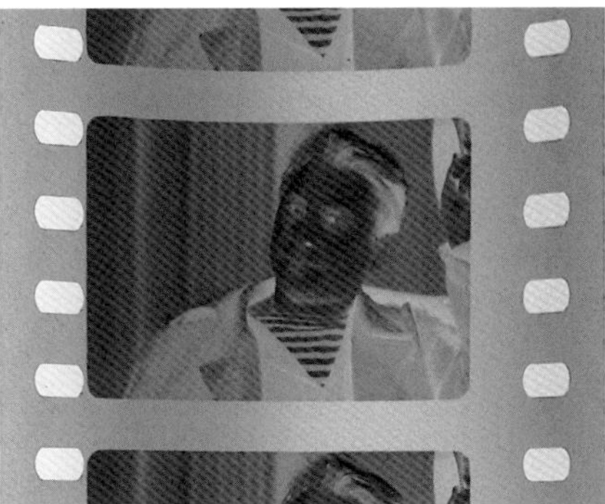

Velikii grazhdanin [The Great Citizen], Part 1 (Fridrikh Ermler, 1937).
Camera negative. Film stock SChS-1 with BH perforations – GF

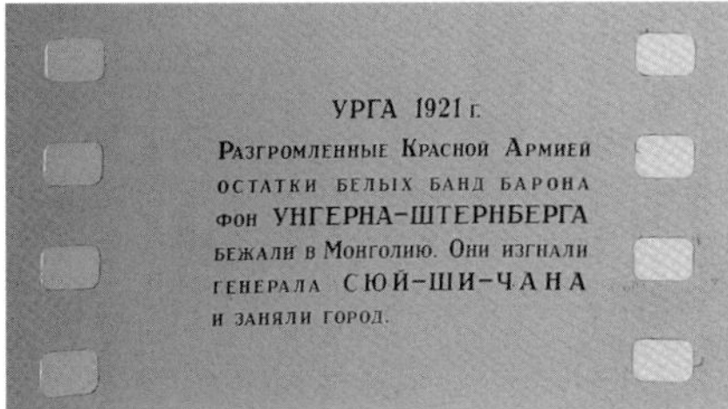

Ego zovut Sukhe-Bator [His Name Is Sukhe-Bator] (Aleksandr Zarkhi and Iosif Kheifits, 1942).
Camera negative with KS perforations – GF

39 Vsevolod Tolmachiov, "Novye mezhdunarodnye kino-standarty" ["New International Film Standards"], *Kino-foto-khim promyshlennost'* [Cinema-Photo-Chemical Industry], No. 9 (1937), p.57.
40 Anoschenko, *Obschaia kinotekhnika* [General Film Technology], p.70.

The aforementioned "special requests" might have included technical experiments – such as colour (until the post-World War II years the production and distribution of colour films in the USSR was still very limited, and practically each of them was considered experimental). Half of the existing Soviet colour prints from the 1930s to the early 1940s have BH perforations, including *Koniok-Gorbunok* [The Little Humpbacked Horse], which was completed and released only in 1941.

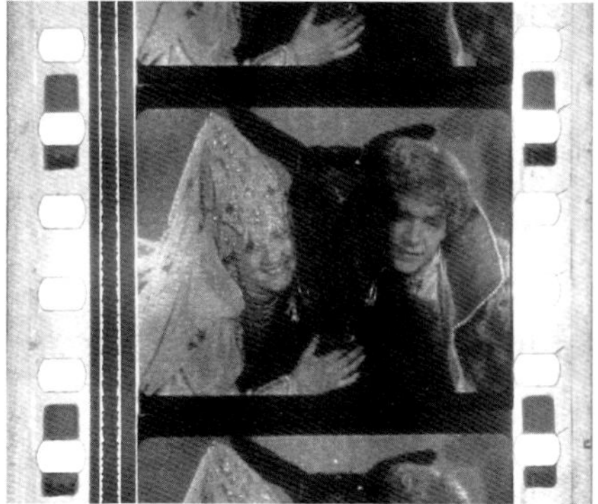

Koniok-Gorbunok [The Little Humpbacked Horse] (Aleksandr Rou, 1941). Colour print – BFI

Reel Lengths

Another requirement of OST-KINO 1 which is worth mentioning deals with the standard reel length. Negative 35mm film had to be released in reels of 30, 120, or 300 metres; other types of 35mm film were released in reels of 300 metres.[41] This indeed was the rule until the dissolution of the Soviet Union and even later. All Soviet 35mm prints were released in 300-metre reels; 600-metre reels were introduced in Russia only in the 21st century. (16mm prints did exist in larger reels.)

Film Stock and Edge Marks / Edge Codes

The question of edge marks is much more complicated. It seems that until the late 1940s Russian film stock had no edge marks whatsoever. In fact, at the Russian archives the lack of an edge code on a Soviet nitrate film has always been one of the main indicators of Soiuz film stock, whether negative or positive.

In spite of endless pathetic speeches and public affirmations, Soviet filmmakers were loath to work with domestic negatives, and preferred to cobble together any foreign film stock that was at hand. An interesting case study is *Chapaev* (1934), the number-one Soviet blockbuster of the 1930s. It is known that one of the reasons for the creation of Gosfilmofond was Stalin's wish to safeguard the negative of *Chapaev*. This negative still exists. It is made up of bits and pieces of Gevaert Belgium, Agfa, Kodak, and Soiuz stock. The filming and grading of such a blend must have been quite a challenge.

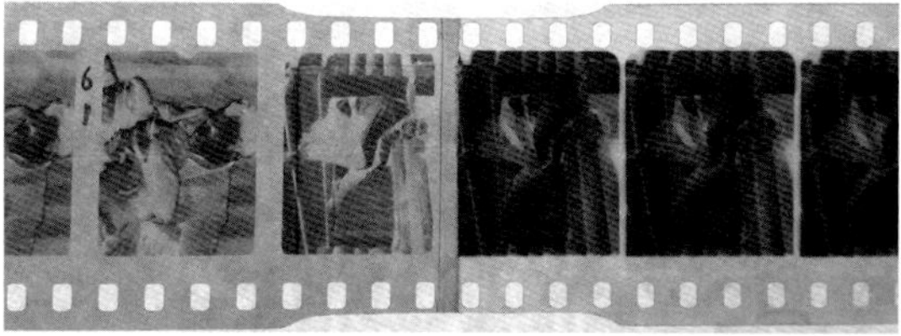

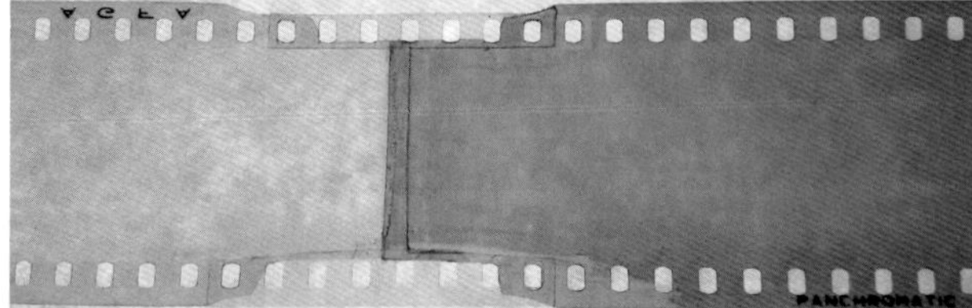

Chapaev (The Vasiliev Brothers, 1934). Camera negative – GF

41 Anoschenko, *Obschaia kinotekhnika* [General Film Technology], p.70.

Yet, several Soviet films from the 1930s are definitely known to have been filmed entirely on Soiuz stock. Among these are *Velikii uteshitel'* [The Great Consoler] (1933),[42] *Velikii grazhdanin* [The Great Citizen] (1937-39), and *Noch' v sentiabre* [A Night in September] (1939).[43] All three negatives are now at Gosfilmofond; I have inspected them, and found no edge codes.

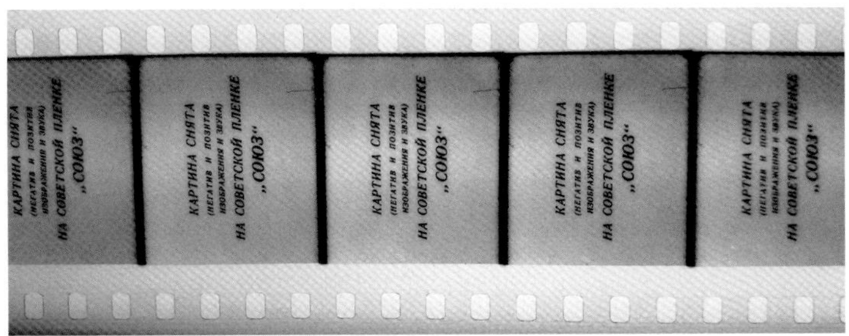

Velikii uteshitel' [The Great Consoler] (Lev Kuleshov, 1933). Camera negative.
The title reads: "This picture was filmed (negative and positive of image and sound) on the Soviet stock Soiuz." – GF

On all the Soviet nitrate prints from the period 1931 to 1949 that I have inspected – which total about 50 – there were no edge codes. Occasionally one might find squares, circles, or semi-circles that bear a slight resemblance to Kodak edge marks, but in reality all of these must be laboratory quality-control or printer marks.

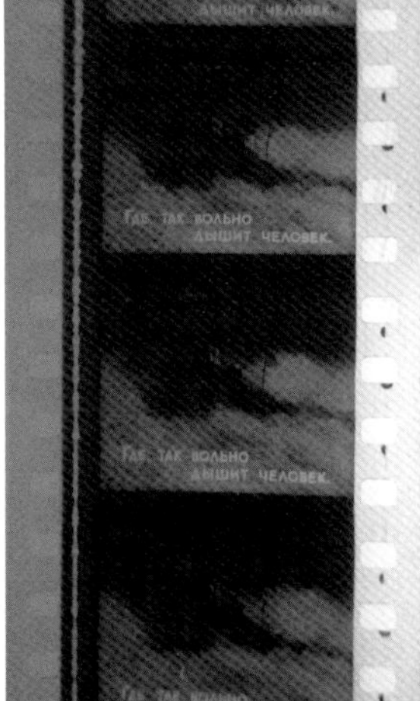

Pesnya o Rodine [Song about the Motherland] (1937) – RGAKFD

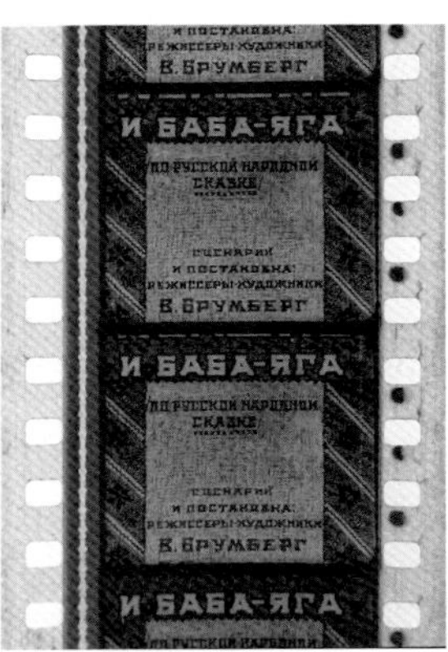

Ivashko i Baba Yaga [Ivashko and Baba Yaga] (Valentina and Zinaida Brumberg, 1938) – BA

42 Among other sources, the main evidence is a special title clearly mentioning the film stock (Soiuz) in the opening credits of the film itself (see *Velikii uteshitel'*, on this page).

43 Kozlov and Tolstoguzov, "Kinoplionochnaia promyshlennost'" ["Film Stock Industry"], p.43.

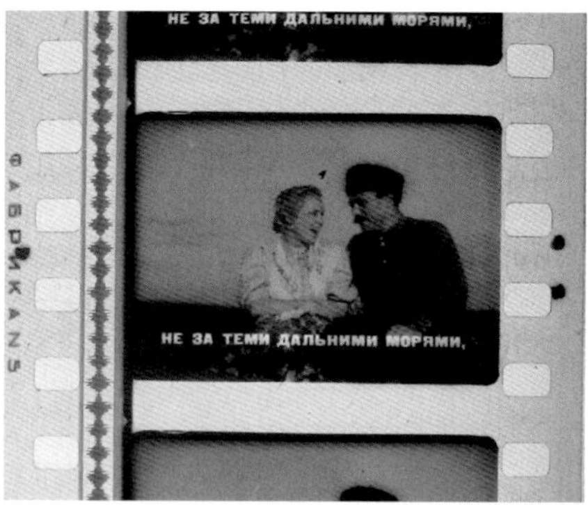

Dinamicheskii subtitr [Dynamic Subtitle] (Vladimir Kasianov, ca. 1952): an experimental reel using footage from *Kubanskie kazaki* [The Cossacks of Kuban] (Ivan Pyriev, 1949) – GF

However, I did run across a single strip of film, from *Vysokaia nagrada* [High Reward] (1939), with the words "ПОЗИТИВ НИТРО" ["pozitiv nitro" / "nitrate positive"] printed along the edge. This being the only example known, it is hard to say how common such an inscription was.

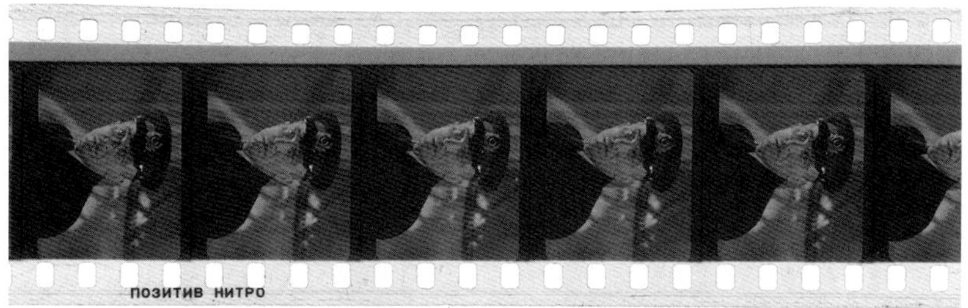

Vysokaia nagrada [High Reward] (Evgenii Shneider, 1939). The edge code reads "nitrate positive". – GTsMK

In any case, there seems to be no system.

The number of surviving acetate 16mm prints from the 1930s and 1940s is absolutely negligible, but none of the existing copies have edge codes either.

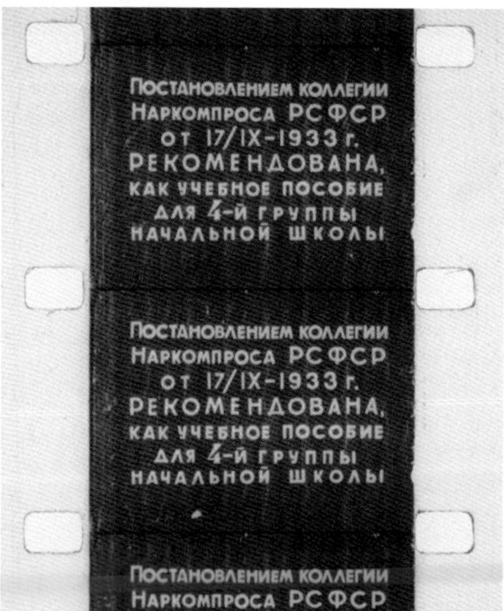

Kak ustroeno i rabotaet telo cheloveka [How a Human Body Is Structured and Functions] (Noi Galkin, 1933). This is one of the first Soviet 16mm prints (diacetate). – GF

The system finally appeared around 1950. This might have to do with the introduction of a new national standard for film stock, replacing OST-KINO 1. GOST 4896-49 laid down the dimensions and control methods for 35mm stock, and GOST 4897-49 and GOST 4898-49 did the same for 16mm, both regular and double-sized (there were no established Soviet standards for small-gauge film before this). Neither of these mentioned edge codes. But the new documents were approved by the All-Union Committee of Standards on 31 May 1949 and had to enter into service on 1 January 1950,[44] whereas the earliest print with an edge code I am aware of was released in 1950. There is probably a connection here. Though I have to mention that David Zolotnitskii's *Kontrol' protsessov obrabotki kinoplionki* [Methods of Film Processing Control], an essential book – basically an instruction manual – on film stock inspection published at the end of the year, doesn't say a single word about edge codes. According to Zolotnitskii, the name of the manufacturer had to be listed only on the label of the can containing raw stock.[45]

From 1950 onwards one could find distinctive edge codes: an indication of the film stock manufacturer, and often the year and month of its manufacture, e.g., "ФАБРИКА № 8. ИЮЛЬ 50 г. ПЕРФ 64", which means "Factory No. 8. July 1950. Perf 64" (the latter indicates the number of the perforating machine).

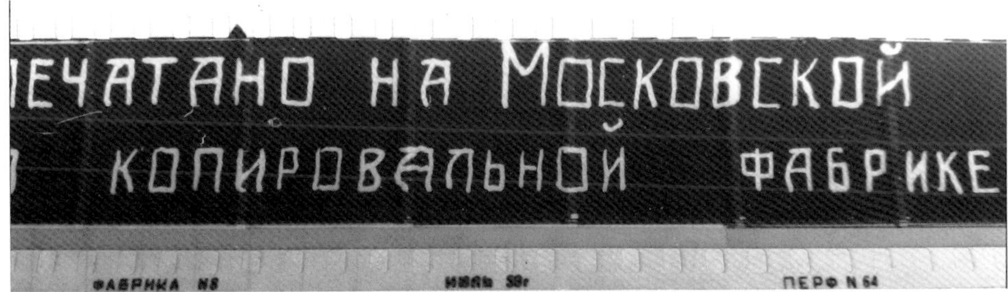

Deputat Baltiki [Baltic Deputy] (Aleksandr Zarkhi, Iosif Kheifits, 1936). A post-World War II re-release (Factory No. 8, July 1950, Perf 64) – GF

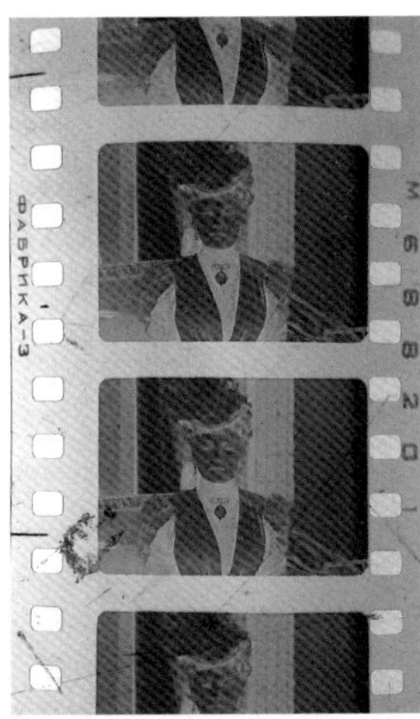

Vol'nitsa [Freemen] (Grigorii Roshal, 1955). Outtakes.
Camera negative (Factory No. 3) – GTsMK

Around 1952 or 1953 the style of the indication changed slightly: from then onwards, the word "ФАБРИКА" ["FABRIKA", "Factory"] was often abbreviated to "Ф-КА", or simply "Ф", and the month was indicated in Roman numerals (e.g., "Ф-ка № 5 XII-1953" or "Ф-8-XII-56"). The negatives in the 1950s contained only the name of the manufacturing factory, but not the date.

44 *Kinoplionka: Osnovnye razmery, metody proverki* [Film Stock: Main Dimensions, Control Methods] ([Moscow:] Standardgiz, 1949).
45 David Zolotnitskii, *Kontrol' protsessov obrabotki kinoplionki* [Methods of Film Processing Control] (Moscow: Goskinoizdat, 1950), p.229.

Vyborgskaia storona [The Vyborg Side] (Grigorii Kozintsev, Leonid Trauberg, 1938). Sound negative – most likely for a post-World War II re-release. (Factory No. 5) – GF

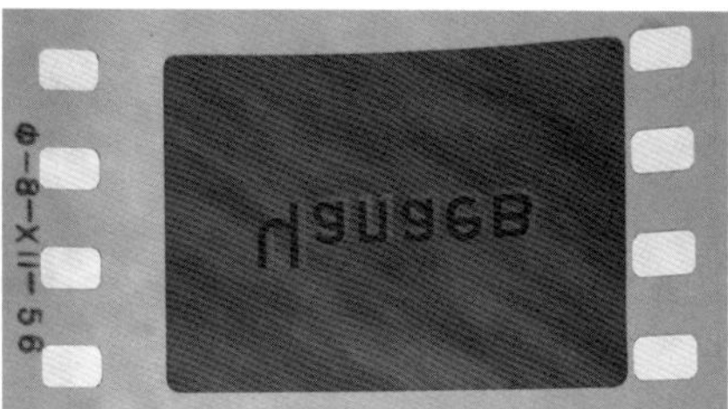

Chapaev (The Vasiliev Brothers, 1934). Negative of the new credits for a 1956 re-release (Factory No. 8, December 1956) – GF

I was not able to find the exact date of the first large output of safety 35mm film, but the earliest prints I am aware of were made in the second half of the 1950s. By this time everything had to be standardized and legitimized, and it wasn't long before a new Standard appeared. GOST 8449-57 ("Safety Film. Control Methods. Coding") was approved by the All-Union Committee of Standards on 3 May 1957, and had to enter into service on 1 July 1958.[46] This document finally states in writing that there should be an edge code, but we know nothing about it except that it must have contained the word "БЕЗОПАСНАЯ" ["Bezopasnaia" / "safety"]. Only the next two standards – GOST 11271-65, developed for 35mm anamorphic film (known abroad as "SovScope") and approved on 1 July 1965, and GOST 11272-65, developed for 70mm and approved on 3 June 1965 – provided detailed guidelines for edge codes. They had to contain the word "БЕЗОПАСНАЯ", or its first letter, "Б", the name or symbol of the manufacturing factory, the month of manufacture, and the number of the perforating machine. It was also specified that both of these widescreen formats had to be printed exclusively on safety stock.[47]

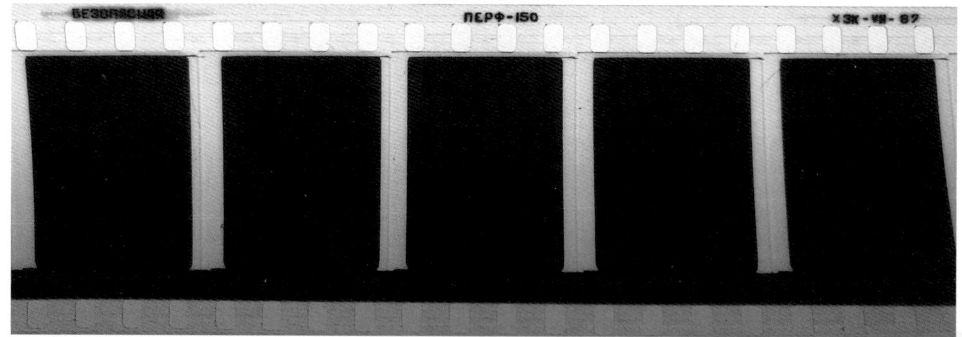

Putiovka v zhizn' [The Road to Life] (Nikolai Ekk, 1931). Triacetate print (KhZK, September 1967) – MoMA

Sometimes, instead of "БЕЗОПАСНАЯ", safety film would be marked with the letters "ТРИАЦЕТАТНАЯ" ["triatsetatnaia" / "triacetate"].

46 *Kinoplionka bezopasnaia: Metody ispytanii; Markirovka* [Safety Film Stock: Control Methods; Coding] ([Moscow: Standardgiz, 1957]).
47 *Kinoplionka 35-mm dlia shirokoekrannykh stereofonicheskikh kinofilmov: Osnovnye razmery, markirovka* [35mm Film Stock for Widescreen Stereophonic Motion Pictures: Main Dimensions, Coding] (Moscow: Izdatel'stvo standartov, 1977); *Kinoplionka 70-mm dlia shirokoformatnykh kinofilmov: Osnovnye razmery, markirovka* [70mm Film Stock for Widescreen Motion Pictures: Main Dimensions, Coding] (Moscow: Izdatel'stvo standartov, 1965).

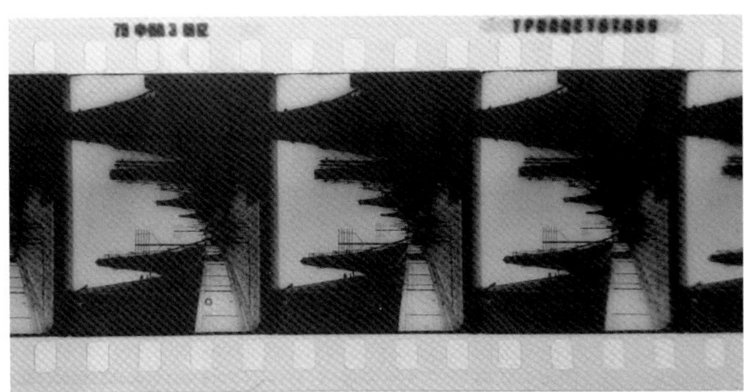

Prividenie, kotoroe ne vozvraschaetsia [The Ghost That Doesn't Return] (Abram Room, 1929).
Triacetate print (Factory No. 3, late 1950s) – GEM

But nitrate was still produced, and this "interregnum" lasted for nearly a decade. The latest Soviet nitrate print I came across was made in 1964; the latest nitrate master element I am aware of was made as late as 1971.

As Dmitrii Masurenkov and Mark Osepian, two cameramen who started their careers as cinematographers in the late 1950s, recall, the mass release of all the new films was supposed to be on triacetate, but filmmakers were eager to get nitrate prints. Early Soviet triacetate base wasn't as transparent as nitrate; it was somewhat dim, and thus led to an overall gray image quality. It lacked the "velvet colour" (Osepian's description) of the black & white nitrate. Striking triacetate prints of a nitrate negative was a special challenge both for the cameraman and the lab – it was extremely difficult to raise the level of contrast. That's why, while most of the copies were on safety film, several fine-quality nitrate prints were usually made for the premiere at the Moscow House of Cinema, for the officials at Goskino, and for private screenings at the *dachas* of the party elite. This went on for a number of years.

The words "БЕЗОПАСНАЯ", "ТРИАЦЕТАТНАЯ", and the letter "Б" disappeared in 1966 – by that time the general transition from nitrate to safety was over. But we have to watch out for those late nitrates: they do exist, if only in small numbers. As well as some of the triacetate prints released before 1957, which did not have respective edge marks – they were specifically mentioned in the "Instructions for the Safeguarding of Film and Photo Documents on Nitrate Stock in the State Archives of the USSR", which were developed in 1983 by the Chief Archival Directorate of the Council of Ministers of the Soviet Union.[48]

Later on, two more new standards for regular 35mm were introduced (in 1973 and 1980), but neither of these contained requirements for edge codes.

However, in Evsei Iofis's 1967 manual on film-stock processing, the bible for all Soviet film technicians for the next 2 to 3 decades, it is specified that all 35mm and 70mm stocks should have an edge code containing "for negatives – factory name, footage counter, number of the marking machine; for positives and duplicate negatives – factory name, number of perforating machine, release date (month and year); for safety film – the letter 'Б'; for positive and reversal film with magnetic soundtrack the standard codes are supplemented with the letter 'M'".[49] These rules were not followed strictly. For example, the month and year did not appear every time, and as far as I have found, there is no consistency.

It might be useful to know the edge code abbreviations of more recent times – some of them could be found on nitrate.

From 1959 to 1964, while the factory in Shostka was called Shostkinskii khimzavod ("Shostka Chemical Factory"), the film stock manufactured there was marked with the letters "ШХЗ" ("ShKhZ"), and from 1964 onwards the abbreviation "СВЕМА" ("SVEMA") was applied.

48 Instruktsiia po obespecheniiu sokhrannosti kinofotodokumentov na nitroosnove v gosudarstvennykh arkhivakh SSSR. Retrieved 26 August 2019, from http://www.gosthelp.ru/text/Instrukciyapoobespecheniy.html

49 Evsei Iofis, *Praktika obrabotki kinoplionki (dlia massovykh rabochikh professii)* [Film Stock Processing Practice (for Mass Service Jobs)] (Moscow: Iskusstvo, 1967), p.193.

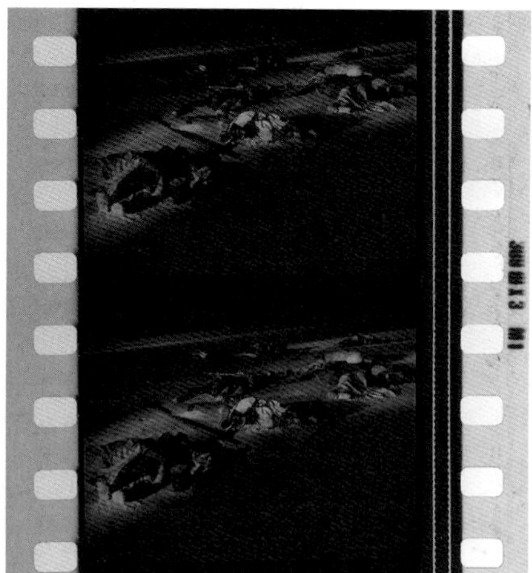

Aleksandr Nevskii [Alexander Nevsky] (Sergei Eisenstein, 1938). Nitrate print (ShKhZ, late 1950s–early 1960s) – ÖFM

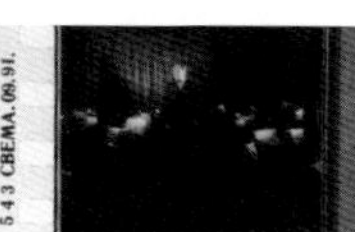

Pasifik-231 [Pacific-231] (Mikhail Tsekhanovskii, 1931). Acetate print (SVEMA, September 1991) – GEM

From 1958 to 1974, when the factory in Kazan was called Khimicheskii zavod imeni V.V. Kuibysheva ("V.V. Kuibyshev Chemical Factory"), the edge code letters for its production were "ХЗК" ("KhZK") (see *Putiovka v zhizn'*, p. 296); starting in 1974 all the Kazan stock was marked with the letters "ТАСМА" ("TASMA").

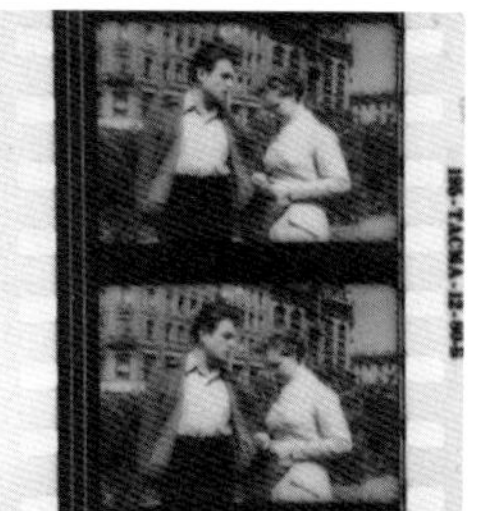

O moiom druge [About My Friend] (Iurii Erzinkian, 1959). Triacetate print (TASMA, December 1980) – GF

After the Pereslavl-Zalessky factory changed its name to Pereslavskii khimicheskii zavod ("Pereslavl Chemical Factory") in 1964, the letters "ПХЗ" appeared on its photography stock, but I have never seen them on motion picture elements.

A summary of the chronology of the factories and their marks follows.

List of Soviet Film Stock Manufacturers and Their Marks

1. The Factory in Pereslavl-Zalessky
    - From 1 July 1931: Fabrika No. 5 [Factory No. 5] (ФАБРИКА № 5)
    - From 1964: Pereslavskii khimicheskii zavod [Pereslavl Chemical Factory] (ПХЗ)

2. The Factory in Shostka
    - From 1 October 1931: Fabrika No. 6 [Factory No. 6] (ФАБРИКА № 6)
    - From December 1943: Fabrika No. 3 [Factory No. 3] (ФАБРИКА № 3)
    - From October 1959: Shostkinskii khimzavod [Shostka Chemical Factory] (ШХЗ)
    - From 1964: SVEMA (СВЕМА)

3. The Factory in Kazan
    - From 1935: Fabrika No. 8 [Factory No. 8] (ФАБРИКА № 8)

From 1958: Khimicheskii zavod imeni V.V. Kuibysheva [V.V. Kuibyshev Chemical Factory] (ХЗК)
From 1974: TASMA (ТАСМА)

4. The Factory in Leningrad (X-ray film)
From 1935 to 17 January 1939: Fabrika No. 9 [Factory No. 9] (ФАБРИКА № 9)

3. "Non-Physical" Characteristics

In order to be consistent with Harold Brown's work and to provide some true aid to film identification, rather than just a history of Soviet film stock, it makes sense to mention such crucial characteristics as trademarks, production serial numbers, and other attributes which do not really answer the definition of "physical".

We could start by eliminating producers' edge marks, embossed and punch marks, and trademarks in scenes: early Russian (let alone Soviet) cinema knew practically none of these.

We should make an obvious exception for Gaumont and Pathé, two French companies who not only distributed but also produced films in Russia before the Revolution (Pathé, in 1909-1914; Gaumont, in 1910-1914). *Kniazhna Tarakanova* [Princess Tarakanova] (1910) is one of the early surviving Pathé Russe productions, and the famous Pathé rooster does appear in some of the sets. The film was released in November 1910, and it is probably one of the latest examples of Pathé's trademark in the image. This was the end of the tradition, and there seems to be no system: the earliest surviving film produced by Pathé Russe, *Ukhar-kupets* [The Happy-go-lucky Merchant], was released more than a year earlier, in October 1909, and does not have any evident marks.

All the Russian Gaumont films seem to be lost, with the exception of *Zhizn i smert' A.S. Pushkina* [The Life and Death of Alexander Pushkin] (1910). This contains no trademarks in scenes, and it is now impossible to tell if the Gaumont logo was present anywhere in the sets of their other Russian films.

Production Serial Numbers and Censorship Numbers

Though almost never mentioned by film historians and (very unwisely) by filmographers, production serial numbers did exist in early Russian cinema.

For one thing, Pathé and Gaumont adopted the same practice in Russia as they did in the rest of the world. And although the serial numbers for Pathé productions were listed in the trade press, to my knowledge they were never put in the intertitles.

Gaumont was one of the leading distribution companies in Russia from 1905 until 1917, and foreign films released by Gaumont had a serial number placed in the main title and intertitles, as customary. We should assume that the same rule was applied to Gaumont's own Russian productions. But out of the 14 films that are known to have been produced by Gaumont in Russia, only one – the aforementioned *Zhizn i smert' A.S. Pushkina* (1910) – is known to exist, and there are no serial numbers in its intertitles.

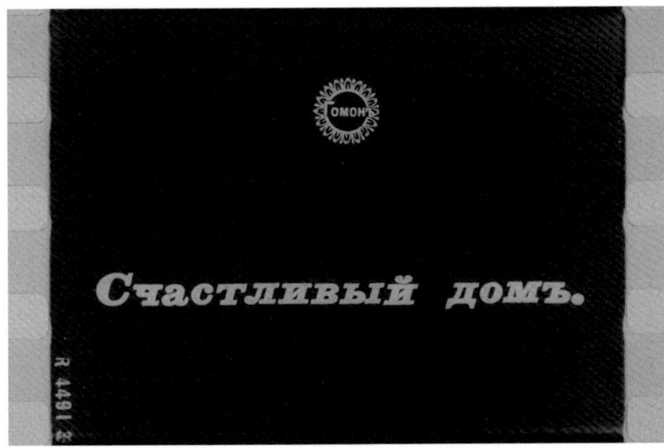

L'Aumône de l'amour (France, released in Russia in 1913).
Safety print of the Russian release version (serial number 4491 in the lower-left corner) – GF

Whether all the Russian-born companies had production serial numbers is hard to tell. Around 2500-2600 feature films were made in Russia between 1907 and 1919. Of these, 352 are known to exist in some form, but only 45 have original Russian intertitles. Judging by these, we may assume that only one company was constantly including serial numbers in its intertitles, that of Aleksandr Khanzhonkov. These are, in fact, not production numbers but censorship numbers. A major part of the archives of Russian censorship has been lost; there are no lists, and we do not know how these numbers were assigned. But, judging by the fact that titles released by different companies within the same year sometimes ended up having identical numbers, there must have been a separate numbering system for each of the companies. Most likely, each company assigned its own numbers before submitting films to the censorship committees. Which makes perfect sense, since there were different censorship authorities in each town, but a given picture would have the same serial number everywhere.

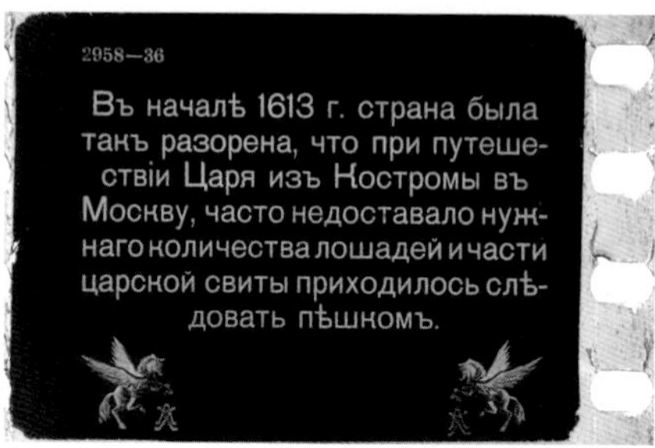

Votsarenie doma Romanovykh [Enthronement of the House of Romanov] (Vasilii Goncharov, Piotr Chardynin, 1913). Nitrate print (serial number 2958 in the upper-left corner) – GF

Aleksandr Khanzhonkov founded his distribution company in 1906, and began producing films in 1909. But the first time we find information about his films accompanied by a serial number (No. 630) is only in September 1910.[50] He stopped publishing serial numbers in the Summer of 1914 (the last number mentioned is 2774)[51] – which, one way or another, must have been connected to the outbreak of World War I.

Until someone compiles a definite filmography with all the serial numbers included, our main reference would be the trade press of the time, first and foremost the Moscow periodicals *Sine-fono* [Cine-Phono] (1907-1918), *Vestnik kinematografii* [Cinematograph Herald] (1910-1917), and *Kine-zhurnal* [Cinema-Journal] (1910-1917).

The serial number was always followed by the number of the intertitle, and, to my knowledge, was never provided with a letter (unlike Gaumont and other worldwide distributing countries). One must keep in mind, though, that the numbers refer not only to films produced by Khanzhonkov, but also to those foreign films which his company released in Russia.

It is natural to assume that if the Russian film industry was controlled by censorship, every single picture had to be assigned a censorship number. And the information found in the trade press confirms that. For example, Thiemann and Reinhardt, another leading production and distribution company, founded in March 1909, started publishing the censorship numbers of their films in mid-February of 1910,[52] and continued to do so until the October Revolution of 1917. However, the very few existing prints with original intertitles show no sign of serial numbers or censorship numbers.

Therefore, a serial number in a Russian intertitle most likely indicates a Khanzhonkov production. In most cases this can be confirmed by the presence of the company's trademark in the very same title. Which brings us to the next key tool of film identification.

50 "Novyia lenty" ["New Pictures"], *Sine-fono* [Cine-Phono], Year III, No. 24 (15 September 1910), pp.17-19.
51 "Novyia lenty" ["New Pictures"], *Sine-fono* [Cine-Phono], Year VII, No. 20 (5 July 1914), pp.68-70.
52 "Spisok novykh lent" ["A List of New Pictures"], *Sine-fono* [Cine-Phono], Year III, No. 10 (15 February 1910), p.22.

Trademarks in Intertitles

Before the nationalization of the film industry in 1919, intertitles in Russian films often contained a trademark. A limited amount of prints with original intertitles (let alone original nitrate prints) makes it difficult to talk about statistics today, and, in some cases, to trace the evolution of a trademark. However, the practice was so common that the lack of a trademark in an intertitle usually indicates that a print was either produced by one of the smaller companies or represents a later re-release.

Without going into too much detail and providing images of all the known trademarks of pre-revolutionary Russian film companies, I shall focus here on several major companies that definitely used trademarks in their intertitles.

Aleksandr Drankov, a producer whose career lasted from 1907 to 1918, started by furnishing each intertitle with his signature and the letters СПБ (SPB – St. Petersburg) – several existing original prints from 1908 have this trademark.

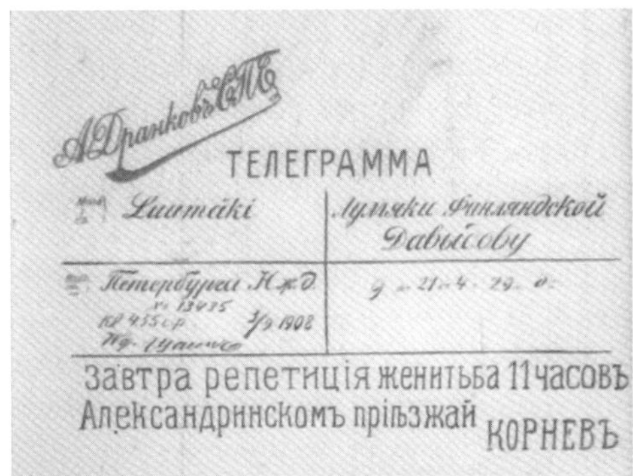

V.N. Davydov u sebia na dache [Vladimir Davydov at His Country House] (1908).
Produced by Aleksandr Drankov. Acetate print – GF

In 1909 the signature was complemented with a complex allegorical image which included a globe, a laurel, a film strip, a girl, a rising sun, and several other objects. I have also encountered prints from 1908-1910 where the signature was substituted with typed letters "Дранковъ" [Drankov] or "фот. А. Дранковъ" [A. Drankov's photography]:

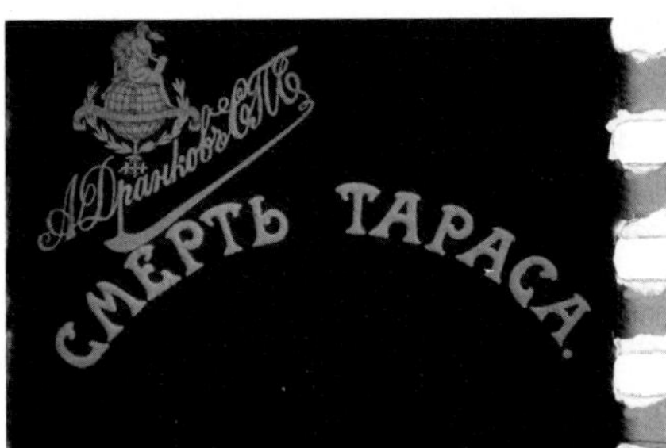

Taras Bul'ba [Taras Bulba] (Aleksandr Drankov, 1909).
Produced by Aleksandr Drankov. Nitrate print – GF

Several years later the garish allegory was replaced by Drankov's name in an elegant Art Nouveau frame:

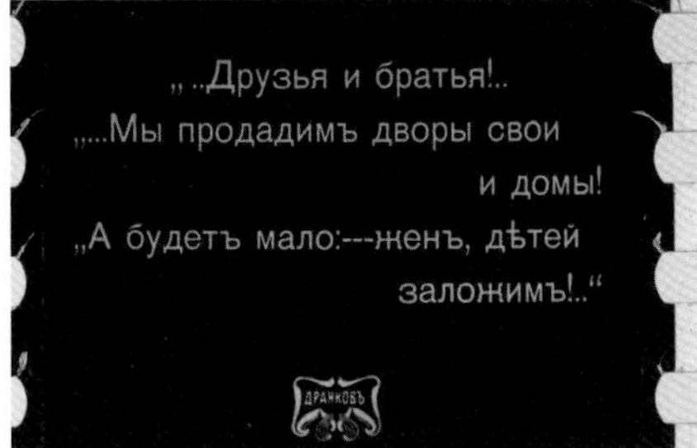

Triokhsotletie tsarstvuiuschego doma Romanovykh (1613-1913)
[Tercentenary of the Ruling House of the Romanovs (1613-1913)]
(Aleksandr Uralskii, 1913). Produced by Aleksandr Drankov. Nitrate print – GF

Drankov's final trademark, adapted sometime in the mid-1910s, was much more laconic: it contained two peacocks facing each other:

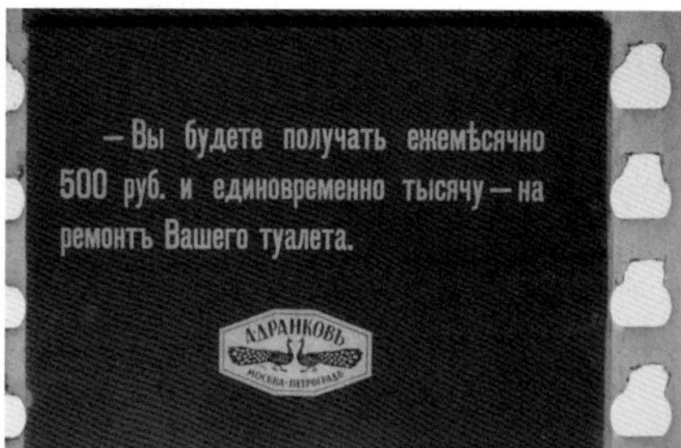

Shiolkovaia pautina [Silken Weed] (Iurii Iurievskii, 1916).
Produced by Aleksandr Drankov. Nitrate print – GF

I have also once encountered a newsreel produced (or just released) by Drankov's company with yet another trademark – a three-horse carriage (*troika*) driven by a coachman with a long whip, and Drankov's signature on top. Judging by the signature and by the font of the intertitle this must have been a relatively early production, of circa 1910. Unfortunately, I do not have a copy of the image to reproduce here.

Aleksandr Khanzhonkov's company, very conveniently for future film scholars, seems to have had only one trademark throughout its entire history. It is simple and memorable: an image of Pegasus, accompanied by Khanzhonkov's monogram, the Russian letters A X (which can be seen in an intertitle from *Votsarenie doma Romanovykh*, p.300). It is seen in the majority of Khanzhonkov's productions. However, I have encountered prints and fragments of his films with no trademarks on the intertitles:

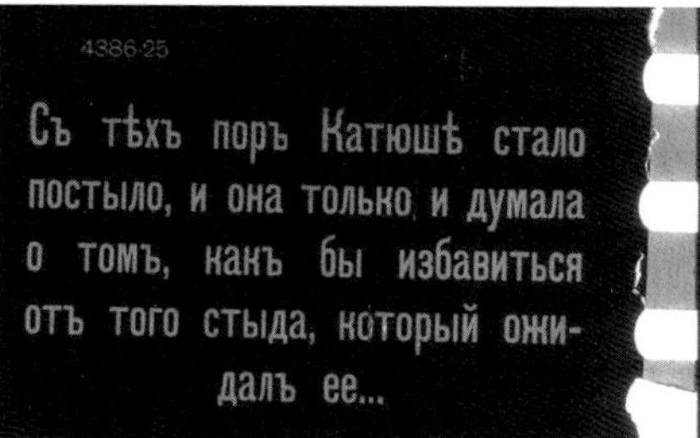

Katiusha Maslova [Katiusha Maslova] (Piotr Chardynin, 1915).
Produced by Aleksandr Khanzhonkov. Nitrate frame – Marina Umanskaia collection

Thiemann and Reinhardt were consistent as well. Even when their company was renamed "Russkaia zolotaia seriia" ["Russian Golden Series" – the name of a series was given to the whole company], in order to avoid German names during World War I, the lettering in their trademark changed but the image remained intact: a swan floating on turbulent waves.

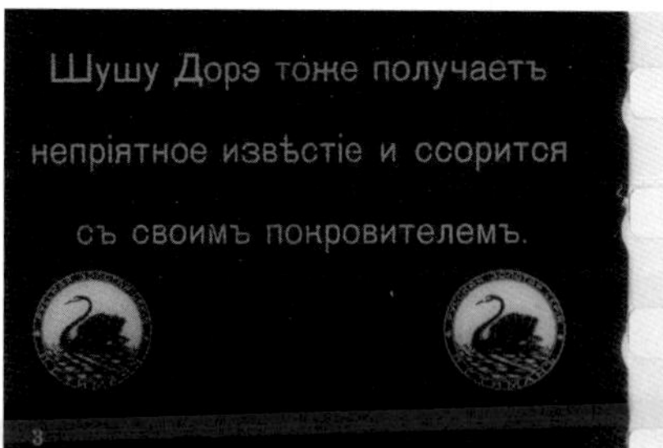

Odin nasladilsia, drugoi rasplatilsia [One Plays, Another Pays] (Yakov Protazanov, 1913).
Produced by Thiemann and Reinhardt. Safety print – GF

This being said, I have to point out that at least two surviving films with original titles that were produced by Thiemann and Reinhardt in 1914 (*Beglets* [The Fugitive] and *Kreitserova sonata* [The Kreutzer Sonata]) have no trademarks. But *Otsy i deti* [Fathers and Sons], produced in 1915, has the traditional swan again.

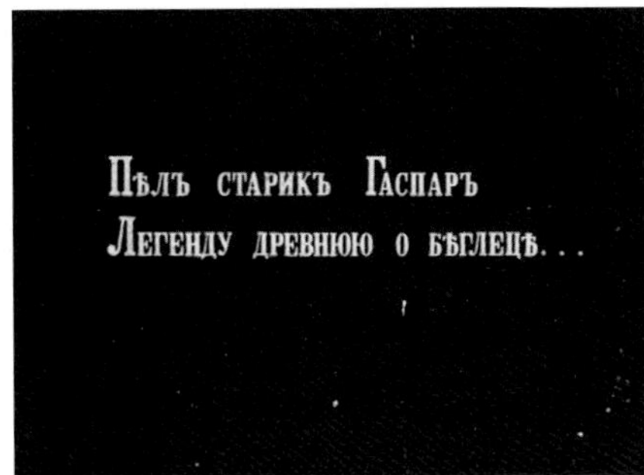

Beglets [The Fugitive] (Aleksandr Volkov, 1914).
Produced by Thiemann and Reinhardt. Safety print – GF

Around 1916 the company was often referred to as "Era", and since its trademark in the press was modified, one assumes that the trademark in the intertitles was also. However, the lack of original prints doesn't allow us to confirm this.

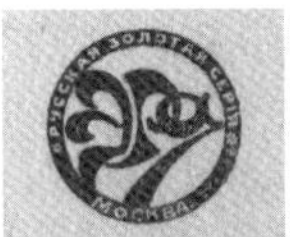

Trademark of the Russian Golden Series
(Era) in 1916-1918 – VGIK

I was not able to locate any trademarks in the intertitles of some of the major film companies of pre-revolutionary Russia: they are lacking in the films produced by Iosif Ermoliev, Dmitrii Kharitonov, Grigorii Libken, Robert Perskii, Aleksandr Taldykin, and the film department of the Skobelev Committee. However, considering the low survival rate of original prints, I would not risk drawing any conclusions. I'd rather provide the images of respective trademarks: these might turn out to be of some help to archivists and film historians.

Iosif Ermoliev's company

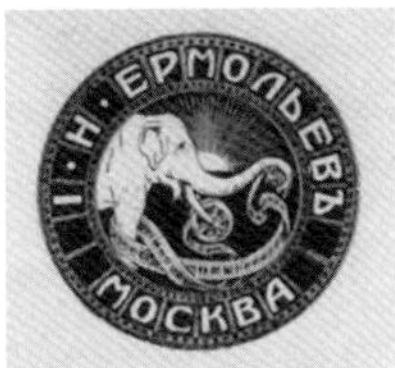

Ts. Iu. Suliminskii (ed.), *Vsia kinematografia: Nastol'naia adresnaia i spravochnaia kniga: 1916* [All Cinematography: A Tabletop Address and Reference Book: 1916] (Moscow: Izdanie Zh. Chibrario de Goden, 1916), ill. between pp.32 and 33.

Grigorii Libken's company

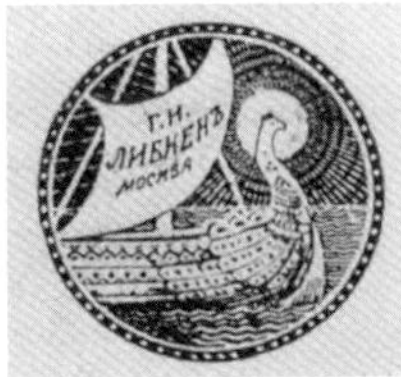

Suliminskii (ed.), *Vsia kinematografia: Nastol'naia adresnaia i spravochnaia kniga: 1916* [All Cinematography: A Tabletop Address and Reference Book: 1916], ill. between pp.32 and 33.

Robert Perskii's company

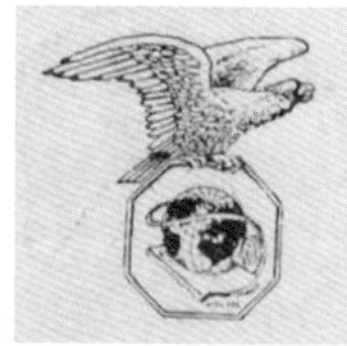

Suliminskii (ed.), *Vsia kinematografia: Nastol'naia adresnaia i spravochnaia kniga: 1916* [All Cinematography: A Tabletop Address and Reference Book: 1916], ill. between pp.32 and 33.

T.K.Iu. (Aleksandr Taldykin, Nikolai Kozlovskii and Stepan Iuriev) company

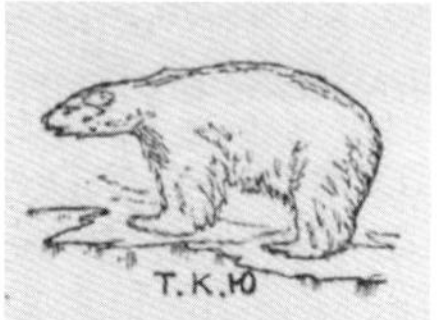

Suliminskii (ed.), *Vsia kinematografia: Nastol'naia adresnaia i spravochnaia kniga: 1916* [All Cinematography: A Tabletop Address and Reference Book: 1916], ill. between pp.32 and 33.

Dmitrii Kharitonov's company

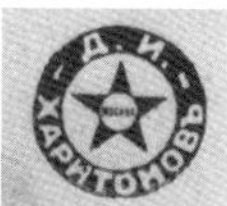

VGIK

Skobelev Committee

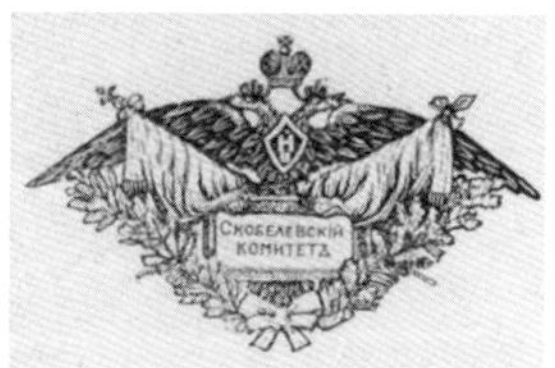

Suliminskii (ed.), *Vsia kinematografia: Nastol'naia adresnaia i spravochnaia kniga: 1916* [All Cinematography: A Tabletop Address and Reference Book: 1916], ill. between pp.32 and 33.

Very little is known about the use of trademarks by minor companies. However, one surviving example from a relatively small production company, Kino-Alfa, suggests that the practice was followed.

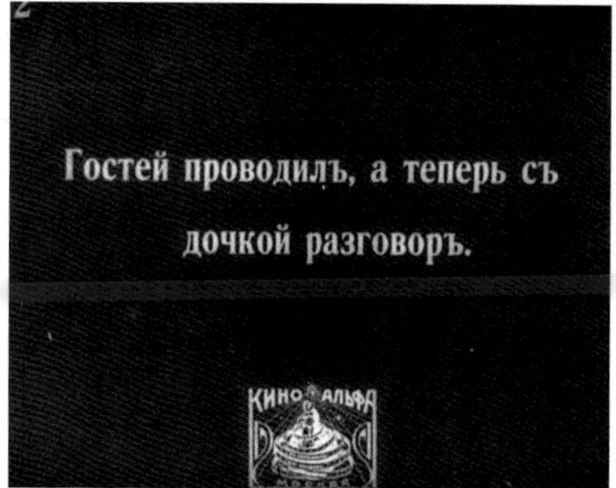

Vskolykhnulas' Rus' sermiazhnaia i grudiu vstala za sviatoe delo [Lumbering Russia Has Stirred to Defend the Sacred Cause] (Boris Svetlov, German Insarov, 1915). Produced by Kino-Alfa. Acetate print – GF

Thus, a complete inventory of Russian pre-revolutionary trademarks would be of much help.

Title Fonts

This section might be the shortest, quite paradoxically, though, as the variety of fonts in Russian and Soviet titles is very large. Listing all of them would require a separate book. But it is particularly important to indicate the "Gosfilmofond font", which has misled and keeps misleading generations of scholars and archivists.

From the very first years of its existence (perhaps as early as 1949 or 1950), Gosfilmofond has been involved in film restoration. Not only was it standard practice to create new opening credits and intertitles, but often original titles were replaced with new ones. This occurred most often when a print was in fact composed of two or three different prints, and each of them had different styles and fonts of intertitles (sometimes, in different languages). For about two decades Gosfilmofond used the same font, adapted from that of the Mosnauchfilm studio.

These newly created titles – often still made on nitrate stock – are often being mistaken for original ones. Particularly since the practice of the time was to insert them in the master element (even in the camera negative, if it still existed).

For clarity I'll provide examples from two films that have a mixed set of titles – both the original ones and the ones created at Gosfilmofond. It is worth mentioning that one of them is a sound film – for this practice was applied to a wide range of motion pictures, including early sound ones – in fact, to any film that lacked opening credits or intertitles.

Note that the new opening credits and intertitles for silent films were always filmed on a solid black background, whereas those for sound films were placed onto a light wavy cloth ground, unless an image from the film was used as a background (see also *Chapaev*, p.296).

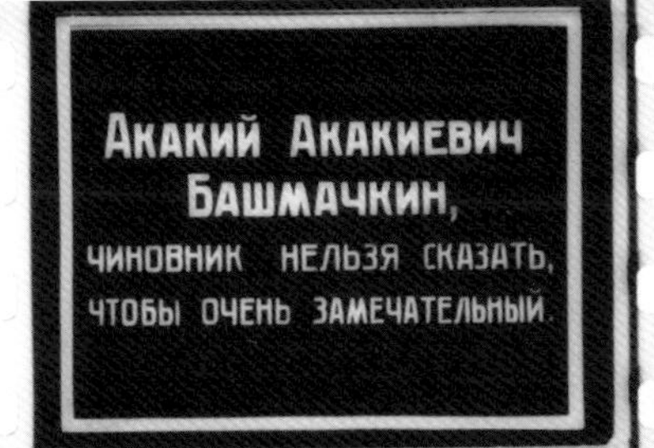

Shinel [The Overcoat]
(Grigorii Kozintsev, Leonid Trauberg, 1926).
Acetate print. 1929 re-release intertitle – GF

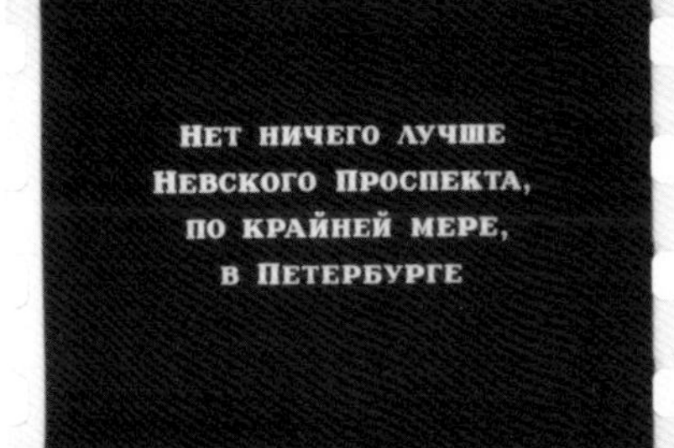

Shinel [The Overcoat] (Grigorii Kozintsev, Leonid Trauberg, 1926). Acetate print.
Intertitle re-created at Gosfilmofond – GF

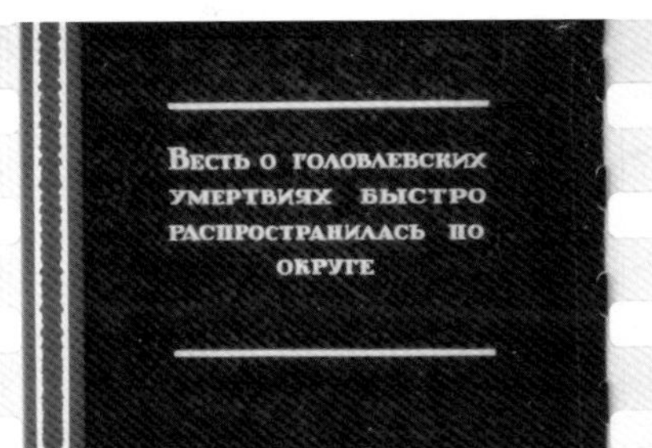

Iudushka Golovliov [Judas Golovlev]
(Aleksandr Ivanovskii, 1933). Acetate print.
Original intertitle – GF

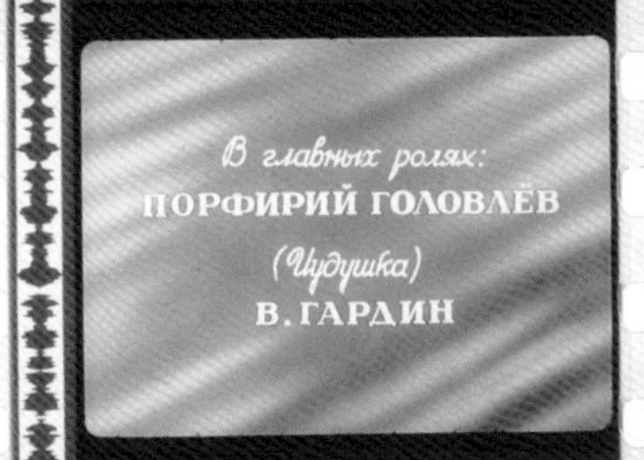

Iudushka Golovliov [Judas Golovlev]
(Aleksandr Ivanovskii, 1933). Acetate print.
Title re-created at Gosfilmofond – GF

13
Some Information on Other Film Stock Manufacturers

Brian Pritchard, with contributions by Camille Blot-Wellens

Please note that unless the description states otherwise, all samples are 35mm.

3M

The Minnesota Mining and Manufacturing Company (3M) was created in 1902, and settled in St. Paul (Minnesota) in 1910. The company is mainly known for adhesive tape; during World War II, they also produced products for the U.S. Army. After the war, the company expanded, including internationally. They took over Ferrania in 1964.[1]

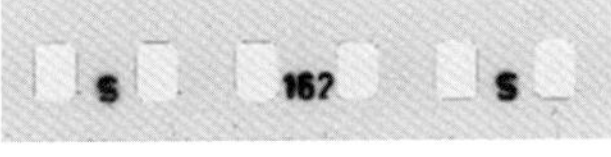

Black & white Positive film 1971

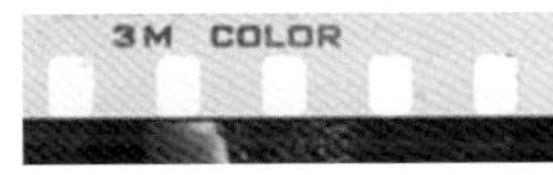

Colour Print

On the black & white positive film, the number 162 is the perforating machine number, and the letter "S" means "Safety".

Ansco

The company was founded in 1842 as E. Anthony & Co. (later E. and H.T. Anthony & Company, when Edward Anthony's brother officially joined the business) and became the Anthony & Scovill Co. in 1901, after a merger with the camera business of Scovill Manufacturing (of Waterbury, Connecticut). That year the company headquarters relocated to Binghamton, New York. In 1928 Ansco merged with the German firm Agfa to become Ansco-Agfa. After World War II, Ansco became independent, but was able to obtain the physical and intellectual property associated with Agfacolor.[2]

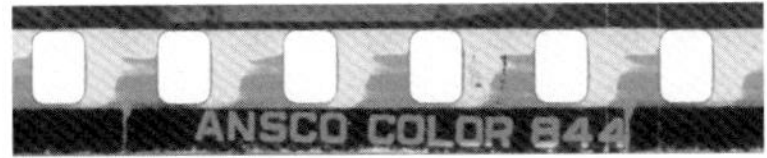

Print (edge printing printed through)

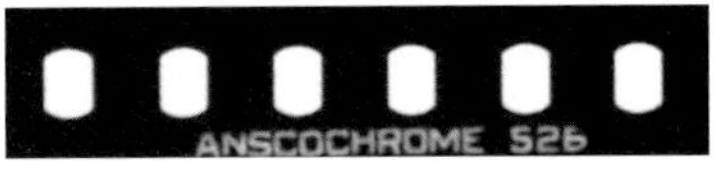

Anscochrome Reversal

Anscocolor was launched in 1948 and Ansochrome in the mid-1950s. It seems that Ansco adopted Agfa's edge printings system, with the indication of the emulsion batch.

1 See more information in the entry about Ferrania, below.
2 Leo Enticknap, *Moving Image Technology. From Zoetrope to Digital*, London & New York: Wallflower Press (2005), p.92.

Deko[3]

Kodak AG VEB Köpenick-Berlin.[4] This was formerly the Kodak AG factory, which was confiscated in 1941 as enemy property. Despite this, Kodak AG went on manufacturing film until 1956, when the factory became VEB Fotochemische Werke Berlin Köpenick, where X-ray films, black & white motion picture film stock, and photographic paper, as well as chemicals, were produced. The factory was returned to Kodak in 1992.

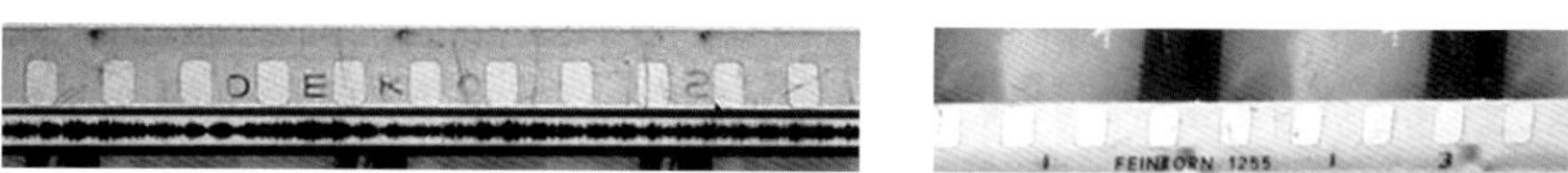

Different edge printings on two edges of film. "Feinkorn" means "Fine Grain" in German; 1255 probably denotes Duplicate Positive emulsion.

Ensign

George Houghton and Sons, founded in 1834 under another name, is mainly known for manufacturing still-photography cameras and accessories. In 1903 the firm started making roll film, sold as Ensign, and in 1904 they changed their name to Houghton's Limited.[5] In 1908 the firm was the biggest British camera factory. The company became Ensign again in 1930; later it was known as Ross Ensign, and operated under that name until 1961.

The company also offered equipment for moving-picture amateurs (cameras, splicers, title-makers, projectors, screens, etc.). Logically, it also supplied raw motion-picture film stock. In its 1930[6] catalogue, the firm provided panchromatic "Ensign 16mm Safety film", black & white or tinted, manufactured by Austin Edwards Ltd. in Warwick, England.

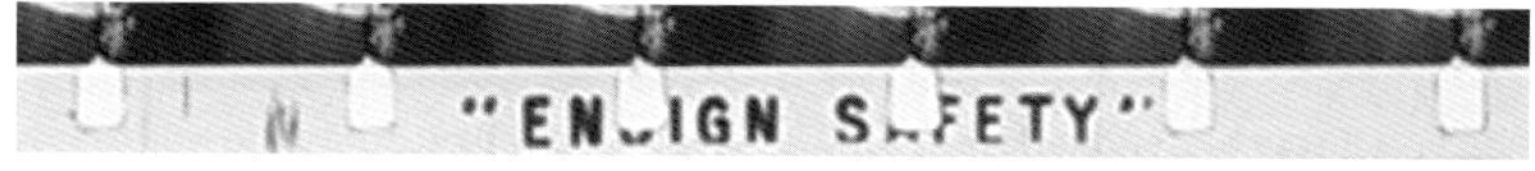

16mm "Ensign Safety" film

Ferrania

FILM (taking its name from the initial letters of Fabbrica Italiana Lamine Milano) was created in 1917 from the Societá Italiana Prodotti Esplodenti.[7] They started collaborating with the French firm Pathé and began manufacturing motion-picture film in the mid-1920s. The company became Ferrania Limited in 1938. In 1954 the Italian firm was acquired by the American 3M Company, who spun it off to their Imation Division. The film factory stopped manufacturing film stock in 2009 and closed down in 2010. The company has since been taken over, and is once more called Ferrania Film.

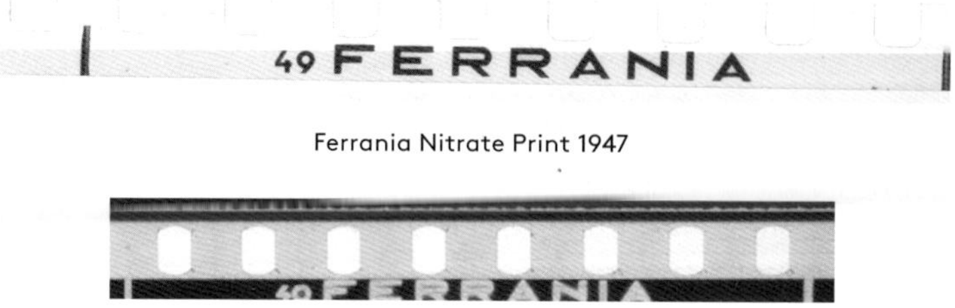

Ferrania Nitrate Print 1947

Ferrania Nitrate Negative (before 1952) printed through on a Cinecolor print

3 The authors thank Martin Koerber for his help.
4 For more information on Kodak AG, see the chapter on Eastman Kodak, pp.246-247.
5 Research by Adrian Richmond, and available on his website (ensign.demon.uk).
6 Made available online by Pacific Rim Camera (pacificrimcamera.com), dealers in vintage and film cameras located in Salem, Oregon (United States). They provide original catalogues of many companies in pdf format.
7 All the information on Ferrania was found on the website of Film Ferrania (filmferrania.it), the company created in 2012 by Nicola Baldini and Marco Pagni. The editor thanks Alice Rispoli, who is currently investigating Ferrania, for her help.

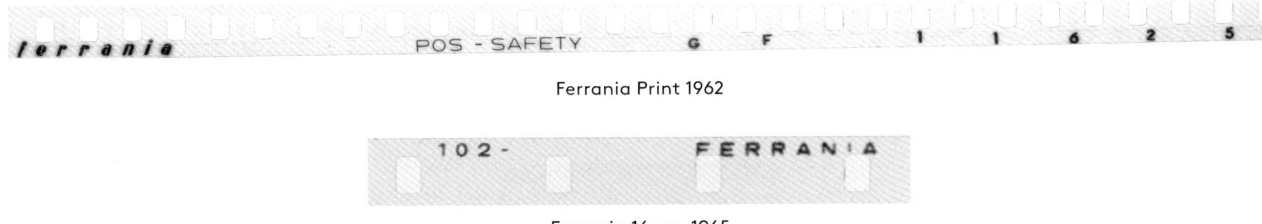

Ferrania Print 1962

Ferrania 16mm 1965

Foton

WARSZAWSKIE ZAKŁADY FOTOCHEMICZNE FOTON [Warsaw Photochemical Factory FOTON]: In 1936 George and Casimir Franaszkowie, the owners of the factory "J. Franaszek SA", began the production of photographic paper and film negatives. The factory was destroyed during the Warsaw Uprising. In 1949, after the end of World War II, the factory was rebuilt. They signed a technology deal with the British company Ilford Ltd. in 1969. If the chemicals and film base were largely produced in Poland in the 1970s and 1980s, towards the end of the 1970s and throughout the 1980s imports gradually increased, such as polyester base (from the factories of Bexford Ltd. in England, DuPont in Luxembourg, and Agfa-Gevaert in Belgium) and gelatine (notably from Rousselot in France and Croda in England). By the 1990s, the majority of the company's production focused on healthcare products.[8]

Foton Positive

Foton 16mm

Ilford

The company was founded in 1879 by Alfred Hugh Harman as Britannia Works and was sold in 1897. In 1898 the firm was known as The Britannia Works Limited. Initially making photographic plates, it grew to occupy a large site in the centre of Ilford, in east London. In 1902 the company took the name of the town, to become Ilford Limited. In 1920, together with Imperial, Gem, and Amalgamated Photographic Manufacturers, they created the company Selo. Their first film stock for motion pictures was launched in 1923.[9] It was decided that all roll films would be labelled "Selo" (apparently from 1928 for motion picture film stock),[10] while the brand "Ilford" was used for plates and paper. In 1932 Ilford opened facilities in London, near the main motion picture companies, while a cine studio and testing department were based in Brentwood.[11] From 1935 the brand Selo was replaced by Ilford,[12] and a laboratory specialized in motion picture film stock was completed in Warley. Initially intended to work mainly on Dufaycolor films, the firm also worked with black & white film stock (camera negative and positive for prints).[13]

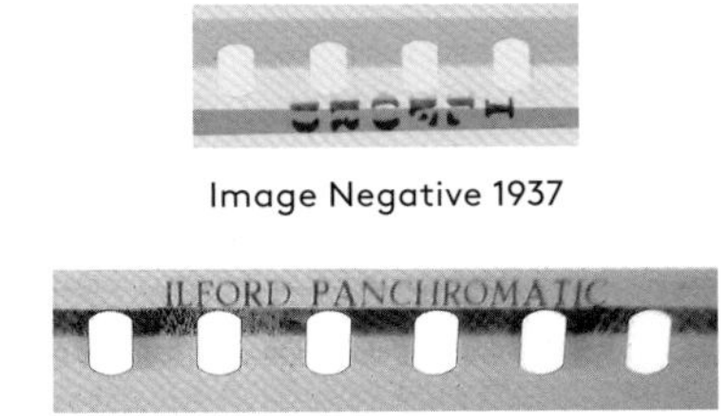

Image Negative 1937

Panchromatic Image Negative from the late 1930s

8 For more information on early Foton (1945-1988), see [Tadeusz Korecki], *Odbudowa i rozwój Warszwskich Zakładów Fotochemicznych* [Reconstruction and Development of the Warsaw Photochemical Factory], brochure, published 1988 by PWiWPChiL CHEMIL. On late Foton (1990-2005), see Jerzy Kuźnicki, "Warszawskie Zakłady Fotochemiczne ORGANIKA – FOTON, Warszawskie Zakłady Fotochemiczne FOTON S.A., Warszawskie Zakłady Fotochemiczne FOTON S.A. w likwidacji w latach 1990-2005" [Warsaw Photochemical Factory ORGANIKA – FOTON, Warsaw Photochemical Factory FOTON, Warsaw Photochemical Factory FOTON, in liquidation in the years 1990-2005], unpublished article. The authors thank Elżbieta Wysocka for sharing these sources.

9 Robert J. Hercock and George A. Jones, *Silver by the ton. The History of Ilford Limited, 1879-1979*, London: McGraw-Hill Book Company (UK) Limited (1979), p.61.

10 See also "Selo" in "Stock Manufacturers' Edge Marks", p.72 in this volume; and Hercock and Jones, *Silver by the ton*, p.62.

11 Hercock and Jones, *Silver by the ton*, p.121.

12 *Kinematograph Year Book* (1936), p.260.

13 Hercock and Jones, *Silver by the ton*, p.66.

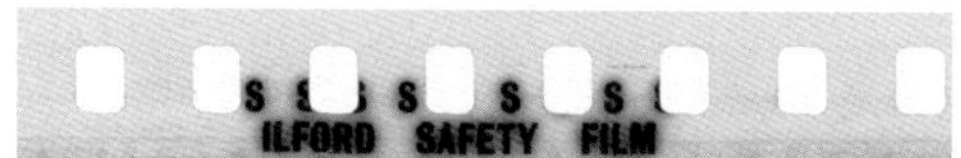

Safety Positive 1951

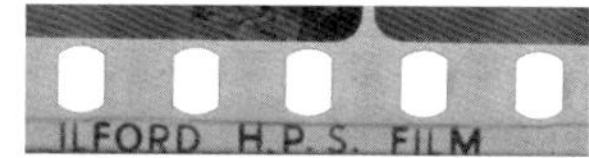

HPS Image Negative 1958

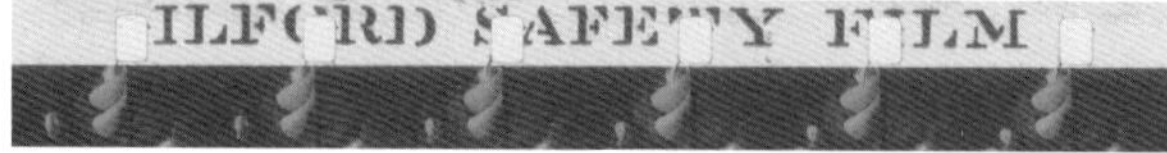

16mm Positive 1956

Towards 1950, Ilford introduced a footage number also on 16mm negative film stock:

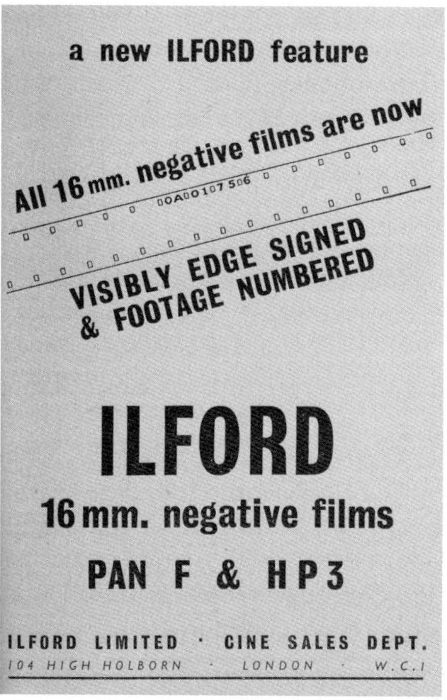

Advertisement published in *British Kinematography*, Vol. 17, No. 2 (August 1950)
– Coll. Frank Wylie (mediahistoryproject.org)

From at least the early 1960s, footage numbers introduced by Ilford specify the emulsion type with letters:

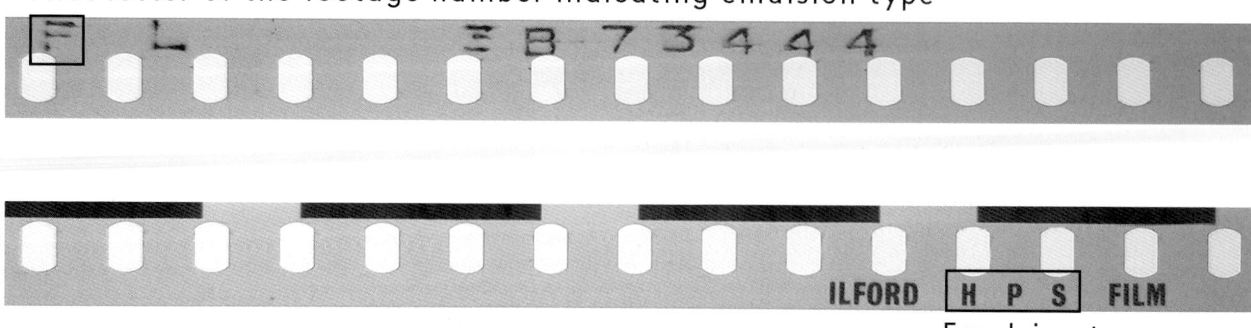

HPS Image negative ca. 1965 – CF

Changes in these footage numbers can help situate the period of manufacture of negative film stock. Unfortunately, the documentation[14] is incomplete and only gives partial information. Nevertheless, it is possible to date 3 changes in the edge printings of Ilford 35mm film stock:

Emulsion	Nature	Code	Example and Date	
FP4	Negative Motion Picture	A-B-C	AE 2A12345	????
HPS	Negative Cine Film	F-G	FJ 3A43215	1962
HP3	Negative Cine Film	L-M	LA 1A64373	1962
HP3	Negative Cine Film	A-B-C	AE ZA 12345	1966
HPS	Negative Motion Picture	F-G	FE AZ 12345	1966
Pan F	Negative Motion Picture	J-K	JE ZA 12345	????
FP3	Negative Motion Picture	L-M	LE ZA 12345	1966
Mark V	Negative Motion Picture	S-T	SE AZ 12345	1966
FP4 +	Negative Motion Picture	A-B-C-D-E	AB2 C76024	1990?
FP5+	Negative Motion Picture	F-G-H	FA2 B75028	1991?

Dates of some Ilford motion picture film emulsions:

Emulsion Type	Beginning Date Motion Picture[15]	End Date Still Photography[16]
HP3	1943	1968
Pan F	1946	1992
FG Pan Positive	1953	
Hyperpan	1953	
N	1957	
FP3	1958	1968
HR	1959	
Mark V	1965 (until ca. 1976)	
HRT	1965	
A	1965	
FP4	1969	
L	1975	

In 2003, Ilford stopped manufacturing motion-picture film stock, and in 2004 the company became Harman Technology.[17]

Indu

Hindustan Photo Films Manufacturing Company (HPF) was an initiative of the Indian Government in 1960. The manufacturing of film stock started in January 1967. In her inaugural speech, then-Prime Minister Indira Gandhi declared: "on one side the Ooty factory would produce the raw film required by the Motion Picture industry for entertainment of the people, on the other, it would produce X-ray Film, which is needed for saving human lives."[18]

14 Documents gathered by Brian Pritchard and consultable in the Madrid Project.
1962: *Ilford. Technical Information Sheet A 50-6. HP3 Negative Motion Picture Film, Ilford. / Technical Information Sheet A 50-8. HPS 35mm Negative Cine Film.*
1966: *Ilford. Technical Information Sheet A 50-3. FP3 Negative Motion Picture Film, Ilford. / Technical Information Sheet A 50-5. HP3 Negative Motion Picture Film, Ilford. / Technical Information Sheet A 50-7. Mark V Negative Motion Picture Film, Ilford. / Technical Sheet A 50-9. HPS Negative Motion Picture Film.*
No date: *Ilford. A50.1. Pan F negative Motion Picture Film, Ilford. / A50.35. FP4 Negative Motion Picture Film, Ilford. / Fact Sheet FP4 Plus Black and White Negative Motion Picture Film, Ilford. / Fact Sheet HP5 Plus Black and White Negative Motion Picture Film.*

15 Hercock and Jones, *Silver by the ton*, pp.158-159. The authors specify that the list is not exhaustive. Note that the list stops in 1979 (the year of the book's publication).

16 Indicated only for information, the dates of the still photography emulsions proceed from the Historische Kleinbild Database established by Erwin Zbinden (bilderdienst.ch). Please be aware that the dates of manufacture of motion-picture emulsions can differ from still-photography emulsions.

17 ilfordphoto.com/about-us/history/

18 *The Silver Years. Hindustan Photo Films (1961-1986)*. Camille Blot-Wellens wants to thank N. Ramesh from AVM Cine Lab (Chennai) for providing her with a digital copy of this book.

HPF developed several emulsions: "For the Indian feature film industry, the largest in the world, HPF offers INDU[19] black & white cine positives, colour cine positive, and cine sound negative — in 35mm, as well as in 16mm — which is widely used for documentaries and educational films."[20] Film stocks were manufactured in Ootacamund (Nilgiris) and Ambattur[21] (near Madras [Chennai], the most important place in the Indian motion picture industry). The introduction of digital technologies in the film industry provoked serious losses to the firm from 1992.[22] The company was declared unsustainable in 1996, and it only sold limited footage of motion-picture colour film stock in 1998-1999.[23] The firm was finally closed in 2018.[24]

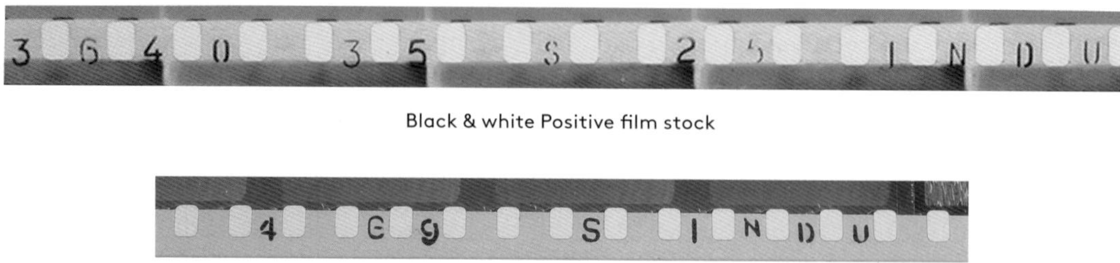

Black & white Positive film stock

Colour positive film stock - Prasad

Even if the meaning of the numberings in the Indu edge printings is unknown, it seems that it might contain the same information as other brands: emulsion number, master roll number, S (for Safety), and maybe the perforation machine number. Unfortunately, not enough materials could be studied to be able to ascertain if the edge printings could help in dating Indu film stocks.

Lucky Film Corporation

Founded in 1958 in Baoding, China. In July 1959 the firm started manufacturing black & white positive film stock for motion pictures.[25] In 1969, the China Lucky group manufactured image positive and image and sound negative black & white film stocks. The firm changed names several times; the latest name, "Lucky", began to be used in 1985. It appears that the manufacture of photographic and/or film products ended in 2012.

Positive 1960 (the Chinese characters say "China, Baoding")

19 INDU, the name appearing on the edge printings, means "Silver" in Sanskrit.

20 *The Silver Years*, p.35. According to Rajesh K. Pillania and Subhojit Banerjee, "Entertainment and economic contributions of the Indian Hindi movie Industry", in Samuel Cameron (ed.), *Handbook on the Economics of Leisure* (Cheltenham: Edward Elgar Publishing, 2011), if HPF made India self-sufficient in black & white positive stock and sound negative stock, all colour film was imported and perforated in India until 1974 (p.305). An article by Arun Ram, "Hindustan Photo Films does not have the technology for colour film", in *India Today* (5 February 2001), also states that "While the Indu brand, HPF's answer to international film brands, was a washout, its real business, of being the monopoly distributor of raw film, went under with import liberalization since 1992."

21 According to Arun Ram's article in *India Today* (5 February 2001) the Ambattur factory in Chennai was "a conversion plant to cut imported jumbo film rolls into retail packs."

22 "After a prolonged illness, Hindustan Photo Films breathes its last", *The Times of India* (26 April 2018).

23 Arun Ram, *India Today* (5 February 2001).

24 "After a prolonged illness, Hindustan Photo Films breathes its last", *The Times of India* (26 April 2018).

25 luckyfilm.com.cn/html/MasterSite/EN/aboutus/gonpn.html

Mafe

Mafe (Manufacturas Fotográficas Españolas / Spanish Photographic Manufactures) was a photographic film factory founded in 1949, near Madrid at Aranjuez, taking advantage of facilities created in 1935, during the Second Spanish Republic; these facilities were abandoned during the Civil War (1936-39). Mafe mainly manufactured material for X-rays. In the 1950s, they started manufacturing film for copying and sound, and also, although in small quantities, negative film for cinematography.[26] Mafe films for copying were a fundamental material in Spanish cinema in that period. They never manufactured colour film stock. From 1962 Mafe manufactured stock using the patents of Perutz (see the next entry); in 1967 this German manufacturer would take control of Mafe. In 1992, Agfa-Gevaert absorbed Mafe, to continue with the production of X-ray materials. Around 2005, Mafe closed down completely.

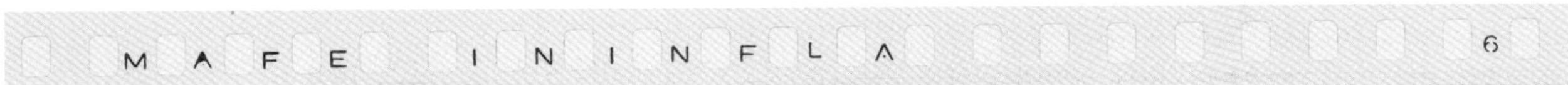

MAFE Positive black & white

Perutz

Perutz-Photowerke was founded in 1880 by Otto Perutz in Munich.[27] It seems that its production of 35mm motion-picture film stock began in 1913. In 1933, the firm started offering small-gauge film stock (16mm, 9.5mm, and 8mm). In the 1950s and early 1960s, they also provided 16mm film stock for television and documentaries. The company was owned by Agfa since 1964, and closed down in 1994.

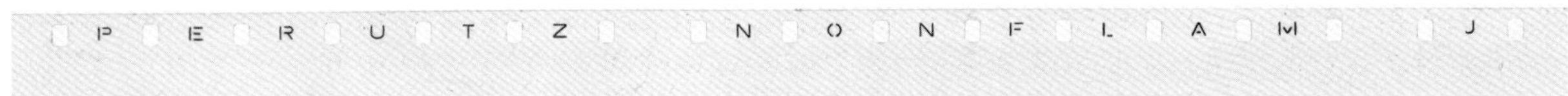

Perutz 16mm

26 See also Jennifer Gallego Christensen and Encarni Rus Aguilar, "La catalogación de las marcas marginales de fábrica como medio para la identificación y conservación de materiales fílmicos", in *Los soportes de la cinematografía 1*, Coll. Cuadernos de la Filmoteca 5 (1999), pp.120-133. The authors present an analysis of the typographies of the edge printings of the manufacturer's products.

27 It was not possible to consult directly the original literature published on Otto Perutz and the firm (such as *Perutz-Kinefilm: 35mm, 16mm*, published by the company in 1956), so part of the information in this entry is derived from the Wikipedia entry on "Perutz-Photowerke".

Contributors to the 2020 Edition

Peter Bagrov is a film historian and archivist. From 2005 to 2013 he was a Research Associate at the Russian Institute of Art History in St. Petersburg. From 2013 to 2017 he was the Senior Curator at Gosfilmofond of Russia, and from 2013 to 2019 he served as the artistic director of the Belye Stolby archival film festival. Since August 2019 he has been the Curator-in-Charge of the Moving Image Department at the George Eastman Museum. His main area of interest is early Russian and Soviet film history of the 1910s through the 1960s.

Luciano Berriatúa is a director and producer of fiction films, documentaries, and animation. He is an authority on F.W. Murnau, about whom he has published three books: *Los proverbios chinos de F.W. Murnau*, *Apuntes sobre las técnicas de dirección cinematográfica de F.W. Murnau*, and *Nosferatu, un film erótico, ocultista, espiritista, metafísico*. He has also made documentaries and written articles about Murnau, and curated the restoration of many of Murnau's films, as well as those of other filmmakers, for the F.W. Murnau Stiftung, the Filmoteca Española, and the Filmoteca de Catalunya. As a professor, he has trained several generations of restorers, at the Université Paris 8 for 15 years and currently at the Universidad de Córdoba and the Elias Querejeta Zine Eskola in San Sebastián.

Camille Blot-Wellens is an independent film historian, researcher, and archivist. She started collaborating with film archives in 2000 on identification, research, restoration, and training projects, and more notably worked for the Filmoteca Española (2000-2007), the Cinémathèque française (2007-2011), and the Svenska Filminstitutet (2016-2019). Specializing in Early Cinema, she is the author of two books and numerous articles. She is a member of the FIAF Technical Commission and on the Board of Domitor, and currently teaches at Université Paris 8 and the Université de Lausanne. In 2018, she was the recipient of the Jean Mitry Award and the Outstanding Achievement Award for Film Preservation.

Christophe Dupin is the Senior Administrator of the International Federation of Film Archives (FIAF), Executive Publisher of the *Journal of Film Preservation*, and a film historian. Previously he worked for the British Film Institute while writing a PhD thesis on the history of the BFI as a film producer, before carrying out extensive doctoral and post-doctoral research on the history of the BFI, which resulted in the publication of *The British Film Institute, the Government and Film Culture, 1933-2004* (Manchester University Press, 2012), co-edited with Geoffrey Nowell-Smith. His current research topic is the history of FIAF and the international film archive movement.

Martin Koerber is head of Audiovisual Heritage – Film at the Stiftung Deutsche Kinemathek in Berlin, and professor for the Restoration of Audiovisual Media at the Hochschule für Technik und Wirtschaft in the same city. After studies in Media, Art History, and Musicology he dabbled in filmmaking in the 1980s, and from 1986 worked for the Deutsche Kinemathek, Nederlands Filmmuseum, and other film archives as a free-lance, followed by a permanent position at the Deutsche Kinemathek from 1999 to 2003, and again since 2007. Since 1988 he has been involved in numerous film restoration projects.

James Layton is a film historian and archivist specializing in the history of motion picture technology. He is the Manager of The Museum of Modern Art's Celeste Bartos Film Preservation Center, and is co-author of two books with David Pierce, *The Dawn of Technicolor, 1915-1935* (2015) and *King of Jazz: Paul Whiteman's Technicolor Revue* (2016). Prior to working at MoMA, Layton was an archivist at the George Eastman Museum, where he curated two gallery exhibitions, celebrating CinemaScope and Technicolor, and the informational website *Technicolor 100*. Layton also contributed to the Image Permanence Institute's *Knowing and Protecting Motion Picture Film* informational poster (2010).

Pierrette Lemoigne was initially drawn to the fields of dance and architecture, but turned to film heritage after working on a film with Patrice Hamel. She holds a Master's degree in film and an undergraduate degree in fine art, and is a former student of the École Louis-Lumière. She is now a research officer in the Documentation Department of the Centre national du cinéma et de l'image animée (CNC) – Direction du patrimoine cinématographique. As a cultural mediator, she assists researchers in the audiovisual room of the Bibliothèque Nationale de France, where the CNC's digitized collections can be consulted.

Eric Loné works as a documentalist at the CNC's Heritage Department. In particular, he has collaborated on the identification and restoration of two silent films previously considered lost, *Bucking Broadway* (1917) by John Ford and *The Deciding Kiss* (1918) by Tod Browning. He is also the author of several articles, including one on Lux, an early French film production company.

Jacques Malthête is the author of numerous contributions on early cinema, in particular on Georges and Gaston Méliès. With Laurent Mannoni, he was co-director of *Méliès, magie et cinéma* (Paris-Musées, 2002) and co-author of *L'œuvre de Georges Méliès* (La Cinémathèque française/Éditions de La Martinière, 2008). More recently, with Stéphanie Salmon he co-directed *Recherches et innovations dans l'industrie du cinéma. Les cahiers des ingénieurs Pathé, 1906-1927* (Fondation Jérôme Seydoux-Pathé, 2017), and with Réjane Hamus-Vallée and Stéphanie Salmon, *Les Mille et un visages de Segundo de Chomón* (Fondation Jérôme Seydoux-Pathé/ Presses universitaires du Septentrion, 2019). He has also participated in collective works on Léon Gaumont, Étienne-Jules Marey, Jean Comandon, and Karel Zeman.

Hidenori Okada is a curator at the National Film Archive of Japan (NFAJ), which he joined in 1996 when it was still the National Film Center (NFC) of the National Museum of Modern Art, Tokyo. Over the years he has been involved in various areas of film archiving, including the collection, preservation, and film programmes. Since 2007 he has been responsible for non-film collections, as well as the curation of exhibitions on film culture. He has also published numerous papers and essays on Japanese and international cinema. His works include *Eiga to iu buttai X* [The Thing from Planet Film] (2016), a book written about the archival activities at the former NFC.

Brian Pritchard is a Motion Picture and Film Archive Consultant. He started his career in 1962 with the Kodak Research Laboratory, and then Motion Picture Sales in 1969. From there he became Technical Director of the Filmatic, Humphries, and then Hendersons Film Laboratories. He worked with the Gamma Group in the 1990s, and became a consultant in 2002. He has worked with the BFI/National Film Archive, BBC, Digital Film Lab, and many film archives. In 2011 he worked on the preservation of the world's first colour moving pictures. He is the co-author with David Cleveland of the book *How Films Were Made and Shown*, published in 2015.

Index

This book is conceived mainly as a working tool. Its contents are presented in a logical and efficient manner. The Index is organized in a way which should assist the reader when examining a film element that needs to be identified. It thus consists of 4 sections:

1. General

This section covers people, companies, and subjects, and includes relevant vocabulary and nouns. Personal names and companies are listed alphabetically. Technical entries include variant name styles, and are grouped by shared terms: e.g. B.H., BH, B&H (Bell & Howell), *see* perforation

2. Trademarks

This section presents trademarks and logotypes that appear onscreen (visible in settings and in intertitles). When text is part of a logo, it is indicated between French quotation marks («...»): e.g. « MARQUE DÉPOSÉE / [black 5-pointed star] / TRADE MARK »

3. Edge Printings

This section gathers the text of edge printings introduced by production companies and film stock manufacturers, as transcribed directly from film elements: e.g. S GEVAERT BELGIUM S

4. Film Titles

This section lists the original titles of films for which an image frame has been reproduced. It does not include films for which only main titles, intertitles, or inserts are reproduced.

Style of references within the Index (page and footnote numbers)

Boldface page numbers: The entry is the main subject.

***Italic* page numbers:** These indicate specific illustrations, charts, or tables.

Footnotes: Index entries appearing only in footnotes are specified by "n": e.g., 60n means page 60, footnote (when there is only one on that page).
When there are several footnotes on a page, "n" is followed by its number: e.g., 72n18 means page 72, footnote 18.
When an entry appears in several separate footnotes on the same page, "n" is repeated several times, followed by its respective number, each separated by a comma: e.g., 312n20,n22 means page 312, footnotes 20 and 22.
In the case of consecutive footnotes on the same page, the numbers are separated by a dash: e.g., 197n3-4 means page 197, footnotes 3 and 4.
If a citation is repeated in several footnotes on the same page (e.g., a reference to the same publication), only the reference to the first footnote is indicated.

Notes

Film archives are not referenced in the Index; e.g., "Gem" refers to the company, not to the George Eastman Museum (GEM).

Illustrations alone are listed only for edge printings, film titles, and trademarks. (*NB: Illustrations of duplicate elements are not indexed under "duplicate".)

1. General

People, companies, subjects.

H

I

J

K

L

M

R

S

2. Trademarks

Texts and descriptions of visual logotypes appearing on the screen, in settings and titles. When the same illustration appears several times, all page references are given, since the illustrations are presented in different contexts that can aid identification.

*NB. More logotypes of Russian Production companies are reproduced on pp.303-304. (These were not observed on film materials but found in contemporary publications.)

3. Edge Printings

Transcriptions of edge printings introduced on film materials by film production companies and film stock manufacturers.

V

W

CHINESE CHARACTERS

CYRILLIC CHARACTERS

Б

П

С

Т

Ф

Х

Ш

4. Film Titles

This section includes only films illustrated with a reproduction of an image frame.
Films are not listed here if only main titles, intertitles, and inserts are reproduced.

I

K

L

M

N

O

P

R

S

T

U

V

W

NOTES

NOTES

NOTES

NOTES